Metabolism of Human Diseases

Eckhard Lammert • Martin Zeeb

Editors

Metabolism of Human Diseases

Organ Physiology and Pathophysiology

Second Edition

Springer

Editors
Eckhard Lammert
Institute of Metabolic Physiology
& German Diabetes Center
Heinrich Heine University Düsseldorf
Düsseldorf, Germany

Martin Zeeb
Corporate Publishing
Springer Medizin Verlag
Neu-Isenburg, Germany

ISBN 978-3-031-96018-5   ISBN 978-3-031-96019-2   (eBook)
https://doi.org/10.1007/978-3-031-96019-2

This Springer imprint is published by the registered company Springer Nature Switzerland AG
The registered company address is: Gewerbestrasse 11, 6330 Cham, Switzerland

If disposing of this product, please recycle the paper.

Contents

Fat Tissue

Lung

Immune System

Kidney

Cancer

Introduction

Eckhard Lammert and Martin Zeeb

The scientific community has increasingly recognized metabolic alterations as being critical components or even drivers of human disease. *Metabolism of human diseases* discusses the metabolism and signaling pathways in tissues and organs known to be relevant for common human diseases. It thus bridges the existing gap between biochemistry and physiology textbooks on the one hand, and pathology textbooks on the other hand.

Metabolism of human diseases is directed at advanced students, doctors, and scientists from all categories of life sciences and medicine (e.g., biochemists, biologists, physiologists, pharmacologists, pharmacists, toxicologists, and physicians) with an interest in the metabolism and molecular mechanisms of human diseases, irrespective of their specialization. The demand for this book became apparent during the first half of 2020 (the starting year of the coronavirus disease 2019 (Covid-19) pandemic): In the first 6 months of this year, more than 9 million chapters of this book were downloaded, and this motivated us to work on a second edition of the book.

The book is divided into sections, each related to a human organ or tissue. Each section begins with an *overview chapter* presenting the anatomic and physiological properties of the organ or tissue in question relevant for the subsequent disease chapter/s of the section (Fig. 1). The overview introduces organ- or tissue-specific metabolism and signaling pathways as well as intra- and inter-organ communication (i.e., "inside-in," "inside-out," and "outside-in" signaling). The *disease chapters* discuss pathomechanisms of the diseases with emphasis on metabolic alterations and affected signaling pathways. In addition, they often introduce major treatments currently in use and in clinical trials as well as their influence on the patient's metabolism.

The diseases have been selected to cover a wide spectrum of human diseases in the industrialized world (as described in the tenth edition of the International Classification of Diseases, ICD-10). They include several of the most common (based on diagnosis), most deadly (based on numbers of deaths), and most expensive (based on treatment costs) illnesses.

Each chapter of *Metabolism of human diseases* contains figures and/or tables that illustrate important elements of anatomy, physiology, metabolism, signaling pathways, or treatment. Virtually all figures are presented in a similar layout to facilitate understanding of the contents of each chapter (Fig. 2). Since Yousun Koh provided the layout and final presentation, we would like to express our gratitude to her.

E. Lammert (✉)
Institute of Metabolic Physiology & German Diabetes Center, Heinrich Heine University Düsseldorf, Düsseldorf, Germany
e-mail: lammert@hhu.de

M. Zeeb
Springer Medizin Verlag, Neu-Isenburg, Germany

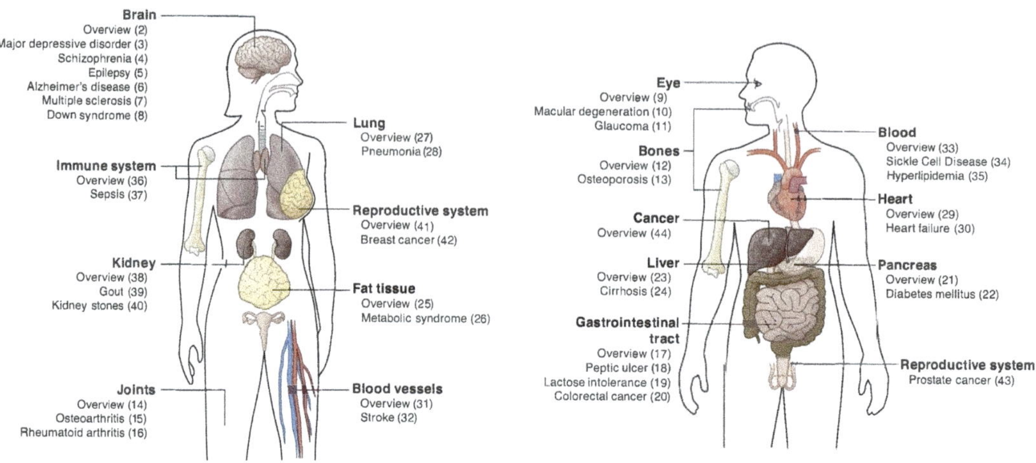

Fig. 1 Summary of *overview chapters* and *disease chapters*

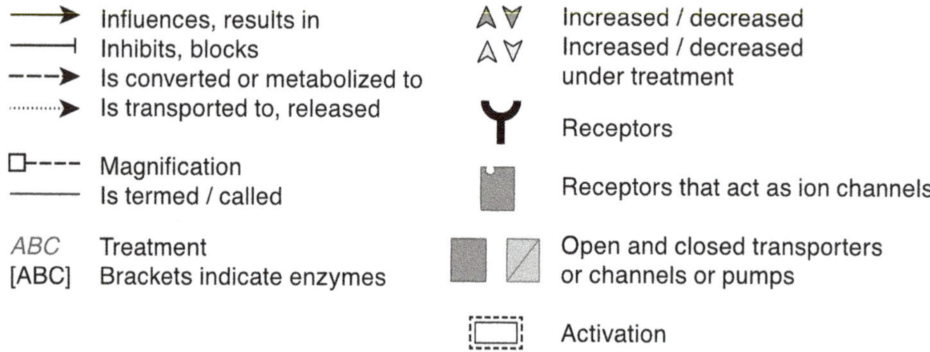

Fig. 2 Explanation of symbols wherever used in this book

Finally, many international experts contributed chapters to the second edition of this book, and we would like to thank all of them for their work and dedication.

We wish the owner of the book a pleasant read! Best regards, Martin Zeeb and Eckhard Lammert

Brain

Brain: Overview

Lorraine V. Kalia and Minesh Kapadia

Anatomy and Physiology of the Brain

The brain is a remarkably complex organ both in its structure and function. At the macroscopic level, it can be divided into three major components: (1) brainstem (which includes medulla, pons, and midbrain), (2) cerebellum (with its cortex and deep nuclei), and (3) cerebral hemispheres (which are composed of cerebral cortex, subcortical white matterSubcortical white matter, basal ganglia and thalamus, limbic system, and hypothalamus and pituitary). The cerebral cortex itself is divided into frontal, parietal, temporal, and occipital lobes (Fig. 1). At the microscopic level, there are two primary cell typesPrimary cell types: (1) neurons (which receive, process, and transmit information by electrical and biochemical changes mediated, in part, by neurotransmitters) and (2) glia (which are a diverse group of cells with expanding roles in brain development, function, and disease).

The various macroscopic regions of the brain are responsible for different physiological functions. The brainstem contains nuclei required for autonomic functions, such as regulation of heart rate and respiration. Most cranial nerves, which provide motor and sensory function to structures of the cranium (or skull), are also located within the brainstem. These include the trigeminal nerve (cranial nerve V); its sensory portion supplies touch, temperature, and pain sensation to the face, as well as innervates the cerebral vessels to form the trigeminovascular system. The cerebellum functions to coordinate movements. The cerebral cortex contains areas important for motor and sensory functions, as well as association areas, which are required for more complex functions, such as language and executive function. The basal ganglia, including the substantia nigra, are responsible for the control of motor activity. The limbic system supports a variety of functions including memory and emotion. It receives inputs from diverse areas of the brain; for example, the mesolimbic system, which plays important roles in reward, motivation, and addiction, is composed of projections from the midbrain to limbic areas (see chapter "Major Depression and Metabolic Syndrome"). The thalamus plays a critical relay function by mediating all motor output from and nearly all sensory input to the cortex. The hypo-

L. V. Kalia (✉)
Krembil Research Institute, Toronto Western Hospital, University Health Network, Toronto, ON, Canada

Division of Neurology, Department of Medicine, University of Toronto, Toronto, ON, Canada
e-mail: lorraine.kalia@utoronto.ca

M. Kapadia
Krembil Research Institute, Toronto Western Hospital, University Health Network, Toronto, ON, Canada

E. Lammert, M. Zeeb (eds.), *Metabolism of Human Diseases*, https://doi.org/10.1007/978-3-031-96019-2_2

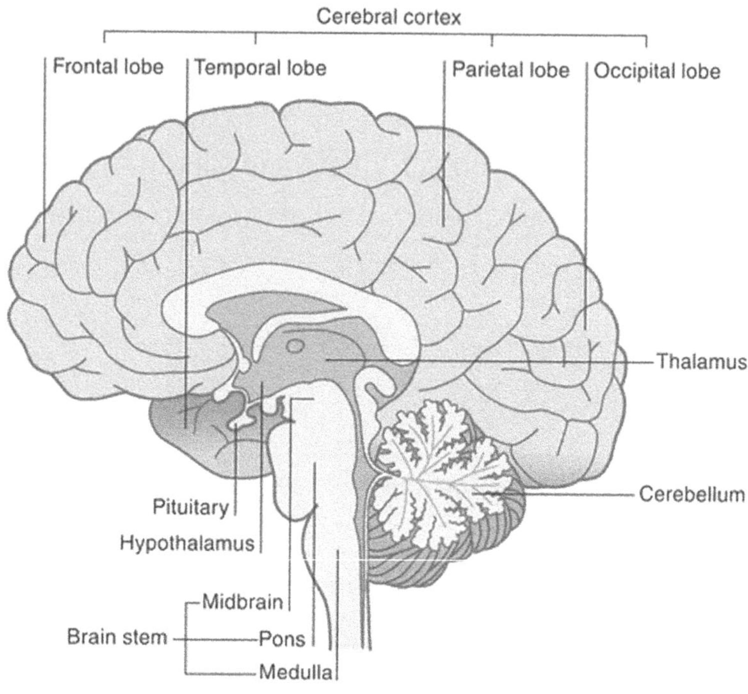

Fig. 1 Basic macroscopic anatomy of the human brain

thalamus is mainly involved in the regulation of visceral and endocrine activities, with the hypothalamus and pituitary being the major hormonal regulators (see chapters "Major Depression and Metabolic Syndrome," "Rheumatoid Arthritis," "Reproductive System: Overview").

Optimal brain function requires a well-regulated metabolic environment involving a complex interplay among different types of neurons and glia. Extracerebral systems remove metabolic products and protect the brain from sudden metabolic perturbations. Necessary nutrients are delivered via the circulation and must cross the blood–brain barrier (BBB), which is formed by endothelial cells lining cerebral microvessels, their basal lamina, and the end-feet of specialized glial cells called astrocytes. Astrocytes and the entire BBB protect neurons from toxic metabolites by preventing their transport into the brain via exclusion or efflux and by neutralizing harmful compounds via uptake or enzymatic inactivation [1]. Oligodendrocytes are another type of glial cell found in the brain, which generate myelin, an extended membrane of the oligodendrocyte that wraps tightly around the axons of neurons (see below). Radial glia serve as scaffolds for developing neurons as they migrate to their end destinations. Microglia, which are derived from the myeloid hematopoietic lineage, are the immune-competent cells of the brain. They migrate into the brain early in development and transform into a ramified state with small somata and long highly motile processes to constantly surveil their environment for perturbations and participate in the removal of cellular debris [2]. Specialized ependymal glial cells line fluid-filled ventricles of the brain and are involved in the production of cerebrospinal fluid (CSF). Functionally, CSF provides buoyancy, nourishment (e.g., vitamins), and endogenous waste product removal for the brain by bulk flow into the venous and lymphatic systems. Although the brain lacks lymphatic vasculature, it contains a network of lymphatic vessels that run beside the dural venous sinuses and extends upon the surrounding dura mater [3, 4]. These meningeal lymphatics drain CSF to deep cervical lymph nodes, clear macromolecules and waste (including β-amyloid; see chapter "Alzheimer's Disease"), and can traffic immune cells under normal and inflammatory conditions (see chapter "Multiple Sclerosis") [5].

Neuronal dysfunction and associated neurological diseases can occur when the brain's stable metabolic environment is disrupted.

Brain-Specific Metabolic/Molecular Pathways and Processes

Neurons are uniquely designed to receive information from the environment or other neurons, process this information, and send information to other neurons or effector tissues (i.e., neurotransmission). The cell body, or soma, contains the nucleus and additional organelles required for protein synthesis and metabolic maintenance. In most neurons, several dendrites and a single axon extend from the soma. Information is typically transported from the dendrites to the soma to the axon within a neuron by means of electrical events, which are mediated by the opening and closing of specific ion channels. Regulated transport of Na^+ and K^+ through the cell membrane is critical. Depolarization of the cell membrane occurs if positive ions (Na^+) enter the cell, whereas hyperpolarization results from exiting of positive ions (K^+) from the cell. When the cell reaches a threshold of depolarization, an electrical signal called an action potential is generated, and this signal is propagated along the length of the axon. Gaps in the myelin sheath coating the axon, where voltage gated sodium and potassium channels cluster, allow for ions to diffuse in and out of the neuron, facilitating the rapid conduction of the signal down the axon (see chapter "Multiple Sclerosis"). At the axon terminal, the neuron communicates with another neuron or an effector tissue within a specialized structure called a synapse. A typical synapse consists of a presynaptic axonal bouton, a postsynaptic dendritic spine, and the intervening space called the synaptic cleft (Fig. 2). Arrival of the action potential at the axonal bouton triggers the release of neurotransmitter from presynaptic vesicles into the synaptic cleft, thereby transforming the electrical signal into a chemical one.

A variety of molecules can act as neurotransmitters: amino acids, such as glutamate (Glu) and γ-aminobutyric acid (GABA), monoamines (including histamine, serotonin, and the catecholamines dopamine, epinephrine, norepinephrine) and nonmonoamines (acetylcholine), nucleotides (e.g., adenosine), neuropeptides (e.g., substance P), and even gases (e.g., nitric oxide). Neurotransmitters act either directly or indirectly in controlling the opening of ion channels in the postsynaptic neuron or effector tissue (see below). They can be classified based on their effects on the postsynaptic cell; those neurotransmitters that cause depolarization are classified as excitatory, and those that cause hyperpolarization are classified as inhibitory. The major inhibitory neurotransmitter in the brain is GABA, whereas Glu represents the major excitatory neurotransmitter.

To exert their effects on the postsynaptic target, neurotransmitters first traverse the synaptic cleft and then bind to specific postsynaptic receptors. There are two major classes of receptors: ionotropic and metabotropic. Ionotropic receptors are transmembrane proteins with an intrinsic ion channel, which opens upon binding of the neurotransmitter to the receptor's extracellular domain. Metabotropic receptors do not contain their own ion channel, but neurotransmitter binding can activate intracellular signaling cascades, which produce second messengers that indirectly gate ion channels. Many neurotransmitters utilize postsynaptic receptors from both classes. As an example, Glu receptors include metabotropic Glu receptors, as well as *N*-methyl-*D*-aspartate (NMDA) receptors, α-amino-3-hydroxy-5--methyl-4-isoxazolepropionic acid (AMPA) receptors, and kainate receptors, which are ionotropic receptors.

The actions of neurotransmitters are terminated by their removal from the synaptic cleft (Fig. 2). Three mechanisms are involved in neurotransmitter removal: diffusion, enzymatic degradation, and reuptake into neurons or uptake into astrocytes. For example, serotonin is removed from the synaptic cleft by a reuptake mechanism as well as degrading enzymes such as monoamine oxidase A (see also chapters "Major Depression and Metabolic Syndrome"). Similarly, dopamine is cleared from the synaptic cleft by dopamine transporters on the presynaptic

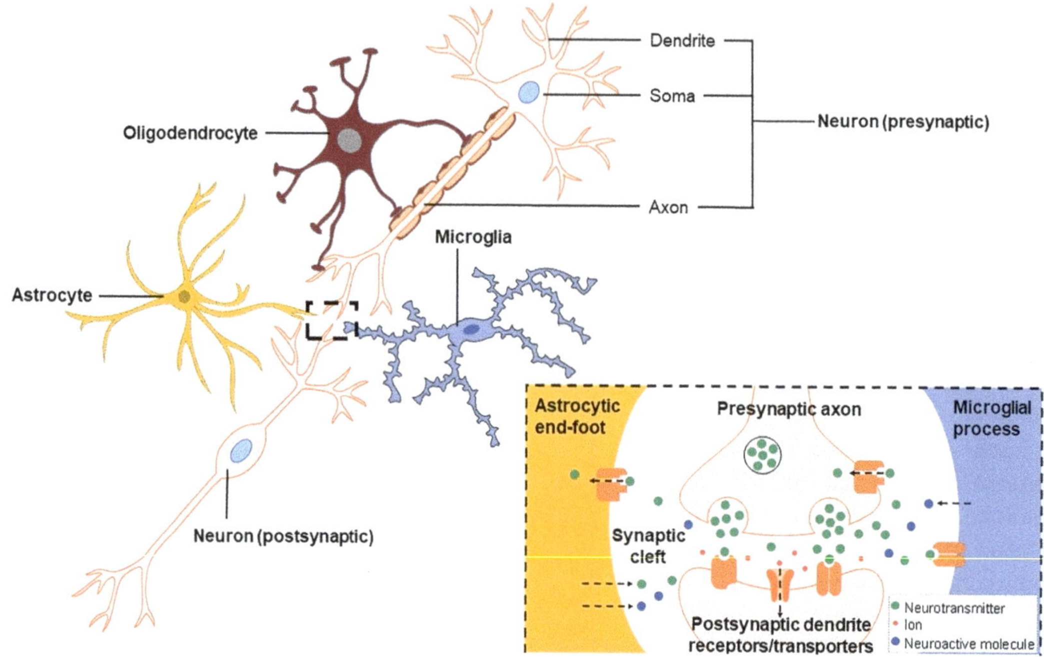

Fig. 2 Basic microscopic anatomy of the functional unit of the brain, the chemical synapse. Synapses exist as tri- or even quad-partite structures with glial end-feet and processes in direct contact with neuronal components. The propagation of an action potential in gaps created by the myelin sheath culminates at the axonal bouton, where the electrical signal is converted into a chemical signal through the release of neurotransmitters from presynaptic vesicles into the synaptic cleft. Neurotransmitters released from presynaptic nerve terminals bind to postsynaptic receptors that are often ion channels or are associated with ion channels. Neurotransmitters are also detected by glial cells that help to regulate efficient neurotransmitter release and clearance, as well as provide trophic factors to ensure healthy neuronal networks during development and adulthood

neuron or on astrocytes and then degraded by monoamine oxidase B and catechol-O-methyltransferase. The latter also acts to degrade dopamine within the synaptic cleft (see also chapters "Major Depression and Metabolic Syndrome"). Astrocytes also play an active role in synaptic transmission by releasing synaptically active molecules including Glu, GABA, and adenosine triphosphate (ATP) in response to neurotransmitter binding [6]. Microglia too express a variety of neurotransmitter receptors, including those for GABA and Glu, and may increase or decrease their release of neuroactive molecules as part of a positive or negative feedback loop, respectively. Microglia play a role in the phagocytic elimination of synapses as part of the widespread pruning of excess synaptic connections during development and activity-dependent synaptic stripping of less active inputs in adulthood [7].

Inside-In: Metabolites of the Brain Affecting Itself

The fundamental metabolic pathways of brain function have been uncovered studying diseases associated with inborn errors of metabolism. These inherited disorders can be classified into three major categories: (1) intoxication disorders (in which there is an acute or progressive accumulation of toxic compounds due to a metabolic block [e.g., phenylketonuria]), (2) storage disorders due to abnormal synthesis or degradation of complex molecules (e.g., leukodystrophies), and (3) energy production disorders resulting from deficiencies in energy production or utilization (e.g., mitochondrial disorders) [8]. Because these metabolic pathways are so critical to normal brain function, perturbations due to genetic mutations often result in the manifestation of disease in the neonatal period, infancy, or childhood. However, abnormalities in any of these metabolic pathways due to genetic and/or environmental factors can also lead to neurological disorders in adults.

As an example, energy production abnormalities are common to many neurological disorders, which present in adulthood. The lipid-rich content and high metabolic activity of neurons (see below) render the brain particularly susceptible to oxidative damage [9]. Mitochondria are the major source of both ATP, the principal medium of energy exchange in cells, and reactive oxygen species that can be highly deleterious for neurons. It is therefore not surprising that there is a growing list of neurological diseases in which aberrations in mitochondrial function are implicated, such as Down syndrome (see chapter "Down Syndrome") [10] and Alzheimer's disease (see chapter "Alzheimer's Disease") [11].

The brain is also sensitive to aberrations in protein homeostasis, or proteostasis. Similar to other cells, neurons possess a variety of cellular systems to manage abnormal or damaging proteins. For instance, a network of interactive molecules known as the chaperone system handles misfolded proteins by refolding the proteins or directing them toward protein elimination systems including the ubiquitin-proteasome system as well as the autophagy-lysosomal and endo-lysosomal pathways [12]. Impairments in these systems can cause harmful proteins to accumulate within the intracellular or extracellular space resulting in neuronal dysfunction and death. Neurodegenerative diseases are increasingly recognized as disorders of proteostasis and thus are termed proteinopathies. Specific proteins appear to accumulate, aggregate, and potentially spread from cell to cell in different neurodegenerative diseases [13]. For example, α-synuclein is implicated in Parkinson's disease, β-amyloid and tau in Alzheimer's disease (see chapter "Alzheimer's Disease"), and prion protein in Creutzfeldt-Jakob disease.

Inside-Out: Metabolites of the Brain Affecting Other Tissues

The by-products of neuronal metabolism are currently not known to directly cause disease in other organs. However, the brain indirectly influences metabolism in other parts of the body. On the one hand, the synthesis and release of hormones from the hypothalamus and pituitary regulate the endocrine system; and on the other hand, the action of neurotransmitters mediates neuronal circuits and ultimately regulates behaviors of the organism (e.g., dopamine mediates the reward system that regulates motivational behaviors; see chapter "Major Depression and Metabolic Syndrome").

The hypothalamus and pituitary are physically and functionally connected. The pituitary is composed of a posterior lobe and an anterior lobe. The supraoptic and paraventricular nuclei of the hypothalamus synthesize oxytocin and vasopressin (antidiuretic hormone), which are transported to, stored in, and released from the posterior pituitary lobe. The anterior pituitary lobe synthesizes and releases gonadotropins (which influence the gonads; see chapter "Reproductive System: Overview" under part "Reproductive System"), thyrotropin (which influences the thyroid gland), corticotropin (which influences the adrenal glands; see chapters "Major Depression and Metabolic Syndrome" and "Rheumatoid Arthritis"), prolactin (which influences the mammillary glands; see chapter "Reproductive System: Overview" under part "Reproductive System"), and somatotropin/growth hormone (which directly influences adipocytes and the liver; see also chapters "Fat Tissue: Overview" under part "Fat Tissue" and "Liver: Overview" under part "Liver"). Releasing or inhibiting hormones secreted by the hypothalamus act on the anterior pituitary gland to regulate the release of these hormones into the circulation (see also chapter "Reproductive System: Overview" under part "Reproductive System").

One of many examples of how the action of neurotransmitters mediates neuronal circuits and thereby regulates behaviors is the dopaminergic reward system, called the mesolimbic pathway. This circuit includes dopaminergic neurons within the medial substantia nigra and ventral tegmental area of the midbrain, which connect to the nucleus accumbens/ventral striatum, as well as limbic structures such as the amygdala and hippocampus. The dopaminergic mesolimbic pathway is involved in motivational behaviors to diverse environmental stimuli ranging in saliency from reward [14] to aversion [15]. Dysfunction of the mesolimbic dopaminergic circuit is believed to play a central role in schizophrenia (see chapter "Schizophrenia") [16].

Outside-In: Metabolites of Other Tissues Affecting the Brain

The energy demand of the brain is immense and mainly satisfied by consumption of glucose. In humans, the brain accounts for approximately 2% of the body weight, but it utilizes up to 20% of glucose-derived energy making it the main consumer of glucose [17]. The energy consumed by the adult brain is used by neurons for maintenance of the membrane gradient (generating ATP to drive ion pumps necessary for electrical transmission), synthesis and recycling of neurotransmitters, as well as dendritic and axonal transport. Unlike most other tissues, the brain has limited fuel stores and is quite inflexible with regard to substrates for energy metabolism due to the presence of the BBB. However, glucose is able to permeate the BBB, enabling the brain to derive most of its energy from the oxidation of glucose. Glucose delivery is related to local regulation of blood flow by astrocytes and its metabolism is connected to cell death pathways. Thus, neurons are particularly vulnerable to disruptions in glucose availability. For example, hypoglycemia is associated with aberrations of cerebral function, which can cause an altered mental state depending on severity and duration of glucose deprivation. Hence, nutrition critically influences brain function and, while glucose cannot be replaced as an energy source, it can be supplemented, as illustrated by elevated blood lactate levels following physical activity or the use of a ketogenic diet to increase blood levels of ketone bodies as treatment for epilepsy (see chapter "Epilepsy"). Metabolites of other tissues that enter the circulation must be able to cross the BBB to influence brain function. Disorders in which

metabolites lead to brain malfunction resulting in altered mental status ranging from a mild confusion to coma are referred to as metabolic encephalopathies. Hypoglycemia can cause such encephalopathy (see above). Abnormalities due to dysfunction in organs such as the kidney and liver may also lead to metabolic encephalopathies. More specifically, elevated ammonia in the brain due to liver failure is a major factor involved in the pathogenesis of hepatic encephalopathy (see chapter "Cirrhosis").

Certain hormones synthesized and secreted from peripheral tissues can also cross the BBB. Most hormones exert their influence by acting on various nuclei within the hypothalamus. For example, leptin is a hormone secreted by adipocytes, which acts on the arcuate nucleus and lateral hypothalamic area to suppress appetite (see chapter "Major Depression and Metabolic Syndrome"). Gonadal hormones (e.g., estradiol, progesterone, testosterone) and adrenal hormones (e.g., cortisol) circulating in the bloodstream act on the hypothalamus to suppress their own release (negative feedback control of the hypothalamo-pituitary-gonadal and hypothalamo-pituitary-adrenal axis, respectively). Hormones are increasingly found to also have effects on neuronal function in extra-hypothalamic brain areas. This is illustrated by leptin and ghrelin, two peripherally secreted peptide hormones implicated in depression (see chapter "Major Depression and Metabolic Syndrome").

Perspectives

The brain is critical for the survival of the organism, mediating functions and behaviors that range from basic and fundamental to incredibly complex. It accomplishes this through a vast number of connections that functionally link neural pathways in distinct macroscopic regions of the brain to each other. At the microscopic level, neurons and glia work in concert to fine-tune the release and subsequent degradation of neuroactive molecules in response to changes in the environment. This coordinated interplay at both macroscopic and microscopic levels demands a well-regulated and stable microenvironment that is protected from sudden metabolic perturbations. Yet, the unique requirements and complexity of the brain also render it particularly susceptible to defects in metabolism that can promote dysfunction and neurological disease. These include inherited disorders associated with inborn errors of metabolism within the brain that manifest early in life as well as impairments in energy production and proteostasis that develop later in adulthood in the context of neurodegeneration. Beyond the metabolites it produces centrally, the brain is also particularly sensitive to alterations in peripheral metabolites including glucose and hormones in circulating blood. Further, the brain itself can indirectly influence metabolism in other parts of the body through the release of hormones. Some important associations between metabolism and brain disease will be discussed in more detail in the following chapters.

Questions and Answers

Question 1 Name the two primary cell types found in the brain.

Answer 1 Neurons and glia

Question 2 Name three functions of cerebrospinal fluid.

Answer 2 Buoyancy, nourishment, and waste product removal

Question 3 What is the functional unit of the brain?

Answer 3 Chemical synapse

Question 4 Name three molecules that can function as neurotransmitters.

Answer 4 Amino acids (Glu and GABA), monoamines (histamine, serotonin, dopamine, epinephrine, norepinephrine), nonmonoamines (acetylcholine), nucleotides (adenosine), neuropeptides (substance P), and gases (nitric oxide)

Question 5 What is the primary source of energy for the brain?

Answer 5 Glucose

References

1. Daneman R, Prat A (2015) The blood–brain barrier. Cold Spring Harb Perspect Biol 7(1):a020412
2. Wolf SA, Boddeke H, Kettenmann H (2017) Microglia in physiology and disease. Annu Rev Physiol 79:619–643
3. Aspelund A et al (2015) A dural lymphatic vascular system that drains brain interstitial fluid and macromolecules. J Exp Med 212(7):991–999
4. Louveau A et al (2015) Structural and functional features of central nervous system lymphatic vessels. Nature 523(7560):337–341
5. Alves de Lima K, Rustenhoven J, Kipnis J (2020) Meningeal immunity and its function in maintenance of the central nervous system in health and disease. Annu Rev Immunol 38:597–620
6. Araque A et al (2014) Gliotransmitters travel in time and space. Neuron 81(4):728–739
7. Wu Y et al (2015) Microglia: dynamic mediators of synapse development and plasticity. Trends Immunol 36(10):605–613
8. Saudubray J-M, Berghe G, Walter JH (2012) Inborn metabolic diseases. Springer
9. Salim S (2017) Oxidative stress and the central nervous system. J Pharmacol Exp Ther 360(1):201–205
10. Izzo A et al (2018) Mitochondrial dysfunction in down syndrome: molecular mechanisms and therapeutic targets. Mol Med 24(1):2
11. Monzio Compagnoni G et al (2020) The role of mitochondria in neurodegenerative diseases: the lesson from Alzheimer's disease and Parkinson's disease. Mol Neurobiol 57(7):2959–2980
12. Friesen EL et al (2017) Chaperone-based therapies for disease modification in Parkinson's disease. Parkinsons Dis 2017:5015307
13. Soto C, Pritzkow S (2018) Protein misfolding, aggregation, and conformational strains in neurodegenerative diseases. Nat Neurosci 21(10):1332–1340
14. Syed ECJ et al (2016) Action initiation shapes mesolimbic dopamine encoding of future rewards. Nat Neurosci 19(1):34–36
15. de Jong JW et al (2019) A neural circuit mechanism for encoding aversive stimuli in the mesolimbic dopamine system. Neuron 101(1):133–151.e7
16. McCutcheon RA, Abi-Dargham A, Howes OD (2019) Schizophrenia, dopamine and the striatum: from biology to symptoms. Trends Neurosci 42(3):205–220
17. Mergenthaler P et al (2013) Sugar for the brain: the role of glucose in physiological and pathological brain function. Trends Neurosci 36(10):587–597

Major Depression and Metabolic Syndrome

D. Marazziti, G. Cappellato, I. Chiarantini,
S. Palermo, A. Arone, and L. Dell'Osso

Introduction to Major Depression

Major depressive disorder (MDD), generally known as depression, is one of the most severe and prevalent mental diseases worldwide. It is characterized by the presence of feelings of sadness, emptiness, anhedonia, and disturbed sleep and appetite, along with physical and cognitive changes that last at least 2 weeks, that markedly impair the person's ability to perform everyday tasks and psychosocial functions [1–3]. As reported by the World Health Organization, about 350 million people of all ages suffer from depression globally that is recognized as both a major contributor to the global burden of disease and a major cause of disability [4–7]. In the USA, for the year 2022, the economic burden of depression was estimated to be $83.1 billion, whereas in Europe, the annual cost due to MDD was €118 billion [8]. Major depressive disorder recognizes a complex and multifaceted etiology that results from the interactions involving genetic, epigenetic, developmental, and environmental factors [9–11]. In addition, MDD is associated with an increased risk of cardiovascular disease (CVD), obesity, and stroke, and possibly through the so-called metabolic syndrome (MeS) [12, 13]. In the last decades, converging studies explored the relationships between depression and MeS, while underlining a bidirectional association. MeS incidence varies globally and frequently relates to the prevalence of obesity. According to the National Health and Nutrition Examination Survey (NHANES) 24% of men and 22% of women suffer from MeS, affecting at least a fifth of American and around a quarter of European population [14, 15]. South-East Asia has a lower prevalence of MeS, but it is rapidly moving toward similar rates to those of Western countries [16]. Metabolic syndrome (MeS) is more common among African-American women and Hispanic women by, respectively, 57% and 26%, as compared with African-American or Hispanic men. Some symptoms of MeS, specifically, insulin resistance, hypertension, and dyslipidemia, are more frequent among Hispanics, African-Americans, and Whites. It is estimated that even 10.1% of US adolescents suffer from MeS [17], especially if they are men and Hispanics [18]. A meta-analysis explored the frequency and correlates of MeS in subjects with MDD. The results showed that 30.5% of the 5531 depressed participants satisfied the MeS criteria, with an odds ratio of 1.54, when compared with age- and gender-matched control groups. These results were confirmed in a subsequent meta-analysis conducted in 2015 by the same group reporting that the prevalence of the MeS was 32.6% (95% confidence interval

D. Marazziti (✉) · G. Cappellato · I. Chiarantini ·
S. Palermo · A. Arone · L. Dell'Osso
Department of Psychiatry, University of Pisa,
Pisa, Italy
e-mail: dmarazzi@med.unipi.it

E. Lammert, M. Zeeb (eds.), *Metabolism of Human Diseases*,
https://doi.org/10.1007/978-3-031-96019-2_3

[CI]: 30.8–34.3%; N = 198; n = 52,678) among individuals with severe mental illness [8, 19].

The purpose of this chapter was to review and comment on the relationships between depression and MeS in order to better understand their possibly shared pathophysiological mechanisms. Clinically speaking, this might indicate the need to assess metabolic profiles of depressed patients, and the relevance of paying attention to mood changes in patients with diabetes, hypertriglyceridemia, obesity, and hypertension. Again, researchers might find new targets for cutting-edge treatments for both conditions.

Association Between Depressive Symptoms and MeS

Different data indicate that depression and MeS are interrelated. Metabolic syndrome may predict the onset of depressive symptoms, and it is linked to a higher likelihood of chronicity, particularly its components involving obesity and dyslipidemia. Not surprisingly, the severity of depressive symptoms may increase according to lifetime exposure to MeS [20]. The association of MeS and depression in geriatric populations is often called vascular subtype of depression, as it is accompanied by metabolic alterations of blood vessels and circulation. If this association is truly causal, reducing the prevalence of obesity and MeS could potentially lead to a decline in the prevalence and incidence of depression in later life [21]. Most consistent finding is the positive association between depression and abdominal obesity (waist circumference [WC]/body mass index [BMI]), hypertriglyceridemia, and low high-density lipoprotein (HDL) cholesterol [20]. A link between affective symptoms and hypertension was observed in men, but not in women. On the other side, depressed women may be more likely to acquire MeS components of obesity and dyslipidemia. First, the link between MeS and depression in women points to a possible function for the gonadotropins. Second, women are more likely than men to have an increased hunger, eating dysregulation, binge eating, and subsequent weight gain. Men may be more inclined

to use alcohol or illegal substances, whereas depressed women are more likely to increase their caloric intake (hyperphagia). Third, according to epidemiological research, women's physical activity declines significantly with age [22]. It has been proposed that the term "metabolic depression" that refers to depression accompanied by metabolic disorders, may denote a chronic subtype of depression. Different clinical subgroups of depression have recently been linked to particular clinical or laboratory abnormalities. The findings demonstrated that compared to control women, those with atypical and undifferentiated characteristics have higher levels of overall and abdominal fat, and waist to hip ratio. In comparison to control patients, this group consistently showed higher levels of low-density lipoprotein (LDL), triglycerides, and total cholesterol, as well as higher levels of fasting glucose, insulin, and Homeostasis model assessment of insulin resistance (HOMA-IR), an indicator of insulin resistance [23]. An increased prevalence of MeS was reported among subjects with atypical and undifferentiated depressive symptoms compared to those with melancholic features. These findings are consistent with the hyperphagia that distinguishes atypical depression. Depressive episodes, lifestyle variables, co-occurring bulimia, genetic-biological factors, and medication all contribute to the greater frequency of obesity in depressed individuals. Some depressed symptoms seem to be linked to some MeS specifically (energy loss, often leading to decreased physical activity, might lead to elevated waist circumferences). Using a dimensional approach, three symptom dimensions were found: (1) "negative affect" that describes common signs of psychological discomfort, such as loss of attention or pessimism that are present in both depression and anxiety and may explain the significant comorbidity of these two disorders; (2) lack of "positive affect" targeting anhedonic symptoms that are mainly specific to depression; and (3) "somatic arousal", including symptoms of hyperarousal like palpitations, shortness of breath, and dizziness, specific for anxiety. Only the latter was found to be strongly and independently associated with the majority of MeS components (especially WC, triglycerides, and blood pressure). Major depression exhibits a more unfavorable course when it coexists with MeS or one of its components, while leading to a more severe clinical presentation, an increased risk for persistent and recurrent episodes (chronicity), a higher frequency of depressive episodes and suicidal thoughts, as well as poorer social functioning [24].

Pathophysiology

Although the precise relationship between MeS and depression is uncertain, different hypotheses have been proposed [25–27]. One of them suggests that MeS might be the result of the poor lifestyle of depressed subjects. This theory is supported by the existence of a strong connection between undesirable lipoprotein patterns (lower HDL level and elevated triglyceride level) and severe depression after adjusting for lifestyle-related variables, such as education, smoking status, alcohol use, and especially BMI. After controlling for all other variables, it should be underlined that lower HDL cholesterol and higher total and LDL cholesterol levels were independently associated with atypical depression with melancholy trait. According to an alternative view, MeS and depression may both induce changes in the hypothalamus-pituitary-adrenal (HPA) axis, the autonomic nervous system (ANS), the inflammatory system, oxidative process, as well as platelet and endothelial function [24, 28, 29] (Fig. 1).

The Hypothalamic-Pituitary-Adrenal Axis

It is well known that psychosocial stressors like loss-related events, interpersonal conflicts, difficult jobs, and social isolation can cause the hypothalamus to release catecholamines and the stress corticotropin-releasing hormone (CRH), promoting the synthesis of cortisol. Catecholamines interact with adreno-receptors, increasing nuclear factor (NF)-κB DNA binding

and releasing inflammatory mediators [30, 31]. Cytokines penetrate the brain and trigger neuroinflammation by activating inflammatory signaling pathways at the level of microglia [31]. In MDD, the cytokine-induced activation of CRH and subsequently of the HPA axis that in normal condition typically inhibits inflammatory signaling pathways, in this case loses its inhibitory impact. Indeed, glucocorticoids lose their capacity to restrain the production of cortisol and to block the activation of inflammatory pathways [32]. This situation, known as glucocorticoid resistance, would result in the continuation of the neuroinflammation and the hyperactivation of the HPA axis, and it appears to be related, at least in part, to the effects of cytokines on glucocorticoid receptors [31]. These occurrences actually produce excitotoxicity, reduce neurotrophic supports, and change the metabolism of monoamines that might contribute to the brain damage that underlies or at least accompanies MDD [30, 33, 34]. Several researchers found elevated cortisol levels in depressive patients' urine, plasma, and saliva that were subsequently linked to enlarged pituitary and adrenal glands [35]. The HPA axis changes, the presence of hypercortisolemia, the induction of inflammatory patterns (such as an increase in circulating cytokines), the stimulation of particular leukocyte populations, and changes to factors associated with cardiovascular risk (such as CRP and platelet reactivity) may all be components of the same pathogenetic mechanism in depression. Desensitization and lack of inhibition in HPA feedback are two ways through which HPA axis dysfunction in depression may be mediated by mineralocorticoid receptors [35]. The effects of this imbalance and glucocorticoid resistance lead to a persistent release of pro-inflammatory cytokines [36]. These cytokines activate particular metabolic pathways, such as the tryptophan shunt (also known as the kynurenine [KYN] pathway), resulting in the formation of quinolinic acid (QA) in the Nicotinamide adenine dinucleotide $(NAD)^+$ pathway, which ultimately results in the stimulation of the glutamatergic system with a markedly reduced production of serotonin (5-HT) [34].

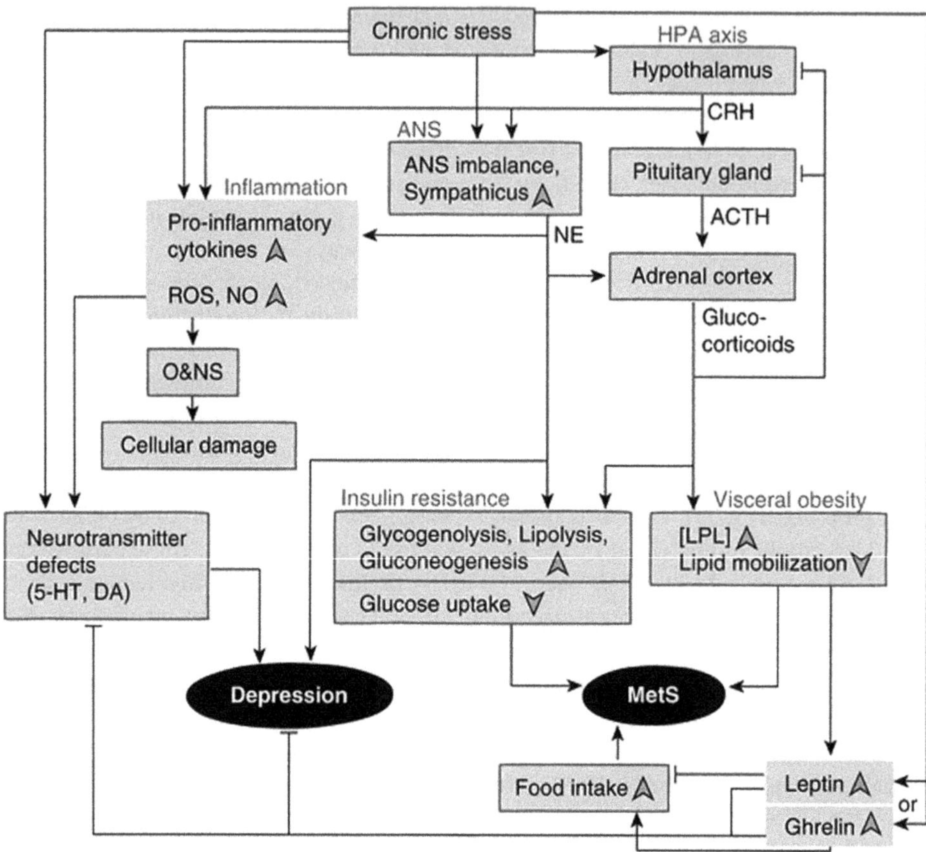

Fig. 1 Pathophysiological overlap between depression and metabolic syndrome. Depression and metabolic syndrome (*MetS*) share common pathways in the stress system, involving an abnormal activation of the hypothalamus-pituitary-adrenal (*HPA*) axis and an imbalance of the autonomic nervous system (*ANS*). In both conditions, a low-grade systemic inflammation manifests, which leads to enhanced oxidative and nitro-sative stress (*O&NS*). An antidepressant efficacy has been demonstrated for leptin and ghrelin, two peripheral hormones classically implicated in the homeostatic control of food intake. *5-HT* serotonin, *ACTH* adrenocorticotrophic hormone, *CRH* corticotropin-releasing hormone, *DA* dopamine, *LPL* lipoprotein lipase, *NE* norepinephrine, *NO* nitric oxide, *ROS* reactive oxygen species

Autonomic Nervous System

Abnormalities in ANS activity are typically observed in MeS, insulin resistance, and depression. Increased QT variability, baroreflex dysfunction, and lower heart rate variability are frequent in depressed individuals and have all been associated with an increased risk of cardiac mortality, including sudden death. Additionally, resting heart rates in depressed individuals are usually greater than in healthy controls. Additional markers of excessive sympathetic activity include increased norepinephrine metab-olites and not suppression of the dexamethasone suppression test. A dysregulated sympathetic/parasympathetic balance has been connected to insulin resistance, elevated blood insulin concentrations, and decreased insulin sensitivity. Sympathetic stimulation may lead to an increase in catecholamine release and circulation of free fatty acid that worsens insulin resistance. It is believed that the sympathetic branch contributes significantly to MeS, by impairing muscle glucose uptake and negatively affecting the function of the heart, large arteries, and skeletal muscle. However, the intra-abdominal compartment of the ANS is turned in favor of the parasympathetic branch that causes an increase in insulin secretion and the development of intra-abdominal adipose tissue [37].

Inflammation

As already mentioned above in the section on HPA axis, stressful stimuli play a role in the development of mood disorders [30, 38–43]. Different data suggest that inflammatory processes and chronic exposure to cytokines are activated in MDD. In fact, cytokines can cross the blood-brain barrier and, once inside the brain, they activate immune and endothelial cells, which in turn cause the release of more inflammatory mediators [41, 44]. The most accurate biological indicators of inflammation in depressed patients are interleukin (IL)-1, IL-6, Tumor necrosis factor (TNF), and C-reactive protein (CRP). Pro-inflammatory cytokine gene polymorphisms, such as those for IL-1, TNF, and CRP, have been linked to depression [30]. Pro-inflammatory cytokines, like interferon (IFN), or their inducers (such as endotoxins or the antityphoid vaccine) induce depressed symptoms in healthy people. On the contrary, individuals with other diseases including rheumatoid arthritis (see chapter "Rheumatoid Arthritis"), psoriasis, and cancer (see chapter "Cancer Metabolism") have considerably fewer depressed symptoms when TNF-type cytokines or other inflammatory factors (such as cyclo-oxygenase 2 [COX-2]) are blocked. Patients who had severe depression experienced the same outcomes [10]. According to a recent study, where nearly half of resistant participants with failure to respond to conventional therapies had a CRP level > 3 mg/L, the presence of inflammation appears to limit the responsiveness to ADs (indicative of elevated inflammatory status) [45]. Antidepressants in depressed mice were associated with a rise in hippocampal neurogenesis and a shift in microglia toward the neuroprotective M2 phenotype [46, 47] with the production of IL-4 by T-cells in the meningeal space that stimulates astrocytes to produce brain-derived neurotrophic factors (BDNFs), a neurotrophin promoting development, survival, and repair of neurons [47]. This process also encourages the transition of meningeal macrophages and monocytes from the pro-inflammatory M1 phenotype to the less inflammatory M2 phenotype. The recent identification of a previously unidentified brain lymphatic system has piqued interest in the mobility of T-lymphocytes within the brain, particularly the meningeal region. In times of stress, T-Reg cells may contribute to the suppression of inflammation and maintenance of neuronal integrity [47]. Depression is associated with a decreased plasma concentration of tryptophan (TRP) and elevated IL-1, TNF, and IFN production [31, 33]. Pro-inflammatory cytokines activate the enzyme indoleamine 2,3-dioxygenase by activating inflammatory signaling pathways such as NF-κB, p38 mitogen-activated protein kinase (MAPK), and STAT1a (that activates IDO). The latter breaks down TRP into TRP catabolites (TRYCATs) that include kynurenine (KYN) along the IDO route [31, 33, 34, 48]. It is also possible that the lower levels of 5-HT found in some MDD patients are due to the pro-inflammatory cytokines' activation of the inflammatory molecular signaling pathways, which result in IDO activation and lower plasma levels of TRP. As a result, the traditional hypothesis of TRP and 5-HT depletion could be replaced by the notion of inflammation-induced TRP degradation and TRYCAT synthesis in the 5-HT hypothesis of MDD ([34, 49]; Catena-Dell'Osso et al. 2016). In fact, more TRP would become TRYCATs after IDO activation, resulting in a drop in both 5-HT and TRP. It has been discovered that TRYCATs can make mice exhibit sad and anxious behaviors [33, 34]. TRYCATs also have neurotoxic side effects. For example, KYN can be metabolized into quinolinic acid (QA), a strong N-methyl-D-aspartate (NMDA) agonist that promotes glutamate release while causing lipid peroxidation [31, 33, 34, 48].

Oxidative Stress

The patients' metabolic and redox status may be impacted by the failure of the HPA axis, monoaminergic, and neurotrophic systems. Additionally, poor lifestyle and food choices may contribute to a dysfunctional response to stress that might continue an antioxidant response as well as the metabolic and nutritional features of depressed patients [50, 51]. Through the modulation of neurotransmitters (glutamate-type), the induction of oxidative stress and an immune-inflammatory response may play a significant role in the pathogenesis of depression by altering crucial brain functions involved in the neurobiology of depression [51–58]. Nonenzymatic antioxidants and enzymatic antioxidants are the two primary elements that make up the antioxidant system [51]. Nonenzymatic antioxidants, which include glutathione, thiols (R-SH), plasma proteins, uric acid (UA), vitamin C, vitamin E, zinc, and coenzyme Q10, exert antioxidant activity and neutralize the free radicals produced and residues of the oxidative metabolism [51]. Enzymes having specialized roles in the neutralization of potentially harmful radical or reactive species, such as catalase, glutathione peroxidase, and reductase, and the thioredoxin system, make up the second part of the antioxidant system [51]. Depression may be associated with altered levels of oxidative stress indicators and lower concentrations of certain nonenzymatic and enzymatic antioxidants, which can be restored with the use of antidepressants. In addition, several antioxidants, including zinc, N-acetylcysteine, and omega-3 free fatty acids, exhibit antidepressant effects [59]. The result of oxidative imbalance appears to be related to an excessive production of reactive oxygen species (ROS), which, at low physiological concentrations, serve as signaling molecules and are crucial for the immune response to control the activity of various cells. However, ROS in high concentrations are known to dramatically harm DNA, lipids, proteins (including receptors and enzymes), and other cellular components, leading to apoptosis and cell death. Indeed, the CNS is very susceptible to oxidative damage [59] causing depression. In a sample of depressed patients, the association between depression and lipid peroxidation was examined showing that there was a drop in PUFAs (polyunsaturated fatty acids) in the lipid membranes of red blood cells, which suggested a rise in long-chain peroxide degradation [60]. In severe forms of depression, DNA damage was seen in the

blood, urine, and brain tissue, as well as an increase in oxidation [60]. The primary defenses against free radical damage to the brain are antioxidant enzymes, which are expressed both peripherally and centrally. The activity of these enzymes, such as copper-zinc superoxide dismutase (SOD), catalase, and glutathione peroxidase, has generally been found to differ from that seen in healthy persons in patients with MDD. Superoxide dismutase (SOD) activity, which is positively correlated with the degree of depression, appears to be affected by ADs. Catalase and glutathione-reductase activity tend to be elevated in MDD and ADs (see chapter "Alzheimer's Disease") would restore the activity of such enzymes [61]. Purine nucleotides are broken down metabolically to produce uric acid, which may also have an impact on MDD [62]. The regulation of the sleep-wake cycle, hunger, cognition, memory, seizure threshold, social interaction, and impulsivity appears to be largely influenced by uric acid and purines [62]. A recent study found that enhancing the type 1 and type 2 purinergic modulators' efficacy appears to significantly alter the clinical picture of the mood disorder [63]. The P1 and P2 trans-membrane receptors, which are distinguished by their pharmacological activation, are part of the purinergic system. P1 is related to protein-G and has four different receptor subtypes (A1, A2a, A2b, and A3). Its activation results in an inhibition of the atrioventricular node, a reduction in cardiac output, coronary vasodilation, bronchospasm, inhibition of neutrophil degranulation, and consequently an anti-inflammatory effect. An increase in secretions, immunological response, platelet aggregation, heart rhythm, vascular tone, and nociception are all made possible by P2, which has two receptor subtypes (P2X, P2Y), once they are activated [64]. The function of ATP as a neurotransmitter on some nerves specifically referred to as "purinergic" was discovered at the beginning of the 1970s. The CNS contains both ATP and adenosine, and neurons can store and release ATP. The purinergic system is present in the cerebral cortex, hypothalamus, basal ganglia, hippocampus, and other limbic system regions of the brain. The purinergic system, which links glial and neuronal cells, produces the intercellular calcium waves that form synapses and support neural plasticity. As a result of changes in the intracellular concentrations of Ca2+ and cAMP, activated synapses and axons release ATP to activate purinergic receptors on glial cells. This circumstance affects the motility, myelination, and both proliferation and differentiation of glial tissue. The gamma-aminobutyric acid (GABA), dopaminergic, glutamatergic, and serotonergic systems, all involved in the pathophysiology of mood, are likewise impacted by the purinergic system's impacts [65]. By activating various purinergic receptors, both adenosine and ATP, as well as their metabolites, may cause cascade effects. The downstream product of adenosine, uric acid (UA), is produced by xanthine oxidase (XO), which cleaves adenosine after it has been broken down into inosine by adenosine deaminase and hypoxanthine by a nucleosidase. The purinergic system may be linked to mood dysregulation and might be potential targets for future drugs [66]. Two inhibitors of the XO enzyme, allopurinol and febuxostat, showed an antidepressant effect like fluoxetine in animal models [67].

Platelet and Endothelial Functions

The classic platelet functions of adhesion, aggregation, and release of prepared mediators make platelets to be a major component that play a role in systemic inflammatory reactions [68]. Numerous factors are produced by the inflammatory process, such as an increase in platelet activation factor (PAF) and the release of thrombin that activates platelets [69]. Through G-protein coupled receptors, PAF and thrombin cause rapid changes in the structure and activity of the platelet, which control the secretome's release (the collection of proteins expressed and secreted into the extracellular space), the translocation of P-selectin on the surface of the platelet cell membrane, which serves as a bond for the leukocytes, and the expression of αIIbβ3 integrin, which binds fibrinogen and permits homotypic plate. In addition to the quick release of inflammatory mediators from intracellular granules during platelet activation, platelets also produce proteins in response to external signals [70] such as cyclooxygenase-2 (COX-2), tissue factor (TF), IL-1, and matrix metalloproteinases (MMP-2 and MMP-9) that prolong inflammation [71]. In the reuptake, storage, and metabolism of 5-HT, platelets share significant biochemical similarities with presynaptic serotonergic neurons [72], and as a result, they are frequently utilized as a peripheral model of neuronal activity in biological psychiatry. According to several studies, depressed people have altered platelet structure, reuptake system or monoamine receptor dysfunction, altered basal and/or induced intracellular calcium levels, variations in the release of bioactive substances, exposure of pro-aggregating molecules on the cell surface, and, in some cases, an increased response to pro-activating stimuli. Research found variations in the amounts of amyloid-(A) and other amyloid-precursor protein (APP) derivatives in the plasma of people who had depression. Since platelets are the primary source of amyloid-(A) in plasma, it is conceivable that people who have depressive symptoms may also have altered APP metabolism and amyloid-(A) generation due to altered platelet responsiveness to specific pro-activating stimuli. There should be a link between depression and MeS, cardiovascular events, and strokes because of changes in APP metabolism and amyloid-(A) formation that cause platelets hyperactivation. It is believed that elective 5-HT reuptake inhibitors (SSRIs) can reduce mortality in depressed individuals by preventing cardiovascular events for their antiplatelet effects in vivo normalizing their function [28].

Coronary Heart Disease

Coronary heart disease (CHD), stroke (see chapter "Stroke"), heart failure (see chapter "Heart Failure"), and a variety of other heart and blood vessel dysfunctions are all considered to be examples of cardiovascular disease (CVD), commonly known as heart disease. For both adult males and females, CVD continues to be the leading cause of death worldwide. Depression and cardiovascular disease are strongly correlated [73–77]. Indeed, it has been demonstrated that depression doubles or triples a healthy person's risk of cardiovascular disease. In addition, depression in people with cardiovascular disease doubles the risk of cardiac morbidity and mortality by a factor of 1.5–2.5 [78]. Additionally, it is clear that individuals with more severe depression have a larger likelihood of suffering from cardiovascular events with obviously an increased morbidity and mortality (Amadio et al. 2020). Despite the comorbidity of CVD and depression, the molecular mechanism between these two conditions is still poorly understood. However, there is evidence that inflammation may play a significant role: elevated levels of systemic inflammatory markers in depression, particularly interleukin 6 (IL-6) and C-reactive protein (CRP), increase the risk of developing CVD. As the body may lose the capacity to regulate the inflammatory response due to a modification of the mechanism of cortisol production, stress-induced inflammation in individuals with depression may show a causative pathway to the development of heart disease [79]. Blood artery constriction and increased blood pressure caused by cortisol increase the circulation of oxygen-rich blood. However, chronic high cortisol levels can lead to blood vessel damage and plaque formation that is the scenario for CVD [80].

Peripheral Hormones

Ghrelin, an endogenous brain-gut peptide, seems essential not only for the regulation of appetite but also for neurohumoral processes such inflammation, cardiovascular function, anxiety, and depression. The stomach's parietal cells release this hormone that enters the circulation, crosses the blood-brain barrier, and influences different brain regions [81, 82]. Elevated levels of ghrelin are supposed to reduce the symptoms of depression enhancing appetite [82], while in pathological obesity and depression, there is an improper ghrelin regulation in different brain areas such as hypothalamus, the dentate gyrus, and hippocam-

pus [83–88]. Leptin, another hormone regulating appetite, is released from adipocytes and through the bloodstream it reaches the brain, where it reduces appetite and promotes weight reduction [43, 89–91]. Leptin and ghrelin exert opposite effects on stress response, with the first inhibiting corticotropin releasing factor (CRF) mRNA expression and glucocorticoid levels, and the second exerting opposite actions. Dopaminergic neurons of the ventral tegmental area (VTA) express both leptin and ghrelin receptors. Leptin acts on the mesolimbic system to limit responses to food rewards, while suppressing the rewarding properties of feeding that are instead promoted by ghrelin [89]. Several converging studies of the last 40 years increasingly suggest that serotonin (5-HT) may control satiety and feeding behaviors through different receptors, specifically 5-HT2C and 5-HT6 at the level of the hypothalamic ventromedial and lateral nuclei. Serotonin interacts with both orexin and melanocortin-stimulating hormone (MSH) in the nucleus of the solitary tract by integrating peripheral satiety inputs [92]. Again, 5-HT is intertwined with ghrelin and leptin [28]. Leptin inhibits appetite by decreasing the 5-HT synthesis or release by neurons of the brainstem, targeting arcuate nucleus 5-HT1A and 5-HT2B receptors, but it also increases appetite depending on the receptor subtype involved. This implies that 5-HT may modulate eating behaviors through a receptor-mediated signaling network in different brain regions and peripheral tissues. The 5-HT neurotransmission and its receptor functions are also influenced by glucocorticoid response in the CNS, affecting feeding behavior and the choice of macronutrient composition. The functioning of the 5-HT network is tightly regulated by the balance between its biosynthesis rate from the essential amino acid tryptophan, its release and catabolism [93], by the degree of occupation/sensitivity of the different 5-HT receptors, and the activity of intracellular 5-HT reuptake via the protein carrier 5-HT transporter (SERT) [28]. Additionally, peripheral 5-HT has been related to adipose tissue formation through modulating energy metabolism [94]. The reduced SERT number seems to constitute a marker of vulnerability toward the onset of depression in obese subjects [28, 95]. High BMI subjects show high levels of leptin and desire for continuing to eat and that increased levels have been associated with 20% prevalence of depression in obese subjects [96]. Leptin triggers depression in obese subjects possibly through a decreased density of the 5-HT transporter (SERT) and a reduced 5-HT uptake. High levels of glycemia, linked to insulin resistance, were positively correlated with SERT affinity, implying a higher affinity for 5-HT under this condition. Insulin stimulated by meal consumption, promotes the synthesis and release of 5-HT (brain) and leptin (adipocytes) [97]. Increased food intake would stimulate peripheral 5-HT generated from the gut and pancreatic cells to induce more insulin secretion, body fat increase [98], further release of leptin, and, ultimately, insulin and leptin resistance, while reinforcing hunger circuits. The normal cooperation between insulin and 5-HT is lost, and insulin, instead of being hampered, continues to be produced by pancreatic islets leading to insulin resistance [28, 90, 96, 99–102].

Diet and Depression

Although the relation between nutrition and depression is not a topic of the current review, a few underlying points should be mentioned. Unhealthy behaviors are strongly related to depression. In particular, the diet of depressed patients is frequently characterized by an excessive intake of calories, carbohydrates, and cholesterol, as well as a deficiency in the consumption of fish, vegetables, cereals, and vitamins [103, 104]. In this regard, studies highlight links between the consumption of specific food groups like seafood or fish and a lower incidence of depression [105, 106]. One of the reasons might be that these foods ensure an adequate intake of vitamins, particularly B vitamins such as B1, B2, and B6, along with folic acid, omega 3-polyunsaturated fatty acids (v3-PUFAs), and monounsaturated fatty acids (MUFAs) that are essential for the homocysteine cycle, which is required for

the synthesis of catecholamines, 5-HT, and other monoamine neurotransmitters [103, 107]. Furthermore, their absence might affect the production of biogenic amines by producing an excess of homocysteine and its metabolites that may cause neurotoxicity through interacting with a glutamate receptor subtype, specifically the *N*-methyl-*D*-aspartate (NMDA) receptor. According to current data, up to 30% of depressive patients show high homocysteine levels; this finding supports the notion that hyperhomocysteinemia is a risk factor for depressed mood [108, 109]. In fact, excitotoxicity, DNA damage, oxidative stress, and inflammation brought on by hyperhomocysteinemia may result in altered neurological functioning [110]. In both cross-sectional and prospective research, eating habits like the Mediterranean diet (MD) and "anti-inflammatory" diet (low in meat and dairy products; high in fruits, nuts, vegetables, legumes, grains, olive oil, and fish; moderate alcohol consumption) are associated with a lower incidence of depression [111, 112]. Increased fish consumption—that is high in v3-PUFA—has been linked to a global decline in the annual prevalence of depression. The biosynthesis of pro-inflammatory cytokines, especially TNF-α and interleukin 1β, appears to be suppressed by v3-PUFA [113, 114]. Furthermore, via altering G-proteins, PUFAs may indirectly influence the plasma membrane. G-proteins go through fatty acylation that redirects the proteins to lipid rafts and changes the signaling that follows downstream [113]. Even olive oil that contains oleic acid and is a strong source of MUFAs, has been proven to decrease the incidence of depression. The D-9 desaturase enzyme activity, essential for the physical and chemical properties of neuronal membranes, is increased by MUFAs, which also has antioxidant qualities and enhances 5-HT binding to its receptor. Additionally, it has been demonstrated that the fatty acid composition of certain brain nuclei varies, as it is related to the amount of gray or white matter present in the nuclei. For instance, cerebral white matter is high in MUFA but low in n-3 and n-6 PUFAs. By contrast, n-3 PUFAs are significantly more abundant in gray matter. It is important to highlight that impaired emotional behaviors seem to be closely related to changes in brain fatty acid composition that are predominantly caused by variations in dietary fat content [115, 116]. The effects of every nutrient component alone might not be as significant as their synergistic combination. Indeed, it has been observed that a Mediterranean diet has an impact on the shared mechanisms between MeS and depression. The MD specifically promotes beneficial changes in insulin/glucose balance, decreases pro-inflammatory cytokines, decreases plasma homocysteine levels, and enhances endothelial function. In particular, the MD (Mediterranean diet) style has been linked to significant improvements in fasting blood glucose, glycated hemoglobin (HbA1c), and insulin levels [117, 118]. In conclusion, there is a reciprocal association between depression and eating patterns. People who are depressed are more likely to follow poor diets that focus mostly on delicious foods with high sugar and fat contents. Furthermore, a good diet should help prevent depression by influencing the vascular, inflammatory, and metabolic pathophysiologic systems positively.

Perspectives

Given the crucial role that modifications of the metabolic networks play in depression, depressive syndromes could be conceived of as MeS. From a therapeutic perspective, clarifying the common pathophysiological mechanisms underlying depression and MeS should allow us to develop more tailored treatment and preventive strategies. Patients with mood disorders should receive treatment from a multidisciplinary, well-coordinated team that include checking for MeS risk factors, such as eating habits, exercise routines, concurrent binge eating disorder, bulimia nervosa, caffeine dependence, smoking, and thyroid dysfunction, as well as baseline and ongoing monitoring of anthropometrics (weight, body mass index, waist-to-hip ratio), fasting blood sugar, and lipid fractionation. The modification of risk variables should be a primary therapeutic objective. While a metabolic comorbidity is present, the "metabolic profile" of antidepressant, or its effect on glucose and lipid homeostasis, should be taken into account when making therapeutic decisions. Several psychotropic drugs have been associated with adverse metabolic effects. The antihistaminergic side effects of tricyclic antidepressants (TCAs) that cause weight gain and dyslipidemia, are principally responsible for the specific metabolic pattern displayed by TCA users. The use of TCAs is also associated with hypertension via peripheral $\alpha1$ adrenergic receptor agonism. It has been demonstrated that the treatment dosage and duration are positively related to the amount of weight gain. On the other hand, selective serotonin (5-HT) reuptake inhibitors (SSRIs) appear to temporarily reduce body weight before increasing it over the first few weeks of treatment. The concept of depression as a "metabolic syndrome of type II" is reinforced by the study of novel and cutting-edge "metabolic therapy" for depressed symptoms. In obese youngsters, a nutrition intervention significantly reduced symptoms of depression and anxiety. Exercise programs, whether used alone or in conjunction with other antidepressant medications, have been demonstrated to reduce the severity of affective symptoms. Future treatments for depression should also take into account other targets like cytokines and their receptors, intracellular inflammatory mediators, glucocorticoid receptors, and neurotrophic factors. It is essential to clarify the impact that currently available anti-inflammatory drugs play in the treatment of depression, and drugs with anti-inflammatory and antioxidant properties have to be investigated as either monotherapy or augmentation strategies.

Disclosures The authors do not have anything to disclose.

Questions and Answers

Question 1 Is there a correlation between depressive symptoms and MeS?

Answer 1 Different evidence sources suggest a connection between depression and MeS, in particularly its components like obesity and dyslipidemia. MeS is linked to an increased risk of chronicity and may predict the onset of depression symptoms. It should not be surprising that the severity of depression symptoms may worsen over the course of a lifetime of exposure to MeS. The correlation between depression and abdominal obesity (waist circumference/BMI), hypertriglyceridemia, and low HDL cholesterol is the most significant. A chronic subtype of depression has been suggested with the name "metabolic depression," which describes depression that is accompanied by metabolic abnormalities. Recently, specific clinical or laboratory abnormalities have been connected to several clinical subgroups of depression. This group consistently displayed higher levels of fasting glucose, insulin, and the insulin resistance indicator HOMA-IR, as well as higher levels of low-density lipoprotein (LDL), triglycerides, and total cholesterol when compared to control patients. Comparatively to those with typical and differentiated depressive symptoms, those with atypical and undifferentiated symptoms had a higher prevalence of MeS.

Question 2 What is the pathophysiology that depression and MeS share?

Answer 2 Different theories have been proposed, even though it is unclear exactly how MeS and depression are related. One of them hypothesizes that MeS may be induced by depressed subjects' inadequate lifestyles. After adjusting for lifestyle-related factors including education, smoking status, alcohol use, and especially BMI, there is a significant correlation between unhealthy lipoprotein patterns (lower HDL level and higher triglyceride level) and severe depression. Another theory suggests that MeS and depression may both affect the hypothalamic-pituitary-adrenal (HPA) axis, the autonomic nerve system (ANS), and the hypothalamus.

Question 3 What inflammatory alterations occur in depression?

Answer 3 Depression is associated with an HPA axis dysfunction that leads to glucocorticoid resistance and so a persistent release of IL-1, IL-6, TNF, and IFN. Pro-inflammatory cytokines cross the blood-brain barrier and trigger neuroinflammation (IDO) by activating inflammatory signaling pathways at the level of microglia, causing the release of more inflammatory mediators. Pro-inflammatory cytokines activate the enzyme indoleamine 2,3-dioxygenase (IDO) by activating inflammatory signaling pathways such as nuclear factor NF-κB, MAPK, and IDO. The latter, breaking down TRP into TRYCATs, can explain the hypothesis that depression is caused by low levels of TRP and 5-HT.

Question 4 What are the peripheral hormones altered in MeS and how are they implicated in depression?

Answer 4 An endogenous brain-gut peptide called ghrelin appears to be crucial for the control of appetite in addition to other neurohumoral processes like inflammation, cardiovascular function, anxiety, and depression. Elevated ghrelin levels are believed to increase hunger and alleviate the symptoms of depression, while pathological obesity and depression have inappropriate ghrelin regulation. Another hormone that controls appetite, leptin, is released from adipocytes and travels through the bloodstream to the brain, where it suppresses hunger and contributes to weight loss. Leptin inhibits the rewarding aspects of feeding that are instead encouraged by ghrelin. Serotonin (5-HT) may affect hunger and eating habits through a variety of receptors. Ghrelin and leptin are linked to 5-HT. According to the receptor subtype implicated, leptin decreases 5-HT synthesis. This suggests that 5-HT may alter eating habits by activating a network of receptor-mediated signals in many parts of the brain and periphery tissues. In obese people, the decreased SERT number appears to represent a marker of susceptibility to the onset of depression.

References

1. American Psychiatric Association (2013) Diagnostic and statistical manual of mental disorders, 5th edn
2. Li Z, Ruan M, Chen J, Fang Y (2021) Major depressive disorder: advances in neuroscience research and translational applications. Neurosci Bull 37(6):863–880
3. Villarroel MA, Terlizzi EP (2019) 2020. Symptoms of Depression Among Adults, United States
4. Friedrich MJ (2017) Depression is the leading cause of disability around the world. JAMA 317(15):1517
5. Li S, Li Y, Li X, Liu J, Huo Y, Wang J, Liu Z, Li M, Luo XJ (2020) Regulatory mechanisms of major depressive disorder risk variants. Mol Psychiatry 25(9):1926–1945
6. World Health Organization (2021) Depression. WHO
7. Young K, Singh G (2018) Biological mechanisms of cancer-induced depression. Front Psych 9:299
8. Al-Khatib Y, Akhtar MA, Kanawati MA, Mucheke R, Mahfouz M, Al-Nufoury M (2022) Depression and metabolic syndrome: A narrative review. Cureus 14(2):e22153
9. Leff-Gelman P, Mancilla-Herrera I, Flores-Ramos M, Cruz-Fuentes C, Reyes-Grajeda JP, García-Cuétara, M.delP., Bugnot-Pérez, M. D., & Pulido-Ascencio, D. E. (2016) The immune system and the role of inflammation in perinatal depression. Neurosci Bull 32(4):398–420
10. Marazziti D, Rutigliano G, Baroni S, Landi P, Dell'Osso L (2014) Metabolic syndrome and major depression. CNS Spectr 19(4):293–304
11. Tayab MA, Islam MN, Chowdhury KAA, Tasnim FM (2022) Targeting neuroinflammation by polyphenols: A promising therapeutic approach against inflammation-associated depression. Biomedicine & pharmacotherapy = Biomedecine & pharmacotherapie 147:112668
12. Bot M, Milaneschi Y, Al-Shehri T, Amin N, Garmaeva S, Onderwater GLJ, Pool R, Thesing CS, Vijfhuizen LS, Vogelzangs N, Arts ICW, Demirkan A, van Duijn C, van Greevenbroek M, van der Kallen CJH, Köhler S, Ligthart L, van den Maagdenberg AMJM, Mook-Kanamori DO, de Mutsert R et al (2020) Metabolomics profile in depression: A pooled analysis of 230 metabolic markers in 5283 cases with depression and 10,145 controls. Biol Psychiatry 87(5):409–418
13. Guerreiro Costa LNF, Carneiro BA, Alves GS, Lins Silva DH, Faria Guimaraes D, Souza LS, Bandeira ID, Beanes G, Miranda Scippa A, Quarantini LC (2022) Metabolomics of major depressive disorder: A systematic review of clinical studies. Cureus 14(3):e23009
14. Fahed G, Aoun L, Bou Zerdan M, Allam S, Bou Zerdan M, Bouferraa Y, Assi HI (2022) Metabolic syndrome: updates on pathophysiology and management in 2021. Int J Mol Sci 23(2):786
15. Swarup S, Goyal A, Grigorova Y, Zeltser R (2022 Jan.) Metabolic syndrome. 2022 may 2. In: StatPearls [internet]. StatPearls Publishing, Treasure Island (FL)
16. Pan WH, Yeh WT, Weng LC (2008) Epidemiology of metabolic syndrome in Asia. Asia Pac J Clin Nutr 17(Suppl 1):37–42
17. Miller JM, Kaylor MB, Johannsson M, Bay C, Churilla JR (2014) Prevalence of metabolic syndrome and individual criterion in US adolescents: 2001-2010 National Health and nutrition examination survey. Metab Syndr Relat Disord 12(10):527–532
18. Rochlani Y, Pothineni NV, Kovelamudi S, Mehta JL (2017) Metabolic syndrome: pathophysiology, management, and modulation by natural compounds. Ther Adv Cardiovasc Dis 11(8):215–225
19. Vancampfort D, Stubbs B, Mitchell AJ, De Hert M, Wampers M, Ward PB, Rosenbaum S, Correll CU (2015) Risk of metabolic syndrome and its components in people with schizophrenia and related psychotic disorders, bipolar disorder and major depressive disorder: a systematic review and meta-analysis. World psychiatry: official journal of the World Psychiatric Association (WPA) 14(3):339–347
20. Milaneschi Y, Simmons WK, van Rossum EFC, Penninx BW (2019) Depression and obesity: evidence of shared biological mechanisms. Mol Psychiatry 24(1):18–33
21. Jellinger KA (2022) Correction: Jellinger, K.A. Pathomechanisms of Vascular Depression in Older Adults. Int J Mol Sci 23(21):12949
22. McHugh RK, Sugarman DE, Meyer L, Fitzmaurice GM, Greenfield SF (2020) The relationship between perceived stress and depression in substance use disorder treatment. Drug Alcohol Depend 207:107819
23. Jiang D, Kong Y, Ren S, Cai H, Zhang Z, Huang Z, Peng F, Hua F, Guan Y, Xie F (2020) Decreased striatal vesicular monoamine transporter 2 (VMAT2) expression in a type 1 diabetic rat model: A longitudinal study using micro-PET/CT. Nucl Med Biol 82-83:89–95
24. Penninx BWJH, Lange SMM (2018) Metabolic syndrome in psychiatric patients: overview, mechanisms, and implications. Dialogues Clin Neurosci 20(1):63–73
25. Nedic Erjavec G, Sagud M, Nikolac Perkovic M, Svob Strac D, Konjevod M, Tudor L, Uzun S, Pivac N (2021) Depression: biological markers and treatment. Prog Neuro-Psychopharmacol Biol Psychiatry 105:110139

26. Roman M, Irwin MR (2020) Novel neuroimmunologic therapeutics in depression: A clinical perspective on what we know so far. Brain Behav Immun 83:7–21
27. Sakamoto S, Zhu X, Hasegawa Y, Karma S, Obayashi M, Alway E, Kamiya A (2021) Inflamed brain: targeting immune changes and inflammation for treatment of depression. Psychiatry Clin Neurosci 75(10):304–311
28. Marazziti D, Betti L, Baroni S, Palego L, Mucci F, Carpita B, Cremone IM, Santini F, Fabbrini L, Pelosini C, Marsili A, Massimetti E, Giannaccini G, Dell'Osso L (2022) The complex interactions among serotonin, insulin, leptin, and glycolipid metabolic parameters in human obesity. CNS Spectr 27(1):99–108
29. Zhang M, Chen J, Yin Z, Wang L, Peng L (2021) The association between depression and metabolic syndrome and its components: a bidirectional two-sample Mendelian randomization study. Transl Psychiatry 11(1):633
30. Miller AH, Raison CL (2016) The role of inflammation in depression: from evolutionary imperative to modern treatment target. Nat Rev Immunol 16(1):22–34
31. Rosenblat JD, McIntyre RS (2017) Bipolar disorder and immune dysfunction: epidemiological findings, proposed pathophysiology and clinical implications. Brain Sci 7(11):144
32. Feng X, Zhao Y, Yang T, Song M, Wang C, Yao Y, Fan H (2019) Glucocorticoid-driven NLRP3 Inflammasome activation in hippocampal microglia mediates chronic stress-induced depressive-like behaviors. Front Mol Neurosci 12:210
33. Catena-Dell'Osso M, Rotella F, Dell'Osso A, Fagiolini A, Marazziti D (2013) Inflammation, serotonin and major depression. Curr Drug Targets 14(5):571–577
34. Dell'Osso L, Carmassi C, Mucci F, Marazziti D (2016) Depression, serotonin and tryptophan. Curr Pharm Des 22(8):949–954
35. Pariante CM (2006) The glucocorticoid receptor: part of the solution or part of the problem? Journal of psychopharmacology (Oxford, England) 20(4 Suppl):79–84
36. Joseph JJ, Golden SH (2017) Cortisol dysregulation: the bidirectional link between stress, depression, and type 2 diabetes mellitus. Ann N Y Acad Sci 1391(1):20–34
37. Stone LB, McCormack CC, Bylsma LM (2020) Cross system autonomic balance and regulation: associations with depression and anxiety symptoms. Psychophysiology 57(10):e13636
38. Beurel E, Toups M, Nemeroff CB (2020) The bidirectional relationship of depression and inflammation: double trouble. Neuron 107(2):234–256
39. Brites D, Fernandes A (2015) Neuroinflammation and depression: microglia activation, extracellular microvesicles and microRNA dysregulation. Front Cell Neurosci 9:476
40. Felger JC (2019) Role of inflammation in depression and treatment implications. Handb Exp Pharmacol 250:255–286
41. Haroon E, Raison CL, Miller AH (2012) Psychoneuroimmunology meets neuropsychopharmacology: translational implications of the impact of inflammation on behavior. *Neuropsychopharmacology: official publication of the American college of.* Neuropsychopharmacology 37(1):137–162
42. Jones BDM, Daskalakis ZJ, Carvalho AF, Strawbridge R, Young AH, Mulsant BH, Husain MI (2020) Inflammation as a treatment target in mood disorders: review. BJPsych open 6(4):e60
43. Liu JJ, Wei YB, Strawbridge R, Bao Y, Chang S, Shi L, Que J, Gadad BS, Trivedi MH, Kelsoe JR, Lu L (2020) Peripheral cytokine levels and response to antidepressant treatment in depression: a systematic review and meta-analysis. Mol Psychiatry 25(2):339–350
44. Brás JP, Pinto S, Almeida MI, Prata J, von Doellinger O, Coelho R, Barbosa MA, Santos SG (2019, 2011) Peripheral biomarkers of inflammation in depression: evidence from animal models and clinical studies. Methods Molecul Biol (Clifton, N.J.):467–492
45. Arteaga-Henríquez G, Simon MS, Burger B, Weidinger E, Wijkhuijs A, Arolt V, Birkenhager TK, Musil R, Müller N, Drexhage HA (2019) Low-grade inflammation as a predictor of antidepressant and anti-inflammatory therapy response in MDD patients: A systematic review of the literature in combination with an analysis of experimental data collected in the EU-MOODINFLAME consortium. Front Psych 10:458
46. Jia X, Gao Z, Hu H (2021) Microglia in depression: current perspectives. Sci China Life Sci 64(6):911–925
47. Maes M, Smith R, Scharpe S (1995) The monocyte-T-lymphocyte hypothesis of major depression. Psychoneuroendocrinology 20(2):111–116
48. Catena-Dell'Osso M, Marazziti D, Rotella F, Bellantuono C (2012) Emerging targets for the pharmacological treatment of depression: focus on melatonergic system. Curr Med Chem 19(3):428–437
49. Kessler RC, Berglund P, Demler O, Jin R, Koretz D, Merikangas KR, Rush AJ, Walters EE, Wang PS, Replication NCS (2003) The epidemiology of major depressive disorder: results from the National Comorbidity Survey Replication (NCS-R). JAMA 289(23):3095–3105
50. de Melo LGP, Nunes SOV, Anderson G, Vargas HO, Barbosa DS, Galecki P, Carvalho AF, Maes M (2017) Shared metabolic and immune-inflammatory, oxidative and nitrosative stress pathways in the metabolic syndrome and mood disorders. Prog Neuro-Psychopharmacol Biol Psychiatry 78:34–50
51. Solleiro-Villavicencio H, Rivas-Arancibia S (2018) Effect of chronic oxidative stress on Neuroinflammatory response mediated by CD4+T

cells in neurodegenerative diseases. Front Cell Neurosci 12:114

52. Hambright WS, Fonseca RS, Chen L, Na R, Ran Q (2017) Ablation of ferroptosis regulator glutathione peroxidase 4 in forebrain neurons promotes cognitive impairment and neurodegeneration. Redox Biol 12:8–17

53. Juszczyk G, Mikulska J, Kasperek K, Pietrzak D, Mrozek W, Herbet M (2021) Chronic stress and oxidative stress as common factors of the pathogenesis of depression and Alzheimer's disease: the role of antioxidants in prevention and treatment. Antioxidants (Basel, Switzerland) 10(9):1439

54. Kim YK, Na KS (2016) Role of glutamate receptors and glial cells in the pathophysiology of treatment-resistant depression. Prog Neuro-Psychopharmacol Biol Psychiatry 70:117–126

55. Pizzino G, Irrera N, Cucinotta M, Pallio G, Mannino F, Arcoraci V, Squadrito F, Altavilla D, Bitto A (2017) Oxidative stress: harms and benefits for human health. Oxidative Med Cell Longev 2017:8416763

56. Weiland A, Wang Y, Wu W, Lan X, Han X, Li Q, Wang J (2019) Ferroptosis and its role in diverse brain diseases. Mol Neurobiol 56(7):4880–4893

57. Wojsiat J, Zoltowska KM, Laskowska-Kaszub K, Wojda U (2018) Oxidant/antioxidant imbalance in Alzheimer's disease: therapeutic and diagnostic prospects. Oxidative Med Cell Longev 2018:6435861

58. Zhao RZ, Jiang S, Zhang L, Yu ZB (2019) Mitochondrial electron transport chain, ROS generation and uncoupling (review). Int J Mol Med 44(1):3–15

59. Siwek M, Sowa-Kućma M, Dudek D, Styczeń K, Szewczyk B, Kotarska K, Misztakk P, Pilc A, Wolak M, Nowak G (2013) Oxidative stress markers in affective disorders. Pharmacological reports: PR 65(6):1558–1571

60. Guu TW, Mischoulon D, Sarris J, Hibbeln J, McNamara RK, Hamazaki K, Freeman MP, Maes M, Matsuoka YJ, Belmaker RH, Jacka F, Pariante C, Berk M, Marx W, Su KP (2019) International Society for Nutritional Psychiatry Research Practice Guidelines for Omega-3 fatty acids in the treatment of major depressive disorder. Psychother Psychosom 88(5):263–273

61. Liu T, Zhong S, Liao X, Chen J, He T, Lai S, Jia Y (2015) A meta-analysis of oxidative stress markers in depression. PLoS One 10(10):e0138904

62. Muti M, Del Grande C, Musetti L, Marazziti D, Turri M, Cirronis M, Pergentini I, Corsi M, Dell'Osso L, Corsini GU (2015) Serum uric acid levels and different phases of illness in bipolar I patients treated with lithium. Psychiatry Res 225(3):604–608

63. Ortiz R, Ulrich H, Zarate CA Jr, Machado-Vieira R (2015) Purinergic system dysfunction in mood disorders: a key target for developing improved therapeutics. Prog Neuro-Psychopharmacol Biol Psychiatry 57:117–131

64. Krügel U (2016) Purinergic receptors in psychiatric disorders. Neuropharmacology 104:212–225

65. Malewska-Kasprzak MK, Permoda-Osip A, Rybakowski J (2019) Disturbances of purinergic system in affective disorders and schizophrenia. Zaburzenia układu purynergicznego w chorobach afektywnych i schizofrenii. Psychiatr Pol 53(3):577–587

66. Brooks SC, Linn JJ, Disney N (1978) Serotonin, folic acid, and uric acid metabolism in the diagnosis of neuropsychiatric disorders. Biol Psychiatry 13(6):671–684

67. Karve AV, Jagtiani SS, Chitnis KA (2013) Evaluation of effect of allopurinol and febuxostat in behavioral model of depression in mice. Indian journal of pharmacology 45(3):244–247

68. Felger JC (2018) Imaging the role of inflammation in mood and anxiety-related disorders. Curr Neuropharmacol 16(5):533–558

69. Chehab FF (2000) Leptin as a regulator of adipose mass and reproduction. Trends Pharmacol Sci 21(8):309–314

70. Sinha R, Jastreboff AM (2013) Stress as a common risk factor for obesity and addiction. Biol Psychiatry 73(9):827–835

71. Ohta Y, Kosaka Y, Kishimoto N, Wang J, Smith SB, Honig G, Kim H, Gasa RM, Neubauer N, Liou A, Tecott LH, Deneris ES, German MS (2011) Convergence of the insulin and serotonin programs in the pancreatic β-cell. Diabetes 60(12):3208–3216

72. Stahl SM (1985) Platelets as pharmacologic models for the receptors and biochemistry of monoaminergic neurons. In: The platelets: physiology and pharmacology, vol 13, New York, NY, USA, pp 307–335

73. Almeida OP, Ford AH, Hankey GJ, Golledge J, Yeap BB, Flicker L (2019) Depression, antidepressants and the risk of cardiovascular events and death in older men. Maturitas 128:4–9

74. Graham N, Ward J, Mackay D, Pell JP, Cavanagh J, Padmanabhan S, Smith DJ (2019) Impact of major depression on cardiovascular outcomes for individuals with hypertension: prospective survival analysis in UK biobank. BMJ Open 9(9):e024433

75. Jee YH, Chang H, Jung KJ, Jee SH (2019) Cohort study on the effects of depression on atherosclerotic cardiovascular disease risk in Korea. BMJ Open 9(6):e026913

76. Okunrintemi V, Valero-Elizondo J, Michos ED, Salami JA, Ogunmoroti O, Osondu C, Tibuakuu M, Benson EM, Pawlik TM, Blaha MJ, Nasir K (2019) Association of Depression Risk with patient experience, healthcare expenditure, and health resource utilization among adults with atherosclerotic cardiovascular disease. J Gen Intern Med 34(11):2427–2434

77. Xue Y, Liu G, Geng Q (2020) Associations of cardiovascular disease and depression with memory related disease: A Chinese national prospective cohort study. J Affect Disord 260:11–17

78. Carney RM, Freedland KE (2017) Depression and coronary heart disease. Nat Rev Cardiol 14(3):145–155

79. Chávez-Castillo M, Nava M, Ortega Á, Rojas M, Núñez V, Salazar J, Bermúdez V, Rojas-Quintero J (2020) Depression as an Immunometabolic disorder: exploring shared Pharmacotherapeutics with cardiovascular disease. Curr Neuropharmacol 18(11):1138–1153

80. Shao M, Lin X, Jiang D, Tian H, Xu Y, Wang L, Ji F, Zhou C, Song X, Zhuo C (2020) Depression and cardiovascular disease: shared molecular mechanisms and clinical implications. Psychiatry Res 285:112802. Advance online publication

81. Bouillon-Minois JB, Trousselard M, Thivel D, Gordon BA, Schmidt J, Moustafa F, Oris C, Dutheil F (2021) Ghrelin as a biomarker of stress: A systematic review and meta-analysis. Nutrients 13(3):784

82. Harmatz ES, Stone L, Lim SH, Lee G, McGrath A, Gisabella B, Peng X, Kosoy E, Yao J, Liu E, Machado NJ, Weiner VS, Slocum W, Cunha RA, Goosens KA (2017) Central ghrelin resistance permits the Overconsolidation of fear memory. Biol Psychiatry 81(12):1003–1013

83. Duan K, Gu Q, Petralia RS, Wang YX, Panja D, Liu X, Lehmann ML, Zhu H, Zhu J, Li Z (2021) Mitophagy in the basolateral amygdala mediates increased anxiety induced by aversive social experience. Neuron 109(23):3793–3809.e8

84. Esteban-Cornejo I, Stillman CM, Rodriguez-Ayllon M, Kramer AF, Hillman CH, Catena A, Erickson KI, Ortega FB (2021) Physical fitness, hippocampal functional connectivity and academic performance in children with overweight/obesity: the ActiveBrains project. Brain Behav Immun 91:284–295

85. Huang HJ, Chen XR, Han QQ, Wang J, Pilot A, Yu R, Liu Q, Li B, Wu GC, Wang YQ, Yu J (2019) The protective effects of ghrelin/GHSR on hippocampal neurogenesis in CUMS mice. Neuropharmacology 155:31–43

86. Huang YQ, Wang Y, Hu K, Lin S, Lin XH (2021) Hippocampal Glycerol-3-phosphate acyltransferases 4 and BDNF in the Progress of obesity-induced depression. Front Endocrinol 12:667773

87. Jiao ZT, Luo Q (2022) Molecular mechanisms and health benefits of ghrelin: A narrative review. Nutrients 14(19):4191

88. Pierre A, Regin Y, Van Schuerbeek A, Fritz EM, Muylle K, Beckers T, Smolders IJ, Singewald N, De Bundel D (2019) Effects of disrupted ghrelin receptor function on fear processing, anxiety and saccharin preference in mice. Psychoneuroendocrinology 110:104430

89. Field BC (2014) Neuroendocrinology of obesity. Br Med Bull 109:73–82

90. Huang HJ, Zhu XC, Han QQ, Wang YL, Yue N, Wang J, Yu R, Li B, Wu GC, Liu Q, Yu J (2017) Ghrelin alleviates anxiety- and depression-like behaviors induced by chronic unpredictable mild stress in rodents. Behav Brain Res 326:33–43

91. Makris AP, Karianaki M, Tsamis KI, Paschou SA (2021) The role of the gut-brain axis in depression:

92. Voigt JP, Fink H (2015) Serotonin controlling feeding and satiety. Behav Brain Res 277:14–31

93. Höglund E, Øverli Ø, Winberg S (2019) Tryptophan metabolic pathways and brain serotonergic activity: A comparative review. Front Endocrinol 10:158

94. El-Merahbi R, Löffler M, Mayer A, Sumara G (2015) The roles of peripheral serotonin in metabolic homeostasis. FEBS Lett 589(15):1728–1734

95. Rajan TM, Menon V (2017) Psychiatric disorders and obesity: A review of association studies. J Postgrad Med 63(3):182–190

96. Milaneschi Y, Lamers F, Bot M, Drent ML, Penninx BW (2017) Leptin dysregulation is specifically associated with major depression with atypical features: evidence for a mechanism connecting obesity and depression. Biol Psychiatry 81(9):807–814

97. Tomiyama AJ (2019) Stress and obesity. Annu Rev Psychol 70:703–718

98. Wyler SC, Lord CC, Lee S, Elmquist JK, Liu C (2017) Serotonergic control of metabolic homeostasis. Front Cell Neurosci 11:277

99. Ge T, Fan J, Yang W, Cui R, Li B (2018) Leptin in depression: a potential therapeutic target. Cell Death Dis 9(11):1096

100. Ozsoy S, Besirli A, Unal D, Abdulrezzak U, Orhan O (2015) The association between depression, weight loss and leptin/ghrelin levels in male patients with head and neck cancer undergoing radiotherapy. Gen Hosp Psychiatry 37(1):31–35

101. Webb M, Davies M, Ashra N, Bodicoat D, Brady E, Webb D, Moulton C, Ismail K, Khunti K (2017) The association between depressive symptoms and insulin resistance, inflammation and adiposity in men and women. PLoS One 12(11):e0187448

102. Zou X, Zhong L, Zhu C, Zhao H, Zhao F, Cui R, Gao S, Li B (2019) Role of leptin in mood disorder and neurodegenerative disease. Front Neurosci 3(13):378. https://doi.org/10.3389/fnins.2019.00378

103. Bender A, Hagan KE, Kingston N (2017) The association of folate and depression: A meta-analysis. J Psychiatr Res 95:9–18

104. Kris-Etherton PM, Petersen KS, Hibbeln JR, Hurley D, Kolick V, Peoples S, Rodriguez N, Woodward-Lopez G (2021) Nutrition and behavioral health disorders: depression and anxiety. Nutr Rev 79(3):247–260

105. Molendijk M, Molero P, Ortuño Sánchez-Pedreño F, Van der Does W, Angel Martínez-González M (2018) Diet quality and depression risk: A systematic review and dose-response meta-analysis of prospective studies. J Affect Disord 226:346–354

106. Quirk SE, Williams LJ, O'Neil A, Pasco JA, Jacka FN, Housden S, Berk M, Brennan SL (2013) The association between diet quality, dietary patterns and depression in adults: a systematic review. BMC Psychiatry 13:175

107. Parletta N, Zarnowiecki D, Cho J, Wilson A, Bogomolova S, Villani A, Itsiopoulos C, Niyonsenga

T, Blunden S, Meyer B, Segal L, Baune BT, O'Dea K (2019) A Mediterranean-style dietary intervention supplemented with fish oil improves diet quality and mental health in people with depression: A randomized controlled trial (HELFIMED). Nutr Neurosci 22(7):474–487

108. Young LM, Pipingas A, White DJ, Gauci S, Scholey A (2019) A systematic review and meta-analysis of B vitamin supplementation on depressive symptoms, anxiety, and stress: effects on healthy and 'at-risk' individuals. Nutrients 11(9):2232

109. Zaric BL, Obradovic M, Bajic V, Haidara MA, Jovanovic M, Isenovic ER (2019) Homocysteine and Hyperhomocysteinemia. Curr Med Chem 26(16):2948–2961. https://doi.org/10.2174/092986 7325666180313105949

110. Moradi F, Lotfi K, Armin M, Clark CCT, Askari G, Rouhani MH (2021) The association between serum homocysteine and depression: A systematic review and meta-analysis of observational studies. Eur J Clin Investig 51(5):e13486

111. Lassale C, Batty GD, Baghdadli A, Jacka F, Sánchez-Villegas A, Kivimäki M, Akbaraly T (2019) Healthy dietary indices and risk of depressive outcomes: a systematic review and meta-analysis of observational studies. Mol Psychiatry 24(7):965–986

112. Tolkien K, Bradburn S, Murgatroyd C (2019) An anti-inflammatory diet as a potential intervention for depressive disorders: A systematic review and meta-analysis. Clinical nutrition (Edinburgh, Scotland) 38(5):2045–2052

113. Burhani MD, Rasenick MM (2017) Fish oil and depression: the skinny on fats. J Integr Neurosci 16(s1):S115–S124

114. Deacon G, Kettle C, Hayes D, Dennis C, Tucci J (2017) Omega 3 polyunsaturated fatty acids and the treatment of depression. Crit Rev Food Sci Nutr 57(1):212–223

115. Assies J, Lok A, Bockting CL, Weverling GJ, Lieverse R, Visser I, Abeling NG, Duran M, Schene AH (2004) Fatty acids and homocysteine levels in patients with recurrent depression: an explorative pilot study. Prostaglandins Leukot Essent Fatty Acids 70(4):349–356

116. Fernandes MF, Mutch DM, Leri F (2017) The relationship between fatty acids and different depression-related brain regions, and their potential role as biomarkers of response to antidepressants. Nutrients 9(3):298

117. Beyer JL, Payne ME (2016) Nutrition and bipolar depression. Psychiatr Clin North Am 39(1):75–86

118. Mirabelli M, Chiefari E, Arcidiacono B, Corigliano DM, Brunetti FS, Maggisano V, Russo D, Foti DP, Brunetti A (2020) Mediterranean diet nutrients to turn the tide against insulin resistance and related diseases. Nutrients 12(4):1066

Schizophrenia

Adriana Foster and Jordanne King

Introduction to Schizophrenia

Schizophrenia is a serious and disabling mental disorder, affecting just under 1% of the population. While its etiological bases remain obscure and consequently its nosological boundaries are uncertain, the condition classically has its clinical onset in childhood or early adolescence following brain alterations that take place early in development [1]. It is characterized by (1) "positive" psychotic symptoms like delusions (fixed false ideas that are held with unshakable conviction), hallucinations (perceptions apparently without a relevant stimulus), and thought disorder (difficulty in assembling a coherent stream of speech), (2) "negative" symptoms including lack of motivation and pleasure, inability of expressing the full range of emotions, neglect of personal appearance, diminished speech output, and disinterest in life events, and (3) disorganized, abnormal motor behavior ranging from unpredictable agitation to limited motor and verbal responses to environment. Associated features are cognitive impairment (memory and attention difficulties), mood disturbance, particularly depression and anxiety, sleep disturbance, and lack of insight about the disorder, leading to treatment non-adherence and poor prognosis [2]. All of these attributes, persistent over time, culminate in a decline in social and occupational performance. These features—coupled with the consequences of sustained impairment—result in comorbid depression among people with schizophrenia. Approximately 20% of patients attempt and about 5–6% of patients with schizophrenia die by suicide [2]. Patients with schizophrenia frequently have comorbid substance use disorders, anxiety, obsessive compulsive disorder, as well as other medical conditions like diabetes and cardiovascular and pulmonary disorders, which all contribute to poor outcomes [2]. Schizophrenia is poorly understood by the public, and it is often highly stigmatizing [3].

Pathophysiology of Schizophrenia and Metabolic Alterations

The causes of schizophrenia are largely unknown and the exact mixture and confluence of etiological factors that result in schizophrenia most likely differ from one patient to another [4, 5]. Notably, people with schizophrenia have a median prevalence of type-2 diabetes of 13% (vs. 8.3% worldwide prevalence of diabetes), which occurs not only in those treated with antipsychotic drugs but also in drug-naïve patients [6]. Furthermore, patients with non-affective psychoses, including

A. Foster (✉)
HCA Florida Woodmont Hospital, Tamarac, FL, USA

J. King
Presbyterian Healthcare Services,
Albuquerque, NM, USA

© The Author(s), under exclusive license to Springer Nature Switzerland AG 2026
E. Lammert, M. Zeeb (eds.), *Metabolism of Human Diseases*,
https://doi.org/10.1007/978-3-031-96019-2_4

schizophrenia, have a fourfold risk of developing type-2 diabetes (see chapter "Diabetes Mellitus") if they have family history of diabetes, vs. patients with non-affective psychoses without such family history [7]. Thus, we will focus below not only on schizophrenia but also on its possible shared pathophysiological mechanisms with type-2 diabetes.

Genetics

Genetic influence is present in schizophrenia as evidenced in high heritability and monozygotic twin concordance [8]. From studies that detected the association of susceptibility genes like COMT (catechol-O-methyltransferase), neuregulin-1 (implicated in neurotransmitter receptors' expression and activation), dysbindin, disrupted in schizophrenia 1 (DISC1), dopamine type-2 (D2) receptor, and transporter genes with schizophrenia, the field has evolved to multinational genetic studies pointing to subtle yet multiple and reproducible genetic findings [9–11]. People with schizophrenia have a higher rate of copy number variations (CNVs) than controls, with affected genes encoding N-methyl-D-aspartate (NMDA) glutamate receptors and genes involved in synaptic plasticity [11, 12]. Genome-wide studies reveal single nucleotide polymorphisms (SNPs) comprising loci on genes involved in protein coding, glutamatergic synaptic neurotransmission, dopamine D2 receptor, calcium channel function, synaptic plasticity, and immunity [11–13]. Some of the risk alleles found in schizophrenia are shared by other psychiatric disorders like bipolar disorder, major depression, attention deficit hyperactivity disorder, autism spectrum disorder, and intellectual disability [12]. Furthermore, schizophrenia is believed to share susceptibility genes with type-2 diabetes through inflammation-associated genes (alleles on the major histocompatibility complex C4 and interleukin 1-beta [IL-1ß], interleukin 6 [IL-6], and IL-6-receptor genes) and genes associated with oxidative stress (e.g., glutathione S-transferase [GST] and methyl tetrahydrofolate reductase [MTHFR]) [14].

Nongenetic Influences

Nongenetic influences include (1) obstetric events like hypoxia and maternal malnutrition, birth during late winter and spring, and advanced paternal age, (2) chronic psychosocial stress including urban dwelling, migration, and ethnic minority, and (3) prenatal infections (such as rubella and influenza), head injury, and use of cannabis [15]. Additionally, acute stress appears to trigger acute psychotic exacerbations in people with established illness [16, 17].

Brain Morphology in Schizophrenia

On whole brain imaging studies and postmortem brain studies, patients with schizophrenia reveal a series of macroscopic and histological abnormalities. The overall brain volume is reduced (by around 5%, in both white and gray matter), and the temporal and frontal lobes are smaller. In addition, the hippocampus, insula, anterior cingulate, and amygdala volumes are smaller. In contrast, the lateral ventricles are enlarged. On a cellular level, changes of attuned arrangement and altered white matter integrity are noted rather than an actual decrease in number of neurons [18]. Genetic and environmental factors likely converge, leading to abnormal neuronal connectivity, synaptic plasticity, synaptic signaling, and altered dopaminergic, gamma-aminobutyric acid (GABA)-ergic, and glutamatergic pathways of neurotransmission in the brain [19].

Neurotransmitter Dysfunction in Schizophrenia

Dopamine Altered dopaminergic function is considered fundamental in understanding schizophrenia [1, 17, 20]. The dopamine (DA) hypothesis has most influenced antipsychotic drug development and treatment of schizophrenia, and it also still represents the predominant (yet still inadequate) explanatory model for schizophrenia and its treatment, although influence of multiple other neurotransmitters has been explored more recently. Dopamine dysfunction in schizophrenia includes increased presynaptic DA synthesis, DA release, and dopamine type-2 (D2) receptor density in the striatum. The DA role in attributing salience to stimuli and predicting reward is impaired in schizophrenia, with DA cells firing unpredictably in response to stimuli of low significance thus attributing unwarranted meaning to such stimuli. This mechanism is postulated to underlie positive symptoms (delusions) [19]. Dopamine cell response is decreased in anticipation of reward, possibly underlying negative symptoms. Dopamine is generated in the midbrain in the ventral tegmental area (VTA) and substantia nigra, from where it projects throughout the basal ganglia-thalamo-cortical network in distinct pathways, as follows:

1. *The mesocortical pathway*, originating in the VTA, projects to the frontal cortex; underactivity of dopamine in this pathway is linked to cognitive, negative, and depressive symptoms in schizophrenia.
2. *The mesolimbic pathway*, from the VTA to the ventral striatum and limbic areas with dopamine underactivity in this pathway also being implicated in negative symptoms.
3. *The nigrostriatal pathway*, originating in substantia nigra with fibers projecting distinctly to the dorso-medial striatum and associative cortex and overactivity of dopamine in this pathway thought to be associated with psychosis, while fibers projecting to the "sensorimotor" striatum and sensorimotor cortex form a movement pathway, implicated in extrapyramidal symptoms and tardive dyskinesia [19, 21].
4. *The tuberoinfundibular pathway* from the hypothalamus to the pituitary gland controls prolactin secretion, and DA hyperactivity in this pathway can lead to increased serum prolactin, amenorrhea, galactorrhea, and sexual dysfunction [19].

Glutamate The balance between the neuroexcitatory glutamate system and the inhibitory GABA-ergic system in the central nervous system is thought to be altered in schizophrenia.

Ionotropic *N*-methyl-*D*-aspartate (NMDA) glutamate receptors influence the dopaminergic neuron firing [19, 22]. In schizophrenia, inhibition of NMDA receptors on GABA interneurons leads to decreased amounts of extracellular GABA and increased extracellular glutamate, with resultant increased cortical activity and psychosis [19, 20].

The Cholinergic System The cholinergic system is also thought to contribute to schizophrenia etiology by stimulating the DA neurons and thus facilitating DA release in the striatum and affecting the excitatory/inhibitory balance of the glutamate and GABA systems [19].

Oxidative Stress and Inflammation

Oxidative stress and inflammation are responses to intrinsic and environmental insults implicated in pathophysiology of many disease processes (cancer, diabetes, and cardiovascular disease) and thus are not specific for schizophrenia. However, both mechanisms have emerged as foci of research in schizophrenia psychopathology and are thought to be shared with diabetes.

Inflammation It is thought to be a central nervous system response to harmful stimuli generated by genetic and environmental factors during development in schizophrenia (e.g., prenatal exposure to infections, obstetric complications, psychological trauma, substance use), resulting in changes in neurogenesis, brain connectivity, and neurotransmitter function [14, 23]. Inflammation markers like cytokines, IL-6, IL-1ß, and tumor necrosis factor (TNF) levels are elevated in blood and cerebrospinal fluid of patients with schizophrenia and are associated with first episode of illness as well as illness exacerbations and negative symptoms [14]. Same inflammatory markers (IL-6, IL-1ß, and TNF) have been associated with insulin resistance and thus risk of development of type-2 diabetes (see chapter "Diabetes Mellitus"). Diabetes is also associated with allelic variants in IL-6 and IL-1ß [14].

Oxidative Stress It results from an imbalance between overproduction of reactive oxygen species (ROS) and a deficiency in ROS elimination by antioxidants, resulting in chronic inflammation. Oxidative stress is thought to affect interneurons, myelin, and cellular signaling cascades in schizophrenia [24]. Type-2 diabetes and schizophrenia share associations with polymorphisms of genes that protect against oxidative stress and affect DNA methylation [14], again raising the suspicion that inflammation could be a common underlying mechanism for schizophrenia and diabetes, which are frequently comorbid [7, 14, 25]. Figure 1 illustrates the mechanisms that underlie the association between schizophrenia and diabetes, with the lifestyle and diet, as well as the contribution of antipsychotics being represented once the disease is clinically apparent [14].

Treatment of Schizophrenia

Antipsychotic medications form the bedrock of treatment for schizophrenia and other psychoses [19, 26, 27]. All these medications block dopamine (D2) receptors to some extent, although newer antipsychotic medications tend to have broader effects that extend to other (e.g., cholinergic, adrenergic, serotonergic) neurotransmitters [26, 27]. At each stage of medication decision-making, psychosocial interventions and assessment of adherence to the medications should be integral.

Antipsychotic drugs are grouped into a first- and second-generation antipsychotics (FGAs and SGAs), based on their receptor and adverse effect profiles. FGAs' (e.g., haloperidol and perphenazine) mechanism of action is based primarily on the inhibition of dopamine D2 receptors. Clozapine, the initial SGA drug, only partially binds to dopamine D2 receptors. However, it also binds dopamine D4 receptors with high affinity. Other targets include dopamine D1, D3, and D5 receptors as well as serotonin, adrenergic, and histaminergic receptors. Other SGAs (like risperidone, quetiapine, and olanzapine) maintain the combined inhibition of dopamine D2 receptors and serotonin 5-HT$_{2A}$ receptors. Both groups of medications also have actions on muscarinic, histaminic, and adrenergic receptors [28].

Treatment for schizophrenia usually begins with SGA at the lowest effective dose. Treatment response, tolerability, and side effects of any given drug are highly variable between patients [26, 27], and side effects should thus be monitored closely. Drug choice is often individual. It is also integral that each stage of medication decision-making includes psychosocial interventions and assessment of adherence to the medications. Changes in antipsychotic medications may be indicated either due to a lack of sufficient effect or due to adverse effects. Switches should be gradual and involve a cross taper of the medications. After two trials of monotherapy, use of more than one antipsychotic may have utility in certain cases. Beyond this stage, treatment may involve the addition of adjunctive medications in other classes or neuromodulation [29].

When starting schizophrenia treatment (usually with a second-generation antipsychotic, SGA), the lowest effective drug dose should be used. Antipsychotic drugs often appear to work in about 48 hours, although it may take up to 4 weeks at adequate dose to determine whether the drug is ultimately effective. Treatment response, tolerability, and side effects of any given drug are highly variable between patients [14], and side effects should thus be monitored closely. Switching antipsychotic medications may be indicated, for either lack of effect or presence of side effects on the present medications, although it is a complicated process and the switch to a new medication should be gradual and phased-in with a cross taper. Antipsychotic polypharmacy is common although probably not justified. At present, treatment remains a clinical approach of "trial and error" with the selection of each antipsychotic medication.

Only one medication is currently approved by U.S. Food and Drug Administration (FDA) to treat psychosis (in Parkinson's disease only) without blocking postsynaptic D2 receptor, postulated to act through antagonism of serotonin receptors and others are in development, target-

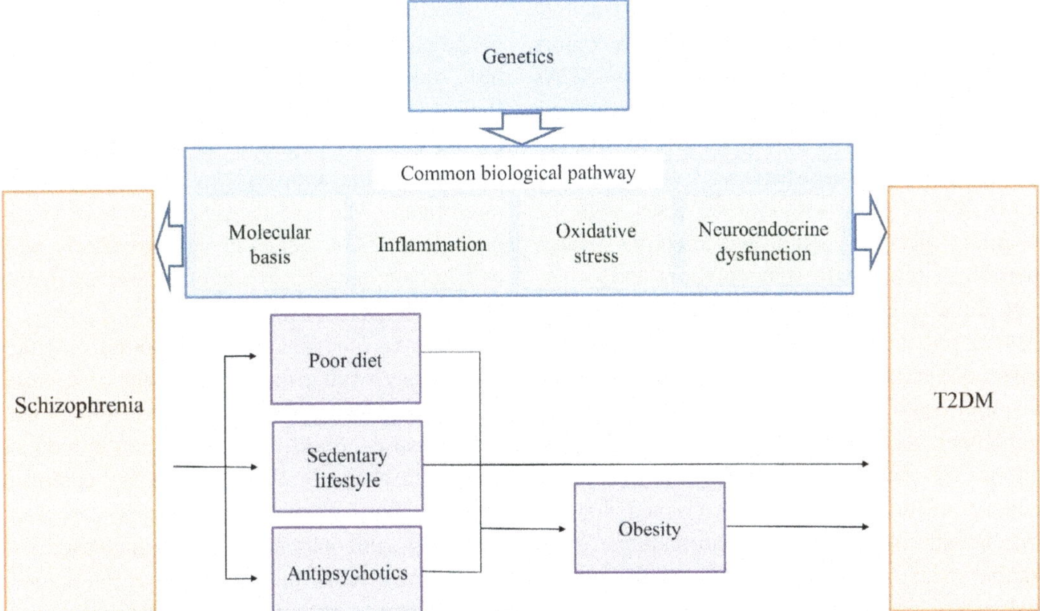

Fig. 1 Mechanisms that underlie the association between schizophrenia and type-2 diabetes mellitus (T2DM). The mechanisms of the increasing prevalence of T2DM in patients with schizophrenia are multifactorial. Poor diet and sedentary lifestyle are included in the traditional risk factors. Iatrogenic risk during treatment with antipsychotics is included in risk factors unique to schizophrenia [25]. Accumulating evidence suggests shared genetic susceptibility and biological common pathway of both schizophrenia and T2DM. (From Mizuki et al. 2021)

ing cholinergic muscarinic receptors among other mechanisms [30, 31].

Psychosocial interventions are gaining broader acceptance in schizophrenia, are adopted in treatment guidelines, and are being broadly implemented in various countries. Family involvement and social skills training applied by multidisciplinary teams at the onset of the illness, during the first episode of psychosis, yielded favorable outcomes including decreased mortality, number and duration of hospital admissions, and outpatient visits in large studies [32, 33].

Influence of Treatment on Metabolism

As dopamine affects many neurological systems in the brain, antipsychotic treatments have broad therapeutic and adverse effects. However, there is some evidence that antipsychotic medications may have neuroplastic effects [34]. On the other hand, there is lingering concern that antipsychotic medications might be neurotoxic and thereby contribute to progressive neurodegeneration in schizophrenia [35]. Understanding these correlations is complicated by a wide range of factors from age and duration of illness to how metabolic side effects can affect cognition [36].

The FGAs have a side effect profile largely characterized by acute and long-term motor impairments through their effect on the dopaminergic nigrostriatal pathway. In addition, FGAs antagonize the action of dopamine (also known as prolactin-inhibiting hormone) and thus induce hyperprolactinemia by releasing the dopamine block of prolactin secretion in the pituitary gland. This can lead to gonadal failure and subsequently to bone loss and osteoporosis [37, 38]. Aripiprazole, an SGA, which has been shown to decrease prolactin levels, can be used as possible treatment for antipsychotic-induced hyperprolactinemia [29].

SGAs show a more complex side effect profile that is increasingly characterized by metabolic disturbances [28]. While they yield less risk for movement disorders and hyperprolactinemia, due to lower blockade of dopamine D2 receptors and concomitant serotonin antagonism, SGAs are known to induce dangerous metabolic effects [39]. These include weight gain, glucose dysregulation, and lipid disturbances, which can affect up to 69% of those with chronic cases, with the resulting metabolic syndrome affecting a large portion of those with schizophrenia and occurring three to five times more frequently than in general population. This is particularly important given that cardiovascular disease is a significant cause of death in this population [39]. Therefore, guidelines on monitoring patients on antipsychotics for obesity, diabetes, lipid abnormalities, and cardiovascular risk have been issued [40, 41]. When factoring metabolic abnormalities, olanzapine and clozapine are associated with the most risk while aripiprazole, brexpiprazole, cariprazine, lurasidone, and ziprasidone are considered to have the lowest risk [39].

The mechanism for the development of these metabolic conditions is multifactorial with serotonin 5-HT$_{2c}$, histamine, and muscarinic receptors as well as the hormones leptin and insulin being heavily implicated in those mechanisms (Fig. 2) [42]. Activation of histamine H1 and serotonin 5-HT$_{2c}$ receptors by SGAs increases appetite [42]. H1 antagonism not only affects feeding behavior but also through its proportional activation of adenosine monophosphate-activated (AMP) activated protein kinase (AMPK) can reduce the effect of leptin to decrease appetite [43].

Leptin, a hormone implicated in the pathophysiology of food and energy regulation is known to increase with the use of certain SGAs and there is a positive correlation between high leptin levels, hyperinsulinemia, and weight gain [44–46]. Leptin and insulin levels also are partnered as insulin secretes leptin and leptin has a role in regulating insulin. Insulin also has an anorexigenic effect in the brain, and antipsychotics through their actions on histamine and muscarinic receptors as well as leptin can reduce the secretion of insulin. Other neuro hormones such as adiponectin, melatonin, and ghrelin have been implicated in mechanisms of metabolic adverse effects but further research is needed [43].

Another upcoming area of interest is the role of the gut microbiome in schizophrenia, particularly given its placement at the interface of biological mechanisms with environmental stressors during neurodevelopment, as well as the potential role of the antipsychotics in changing gut microbiota, which is thought to at least partly mediate the SGA metabolic adverse effects, with weight gain, increase in adiposity, and inflammation [43, 47].

However, medication alone is not the only factor that may result in these disturbances. Increased baseline weight, gender, and ethnicity may also predict susceptibility to antipsychotic-induced metabolic change [39, 44]. Data are emerging showing that patients with psychotic disorders, including schizophrenia, have an increased risk of hyperprolactinemia and glucose intolerance independent of antipsychotic medication use, as evidenced by changes in these metabolic factors during a first episode of psychosis for drug-naive patients [14, 48].

Management of the metabolic syndrome (see chapter "Metabolic Syndrome") in schizophrenia can be pharmacological or nonpharmacological. Historically, nonpharmacological interventions focused on weight reduction instead of prevention. These interventions are effective and include cognitive behavioral interventions, nutritional interventions, and exercise [49]. Pharmacologic interventions have included studies with multiple medications, with the most evidence for clinical use having been for metformin and topiramate.

Metformin, a hypoglycemic agent broadly prescribed for type-2 diabetes (see chapter "Diabetes Mellitus"), has been used as an adjunctive treatment for weight gain and dyslipidemia occurring as a result of treatment with SGAs in schizophrenia. In a meta-analysis of randomized controlled trials, patients with schizophrenia taking antipsychotics, treated with adjunctive metformin, had significantly lower total cholesterol, triglycerides, body weight, body mass index, glycosylated hemoglobin, and fasting insulin than patients treated with antipsychotics without adjunctive metformin, while high-density lipoprotein cholesterol was significantly higher than in the control group. No significant differences

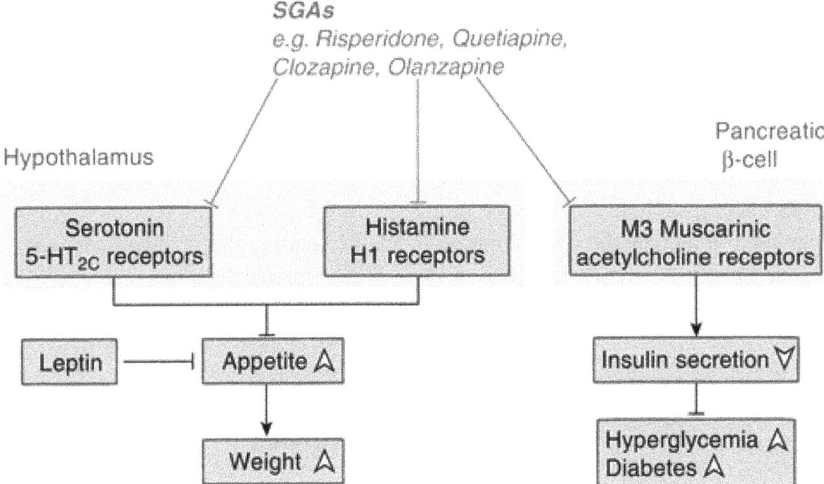

Fig. 2 Side effect profile of second-generation antipsychotics (SGAs). These drugs (*SGAs*) show side effects different from the first generation antipsychotics (FGAs) due to their binding to 5-HT$_{2C}$ dopamine (*5-HT*) and H1 histamine receptors in the hypothalamus (*left side*) and M3 muscarinic acetylcholine receptors on pancreatic β-cells (*right side*). Blockade of 5-HT$_{2C}$ and H1 receptors releases the inhibitory effect on appetite that these receptors share with leptin. Increased appetite results in weight gain and associated effects. Blockade of M3 receptors may reduce insulin secretion and subsequently cause hyperglycemia and diabetes

between groups were noted on low-density lipoprotein cholesterol, fasting serum glucose, blood pressure, or waist circumference [50].

Perspectives

In addition to the use of FGAs and SGAs in the treatment of schizophrenia, there are emerging approaches targeting specific aspects of the illness, for example, psychological and cognitive approaches (such as cognitive therapy adapted for schizophrenia) and cognitive remediation training, particularly when combined with functional adaptation skills training, are important and impactful treatment modalities [51]. The field is still awaiting a third generation of drugs for schizophrenia, which can exercise therapeutic effect without considerable metabolic and endocrine risk [19, 31]. Pharmacogenetic analyses have offered some insights into effectiveness, tolerability, and side effects in individual patients. However, with 30% of patients considered to be "refractory" to treatment with most antipsychotics, and over 70% of patients discontinuing antipsychotic treatment within 1 year, the need for personalized approaches to treatment, although acute, is still a long way off [8, 52]. Early focus on environmental factors like nutrition and sedentary lifestyle targeting the potential vulnerability to type-2 diabetes would enhance research and clinical approaches to treatment in schizophrenia [6, 49]. Although medications form the basis of treatment, medications alone are insufficient for treating schizophrenia with psychosocial interventions early in the course of illness being broadly adopted and implemented in some countries [32, 33].

Research also focuses on earlier diagnosis and treatment of schizophrenia, thereby intuitively leading to better results and less secondary consequences of protracted psychosis. Indeed, research points to subtle signs of psychosis in advance of more florid psychotic manifestations [53]. Moreover, many of the neurobiological hallmarks of schizophrenia such as decreased gray matter in the temporal, frontal, and cingulate cortex as well as subtle clinical symptoms like attention difficulty, cognitive decline, social withdrawal, and affective flattening exist in early states, albeit in much more attenuated forms [54]. Unfortunately, identification and clinical incorporation of disease biomarkers that could inform and reliably predict treatment outcomes prove difficult [55]. Initiating psychopharmacological treatment prior to the development of the full clinical picture of illness with its functional consequences, is limited by exposure to possible metabolic risks of antipsychotics. Moreover, the lack of fundamental understanding of the pathobiology of schizophrenia greatly hampers this quest.

Questions and Answers

Question 1 A patient with schizophrenia on clozapine following a nutritionist-guided strict diet and exercise regimen may still develop type-2 diabetes due to:

A. A combination of various genetic factors that increase risk
B. Not following a strict-enough diet regimen
C. An inevitable adverse effect of the medication regimen
D. An inevitable progression of schizophrenia

Answer 1 A

Question 2 Which of the following changes correctly represent how some second-generation antipsychotics (SGAs) may affect leptin:

A. Decrease in leptin, increase in appetite, increase in insulin
B. Increase in leptin, increase in appetite, decrease in insulin
C. No change in leptin, increase in appetite, decrease in insulin
D. Decrease in leptin, decrease in appetite, increase in insulin

Answer 1 B

Question 3 A patient with both schizophrenia and type-2 diabetes wants to know if a diet change alone is enough to manage the metabolic risk. Which of the following is the best response as to why diet alone is likely to be insufficient?

A. Diet has to be combined with metformin.
B. Diet has to be combined with exercise.
C. Diet has to be combined with stopping the antipsychotic.
D. Metabolic risk management, as it addresses multiple factors.

Answer 3 D

References

1. Van Os J, Kapur S (2009) Schizophrenia Lancet 374:635–645
2. The diagnostic and statistical manual of mental disorders, 5th, Text Revision (DSM-5-TR), American Psychiatric Association Publishing, 2022
3. Pescosolido BA, Medina TR, Martin JK, Long JS (2013) The "backbone" of stigma: identifying the global core of public prejudice associated with mental illness. Am J Public Health 103:853–860
4. Tandon R, Nassrallah HA, Keshavan MS (2012) Schizophrenia: just the facts 5. Treatment and prevention. Past, present, and future. Schizophr Res 122:1–23
5. McGrath J, Meyer-Lindenberg A (2013) Is it time schizophrenia research left the museum? Clin Schizophr Relat Psychoses 6:170–171
6. Gragnoli C, Reeves GM, Reazer J, Postolache TT (2016) Dopamine–prolactin pathway potentially contributes to the schizophrenia and type 2 diabetes comorbidity. Transl Psychiatry 6(4):e785–e785
7. Chung J, Miller BJ (2020) Meta-analysis of comorbid diabetes and family history of diabetes in nonaffective psychosis. Schizophr Res 216:41–47
8. Lisoway AJ, Chen CC, Zai CC, Tiwari AK, Kennedy JL (2021) Toward personalized medicine in schizophrenia: genetics and epigenetics of antipsychotic treatment. Schizophr Res 232:112–124
9. Owen MJ (2012) Implications of genetic findings for understanding schizophrenia. Schizophr Bull 38:904–907
10. Cross-Disorder Group of the Psychiatric Genomics Consortium (2013) Genetic risk outcome of psychosis (GROUP) consortium. Identification of risk loci with shared effects on five major psychiatric disorders: a genome-wide analysis. Lancet 381(9875):1371–1379
11. Marshall CR, Howrigan DP, Merico D, Thiruvahindrapuram B, Wu W, Greer DS et al (2017) Contribution of copy number variants to schizophrenia from a genome-wide study of 41,321 subjects. Nat Genet 49(1):27–35

12. Rees E, O'Donovan MC, Owen MJ (2015) Genetics of schizophrenia. Curr Opin Behav Sci 2:8–14
13. Pantelis C, Papadimitriou GN, Papiol S, Parkhomenko E, Pato MT, Paunio T et al (2014) Biological insights from 108 schizophrenia-associated genetic loci. Nature 511(7510):421–427
14. Mizuki Y, Sakamoto S, Okahisa Y, Yada Y, Hashimoto N, Takaki M, Yamada N (2021) Mechanisms underlying the comorbidity of schizophrenia and type 2 diabetes mellitus. Int J Neuropsychopharmacol 24(5):367–382
15. Piper M, Beneyto M, Burne T, Eyles D, Lewis D, McGrath J (2012) The neurodevelopmental hypothesis of schizophrenia: convergent clues from epidemiology and neuropathology. Psychiatr Clin North Am 35:571–584
16. Colizzi M, McGuire P, Pertwee RG, Bhattacharyya S (2016) Effect of cannabis on glutamate signalling in the brain: a systematic review of human and animal evidence. Neurosci Biobehav Rev 64:359–381
17. Howes OD, McCutcheon R, Owen MJ, Murray RM (2017) The role of genes, stress, and dopamine in the development of schizophrenia. Biol Psychiatry 81(1):9–20
18. Kwon E, Soda T, Tsai L (2013) Neurodevelopment and Schizophrenia. In: Charney DS, Buxbaum JD, Sklar P, Nestler EJ (eds) Neurobiology of mental illness. Oxford University Press, pp 327–337.19
19. Correll CU, Abi-Dargham A, Howes O (2022) Emerging treatments in schizophrenia. J Clin Psychiatry InfoPack 1:SU21024IP1
20. McCutcheon RA, Krystal JH, Howes OD (2020) Dopamine and glutamate in schizophrenia: biology, symptoms and treatment. World Psychiatry 19(1):15–33
21. Gremel CM, Lovinger DM (2017) Associative and sensorimotor cortico-basal ganglia circuit roles in effects of abused drugs. Genes Brain Behav 16(1):71–85
22. Jevitt D (2012) Twenty-five years of glutamate and schizophrenia: are we there yet? Schizophr Bull 38:911–913
23. Miller BJ, Goldsmith DR (2019) Inflammatory biomarkers in schizophrenia: implications for heterogeneity and neurobiology. Biomarkers Neuropsychiatry 1:100006
24. Emiliani FE, Sedlak TW, Sawa A (2014) Oxidative stress and schizophrenia: recent breakthroughs from an old story. Curr Opin Psychiatry 27(3):185
25. Ward M, Druss B (2015) The epidemiology of diabetes in psychotic disorders. Lancet Psychiatry 2:431–451
26. Hasan A, Falkai P, Wobrock T, Lieberman J, Glenthoi B, Gattaz WF, Thibaut F, Möller HJ (2012) World Federation of Societies of biological psychiatry (WFSBP). Guidelines for biological treatment of schizophrenia, part 1: update 2012 on the acute treatment of schizophrenia and the management of treatment resistance. World J Biol Psychiatry 13:318–378
27. Buchanan RW, Kreyenbuhl J, Kelly DL, Noel JM, Boggs DL, Fischer BA, Himelhoch S, Fang B, Peterson E, Aquino PR, Keller W, Schizophrenia Patient Outcomes Research Team (PORT) (2010) The 2009 schizophrenia PORT psychopharmacological treatment recommendations and summary statements. Schizophr Bull 36:71–93
28. Leucht S, Cipriani A, Spineli L et al (2013) Comparative efficacy and tolerability of 15 antipsychotic drugs in schizophrenia: a multiple treatments meta-analysis. Lancet 382:951–962
29. Foster A, King J (2020) Antipsychotic polypharmacy. Focus 18(4):375–385
30. Sahli ZT, Tarazi FI (2018) Pimavanserin: novel pharmacotherapy for Parkinson's disease psychosis. Expert Opin Drug Discov 13(1):103–110
31. Weiden PJ, Breier A, Kavanagh S, Miller AC, Brannan SK, Paul SM (2022) Antipsychotic efficacy of KarXT (Xanomeline– Trospium): post hoc analysis of positive and negative syndrome scale categorical response rates, time course of response, and symptom domains of response in a phase 2 study. J Clin Psychiatry 83(3):40913
32. Kane JM, Robinson DG, Schooler NR, Mueser KT, Penn DL, Rosenheck RA et al (2016) Comprehensive versus usual community care for first-episode psychosis: 2-year outcomes from the NIMH RAISE early treatment program. Am J Psychiatry 173(4):362–372
33. Posselt CM, Albert N, Nordentoft M, Hjorthøj C (2021) The Danish OPUS early intervention services for first-episode psychosis: a phase 4 prospective cohort study with comparison of randomized trial and real-world data. Am J Psychiatry 178(10):941–951
34. Ho BC, Andreasen NC, Ziebell S, Pierson R, Vincent M (2011) Long term antipsychotic treatment and brain volumes: longitudinal study of first episode schizophrenia. Arch Gen Psychiatry 68:128–137
35. Ronopaske GT, Dorph-Peterson KA, Sweet RA (2008) Effects of antipsychotic exposure on astrocyte and oligodendrocyte numbers in macaque monkeys. Biol Psychiatry 63:758–765
36. Mackenzie NE, Kowalchuk C, Agarwal SM, Costa-Dookhan KA, Caravaggio F, Gerretsen P et al (2018) Antipsychotics, metabolic adverse effects, and cognitive function in schizophrenia Frontiers media SA, vol 9. https://doi.org/10.3389/fpsyt.2018.00622
37. Howard L, Kirkwood G, Leese M (2007) Risk of hip fracture in patients with a history of schizophrenia. Br J Psychiatry 190:129–134
38. Inder WJ, Castle D (2011) Antipsychotic-induced hyperprolactinemia. Aust N Z J Psychiatry 45:830–837
39. Pillinger T, McCutcheon RA, Vano L, Mizuno Y, Arumuham A, Hindley G et al (2020) Comparative effects of 18 antipsychotics on metabolic function in patients with schizophrenia, predictors of metabolic dysregulation, and association with psychopathology: a systematic review and network meta-analysis. Lancet Psychiatry 7(1):64–77

40. American Diabetes Association, American Psychiatric Association, American Association of Clinical Endocrinologists, North American Association for the Study of Obesity (2004) Consensus development conference on antipsychotic drugs and obesity and diabetes. Diabetes Care 27:596–601

41. De Hert M, Dekker JM, Wood D et al (2009) Cardiovascular disease and diabetes in people with severe mental illness position statement from the European psychiatric association (EPA), supported by the European association for the study of diabetes (EASD) and the European Society of Cardiology. Eur Psychiatry 24:412–424

42. Pramyothin P, Khaodhiar L (2010) Metabolic syndrome with the atypical antipsychotics. Curr Opin Endocrinol Diabetes Obes 17(5):460–466

43. Libowitz MR, Nurmi EL (2021) The burden of antipsychotic-induced weight gain and metabolic syndrome in children. Front Psych 12:623681. https://doi.org/10.3389/fpsyt.2021.623681

44. Coccurello R, Moles A (2010) Potential mechanisms of atypical antipsychotic-induced metabolic derangement: clues for understanding obesity and novel drug design. Pharmacol Ther 127(3):210–251

45. Potvin S, Zhornitsky S, Stip E (2015) Antipsychotic-induced changes in blood levels of leptin in schizophrenia: a meta-analysis. Can J Psychiatr 60(3 Suppl 2):S26–S34. PMID: 25886677; PMCID: PMC4418620

46. Zhang Y, Li X, Yao X, Yang Y, Ning X, Zhao T et al (2021) Do Leptin Play a Role in Metabolism–Related Psychopathological Symptoms? Front Psych 12:710498

47. Kelly JR, Minuto C, Cryan JF, Clarke G, Dinan TG (2021) The role of the gut microbiome in the development-meenia. Schizophr Res 234:4–23

48. Petruzzelli MG, Margari M, Peschechera A, De Giambattista C, De Giacomo A, Matera E et al (2018) Hyperprolactinemia and insulin resistance in drug naive patients with early onset first episode psychosis. BMC Psychiatry 18(1):246. https://doi.org/10.1186/s12888-018-1827-3

49. Gurusamy J, Gandhi S, Damodharan D, Ganesan V, Palaniappan M (2018) Exercise, diet and educational interventions for metabolic syndrome in persons with schizophrenia: a systematic review. Asian J Psychiatr 36:73–85

50. Jiang WL, Cai DB, Yin F, Zhang L, Zhao XW, He J et al (2020) Adjunctive metformin for antipsychotic-induced dyslipidemia: a meta-analysis of randomized, double-blind, placebo-controlled trials. Transl Psychiatry 10(1):1–9

51. Harvey PD, Bowie R (2012) Cognitive enhancement in schizophrenia: pharmacological and remediation approaches. Psychiatr Clin North Am 35:683–698

52. Foster, A., & Buckley, P. F. (2014). Pharmacogenetics and treatment-resistant schizophrenia. In Treatment–Refractory Schizophrenia (pp. 179–193). Springer, Berlin, Heidelberg

53. Fusar-Poli P, Borgwardt S, Bechdolf A, Addington J, Riecher-Rossler A, Schultze-Lutter F et al (2013) The psychosis high-risk state: a comprehensive state-of-the-art review. JAMA Psychiatry 70:107–120

54. Kerns J, Lauriello J (2012) Can structural neuroimaging be used to define phenotypes and course of schizophrenia? Psychiatr Clin North Am 35:633–644

55. Weickert CS, Weickert T, Pillai A, Buckley PF (2013) Biomarkers for schizophrenia: a brief conceptual consideration. Dis Markers 35:3–9

Epilepsy

Detlev Boison

Introduction to Epilepsy

Epilepsy is a complex syndrome comprised of seizures and associated comorbidities affecting around 80 million persons worldwide. The condition is characterized by excessive electrical discharges in hyperexcitable neuronal clusters that result in spontaneous and recurrent seizures. The seizures may be subclinical and thus only apparent on an electroencephalogram (EEG), but more often they fit into two clinical classifications: partial and generalized. Partial seizures have a focused origin in the brain, and, therefore, seizure symptoms may present in a localized manner. Generalized seizures lack a focal origin and instead involve the entire brain. Epileptic seizures can range from altered states of consciousness to those involving motor function with clonic and/or tonic components.

Our understanding of the pathophysiological processes that turn a normally functioning brain into a hyperexcitable one remains limited. Many theories exist to explain the apparent disruption of homeostatic functions that alter neuronal excitation and/or inhibition. Research has thus far examined extracellular ion homeostasis, altered energy metabolism, changes in receptor function,

alterations in transmitter uptake, and epigenetic reprogramming.

Pathophysiology of Epilepsy and Metabolic Alterations

Epileptic seizures result from significant electrical discharges in hyperexcitable groups of neurons in which the normal neurophysiology is pathologically altered to favor excitation over inhibition. This can be caused by decreased inhibitory or increased excitatory signals or altered response to these input signals, all of which probably contribute to the etiology of epilepsy.

Within neurons, altered functions of ion channels, such as those controlling the flux of Ca^{2+}, K^+, or Na^+, a decrease in inhibitory γ-aminobutyric acid (GABA) signaling, or an excess in glutamatergic excitatory signaling, all contribute to the hyperexcitable state of neuronal networks. Those mechanisms generally decrease the threshold for neuronal activation, and, thus, stimuli that normally are subthreshold are now able to trigger excessive excitation. Another feature of epileptogenic groups of neurons is synchrony, which is favored by altered neuronal connectivity leading to recurrent circuitry.

Recent evidence suggests that neuronal hyperexcitability in epilepsy is not only an intrinsic deficiency of neurons but to a large degree

D. Boison (✉)
Department of Neurosurgery, Robert Wood Johnson Medical School, Rutgers University, Piscataway, NJ, USA
e-mail: detlev.boison@rutgers.edu

E. Lammert, M. Zeeb (eds.), *Metabolism of Human Diseases*,
https://doi.org/10.1007/978-3-031-96019-2_5

determined by the disruption of homeostatic and metabolic functions, which are controlled by astrocytes (Fig. 1) [1]. Many forms of epilepsy, particularly those involving the temporal lobe, are characterized by astrogliosis, a macroglial response that leads to increased proliferation and hypertrophy of astrocytes. Astrogliosis in turn is associated with changes in the membrane conductance for ions, which limits the buffering capacity of astrocytes for K$^+$, Ca^{2+}, and H$^+$, and disrupts the homeostasis of neurotransmitters such as glutamate (Glu) and of neuromodulators such as adenosine. As a consequence of those alterations in astroglial physiology, the extracellular concentrations of K$^+$ and Glu are increased, whereas those of adenosine are decreased. In particular, overexpression of adenosine kinase, the key enzyme for the metabolic clearance of adenosine through astrocytes, has been identified as

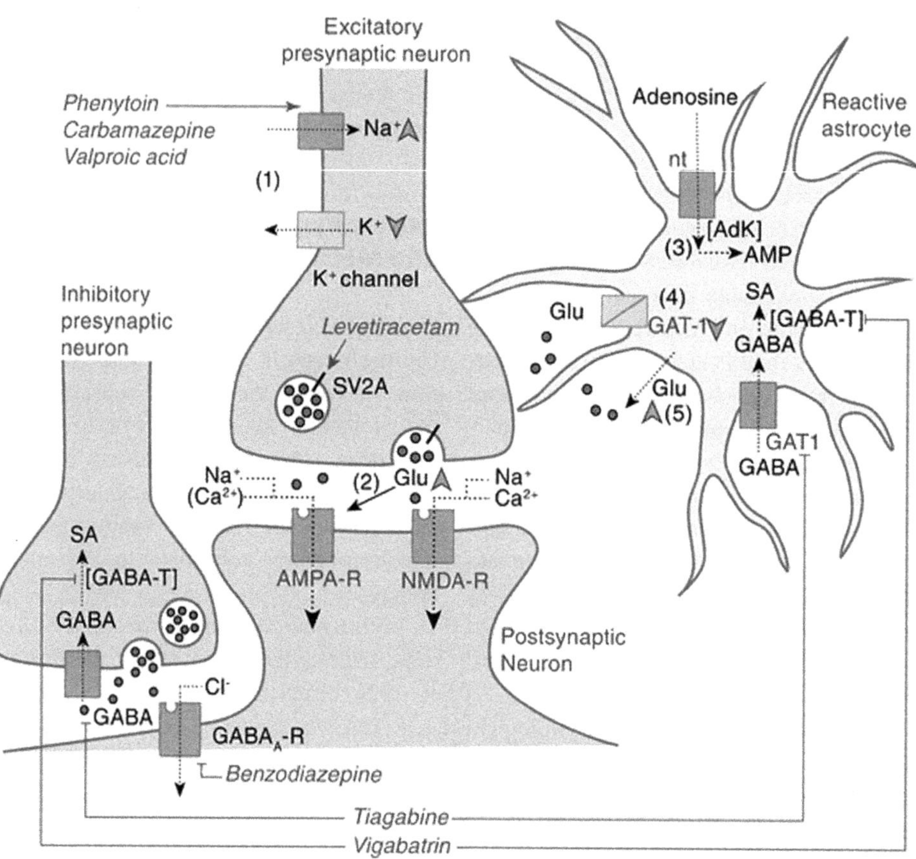

Fig. 1 Disruption of multiple systems in epilepsy involving neurons and astrocytes. (1) Hyperexcitable neurons are associated with overactivation of voltage-gated Na$^+$ channels and reduced activity of K$^+$ channels, which generate an action potential that leads to (2) increased release of glutamate (*Glu*). This activates α-Amino-3-hydroxy-5--methyl-4-isoxazolepropionic acid (AMPA) and *N*-methyl-D-aspartate (*NMDA*) receptors on the postsynaptic membrane, causing excitatory synaptic potentials. (3) Upregulated adenosine kinase (AdK) in reactive astrocytes increases the removal of extracellular adenosine, enhancing hyperexcitability. (4) Decreased membrane transporters for K$^+$ (not shown) and Glu (such as *GLT-1*) on reactive astrocytes further enhance hyperexcitability. (5) Excessive Ca^{2+} intake in astrocytes even induces Glu release. Points of interference of a few antiepileptic drugs are also shown. These drugs generally aim to decrease hyperexcitability by reducing glutamate release or signaling or by increasing glutamate removal or inhibitory signaling. *GABA* γ-aminobutyric acid, *GABA-T* GABA transaminase, *GAT-1* GABA transporter-1, *nt* nucleoside transporter, *SA* succinic semialdehyde, *SV2A* synaptic vesicle glycoprotein 2A

a key pathological hallmark of epilepsy as it results in chronic adenosine deficiency [2].

Adenosine is a key link between energy homeostasis and metabolic activity and has been termed a "retaliatory metabolite" due to its capability to adjust energy consumption to energy supplies [3]. In general, under any conditions of stress or distress, the levels of adenosine rise, and it is this rise in adenosine that limits further neuronal energy consumption [4] by binding to inhibitory adenosine A_1 receptors [5], which couple to inhibitory G_i and G_o proteins and thereby (1) lead to a decrease in the intracellular messenger cAMP (cyclic adenosine monophosphate), (2) induce neuronal hyperpolarization by augmenting G-protein-coupled inwardly rectifying K^+ channels, and (3) induce presynaptic inhibition by limiting Ca^{2+} influx into the presynaptic neuron. More specifically, in epilepsy, seizures consume a large amount of energy, which leads to a rise in adenosine that acts as an endogenous anticonvulsant and seizure terminator [6].

Treatment of Epilepsy

Treatment of epilepsy with antiseizure medications (ASMs) generally attempts to correct the disruption in normal electrical functionality via decreases in excitatory processes or increases in inhibitory processes. ASMs aim to reduce seizures by regulating the nervous system's primary excitatory neurotransmitter, Glu, its primary inhibitory neurotransmitter, GABA, or ion channels that control the conduction of electrical impulses in neurons. It should be noted, however, that the treatment of epilepsy, based on the modulation of neuronal downstream targets, fails to control seizures in more than 30% of patients with epilepsy [7, 8]. The control of neuronal excitability by blockade of ion channels such as those for Na^+, K^+, and Ca^{2+} is a mechanism of commonly used ASMs (e.g., phenytoin, carbamazepine, valproic acid). Newer ASMs act by interfering with the synthesis, function, release, and metabolism of neurotransmitters and their receptors. Levetiracetam binds to the synaptic vesicle glycoprotein SV2A and inhibits presynaptic Ca^{2+} channels. Presynaptic mechanisms are thought to impede impulse conduction across synapses. Vigabatrin blocks GABA transaminase, the major GABA degrading enzyme, and tiagabine blocks reuptake of GABA via GABA transporter 1 (GAT1), thereby increasing the level of extracellular GABA in the brain. Moreover, ASMs can enhance the effect of GABA (e.g., benzodiazepine). However, conventional pharmacological strategies frequently fail due to the development of pharmacoresistance, and the global manipulation of ion channels leads to widespread side effects, particularly in the cognitive domain, at higher and more effective doses. Alternative treatments include surgical resection of a seizure focus. The goal of surgical resection is to remove the seizure origin or disrupt the spread of the seizure throughout the brain. If an epileptogenic focus is found and does not reside in cortical areas responsible for speech, it may be possible to remove without compromising neurological functions. Surgery has been successful in reducing seizures and in some instances may completely alleviate the need for other treatments [9]. While ASMs and surgery have been the classical foundation of epilepsy treatment, they do not provide a satisfactory solution for many patients (see below). In contrast, dietary therapies may provide seizure control in drug-resistant forms of epilepsy [10]. The therapeutic effect of fasting on the control of seizures has been known since Hippocrates. Fasting's suppression of seizures can be mimicked by a high-fat, low-carbohydrate ketogenic diet (Fig. 2). The anticonvulsant effects brought on by metabolic changes during this diet therapy offer great promise for identifying new antiseizure targets as well as providing insight into the pathophysiology of the epileptic brain (see below).

Influence of Treatment on Metabolism

The sharp decrease in carbohydrate intake during ketogenic diet treatment reduces glucose utilization as an energy source. Instead, the liver utilizes fatty acids to produce ketone bod-

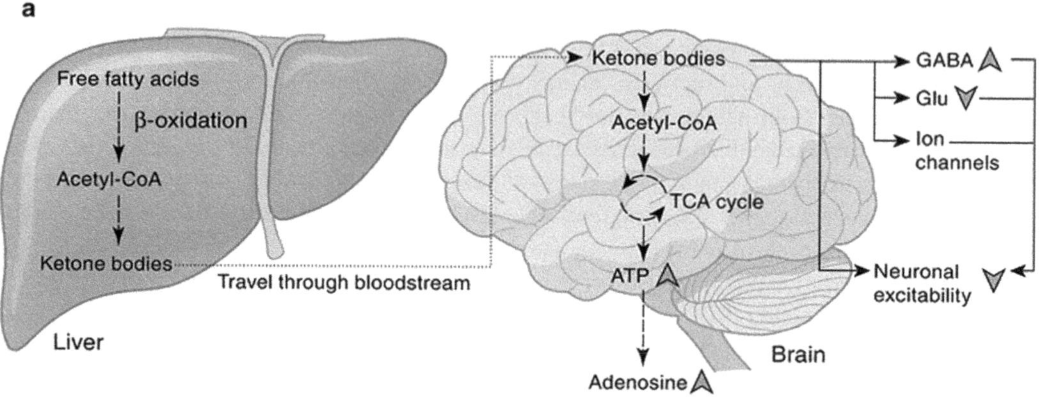

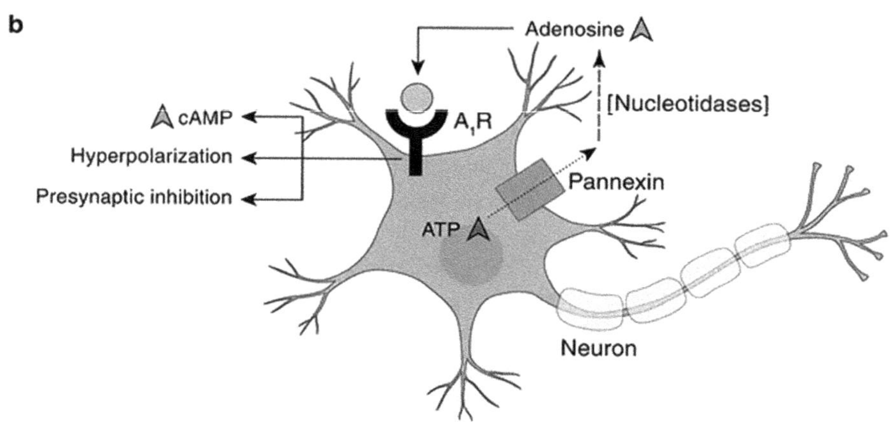

Fig. 2 Mechanism of action for the ketogenic diet. (**a**) Schematic diagram illustrating ketone body formation from fatty acids present in the ketogenic diet and their subsequent use in the brain as fuel for ATP production. Ketogenic diet is reported to increase adenosine and γ-aminobutyric acid (*GABA*), to decrease glutamate (*Glu*), and to directly act on neurotransmitters subsequently reducing neuronal excitability. (**b**) Increased intracellular ATP moves ATP along its concentration gradient through pannexin to the extracellular space. Here, it is degraded by nucleotidases to its core constituent, adenosine, which can act on inhibitory A_1 receptors (*A_1R*) to cause presynaptic inhibition and hyperpolarization to prevent seizures. *cAMP* cyclic adenosine monophosphate

ies, mainly β-hydroxybutyrate and acetoacetate. Neurons in turn will use these ketone bodies for cellular metabolism in place of glucose (Fig. 2). Many hypotheses exist to explain the anticonvulsant action of the ketogenic diet, but despite its name, the ketone bodies themselves may not necessarily be the primary effector, and the underlying mechanisms leading to reduced seizure activity are largely unknown. Among beneficial metabolic changes induced by ketogenic diet therapy are (1) increased levels of adenosine [11], (2) increased GABA, and (3) decreased Glu, which all combine with a direct action of ketone bodies on ion channels to reduce neuronal excitability [10]. Ketosis increases mitochondrial biogenesis and thereby the production of ATP, which is a direct metabolic precursor of adenosine, as it leaves the neuron via the transmembrane channel pannexin and is converted to adenosine extracellularly (Fig. 2) [12]. It is this unique metabolic modulation that builds excitement not only for a deeper understanding of the ketogenic diet but also for metabolic manipulations in general as a novel anticonvulsant strategy with good efficacy and few side effects.

Epilepsy Prevention

A new frontier in epilepsy research is the development of therapies capable to prevent the development or the progression of epilepsy following a precipitating event such as a traumatic brain injury, a stroke, or an infection of the brain. It has now become clear that those precipitating events cause multiple changes in network homeostasis, including inflammation, glial activation, as well as metabolic and epigenetic changes [13]. Those changes that turn a healthy brain into an epileptic brain constitute new therapeutic targets for the development of antiepileptogenic therapies (AEGs). Network pharmacology is one approach for epilepsy prevention. It is based on the rationale to strategically combine several existing Food and Drug Administration (FDA)-approved drugs to target critical nodes in the epileptogenic network in a synergistic fashion; proofs of feasibility studies have shown that this is a workable approach for epilepsy prevention [14, 15]. New findings have shown that adenosine not only has antiseizure properties, but also acts as an AEG. Biochemically, adenosine is a product of DNA methylation reactions and DNA methylation can only proceed when adenosine is removed by the action of an isoform of adenosine kinase expressed in the cell nucleus (ADK-L) [16–18]. It was shown that pathological overexpression of ADK drives the epileptogenic process through an epigenetic mechanism [16, 19] and that therapeutic adenosine augmentation is antiepileptogenic [16, 20]. In line with those findings metabolic therapies, such as the ketogenic diet, have been shown to interfere with the epileptogenic process through a variety of epigenetic mechanisms, which include DNA methylation and histone acetylation [21].

Perspectives

Inefficacy of current ASM treatment in 30% of patients signals the need for a deeper understanding of homeostatic mechanisms that govern the pathophysiology and etiology of epilepsy. Indeed, it is time to redefine epilepsy as a complex syndrome of disrupted network homeostasis, which includes seizures not only as the major pathological trait but also as a wide range of so-called comorbidities including cognitive impairment, sleep dysfunction, depression (see chapter "Major Depression and Metabolic Syndrome"), and psychiatric impairment. Conventional ASMs were solely designed to suppress seizures, which is only a symptom of epilepsy, albeit the most obvious. However, there remains an urgent need for the development of AEGs, which truly interfere with the underlying pathogenic processes. Based on the neurocentric rationale of past drug development, it becomes clear that conventional ASMs are a poor choice to treat epilepsy as a syndrome in a "holistic" sense. Novel therapeutic avenues aimed at reconstructing the homeostasis of network regulation may ultimately provide better treatment and prevention options and hopes for finding a cure for epilepsy. In this regard, metabolic interventions and adenosine augmentation appear to be promising homeostatic therapies, which might not only be effective in pharmaco-resistant epilepsy, but are also poised to prevent the development of epilepsy and its progression and thereby reduce the disease burden of epilepsy.

Acknowledgments The author is supported by National Institutes of Health (NS065957; NS103740; NS127846) and the U.S. Department of the Army (W81XWH2210638).

Questions and Answers

Question 1 What is the difference between seizures and epilepsy?

Answer 1 Epilepsy is a syndrome comprised of seizures, which are the dominant symptom. Seizures are the dominant symptom of epilepsy but can also occur as a consequence of a triggering event independent of epilepsy.

Question 2 Which is a major non-neuronal mechanism of epilepsy?

Answer 2 Astrogliosis. This is a pathological process, which profoundly alters brain metabolism.

Question 3 What is the premise of epilepsy prevention therapies?

Answer 3 For the prevention of epilepsy, it is necessary to target several mechanisms involved in the epileptogenic process synergistically. This can be achieved through network pharmacology, or metabolic strategies (e.g., adenosine), which target multiple mechanisms.

References

1. Steinhauser C, Boison D (2012) Epilepsy: crucial role for astrocytes. Glia 60:1191
2. Boison D (2012) Adenosine dysfunction in epilepsy. Glia 60:1234–1243
3. Newby AC, Worku Y, Holmquist CA (1985) Adenosine formation. Evidence for a direct biochemical link with energy metabolism. Adv Myocardiol 6:273–284
4. Fredholm BB (2007) Adenosine, an endogenous distress signal, modulates tissue damage and repair. Cell Death Differ 14:1315–1323
5. Boison D (2005) Adenosine and epilepsy: from therapeutic rationale to new therapeutic strategies. Neuroscientist 11:25–36
6. Dragunow M (1991) Adenosine and seizure termination. Ann Neurol 29:575
7. Kwan P, Brodie MJ (2000) Early identification of refractory epilepsy. N Engl J Med 342:314–319
8. Loscher W, Schmidt D (2011) Modern antiepileptic drug development has failed to deliver: ways out of the current dilemma. Epilepsia 52:657–678
9. de Tisi J, Bell GS, Peacock JL, McEvoy AW, Harkness WF, Sander JW, Duncan JS (2011) The long-term outcome of adult epilepsy surgery, patterns of seizure remission, and relapse: a cohort study. Lancet 378:1388–1395
10. Lutas A, Yellen G (2013) The ketogenic diet: metabolic influences on brain excitability and epilepsy. Trends Neurosci 36:32–40
11. Masino SA, Li T, Theofilas P, Sandau US, Ruskin DN, Fredholm BB, Geiger JD, Aronica E, Boison D (2011) A ketogenic diet suppresses seizures in mice through adenosine A1 receptors. J Clin Invest 121:2679–2683
12. Masino SA, Geiger JD (2008) Are purines mediators of the anticonvulsant/neuroprotective effects of ketogenic diets? Trends Neurosci 31:273–278
13. Klein P, Dingledine R, Aronica E, Bernard C, Blumcke I, Boison D, Brodie MJ, Brooks-Kayal AR, Engel J Jr, Forcelli PA, Hirsch LJ, Kaminski RM, Klitgaard H, Kobow K, Lowenstein DH, Pearl PL, Pitkanen A, Puhakka N, Rogawski MA, Schmidt D, Sillanpaa M, Sloviter RS, Steinhauser C, Vezzani A, Walker MC, Loscher W (2018) Commonalities in epileptogenic processes from different acute brain insults: do they translate? Epilepsia 59:37–66
14. Schidlitzki A, Bascunana P, Srivastava PK, Welzel L, Twele F, Tollner K, Kaufer C, Gericke B, Feleke R, Meier M, Polyak A, Ross TL, Gerhauser I, Bankstahl JP, Johnson MR, Bankstahl M, Loscher W (2020) Proof-of-concept that network pharmacology is effective to modify development of acquired temporal lobe epilepsy. Neurobiol Dis 134:104664
15. Welzel L, Twele F, Schidlitzki A, Tollner K, Klein P, Loscher W (2019) Network pharmacology for antiepileptogenesis: tolerability and neuroprotective effects of novel multitargeted combination treatments in non-epileptic vs. post-status epilepticus mice. Epilepsy Res 151:48–66
16. Williams-Karnesky RL, Sandau US, Lusardi TA, Lytle NK, Farrell JM, Pritchard EM, Kaplan DL, Boison D (2013) Epigenetic changes induced by adenosine augmentation therapy prevent epileptogenesis. J Clin Inv 123:3552–3563
17. Boison D (2013) Adenosine kinase: exploitation for therapeutic gain. Pharmacol Rev 65:906–943
18. Boison D, Yegutkin GG (2019) Adenosine metabolism: emerging concepts for cancer therapy. Cancer Cell 36:582–596
19. Li T, Ren G, Lusardi T, Wilz A, Lan JQ, Iwasato T, Itohara S, Simon RP, Boison D (2008) Adenosine kinase is a target for the prediction and prevention of epileptogenesis in mice. J Clin Inv 118:571–582
20. Sandau US, Yahya M, Bigej R, Friedman JL, Saleumvong B, Boison D (2019) Transient use of a systemic adenosine kinase inhibitor attenuates epilepsy development in mice. Epilepsia 60:615–625
21. Boison D, Rho JM (2020) Epigenetics and epilepsy prevention: the therapeutic potential of adenosine and metabolic therapies. Neuropharmacology 167:107741

Alzheimer's Disease

Milos D. Ikonomovic and Steven T. DeKosky

Introduction to Alzheimer's Disease

After more than a century since it was first described, there is still controversy regarding the cause(s) of Alzheimer's disease (AD). Dr. Alois Alzheimer published the first clinicopathological report of the disease in 1906 [3], and described the two major histopathological hallmarks, "senile" plaques (also called amyloid plaques) and neurofibrillary tangles [4], composed primarily of fibrillar amyloid-β (Aβ) peptides and over-phosphorylated tau (p-tau) protein, respectively. We know now that pathological aggregates of other proteins (e.g., ubiquitin, TAR DNA-binding protein [TDP-43], and α-synuclein) are also commonly found in brains of patients with AD [5, 6] as well as other proteinopathies such as fronto-temporal lobar degeneration and Parkinson's disease [7]. Prodromal states of AD have been identified, with mild but abnormal cognitive function insufficient to qualify as dementia. Such early symptoms are referred to as mild cognitive impairment (MCI) or prodromal AD. As the disease progresses, performance in other cognitive domains (language, executive function) declines. Along with the clinical symptoms there is activation of the brain's inflammatory responses, as well as degeneration and loss of synapses and neurons [8]. These structural changes, which correlate with cognitive decline [9], are accompanied by atrophy in the medial temporal lobe and eventually diffuse brain atrophy more marked in the association cortices.

While tissue-based analyses of brain biopsy (very rarely performed) or postmortem examination of the brain are the only definitive methods of diagnosing AD, clinical diagnosis is increasingly accurate, using biomarkers based on protein analyses of amyloid-β and p-tau in cerebrospinal fluid (CSF) and more recently in blood (plasma). In vivo imaging of amyloid-β and neurofibrillary (p-tau) pathology with positron emission tomography (PET) also adds significant confidence to the diagnosis and is regarded as sufficient evidence of AD to act as a screen for subjects to enter clinical trials of anti-amyloid medications. Additionally, advances in magnetic resonance imaging (MRI) continue to yield clearer understanding of the locations and rate of brain volume loss and correlation with cognitive decline (Fig. 1) [10].

The most powerful risk factor for sporadic AD is advanced age. Epidemiological studies suggest that the prevalence of AD doubles every 5 years

M. D. Ikonomovic
Departments of Neurology and Psychiatry, University of Pittsburgh School of Medicine and Geriatric Research Education and Clinical Center, VA Pittsburgh Healthcare System, Pittsburgh, PA, USA
e-mail: ikonomovicmd@upmc.edu

S. T. DeKosky (✉)
McKnight Brain Institute and Department of Neurology, University of Florida College of Medicine, Gainesville, FL, USA
e-mail: dekosky@ufl.edu

E. Lammert, M. Zeeb (eds.), *Metabolism of Human Diseases*,
https://doi.org/10.1007/978-3-031-96019-2_6

45

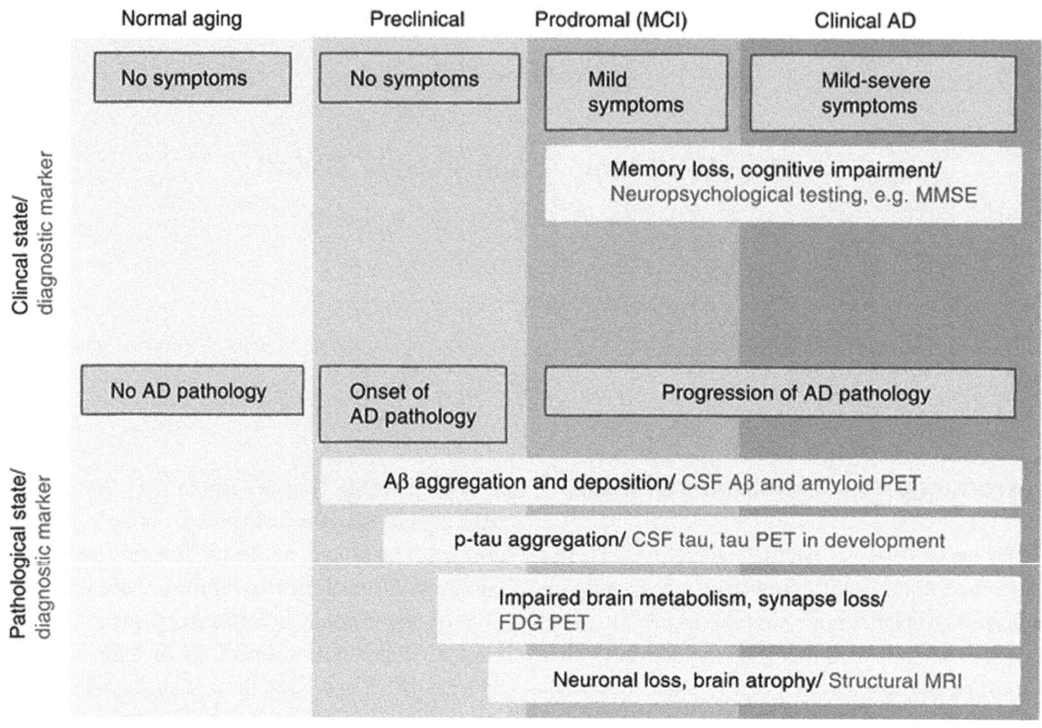

Fig. 1 Clinical and neuropathological continuum of mild cognitive impairment (MCI) and Alzheimer's disease (AD). AD pathology is initiated well over a decade prior to emergence of symptoms. The earliest detectable pathological changes are amyloid-ß (Aß) plaque deposits in the brain parenchyma, clinically detectable using amyloid positron emission tomography (PET) imaging, and decreased Aβ concentration in the cerebrospinal fluid (CSF), as Aβ is retained in the brain (in amyloid plaques) rather than cleared to the blood and CSF. The latter is detectable using enzyme-linked immunosorbent assay (ELISA). Newly developed ultrasensitive immunoassays are emerging to measure plasma concentrations of Aβ and p-tau, as well as the neurofilament light (NfL) protein as a marker for neurodegeneration. Aβ aggregation precedes p-tau aggregation. Impairment of neuronal metabolism and loss of synaptic structure and function may begin in the preclinical stage and the decreased metabolic activity are detectable using fluorodeoxyglucose (FDG)-PET. Finally, neuronal loss and resulting brain atrophy occur, detectable with structural magnetic resonance imaging (MRI). Once initiated, these pathologies advance progressively and contribute to clinical symptoms. *MMSE* Mini-Mental State Examination

after age 65 and approaches 50% by age 90. Other risk factors include cardiovascular and metabolic diseases and related conditions, such as hypertension, hyperlipidemia (see chapter "Hyperlipidemia"), obesity, diabetes (see chapter "Diabetes Mellitus"), and metabolic syndrome (see chapter "Metabolic Syndrome"). Brain injury, such as traumatic brain injury, ischemic stroke (see chapter "Stroke"), brain hemorrhage, substance abuse, e.g., smoking and alcohol abuse, and depression (see chapter "Major Depression and Metabolic Syndrome") also elevate the risk. Genetic contributions to risk of early onset (familial) AD are also significant and are discussed below.

Genetic risk factors for AD can be divided into autosomal dominant familial AD (FAD) genes (mutations), which are completely penetrant and manifest earlier in life, and alleles of genes that elevate the risk of developing the disease but do not cause it. Risk alleles are most commonly those of late-life AD onset [11]. FAD forms are caused by mutations in several genes essential for metabolism of the amyloid precursor protein (APP, see below). Transgenic mice expressing these gene mutations are animal models commonly used to study AD. FAD mutations show earlier onset (usually between ages 35 and 50), yet are responsible for less than 1% of all AD cases. In contrast, nondominant forms of AD are seen in over 99% of cases and are characterized by late onset (usually over age 65) and slow clinical progression. The highest risk elevation is the presence of an apolipoprotein E4 allele (ApoE4, see below).

Pathophysiology of Alzheimer's Disease and Metabolic Alterations

Amyloid-β (Aβ) pathology is detectable in the preclinical phase of AD, prior to detectable changes in any other biomarkers and sometimes a decade or more before the earliest changes in an individual's cognition (Fig. 1) [12, 13]. This is in agreement with the "amyloid cascade" hypothesis of AD, in which altered metabolism of APP results in accumulation of fibrillar Aβ aggregates in brain parenchyma (amyloid plaques) and vasculature (cerebral amyloid angiopathy [CAA]), or diffusible oligomeric forms, driving AD pathogenesis [14]. Additionally, intracellular fibrillar bundles of the microtubule-associated protein tau, which has been highly phosphorylated and crosslinked and thus rendered nonfunctional, accumulate in neurofibrillary tangles that are commonly observed in AD, but also in pathological inclusions which define several other neurodegenerative diseases called "tauopathies." The loss of normal function of tau, as a microtubule-associated protein that stabilizes axons, also contributes to impaired cognition.

Amyloid-β precursor protein (APP) is a putative transmembrane protein, as deduced from its protein sequence, with a short intracellular C-terminus and a long extracellular N-terminus. While much is now known about APP metabolism, its primary function is less clear [15]. Under physiological conditions, the predominant APP metabolic pathway involves the enzyme α-secretase, which cuts the APP within the Aβ peptide sequence, thus preventing its genesis (Fig. 2). An alternative metabolic pathway (the amyloidogenic pathway) produces intact Aβ peptide via proteolytic actions of the enzymes β- and γ-secretase (Fig. 2).

The sequence of full-length Aβ peptides spans 40–43 amino acids from an N-terminal location (the β-secretase cut site) to an intramembranous location (the γ-secretase cut site). The intramembranous γ-secretase is a complex of four proteins, one of which, presenilin 1, is the product of the most frequent gene mutation causing FAD [16].

Aβ is found in many heterogeneous forms, including truncations and modifications at C- and N-terminus. Once released from APP, Aβ monomers self-assemble into diffusible oligomers (Fig. 2), which further assemble into protofibrils and insoluble fibrils that clump together and form deposits in amyloid plaques and CAA. These deposits consist of predominantly Aβ42 or Aβ40, respectively. Amyloid fibrils (and their characteristic β-sheet structure) can be detected using Congo red histochemistry in postmortem tissue sections or PET imaging with specific radiotracers in brains of living patients. PET utilizes the differential uptake of amyloid-binding PET radioligands (e.g., ^{11}C-Pittsburgh compound B, ^{18}F-florbetapir, ^{18}F-florbetaben, ^{18}F-flutemetamol), all of which bind transiently to Aβ fibrils, enabling differential focal positron emission and generating an elevated signal in the region of plaque deposition and/or CAA in the brain. While all the currently available Aβ-amyloid PET radiopharmaceuticals are nonselective markers of Aβ amyloidosis in brain parenchyma and vasculature, CAA-selective PET imaging is currently under development [17].

Aβ plaques are morphologically heterogeneous and in advanced stages contain a dense amyloid core surrounded by dystrophic neurites (i.e., "neuritic" plaques), the latter containing hyper-phosphorylated tau aggregates and associated with glial inflammatory reactions in AD brain (Fig. 2). Diffusible (nonfibrillar) oligomers of Aβ and tau interfere with synapse function (Fig. 2), contributing to cognitive impairment [14, 18].

The relationship between amyloid pathology and the development of phosphorylated tau (p-tau), which leads to neurofibrillary tangles, is

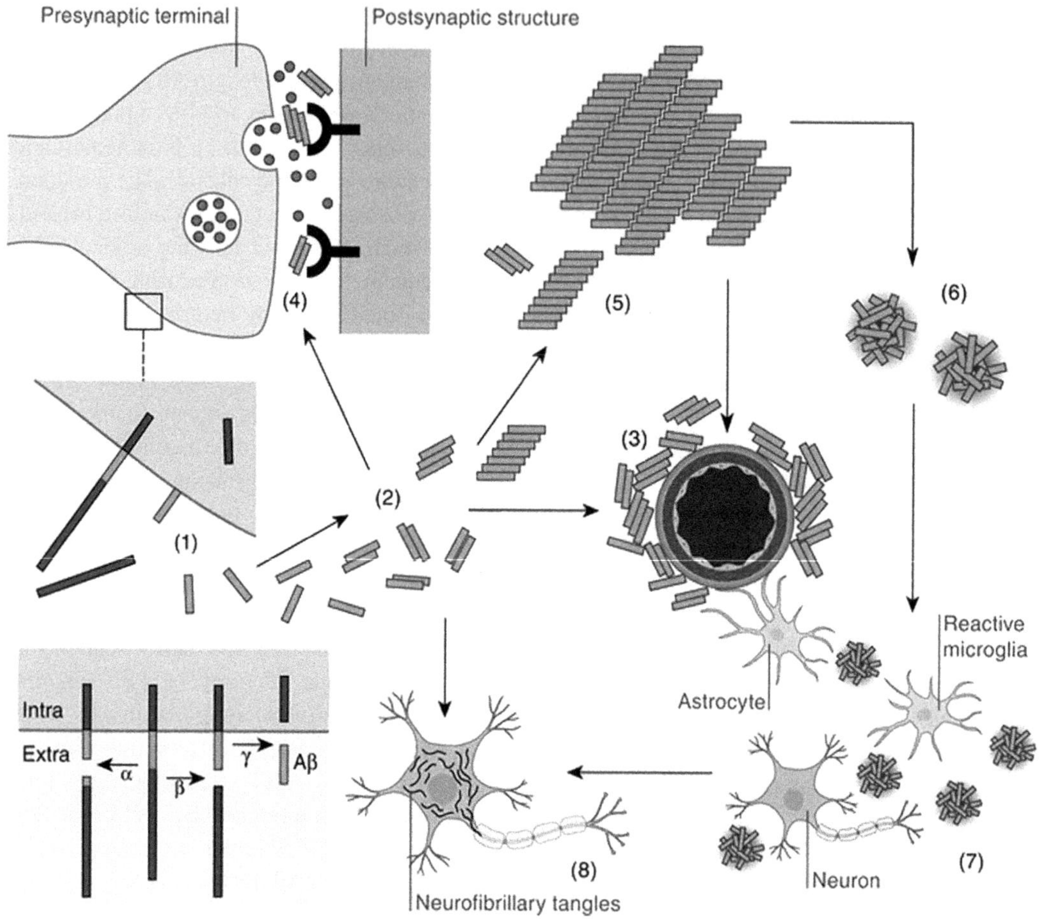

Fig. 2 Generation of Aβ and the pathological cascade in Alzheimer's disease. (*1*) The amyloid precursor protein APP is metabolized mainly by α-secretase (*left arrow*), preventing excessive formation of the Aβ peptide. β- and γ-secretase (*right arrows*) cleave APP to release soluble Aβ. Aβ monomers assemble into soluble Aβ oligomers (*2*). Aβ oligomers can be cleared via the brain vasculature, but in advanced disease stages, Aβ fibrils can deposit in cerebral blood vessel walls (*3*). Soluble oligomers of Aβ (and tau) are believed to interfere with synaptic function, and it is tau deposits that correlate better than Aβ plaques with memory impairment (*4*). The oligomers form Aβ fibrils that comprise the dense β-pleated sheets of amyloid in plaques (*5*). Typical amyloid plaques in the brain parenchyma consist of fibrillar Aβ deposits (*6*) and apolipoproteins. In advanced stages, they are surrounded by dystrophic neuronal processes that contain hyperphosphorylated tau protein and are called neuritic plaques. Amyloid plaques probably initiate an inflammatory process that involves reactive microglia and astrocytes (*7*). Finally, Aβ pathology is believed to precede, and potentially contributes to, the development of neurofibrillary tangles in neocortical areas (*8*) and neuronal cell death

an important question that is not understood fully. Aβ and p-tau fibrils might propagate AD lesions via a common mechanism, allowing the spread of Aβ, tau, and other misfolded fibrillar proteins among interconnected brain areas [19–22]. However, in AD the amyloid deposition begins and is established in the neocortex before there is a great deal of neocortical tau/tangle deposition; thus some change/s that involve the presence of Aβ appears to stimulate tau aggregation and/or spread.

The accumulation of Aβ in the brain might be due to its increased production or decreased degradation/clearance or both. Progressive slowing of Aβ clearance in aging would lead to increased concentrations in the brain and as it begins to aggregate, it would reduce concentrations in cerebrospinal fluid (CSF). ApoE is the lipoprotein carrier that docks with low-density lipoprotein (LDL) receptor-related protein 1 (LRP1) and removes Aβ from the brain, moving it to the blood vessel wall and then to the bloodstream. The E4 allele produces an apolipoprotein that is not as efficient at such removal, providing a rationale for why it is associated with greater risk of AD and adding support to the hypothesis that in late-onset AD the buildup of amyloid plaques in the brain is due to decreased removal of Aβ over the lifespan (rather than increased production).

Implications for Treatment and Influence of Treatment on Metabolism

Medications for AD can be grouped into three categories: (1) symptomatic medications, such as efforts to boost neurotransmitter signaling in affected systems or preventing excitotoxicity, (2) medications directed at nonspecific pathological processes in AD, for example, antioxidants or anti-inflammatory medications, and (3) medications directed at specific AD pathologies, for example, against Aβ or against pathological transformation of tau. There are few approved and mildly effective disease-modifying treatments, although a number of studies are underway.

Reduced acetylcholine (ACh) transmission and metabolism is a well-known feature of AD, stemming from degeneration of the cholinergic neurons in the basal forebrain, the source of all cholinergic input to the cortex and hippocampus. The most common symptomatic treatment is the use of acetylcholinesterase inhibitors (donepezil, rivastigmine, galantamine), which act by preventing the degradation of ACh, thereby prolonging its action at the synapses and improving cholinergic neurotransmission. While there is statistical evidence of their efficacy, the clinical effects are modest. The most common side effects include nausea, vomiting, and diarrhea, due to off-target cholinergic stimulation in the gastrointestinal tract.

Memantine, a noncompetitive antagonist of glutamatergic N-methyl-D-aspartate (NMDA) receptors, has also been shown to correct neurochemical abnormalities, such as excitotoxicity. Side effects are uncommon but include headache, dizziness, and hallucinations. Memantine is only approved for use in patients with moderate to severe AD.

Despite high hopes and suggestions of risk reduction seen in epidemiological studies, antioxidants, anti-inflammatories, and endocrine agents have been unsuccessful in showing positive effects in clinical trials in mild to moderate AD. Decreases in amyloid plaques (based on PET imaging) have been reported in anti-Aβamyloid therapies [23]. However, even with the effects of reducing plaque, the anti-Aβ interventions have had limited success in slowing clinical decline. Thus, increased attention is being directed at intervention in earlier disease stages, including mild cognitive impairment (MCI), a prodromal stage of the disease, in which mildly abnormal cognitive function (usually short-term memory) does not yet meet the clinical criteria for AD dementia (Fig. 2) [24]. Most recent studies utilizing humanized anti-Aβ antibodies directed at a variety of Aβ targets—Aβ monomers, oligomers, fibrils, and/or plaques—have not succeeded to slow disease progression even in the presence of PET evidence of Aβ plaque removal. One recent study of aducanumab (an antibody directed at Aβ oligomers) had a small but statistically significant effect on Aβ plaque reduction in one of two twin studies, but not the other. The U.S. Food and Drug Administration (FDA) approved the medication for use via an accelerated authorization, with the requirement that a subsequent trial be successfully performed. Although the medication was approved for use, its unwanted side effects (including brain swelling and bleeding), high cost, and the decision of U.S. Medicare to pay for it only in placebo-controlled studies has limited its use and it was subsequently removed from the market. Two other anti-Aβ antibodies, lecanemab and donanemab, subsequently showed success in removing Aβ from the brain as well as slowing the cognitive decline and were approved by the FDA. The European Medicines Agency (EMA) approved lecanemab for use but did not approve donanemab over concerns about side effects. The search for improved anti-Aβ medications continues. Like the failed clinical trials with anti-Aβ treatments, administration of humanized tau-targeting antibodies did not slow the rate of tau pathology in a Phase-2 randomized, double-blind, placebo-controlled clinical trial of early AD [25], despite the promising results obtained in preclinical studies using tau transgenic mice [26]. These studies are continuing.

Summary and Perspectives

We have learned from analyses of biological fluids and neuroimaging studies that AD pathology develops more than a decade prior to manifestation of clinical symptoms [12]. Thus, while therapies will continue to be developed for all phases of the disease, an increasing emphasis will be on the earliest symptoms and on preclinical detection and intervention in those with known risk factors. Prevention trials, while expensive and of long duration, are necessary for proof of prevention or disease delay [27, 28].

For over 20 years, detection of Aβ and phospho-tau in the CSF has been used for AD diagnosis. However, only recently have the biomarker assays become sensitive enough for reliable detection of Aβ and tau in the peripheral blood. These analyses will be valuable for screening subjects for drug trials without having to use the more expensive (and geographically limited) PET scans for diagnosis. These assays will improve the subjects' selection process and speed up clinical trials.

Development of therapies for people in all stages will continue and be accompanied by a search for medications that better treat neuropsychiatric symptoms. The currently approved medications, the cholinesterase inhibitors and memantine, have mild effects and do not slow disease progression. Like other complex diseases, for example, cancer or heart disease, more

than one therapy will likely be necessary to treat the symptoms or slow disease progression. More effective behavioral interventions to maintain the dignity and safety of people with dementia will also be necessary. The recognition that minimizing barriers and allowing patients in chronic care facilities to use wandering paths and safe environments has decreased the use of physical and "chemical" restraints and improved quality of life for people with moderate to severe AD.

In sum, AD is a multifactorial disorder with a highly complex pathogenesis that warrants multi-target directed therapeutic approaches [29]. The onset and course of AD likely is influenced by a combined effect of multiple demographic and genetic risk factors, and thus genetics of AD continues to be a major area of research [30].

Acknowledgments University of Pittsburgh Alzheimer's Disease Research Center (NIH grant P30 AG066468).

Questions and Answers

Question 1 Name three methods that can aid the clinical diagnosis of AD.

Answer 1 Biofluids (cerebrospinal fluid and plasma) analyses of amyloid-β and tau, PET imaging of amyloid-β and tau, and MRI analysis of brain volume loss.

Question 2 What are the two characteristics of the most prevalent forms of AD?

Answer 2 Later life clinical onset (over age 65) and slower clinical progression.

Question 3 Which pathology, composed primarily of Aβ or tau, is more specific for AD?

Answer 3 Aβ deposits are specific for AD (although not every person with Aβ deposits will develop the clinical symptoms); tau aggregates are seen in AD and many non-AD tauopathies.

Question 4 Greater risk of AD is associated with which lipoprotein allele?

Answer 4 Apolipoprotein E4 (APOE4).

Question 5 What are the two mechanisms of currently available symptomatic medications for AD?

Answer 5 Cholinesterase inhibition (donepezil, galantamine, rivastigmine) and glutamate receptor modulation (memantine).

References

1. Hebert LE, Weuve J, Scherr PA, Evans DA (2013) Alzheimer disease in the United States (2010–2050) estimated using the 2010 census. Neurology 80:1778–1783
2. Alzheimer's Association (2022) Alzheimer's disease facts and figures report. https://www.alz.org/media/documents/alzheimers-facts-and-figures.pdf
3. Alzheimer A (1907) Über eine eigenartige Erkrankung der Hirnrinde. Allgemeine Zeitschrift fur Psychiatrie und Psychisch-Gerichtliche Medizin 64:146–148
4. Maurer K, Volk S, Gerbaldo H (1997) Auguste D and Alzheimer's disease. Lancet 349:1546–1549
5. Montine TJ, Phelps CH, Beach TG, Bigio EH, Cairns NJ, Dickson DW, Duyckaerts C, Frosch MP, Masliah E, Mirra SS, Nelson PT, Schneider JA, Thal DR, Trojanowski JQ, Vinters HV, Hyman BT, National Institute on Aging; Alzheimer's Association (2012) National Institute on Aging-Alzheimer's Association guidelines for the neuropathologic assessment of Alzheimer's disease: a practical approach. Acta Neuropathol 123(1):1–11
6. Tomé SO, Vandenberghe R, Ospitalieri S, Van Schoor E, Tousseyn T, Otto M, von Arnim CAF, Thal DR (2020) Distinct molecular patterns of TDP-43 pathology in Alzheimer's disease: relationship with clinical phenotypes. Acta Neuropathol Commun 8(1):61. https://doi.org/10.1186/s40478-020-00934-5
7. Jellinger KA (2008) Neuropathological aspects of Alzheimer disease, Parkinson disease and frontotemporal dementia. Neurodegener Dis 5(3–4):118–121
8. Arendt T (2009) Synaptic degeneration in Alzheimer's disease. Acta Neuropathol 118:167–179
9. DeKosky ST, Harbaugh RE, Schmitt FA, Bakay RA, Chui HC, Knopman DS, Reeder TM, Shetter AG, Senter HJ, Markesbery WR (1992) Cortical biopsy in Alzheimer's disease: diagnostic accuracy and neurochemical, neuropathological and cognitive correlations. Ann Neurol 32:625–632
10. Leung KK, Bartlett JW, Barnes J, Manning EN, Ourselin S, Fox NC (2013) Cerebral atrophy in mild cognitive impairment and Alzheimer disease: rates and acceleration. Neurology 80:648–654
11. Tanzi RE (2012) The genetics of Alzheimer disease. Cold Spring Harb Perspect Med 2:pii:a006296
12. Jack CRJ, Knopman DS, Jagust WJ, Petersen RC, Weiner MW, Aisen PS, Shaw LM, Vemuri P, Wiste HJ, Weigand SD, Lesnick TG, Pankratz VS, Donohue MC, Trojanowski JQ (2013) Tracking pathophysiological processes in Alzheimer's disease: an updated hypothetical model of dynamic biomarkers. Lancet Neurol 12:207–216
13. Reiman EM, Quiroz YT, Fleisher AS, Chen K, Velez-Pardo C, Jimenez-Del-Rio M, Fagan AM, Shah AR, Alvarez S, Arbelaez A, Giraldo M, Acosta-Baena N, Sperling RA, Dickerson B, Stern CE, Tirado V, Munoz C, Reiman RA, Huentelman MJ, Alexander GE, Langbaum JB, Kosik KS, Tariot PN, Lopera F (2012) Brain imaging and fluid biomarker analysis in young adults at genetic risk for autosomal dominant Alzheimer's disease in the presenilin 1 E280A kindred: a case-control study. Lancet Neurol 11:1048–1056
14. Hardy JA, Higgins GA (1992) Alzheimer's disease: the amyloid cascade hypothesis. Science 256:184–185
15. Thinakaran G, Koo EH (2008) Amyloid precursor protein trafficking, processing, and function. J Biol Chem 283:29615–29619
16. Steiner H, Capell A, Leimer U, Haass C (1999) Genes and mechanisms involved in beta-amyloid generation and Alzheimer's disease. Eur Arch Psychiatry Clin Neurosci 249:266–270
17. Abrahamson EE, Stehouwer JS, Vazquez AL, Huang GF, Mason NS, Lopresti BJ, Klunk WE, Mathis CA, Ikonomovic MD (2021) Development of a PET radioligand selective for cerebral amyloid angiopathy. Nucl Med Biol 92:85–96
18. Marcatti M, Fracassi A, Montalbano M, Natarajan C, Krishnan B, Kayed R, Taglialatela G (2022) Aβ/tau oligomer interplay at human synapses supports shifting therapeutic targets for Alzheimer's disease. Cell Mol Life Sci 79(4):222
19. Braak H, Del Tredici K (2011) Alzheimer's pathogenesis: is there neuron-to-neuron propagation? Acta Neuropathol 121:589–595
20. de Calignon A, Polydoro M, Suárez-Calvet M, William C, Adamowicz DH, Kopeikina KJ, Pitstick R, Sahara N, Ashe KH, Carlson GA, Spires-Jones TL, Hyman BT (2012) Propagation of tau pathology in a model of early Alzheimer's disease. Neuron 73:685–697
21. Liu L, Drouet V, Wu JW, Witter MP, Small SA, Clelland C, Duff K (2012) Trans-synaptic spread of tau pathology in vivo. PLoS One 7:e31302
22. Iba M, Guo JL, McBride JD, Zhang B, Trojanowski JQ, Lee VM (2013) Synthetic tau fibrils mediate transmission of neurofibrillary tangles in a trans-

genic mouse model of Alzheimer's-like tauopathy. J Neurosci 33:1024–1037

23. Rinne JO, Brooks DJ, Rossor MN, Fox NC, Bullock R, Klunk WE, Mathis CA, Blennow K, Barakos J, Okello AA, Martinez R, de Liano S, Liu E, Koller M, Gregg KM, Schenk D, Black R, Grundman M (2010) 11C-PiB PET assessment of change in fibrillar amyloid-beta load in patients with Alzheimer's disease treated with bapineuzumab: a phase 2, double-blind, placebo-controlled, ascending-dose study. Lancet Neurol 9:363–372

24. Jack CRJ, Albert MS, Knopman DS, McKhann GM, Sperling RA, Carrillo MC, Thies B, Phelps CH (2011) Introduction to the recommendations from the National Institute on Aging-Alzheimer's Association workgroups on diagnostic guidelines for Alzheimer's disease. Alzheimers Dement 7:257–262

25. Teng E, Manser PT, Pickthorn K, Brunstein F, Blendstrup M, Sanabria Bohorquez S, Wildsmith KR, Toth B, Dolton M, Ramakrishnan V, Bobbala A, Sikkes SAM, Ward M, Fuji RN, Kerchner GA, Tauriel Investigators (2022) Safety and efficacy of semorinemab in individuals with prodromal to mild Alzheimer disease: a randomized clinical trial. JAMA Neurol 79(8):758–767

26. Ayalon G, Lee SH, Adolfsson O, Foo-Atkins C, Atwal JK, Blendstrup M, Booler H, Bravo J, Brendza R, Brunstein F, Chan R, Chandra P, Couch JA, Datwani A, Demeule B, DiCara D, Erickson R, Ernst JA, Foreman O, He D, Hötzel I, Keeley M, Kwok MCM, Lafrance-Vanasse J, Lin H, Lu Y, Luk W, Manser P, Muhs A, Ngu H, Pfeifer A, Pihlgren M, Rao GK, Scearce-Levie K, Schauer SP, Smith WB, Solanoy H, Teng E, Wildsmith KR, Bumbaca Yadav D, Ying Y, Fuji RN, Kerchner GA (2021) Antibody semorinemab reduces tau pathology in a transgenic mouse model and engages tau in patients with Alzheimer's disease. Sci Transl Med 13(593):eabb2639

27. Sperling RA, Aisen PS, Beckett LA, Bennett DA, Craft S, Fagan AM, Iwatsubo T, Jack CR Jr, Kaye J, Montine TJ, Park DC, Reiman EM, Rowe CC, Siemers E, Stern Y, Yaffe K, Carrillo MC, Thies B, Morrison-Bogorad M, Wagster MV, Phelps CH (2011) Toward defining the preclinical stages of Alzheimer's disease: recommendations from the National Institute on Aging-Alzheimer's Association workgroups on diagnostic guidelines for Alzheimer's disease. Alzheimers Dement 7:280–292

28. DeKosky ST, Williamson JD, Fitzpatrick AL, Kronmal RA, Ives DG, Saxton JA, Lopez OL, Burke G, Carlson MC, Fried LP, Kuller LH, Robbins JA, Tracy RP, Woolard NF, Dunn L, Snitz BE, Nahin RL, Furberg CD (2008) Ginkgo biloba for prevention of dementia: a randomized controlled trial. JAMA 300:2253–2262

29. Gong CX, Dai CL, Liu F, Iqbal K (2022) Multi-targets: an unconventional drug development strategy for Alzheimer's disease. Front Aging Neurosci 14:837649

30. Jun G, Vardarajan BN, Buros J, Yu CE, Hawk MV, Dombroski BA, Crane PK, Larson EB, Alzheimer's Disease Genetics Consortium, Mayeux R, Haines JL, Lunetta KL, Pericak-Vance MA, Schellenberg GD, Farrer LA (2012) Comprehensive search for Alzheimer disease susceptibility loci in the APOE region. Arch Neurol 69:1270–1279

Multiple Sclerosis

Marc Pawlitzki, Markus Kipp, and Sven G. Meuth

Introduction to Multiple Sclerosis

Multiple sclerosis (MS) is a complex multifactorial polygenic disease, influenced by various factors including age, gender, hormonal, and environmental factors [1]. Despite an unknown etiology, the (histo-)pathological hallmarks of MS lesions are well defined and include demyelination and inflammation in various regions of the central nervous system (CNS) [2, 3]. The leading hypothesis for MS suggests that autoreactive T- and B-cells, along with autoantibodies, initiate myelin damage, neuroinflammation, and neurodegeneration, establishing MS as an autoimmune disease (see also chapters "Lactose and Food Intolerance" and "Metabolic Syndrome") [4]. MS affects persons of all ages, but initial symptoms are most likely to appear in individuals between 20 and 40 years of age, and recent epidemiological evidence indicates an increased incidence after the age of 50 years [5, 6]. The estimated prevalence of MS is about 2.8 million people worldwide and is two to three times higher in women than in men [6]. The diagnosis of MS requires evidence of lesions in separate areas of the brain and spinal cord (dissemination in space), and evidence that new lesions developed at different points in time indicating disease chronification (dissemination in time). Other potential causes of CNS lesions must be ruled out. The diagnostic process involves a thorough medical history, neurological examination, and magnetic resonance imaging (MRI) to confirm dissemination in space and time. Additional tests include visual evoked potentials, cerebrospinal fluid (CSF) analysis to assess immune system protein levels and detect oligoclonal bands—indicative of dissemination in time—and blood tests to exclude alternative conditions with similar symptoms to MS. The current McDonald diagnostic criteria [7] additionally require the first demyelination attack, which is also known as clinically isolated syndrome (CIS), to be clinical, with features typical or suggestive of MS and with objective abnormalities on neurologic examination.

Relapsing remitting MS (RRMS) is the most common type of MS, affecting around 85–90% of MS cases. RRMS means that symptoms appear (i.e., a relapse) and then fade away either partially

M. Pawlitzki · S. G. Meuth
Department of Neurology, University Hospital and Medical Faculty, Heinrich-Heine University, Düsseldorf, Germany
e-mail: MarcGuenter.Pawlitzki@med.uni-duesseldorf.de; svenguenther.meuth@med.uni-duesseldorf.de

M. Kipp (✉)
Faculty of Medicine, Institute of Neuroanatomy, RWTH Aachen University, Aachen, Germany
e-mail: mkipp@ukaachen.de

E. Lammert, M. Zeeb (eds.), *Metabolism of Human Diseases*,
https://doi.org/10.1007/978-3-031-96019-2_7

or completely (i.e., remitting). Secondary progressive MS (SPMS) is characterized by at least one relapse followed by progressive clinical worsening over time. This progressive course commonly follows a period of well-defined RRMS. Finally, primary progressive MS (PPMS) occurs in 10–15% of patients and is characterized by steady worsening of neurologic functioning, without any distinct relapses or periods of remission. A PPMS patient's rate of progression may vary over time, with occasional plateaus or temporary improvements, but the progression is continuous [1]. However, instead of this traditional distinction between RRMS and SPMS as well as PPMS, a more contemporary approach classifies MS into relapsing and progressive disease stages [2, 8]. This model emphasizes the continuous nature of the disease and allows for a more flexible classification based on disease activity and progression.

Pathophysiology of Multiple Sclerosis: Metabolic and Immunological Alterations

On the histopathological level, MS lesions are characterized by oligodendrocyte loss and subsequent demyelination of axons, neuro-axonal loss, astroglia and microglia activation (i.e., gliosis), and, to a certain extent, regeneration of myelin and axons [9]. Both, white and gray matter areas of the brain are affected (Fig. 1). These pathological events are paralleled by the recruitment of peripheral inflammatory cells such as lymphocytes and monocytes.

Several advanced MRI techniques have been developed that, compared with conventional MRI measures, are better able to capture the complexity of the pathological processes occurring in the CNS of MS patients [10]. Among those, proton MR spectroscopy (^{1}H-MRS) has the unique ability to provide chemical-pathological characterization of MR-visible lesions and normal-appearing brain tissues.

^{1}H-MRS brain imaging revealed profound changes at the level of metabolites such as N-acetyl aspartate (NAA, a derivative of aspartic acid acting as an important metabolic precursor and osmolyte in the brain), choline, creatine, myoinositol, glutamate (Glu), glutamine (Gln), macromolecules, lipids, and lactate. A reduction in NAA levels indicates neuronal and axonal loss, while elevated choline and creatine levels suggest gliosis and increased cell membrane turnover, reflecting processes of de- and remyelination. Furthermore, lactate, the end product of anaerobic glycolysis, is also increased in MS lesions [11, 12], along with Gln, fructose, and glucose (at early stages of MS), highlighting the complex interplay between inflammation, neurodegeneration, and repair mechanisms in MS (Fig. 2) [13].

Interestingly, metabolic abnormalities were also found in the "normal appearing" white matter, hence in regions, which are not yet affected by demyelination. Of note, even in patients with MRI findings suggestive of MS—referred to as radiologically isolated syndrome—significant brain metabolic abnormalities indicating axonal damage can be detected at this early stage of the disease [14].

Furthermore, studies have demonstrated a reduction in cerebral blood flow of different white and gray matter regions in MS patients (Fig. 1) [15]. The globally decreased blood perfusion in MS is thought to result from diffuse perivascular inflammation leading to microvascular damage, thrombosis, and fibrin deposition. Furthermore, hypoperfusion might result from widespread astrocyte dysfunction, which contributes to the regulation of vascular tone in the CNS [16].

Metabolic alterations in MS patients can also be detected by ^{1}H-MRS in biofluids, most importantly CSF, instead of imaging the brain. This enables quantification of a much larger range of biochemical compounds. In this context, it is important to note that most MS lesions are located in the periventricular white matter of the brain as well as in superficial areas of the spinal cord, which are close to the CSF space. Since MS lesions are not routinely biopsied, CSF analysis remains an important tool to discern MS pathology. CSF is a clear, colorless fluid surrounding the brain and the spinal cord to protect them from

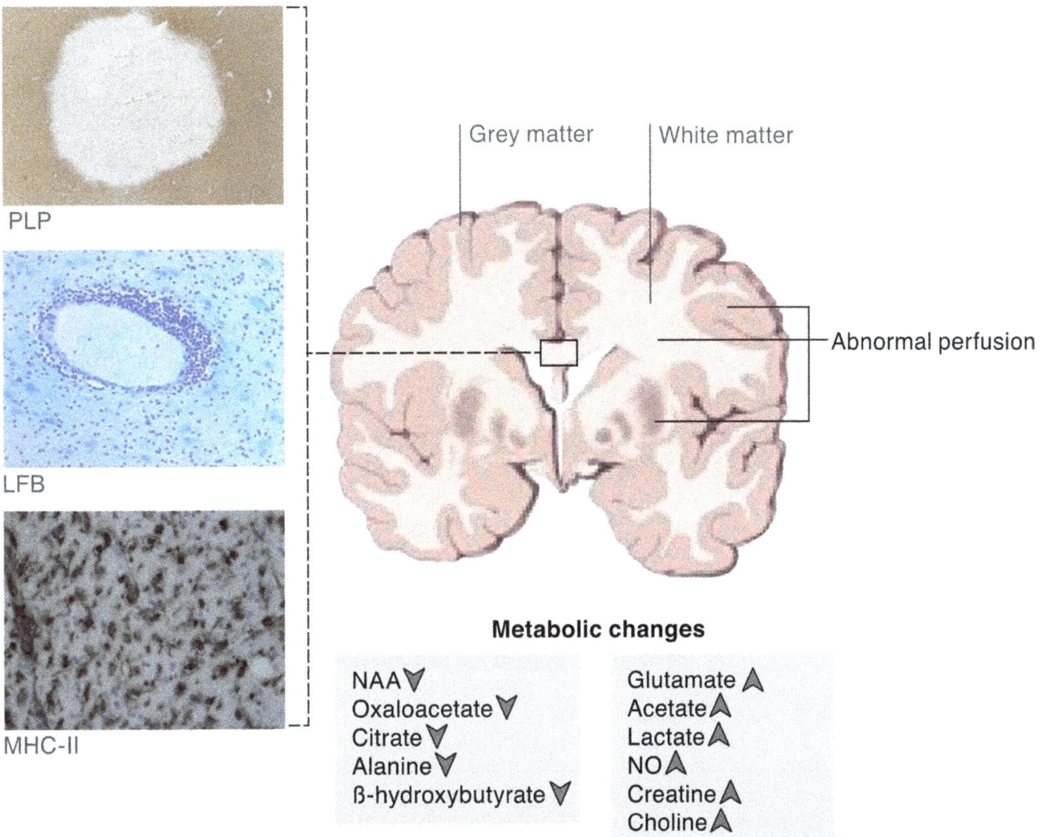

Fig. 1 Pathological and metabolic changes in the brain of an MS patient. Classical active MS lesions [45] can be found in the white and gray matter of the brain. Hallmarks of such lesions are demyelination as indicated by loss of proteo-lipid protein (PLP) staining (top left image) and increased luxol fast blue (LFB) staining (middle left), recruitment of inflammatory cells into the perivascular space, and massive accumulation of major histocompatibility complex (MHC)-II expressing macrophages (bottom left image). Likewise, abnormal perfusion is present at multiple locations in the CNS. Brain and cerebrospinal fluid demonstrate abnormal levels of several metabolites. *NAA* N-acetyl aspartate, *NO* nitric oxide

injury and is predominantly produced by the choroid plexus, a dense network of blood vessels located in each of the four ventricles. CSF contains trace proteins, electrolytes, and nutrients that are needed for the metabolism and normal function of the brain [17]. Remarkably, it also serves to remove waste products from the CNS parenchyma and is thus a vital source of information on physiological and pathological processes occurring within the brain parenchyma [18].

In animal models of MS, CSF metabolite concentrations exhibit significant alterations [19]. For instance, during the early stage of experimental autoimmune encephalomyelitis (EAE), one of the most widely used MS models, arginine levels were found to be significantly reduced [20]. Arginine is the main substrate for nitric oxide (NO) synthesis and its reduction likely results from increased activity of inducible NO synthase (iNOS) in activated immune cells and microglia [21, 22]. This NO can react with the free radical superoxide O_2^-, arising from oxidative phosphorylation in mitochondria to create peroxynitrite anion, a very reactive oxidizing agent, capable of inducing cell death through multiple pathways [23]. Thus, lowered levels of arginine might be an indirect indicator of cell stress and death [24].

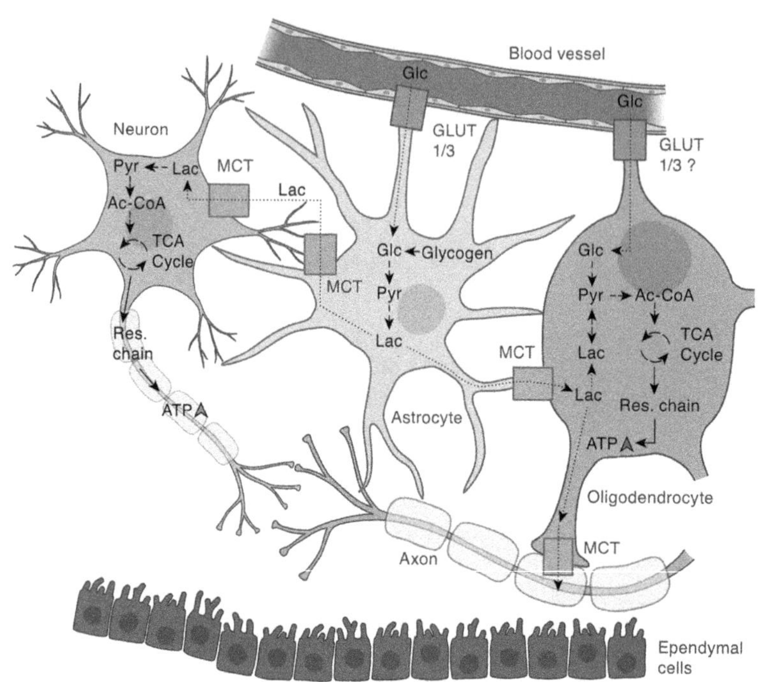

Fig. 2 Movement of metabolites. The brain is mainly fueled by glucose (Glc), transported via the blood and taken up by glucose transporters (GLUT) 1 and 3 (expressed in endothelial and brain cells, respectively). Alternatively, astrocytes are the main reservoir of glycogen in the brain. Therefore, they might serve as supporters for neurons and oligodendrocytes during energy deprivation periods. According to the lactate (Lac) shuttle hypothesis for energy transfer between cells, a heavily glycolytic (i.e., non-oxidative) cell (e.g., an astrocyte) produces large amounts of pyruvate (Pyr) and subsequently lactate, which is then transported out of the cell along its concentration gradient through specific monocarboxylate transporters (MCTs). The extracellular lactate is taken up via distinct MCTs. Intracellularly, lactate is recycled to pyruvate and transformed into acetyl-CoA (Ac-CoA), which is metabolized in the tricarboxylic acid cycle (TCA, also called citric acid cycle) to finally produce ATP via oxidative phosphorylation in the respiratory chain of the mitochondria. This interdependence is hypothesized to occur mainly between astrocytes (center) and neurons (left). Recently, oligodendrocytes have been included in the hypothesis as critical intermediaries for lactate transport to neurons. [46] Oligodendrocytes are also able to take up glucose directly from the endothelial cells of the vessels, probably via GLUT 1 and 3 as well

Furthermore, levels of alanine and branched-chain amino acids (leucine, isoleucine, and valine) are decreased in early EAE and likely MS. These amino acids are utilized as a source of pyruvate for energy metabolism; thus decreased levels indicate increased cell turnover. As the onset of EAE is associated with maximum infiltration of the CNS by blood monocytes and T-cells (see chapter "Anatomy and Physiology of the Immune System"), the observed decrease may suggest these metabolites are utilized for energy metabolism by invading cells [25].

MS patients additionally show increased levels of lactate in the CSF correlating with disease activity [26, 27]. Increased levels of lactate likely arise from increased anaerobic glycolysis, an attempt to compensate for reduced oxidative phosphorylation caused by perturbed microcirculation and hypoxia-like injury during MS plaque formation. Continued perturbation of mitochondrial function results in cellular energy deficit and loss of mitochondrial transmembrane potential ultimately leading to apoptotic cell death. Thus, the increase in lactate in active plaques and CSF can be interpreted as a result of inflammation, local ischemia, and mitochondrial dysfunction (Fig. 3).

Similarly, the concentration of β-hydroxybutyrate (a ketone body and substrate for gluconeogenesis) is increased in the CSF of

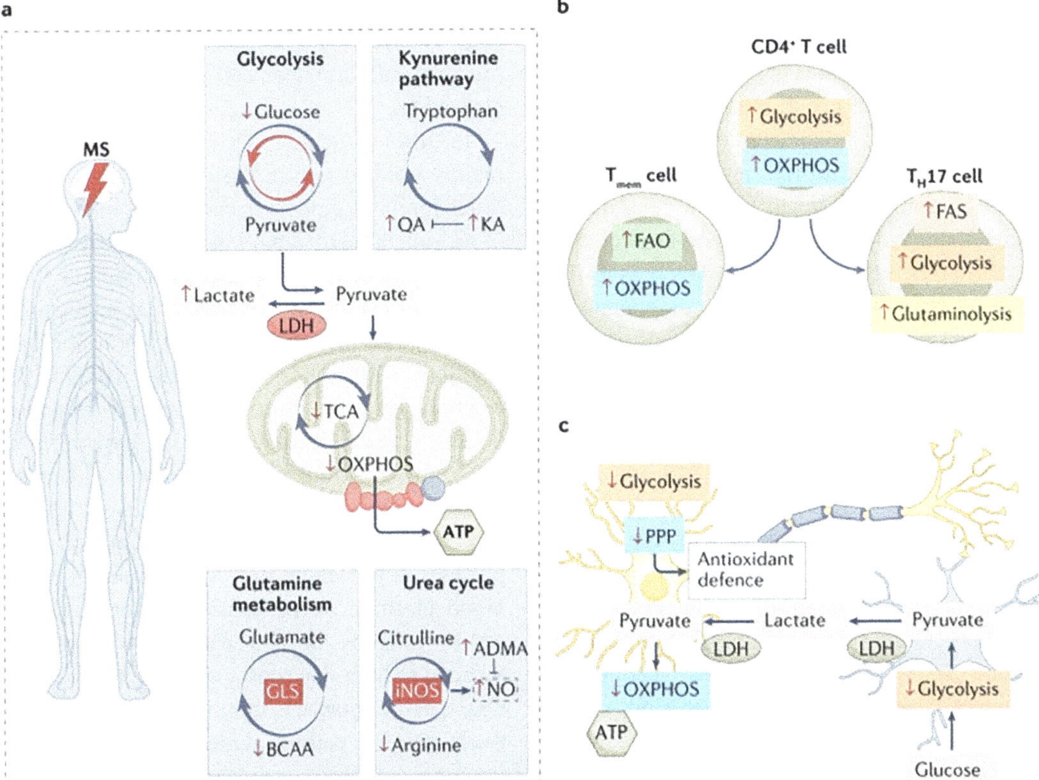

Fig. 3 Overview of metabolic adaptations in MS. (**a**) Multiple sclerosis (MS) is associated with a variety of metabolic alterations, reflected by changes in metabolite levels. MS brain metabolism is characterized by increased glycolysis, which is thought to reduce flux through the tricarboxylic acid (TCA) cycle and oxidative phosphorylation (OXPHOS). Additionally, pathways such as the urea cycle, the kynurenine pathway, and glutamine metabolism are also affected. Changes in metabolite levels or metabolic pathways in MS are indicated by red upward and downward arrows, while dark circular arrows highlight altered pathways. (**b**) T-cell subtypes exhibit distinct metabolic characteristics. In MS, T-cells appear to enhance their metabolic fluxes (marked by red upward arrows). However, after differentiation, their metabolic profiles diverge: T-memory (Tmem) cells primarily utilize fatty acid oxidation (FAO) and OXPHOS, whereas T-helper 17 (TH17) cells exhibit increased fatty acid synthesis (FAS), glycolysis, and glutaminolysis. (**c**) Metabolic activity in neurons and astrocytes follows distinct pathways. Neurons rely on the astrocyte–neuron lactate shuttle for energy supply. Astrocytes take up glucose, metabolize it to pyruvate via glycolysis, and subsequently convert it to lactate through lactate dehydrogenase (LDH). Lactate is then transported extracellularly and can be taken up by neurons, where it fuels ATP synthesis or contributes to redox homeostasis by generating NADPH. In MS, this mechanism appears to be impaired, as indicated by red downward arrows. *ADMA* asymmetric dimethylarginine, *BCAA* branched-chain amino acid, *GLS* glutamine synthetase, *iNOS* inducible nitric oxide synthase, *KA* kynurenic acid, *NO* nitric oxide, *PPP* pentose phosphate pathway, *QA* quinolinic acid. (Source: Bierhansl et al. [32])

early MS patients corresponding to the presence of inflammatory lesions [27]. Interestingly, astrocytes possess the intrinsic ability to perform gluconeogenesis and release glucose [28]. Thus, the increased levels of β-hydroxybutyrate may reflect a perturbation of astrocytic gluconeogenesis potentially due to the presence of inflammatory plaques, or due to reduced cerebral blood flow as a result of a perturbation of microcirculation.

Another important metabolic factor playing a role in MS pathogenesis is nitric oxide (NO). NO products are significantly raised in the CSF of MS patients [22, 29, 30]. NO plays a crucial role in demyelination and axonal degeneration by indirectly disrupting oligodendroglial energy metabolism through mitochondrial DNA damage and lipid membrane impairment. In detail, NO competitively inhibits the binding of oxygen to mitochondrial respiratory complex and thus its increase perturbs ATP synthesis [21]. This condition, in which oxygen is in principle available but cells are unable to use it, mimics hypoxia and is called histotoxic or metabolic hypoxia. Again, subsequent mitochondrial respiratory dysfunction may cause cell death and, in consequence, demyelination and neurodegeneration (Fig. 3).

Moreover, elevation of Glu levels in CSF, a known CNS neurotoxic trigger, compounded by low levels of oxaloacetate (an inhibitor of neuronal cell death, Fig. 1) may contribute to axonal loss and represents an area for further studies [31].

These metabolic changes in MS are closely linked to immune dysfunction and neuroinflammation, as altered energy metabolism in immune cells can drive chronic inflammation and neurodegeneration. Pro-inflammatory immune cells in MS, particularly activated T- and B-lymphocytes, exhibit a shift toward glycolysis, fueling their pathogenic activity and sustaining CNS inflammation and damage [32]. Especially B-cells and plasma cells contribute significantly to CNS pathology by mediating demyelination, inducing neuroinflammation, and promoting neurodegeneration. Within the CNS, B-cells can form ectopic lymphoid structures, secrete pro-inflammatory cytokines, and produce autoantibodies against myelin components, thereby triggering complement activation and attracting further immune cells [33]. As a result, these processes intertwine with T-cell responses, creating a persistent inflammatory environment that drives the progressive nature of MS.

Introduction to Treatment

Despite significant scientific advancements, there is still no cure for MS. Rather than directly targeting neuroprotection or remyelination, most available therapies focus on modulating immune mechanisms through different approaches, including immune cell depletion (e.g., anti-CD20 therapies), inhibition of immune cell migration across the blood-brain barrier (BBB) (e.g., natali-

zumab), pleiotropic immunomodulatory effects (e.g., interferons, glatiramer acetate), and suppression of immune cell proliferation (e.g., cladribine).

In times of acute relapse, corticosteroids are mainly used to reduce the inflammation. These drugs inhibit lymphocyte proliferation, synthesis of pro-inflammatory cytokines, and expression of cell surface molecules required for immune function (see chapter "Cancer Metabolism"). Furthermore, it is believed that corticosteroids stabilize the blood-brain barrier (BBB), for example, by decreasing the expression of angiopoietin-1 and vascular endothelial growth factor A (VEGF-A), both well known to regulate the permeability of the BBB (see chapters "Brain: Overview" and "Gout"). Alternatively, plasmapheresis, a procedure that removes and replaces blood plasma, can be considered for managing severe relapses in MS patients who do not respond to high-dose corticosteroid therapy [34]. One specific approach, immunoadsorption, selectively removes pathogenic immunoglobulins and immune complexes while preserving essential plasma components. This technique may be particularly beneficial in cases where autoantibody-mediated mechanisms contribute to disease activity [35].

As long-term therapy, disease-modifying drugs are prescribed with the aim to reduce relapse rate and slow silent disease progression. Currently, immunotherapies can be categorized into three main groups based on their mode of administration: (i) subcutaneous-based medications, (ii) oral medications, and (iii) intravenous medications [1].

The first group, subcutaneous-based medications, includes glatiramer acetate, interferons, and ofatumumab. Glatiramer acetate is a mixture of randomly polymerized four amino acids that structurally resemble myelin basic protein, a key component of the myelin sheath. By mimicking its antigenic properties, it competes for presentation to T-cells, thereby modulating the immune response and reducing autoimmunity in MS. Interferons (interferon-β1a/b) balance the expression of pro- and anti-inflammatory cytokines in the brain. They downregulate the production of pro-inflammatory cytokines such as interleukin (IL)-12, Interferon-γ, and Tumor necrosis factor-α, while upregulating anti-inflammatory mediators like IL-10 and Transforming growth factor-β, thereby shifting the immune response toward a more regulatory state. Ofatumumab is a monoclonal antibody that binds CD20, a receptor located on the pre-B- and mature B-lymphocytes, resulting in antibody-dependent cellular cytolysis and complement-mediated lysis.

The second group, oral medications, includes cladribine, dimethyl fumarate, diroximel fumarate, fingolimod, ozanimod, ponesimod, siponimod, and teriflunomide. Cladribine increases the expression of deoxycytidine kinase, leading to lymphocyte apoptosis; it disrupts intracellular processes, inhibiting DNA synthesis and repair, ribonucleotide enzymes, and alternating endonuclease activity. Fumarates (dimethyl fumarate, diroximel fumarate) provide an anti-inflammatory immune response through targeting of the Nrf-2 (nuclear respiratory factor 2)-dependent and independent pathways. Fingolimod, siponimod, ozanimod, and ponesimod are modulators of the sphingosine-1-phosphate receptor (S1PR), which is pivotal in pathways regulating lymphocyte egress from lymph nodes. Teriflunomide belongs to a class of drugs called pyrimidine synthesis inhibitors. Its ability to inhibit the mitochondrial enzyme dihydro-orotate dehydrogenase, which is relevant for the de novo synthesis of pyrimidine, is believed to exert the most important therapeutic effect. By inhibiting dihydro-orotate dehydrogenase and diminishing DNA synthesis, teriflunomide has a cytostatic effect on proliferating B- and T-cells.

The third group, intravenous medications, includes mitoxantrone, natalizumab, and monoclonal antibodies targeting B-cells, such as the CD20-depleting agents ocrelizumab and ublituximab, as well as alemtuzumab, which targets CD52 on B- and T-lymphocytes. Mitoxantrone is a type II topoisomerase inhibitor; it disrupts DNA synthesis and DNA repair in both healthy cells and cancer cells. Hence, it suppresses the proliferation of T-cells, B-cells, and macrophages (see chapter "Cancer Metabolism"), impairs antigen presentation, and decreases the secretion of pro-inflammatory cytokines. Natalizumab is a humanized monoclonal antibody against the cell adhesion molecule integrin α4 and reduces the ability of inflammatory immune cells to pass through the BBB. Of note, natalizumab has recently also become available for subcutaneous administration.

Ocrelizumab and ublituximab are recombinant, chimeric, and humanized monoclonal antibodies targeting CD20 positive B-cells. The latter differs from ocrelizumab and ofatumumab in its glycoengineered structure, which may enhance antibody-dependent cellular cytotoxicity. Unlike CD20-targeting therapies, alemtuzumab depletes a broader range of immune cells, including B- and T-cells, monocytes, and natural killer cells. Similar to natalizumab, ocrelizumab is now also available for subcutaneous administration.

Influence of Current Therapies on Metabolism

Evidence supporting the ability of disease-modifying therapies to counteract metabolic changes in MS remains limited. However, early-stage MS patients, including those with clinically isolated syndrome (CIS), who received glatiramer acetate or interferon-β, demonstrated improved neuroaxonal integrity, as reflected by an increased NAA/creatine ratio [36, 37]. With regard to fumarates, they seem to modulate glutathione metabolism [38], improving cellular redox balance and reducing metabolic stress in neurons and glial cells, suggesting a potential neuroprotective effect in MS. Interestingly, S1PR1 signaling in astrocytes has been implicated in oligodendrocyte differentiation and remyelination, suggesting a potential role in CNS repair of S1PR modulators like fingolimod [39]. Despite these findings, our understanding of how these immunotherapies influence metabolic alterations in MS is still in its early stages and requires further investigation. Given that neuroaxonal damage and metabolic dysfunction play a key role in disease progression, understanding these effects is crucial. Current therapies might primarily target immune-mediated demyelination but fail to directly promote remyelination or neuronal repair, leaving neurodegeneration largely unaddressed [9]. Since metabolic disturbances contribute to mitochondrial dysfunction, oxidative stress, and energy deficits in MS, further research is needed to explore whether modulating metabolism could offer neuroprotective or regenerative benefits beyond immunosuppression. Of note, Bruton's tyrosine kinase inhibitors (BTKIs, i.e., Evobrutinib, Fenebrutinib, Orelabrutinib, Remibrutinib, Tolebrutinib) are increasingly being investigated for MS and other autoimmune diseases [40], as they target B-cell and myeloid cell signaling, modulating both adaptive and innate immune responses. By inhibiting BTK, these drugs may reduce neuroinflammation and potentially influence cellular metabolism, particularly in microglia and astrocytes, which play a key role in CNS inflammation and neurodegeneration [41, 42].

Perspectives

Clearly, several metabolic pathways are disturbed in the CNS of MS patients (see above). However, we are far from understanding how these alterations impact cellular physiology and whether a manipulation of these pathways might result in new therapeutic options in the future. Recent findings suggest that neurons, astrocytes, and oligodendrocytes form a so-called "tri-cellular compartmentation" of brain metabolism [32, 43], meaning that complex metabolic pathways are scattered over different cell types and the metabolites are then transported to their target cells. This is, for example, well described for NAA metabolism and catabolism. Aspartate needed for NAA production can only be synthesized de novo in astrocytes and is transported to neurons in the form of glutamine. In neurons, NAA is then assembled and released. Subsequently, NAA is hydrolyzed to acetate and aspartate by aspartoacylase, which is predominantly located in oligodendroglia. Following this theory,

oligodendrocytes and astrocytes are pivotal for neuronal survival and functioning by providing energy metabolites to axons and/or neuronal cell bodies. A more comprehensive understanding of these cellular interactions could pave the way for novel therapeutic strategies to prevent neuronal damage in MS. One potential approach involves modulating tryptophan metabolism [44], which is diminished in the serum and CSF of MS patients and closely linked to disease progression. During MS, the tryptophan-kynurenine pathway undergoes significant shifts: in early stages, increased levels of neuroprotective kynurenic acid may counteract excitotoxicity (Fig. 3), whereas in progressive forms (SPMS and PPMS), rising concentrations of neurotoxic quinolinic acid contribute to neurodegeneration [32]. Additionally, targeting metabolic pathways or key substrates involved in immune cell metabolism may offer a new approach to modulate immune cell differentiation and function, potentially shaping future therapeutic strategies for MS [32] (Fig. 4).

Questions and Answers

Question 1 What are the key pathological hallmarks of MS?

Answer 1 MS is characterized by inflammation, demyelination, axonal loss, and gliosis in both white and gray matter of the CNS. Immune-mediated damage leads to neurodegeneration, contributing to progressive disability.

Question 2 What metabolic changes are observed in MS?

Answer 2 MS patients exhibit altered glucose metabolism, mitochondrial dysfunction, and oxidative stress, contributing to neuroinflammation and energy deficits. Increased lactate and reduced N-acetyl aspartate (NAA) levels in brain and CSF indicate neurodegeneration.

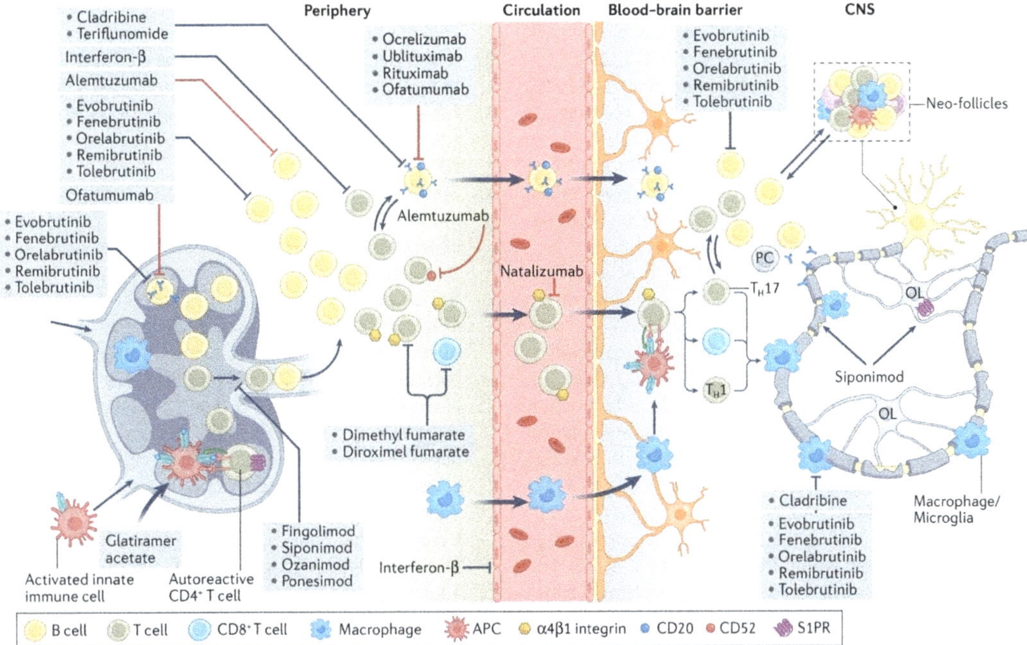

Fig. 4 Current treatment strategies. Overview of the immunopathogenesis and therapeutic targets of available disease-modifying therapies in multiple sclerosis (MS). Current treatment strategies employ diverse mechanisms of action—including pleiotropic immune modulation, targeted immune cell depletion, inhibition of proliferation, and blockade of migration—to modify or suppress key steps of the inflammatory process. These therapies act at different levels, targeting the peripheral immune system, the blood-brain barrier, and inflammatory processes within the central nervous system (CNS). The therapies depicted are subdivided into monoclonal antibodies (red lines) and pharmacological agents (dark lines). *APC* antigen-presenting cell, *OL* oligodendrocyte, *S1PR* sphingosine 1-phosphate receptor, *TH cell* T-helper cell (adapted from Bierhansl et al. [32])

Question 3 What are the treatment options for MS?

Answer 3 Therapies primarily target immune modulation, including:

- Injectable therapies: Interferons, glatiramer acetate, anti-CD20 antibodies and natalizumab
- Oral therapies: Sphingosine-1-phosphate receptor modulators, fumarates, and pyrimidine synthesis inhibitors
- Intravenous therapies: Natalizumab, alemtuzumab, mitoxantrone, and anti-CD20 antibodies

Question 4 How do metabolic changes impact MS progression?

Answer 4 Altered immune cell metabolism sustains inflammation and neurodegeneration. Increased glycolysis in activated T- and B-cells promotes disease activity, while mitochondrial dysfunction in neurons contributes to progressive neurodegeneration.

References

1. Jakimovski D, Bittner S, Zivadinov R et al (2024) Multiple sclerosis. Lancet 403:183–202
2. Kuhlmann T, Moccia M, Coetzee T et al (2023) Multiple sclerosis progression: time for a new mechanism-driven framework. Lancet Neurol 22:78–88
3. Kuhlmann T, Ludwin S, Prat A, Antel J, Brück W, Lassmann H (2017) An updated histological classification system for multiple sclerosis lesions. Acta Neuropathol 133:13–24

4. Charabati M, Wheeler MA, Weiner HL, Quintana FJ (2023) Multiple sclerosis: Neuroimmune crosstalk and therapeutic targeting. Cell 186:1309–1327

5. Prosperini L, Lucchini M, Ruggieri S et al (2022) Shift of multiple sclerosis onset towards older age. J Neurol Neurosurg Psychiatry 93:1137–1139

6. Portaccio E, Magyari M, Havrdova EK et al (2024) Multiple sclerosis: emerging epidemiological trends and redefining the clinical course. Lancet Regional Health – Europe 44:100977

7. Thompson AJ, Banwell BL, Barkhof F et al (2018) Diagnosis of multiple sclerosis: 2017 revisions of the McDonald criteria. Lancet Neurol 17:162–173

8. Yamamura T (2023) Time to reconsider the classification of multiple sclerosis. Lancet Neurol 22:6–8

9. Klotz L, Antel J, Kuhlmann T (2023) Inflammation in multiple sclerosis: consequences for remyelination and disease progression. Nat Rev Neurol 19:305–320

10. John NA, Solanky BS, De Angelis F et al (2024) Longitudinal metabolite changes in progressive multiple sclerosis: a study of 3 potential neuroprotective treatments. Magn Reson Imaging 59:2192–2201

11. Simone IL, Tortorella C, Federico F et al (2001) Axonal damage in multiple sclerosis plaques: a combined magnetic resonance imaging and 1H-magnetic resonance spectroscopy study. J Neurol Sci 182:143–150

12. Sajja BR, Wolinsky JS, Narayana PA (2009) Proton magnetic resonance spectroscopy in multiple sclerosis. Neuroimaging Clin N Am 19:45–58

13. Rovira À, Alonso J (2013) 1H magnetic resonance spectroscopy in multiple sclerosis and related disorders. Neuroimaging Clin N Am 23:459–474

14. Stromillo ML, Giorgio A, Rossi F et al (2013) Brain metabolic changes suggestive of axonal damage in radiologically isolated syndrome. Neurology 80:2090–2094

15. Inglese M, Adhya S, Johnson G et al (2008) Perfusion magnetic resonance imaging correlates of neuropsychological impairment in multiple sclerosis. J Cereb Blood Flow Metab 28:164–171

16. Filosa JA, Iddings JA (2013) Astrocyte regulation of cerebral vascular tone. Am J Phys Heart Circ Phys 305:H609–H619

17. Deisenhammer F, Zetterberg H, Fitzner B, Zettl UK (2019) The cerebrospinal fluid in multiple sclerosis. Front Immunol 10:726

18. Deisenhammer F, Sellebjerg F, Teunissen CE, Tumani H (eds) (2015) Cerebrospinal fluid in clinical neurology [online]. Springer, Cham. Accessed at: https://link.springer.com/10.1007/978-3-319-01225-4. Accessed 26 Feb 2025

19. Haghikia A, Kayacelebi AA, Beckmann B et al (2015) Serum and cerebrospinal fluid concentrations of homoarginine, arginine, asymmetric and symmetric dimethylarginine, nitrite and nitrate in patients with multiple sclerosis and neuromyelitis optica. Amino Acids 47:1837–1845

20. Zahoor I, Rui B, Khan J, Datta I, Giri S (2021) An emerging potential of metabolomics in multiple sclerosis: a comprehensive overview. Cell Mol Life Sci 78:3181–3203

21. Smith KJ, Lassmann H (2002) The role of nitric oxide in multiple sclerosis. Lancet Neurol 1:232–241

22. Förster M, Nelke C, Räuber S et al (2021) Nitrosative stress molecules in multiple sclerosis: a meta-analysis. Biomedicines 9:1899

23. Sonar SA, Lal G (2019) The iNOS activity during an immune response controls the CNS pathology in experimental autoimmune encephalomyelitis. Front Immunol 10:710

24. Rzepiński Ł, Kośliński P, Gackowski M, Koba M, Maciejek Z (2022) Amino acid levels as potential biomarkers of multiple sclerosis in elderly patients: preliminary report. J Clin Neurol 18:529

25. Noga MJ, Dane A, Shi S et al (2012) Metabolomics of cerebrospinal fluid reveals changes in the central nervous system metabolism in a rat model of multiple sclerosis. Metabolomics 8:253–263

26. Albanese M, Zagaglia S, Landi D et al (2016) Cerebrospinal fluid lactate is associated with multiple sclerosis disease progression. J Neuroinflammation 13:36

27. Lutz NW, Viola A, Malikova I et al (2007) Inflammatory multiple-sclerosis plaques generate characteristic metabolic profiles in cerebrospinal fluid. Zimmer J, editor. PLoS ONE 2:e595

28. Yip J, Geng X, Shen J, Ding Y (2017) Cerebral gluconeogenesis and diseases. Front Pharmacol [online serial] 7. Accessed at: http://journal.frontiersin.org/article/10.3389/fphar.2016.00521/full. Accessed 26 Feb 2025

29. Räuber S, Förster M, Schüller J et al (2024) The use of Nitrosative stress molecules as potential diagnostic biomarkers in multiple sclerosis. IJMS 25:787

30. Rejdak K, Petzold A, Stelmasiak Z, Giovannoni G (2008) Cerebrospinal fluid brain specific proteins in relation to nitric oxide metabolites during relapse of multiple sclerosis. Mult Scler 14:59–66

31. Sinclair AJ, Viant MR, Ball AK et al (2010) NMR-based metabolomic analysis of cerebrospinal fluid and serum in neurological diseases—a diagnostic tool? NMR Biomed 23:123–132

32. Bierhansl L, Hartung H-P, Aktas O, Ruck T, Roden M, Meuth SG (2022) Thinking outside the box: non-canonical targets in multiple sclerosis. Nat Rev Drug Discov 21:578–600

33. Cencioni MT, Mattoscio M, Magliozzi R, Bar-Or A, Muraro PA (2021) B cells in multiple sclerosis – from targeted depletion to immune reconstitution therapies. Nat Rev Neurol 17:399–414

34. Rolfes L, Pfeuffer S, Ruck T et al (2019) Therapeutic apheresis in acute relapsing multiple sclerosis: current evidence and unmet needs—a systematic review. JCM 8:1623

35. Pfeuffer S, Rolfes L, Wirth T et al (2022) Immunoadsorption versus double-dose methylprednisolone in refractory multiple sclerosis relapses. J Neuroinflammation 19:220

36. Narayanan S, De Stefano N, Francis GS et al (2001) Axonal metabolic recovery in multiple sclerosis patients treated with interferon β-1b. J Neurol 248:979–986
37. Arnold DL, Narayanan S, Antel S (2013) Neuroprotection with glatiramer acetate: evidence from the PreCISe trial. J Neurol 260:1901–1906
38. Rosito M, Testi C, Parisi G, Cortese B, Baiocco P, Di Angelantonio S (2020) Exploring the use of dimethyl fumarate as microglia modulator for neurodegenerative diseases treatment. Antioxidants 9:700
39. Rothhammer V, Kenison JE, Tjon E et al (2017) Sphingosine 1-phosphate receptor modulation suppresses pathogenic astrocyte activation and chronic progressive CNS inflammation. Proc Natl Acad Sci USA 114:2012–2017
40. Krämer J, Wiendl H (2024) Bruton tyrosine kinase inhibitors in multiple sclerosis: evidence and expectations. Curr Opin Neurol 37:237–244
41. Geladaris A, Torke S, Saberi D et al (2024) BTK inhibition limits microglia-perpetuated CNS inflammation and promotes myelin repair. Acta Neuropathol 147:75
42. Krämer J, Bar-Or A, Turner TJ, Wiendl H (2023) Bruton tyrosine kinase inhibitors for multiple sclerosis. Nat Rev Neurol 19:289–304
43. Amaral AI, Meisingset TW, Kotter MR, Sonnewald U (2013) Metabolic aspects of neuron-oligodendrocyte-astrocyte interactions. Front Endocrinol [online serial] 4. Accessed at: http://journal.frontiersin.org/article/10.3389/fendo.2013.00054/abstract. Accessed 27 Feb 2025
44. Fitzgerald KC, Smith MD, Kim S et al (2021) Multiomic evaluation of metabolic alterations in multiple sclerosis identifies shifts in aromatic amino acid metabolism. Cell Rep Med 2:100424
45. Van Der Valk P, Amor S (2009) Preactive lesions in multiple sclerosis. Curr Opin Neurol 22:207–213
46. Morrison BM, Lee Y, Rothstein JD (2013) Oligodendroglia: metabolic supporters of axons. Trends Cell Biol 23:644–651

Down Syndrome

Maria D. Torres, Isabel Rivera, and Jorge Busciglio

Introduction to Down Syndrome

Down syndrome (DS) or trisomy of chromosome 21 is the most prevalent cause of genetic intellectual disability affecting approximately 1 in 700 live births. Most DS cases are caused by full triplication of chromosome 21, and a small number of cases arise from mosaicism or chromosomal translocations, resulting in multiple medical and physical manifestations. Common characteristics of individuals with DS include skeletal anomalies, craniofacial alterations, hypotonia, increased incidence of congenital heart disease and seizures, abnormalities of the gastrointestinal tract, thyroid dysfunction, and premature aging [1]. Additional clinical features include altered folate metabolism, hormone imbalances, and immune dysregulation [2–4]. Neurological changes include reduced brain mass, impaired neuronal differentiation, aberrant dendritic spine morphology, defects in synaptic plasticity, neuroinflammation, and cerebral amyloid angiopathy [5, 6]. Most middle-aged individuals with DS develop Alzheimer's disease (AD; see chapter "Alzheimer's Disease") due to increased expression of the amyloid precursor protein gene (see chapter "Alzheimer's Disease") located on chromosome 21 [7]. Alterations in reactive oxygen species (ROS) and energy metabolism have long been associated with the development and progression of DS neuropathology [8]. In recent years, significant progress has been made in unraveling the genetic, molecular, and phenotypic conditions associated with Down syndrome, largely due to the gathering and sharing of information from diverse investigators. In large part this has been due to the creation of the Trisomy 21 Research Society (T21RS) that nucleates researchers from multiple disciplines working on DS and associated phenotypes [9–12]. This section focuses on the role of oxidative stress, mitochondrial dysfunction, neuroinflammation, and hypothyroidism in DS.

Pathophysiology of Down Syndrome and Metabolic Alterations

Altered Oxidative Stress and Energy Metabolism in DS

Oxidative stress is a prominent feature associated with DS [13]. Enhanced lipid peroxidation (which can cause DNA [deoxyribonucleic acid] damage) has been documented [14], as well as

M. D. Torres · I. Rivera · J. Busciglio (✉)
Department of Neurobiology & Behavior, Institute for Memory Impairments and Neurological Disorders (UCI MIND), Center for the Neurobiology of Learning and Memory (CNLM), School of Biological Sciences, University of California-Irvine, Irvine, CA, USA
e-mail: irivera1@uci.edu; jbuscigl@uci.edu

E. Lammert, M. Zeeb (eds.), *Metabolism of Human Diseases*,
https://doi.org/10.1007/978-3-031-96019-2_8

differential expression of oxidative stress-related genes [15]. Similarly, DS cortical neurons exhibit intracellular accumulation of ROS, increased lipid peroxidation, and reduced neuronal survival [16]. Another feature of DS pathology closely related to oxidative stress is mitochondrial dysfunction (Fig. 1). DS cells exhibit reduced mitochondrial transmembrane potential, ATP [adenosine triphosphate] production, and oxidoreductase activity [15]. These energy deficits lead to an abnormal pattern of protein processing and secretion in DS including intracellular accumulation of Aβ [17]. Mitochondria not only are the main intracellular source of ROS and free radicals but also play a critical role in apoptotic pathways. Increased levels of ROS and mutations in mitochondrial DNA can initiate a series of molecular events leading to activation of apoptosis [18]. In this regard, reduced mitochondrial activity in DS appears to represent an adaptation to prevent cellular damage and preserve basic cellular functions. In fact, upon stimulation, DS mitochondria possess the capacity to increase ATP production [19]. However, sustained increase of mitochondrial activity leads to lipid peroxidation and increased cell death. Interestingly, downregulation of mitochondrial activity to prevent oxidative damage has been

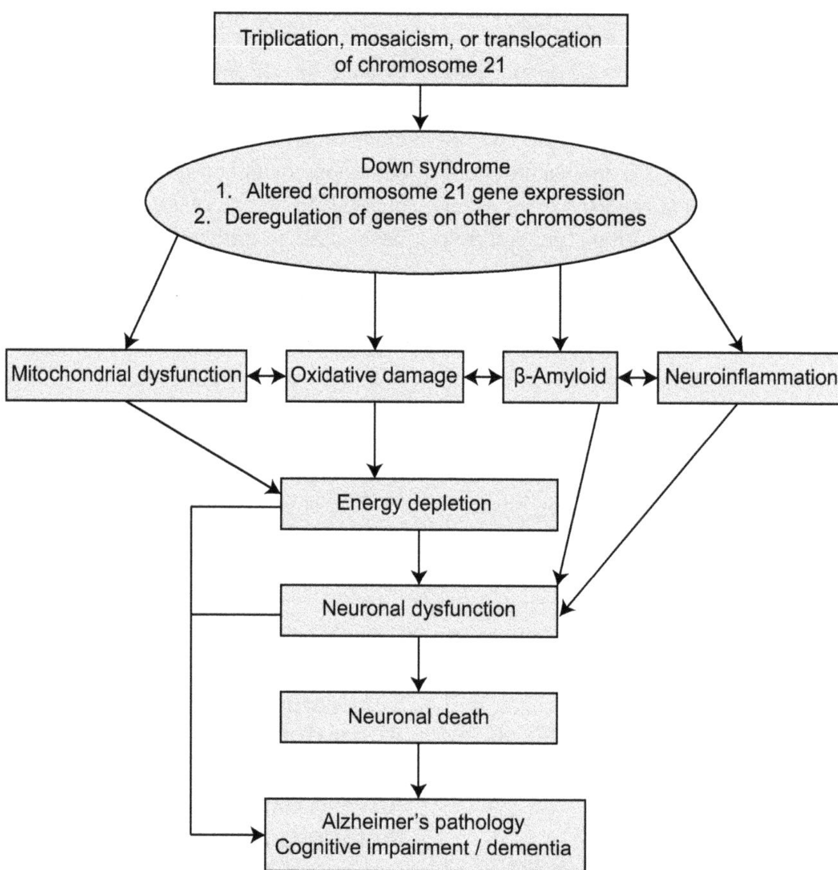

Fig. 1 Schematic representation of the proposed pathological cascade leading to Alzheimer's disease in Down syndrome. Down syndrome is caused by triplication, mosaicism, or translocation of chromosome 21 and characterized by altered expression of genes on chromosome 21, but also other genes. These changes cause multiple derangements (mitochondrial dysfunction, oxidative damage, β-amyloid accumulation, neuroinflammation) that interact and aggravate each other leading to energy depletion, neuronal dysfunction, and neuronal death. The latter factors, along with β-amyloid accumulation, are major contributors to dementia and Alzheimer's disease. (Adapted from Lott et al. [13])

observed in different organisms and cell types [20]. Oxidative stress, mitochondrial dysfunction, and early accumulation of Aβ are likely contributors to nerve growth factor (NGF) metabolic dysfunction and the degeneration of cholinergic neurons in DS [21, 22].

DS neurons and astrocytes also show a reduction in thrombospondin-1 (TSP-1) expression and secretion contributing to dendritic spine abnormalities [23, 24]. TSP-1 is an astrocyte-secreted protein involved in the formation of synapses [25]. Deficits in TSP-1 lead to defects in learning and memory in mice (Torres and Busciglio, unpublished data). A possible explanation is that reduced TSP-1 levels are linked to interferon (IFN) hypersensitivity. Overexpression of and hypersensitivity to IFN ligands have been linked to the presence of four IFN receptor genes on chromosome 21 [26, 27]. IFN-γ inhibits TSP-1 glycosylation and reduces its cellular levels and secretion [28, 29]. Moreover, the role of TSP-1 appears to go beyond synaptogenesis by protecting mitochondria function from Aβ-induced toxicity [30].

Neuroinflammation is mediated primarily by microglia, the immune cells of the central nervous system (CNS). Lifespan studies from brains of individuals with DS show that increases in pro-inflammatory molecules and microglial morphological changes occur as early as childhood and persist through late adulthood [31]. The persistent immune response by microglia in DS contributes to a decrease in neuronal spine density and activity, as well as cellular senescence and cognitive-behavioral deficits, potentially leading to the development of Alzheimer's disease [32, 33].

Endocrine dysfunction is another prominent medical feature of Down syndrome. In particular, the function of the thyroid gland is often compromised. The latter secretes several hormones, that is, thyroxine (T_4), triiodothyronine (T_3), and calcitonin, which regulate metabolism, heart rate, and growth. Alterations in their balance can lead to short stature, skin problems, and intellectual disability. It has been hypothesized that the development of thyroid dysfunction in DS is influenced by oxidative stress [34] or deregulation of

genes on chromosome 21 that participate in the immune response [35]. Individuals with DS are likely to have congenital hypothyroidism or develop thyroid dysfunction later in adulthood. The incidence of hypothyroidism in individuals with DS ranges from 39% to 61%, and recent studies suggest that the prevalence increases (approximately 50%) after the age of 30 [36]. Hypothyroidism results in decreased T_4 and T_3, which are important for physical and intellectual development by affecting protein synthesis and increasing metabolic rate, bone growth, and sensitivity to catecholamines. Hypothyroidism is also associated with delayed sexual maturity and cognitive deficits by altering gonadotropin-releasing hormone (GnRH) pulsatile secretion [37, 38]. A recent pilot study shows that pulsatile GnRH therapy may have potential for enhancing cognitive function in people with DS [39]. Symptoms of hypothyroidism further include hypotonia, enlarged tongue, small stature, skin problems, constipation, and lethargy [40]. Because many of the characteristics involving hypothyroidism overlap with features of DS, it is difficult to get an initial diagnosis.

Thyroid dysfunction is normally diagnosed through blood analysis of thyroid stimulating hormone (TSH). As part of a feedback loop, decreased T_4 and T_3 leads to increased levels of TSH to help stimulate more T_4 and T_3 production. Ultimately, early diagnosis of hypothyroidism is critical for preventing deterioration of physical and mental development in DS individuals. Individuals who develop thyroid autoimmune disease are more likely to develop other endocrine disorders including diabetes mellitus (see chapter "Diabetes Mellitus"). Thus, proper endocrine screening should be routine in individuals with DS.

Treatment and Its Influence on Metabolism

To date, there is no cure for Down syndrome. Existing treatments are directed to alleviate or prevent clinical complications such as congenital heart defects or gastrointestinal blockage.

Additionally, physical, speech, and occupational therapy are available to further assist in providing a better lifestyle for DS individuals. Nutritional therapies have been used to enhance cognition although there is little data supporting their effectiveness [41].

Antioxidants (such as resveratrol, celastrol, vitamins, and Coenzyme Q10) are used to protect cells against oxidative damage through the clearance of free radical intermediates and by delaying lipid peroxidation. Antioxidants increase viability of DS neurons in culture and prevent neuronal death while improving spatial learning in a DS mouse model [16, 42]. Coenzyme Q10, a cofactor of the electron transport chain, has shown promising results to treat mitochondrial dysfunction [43], acting as a scavenger of ROS and preventing lipid peroxide-induced DNA damage (in conjunction with vitamin E) [44]. Recent work indicates that Coenzyme Q10 supplementation restores the antioxidant/oxidant balance in plasma in children with DS [45]. Considering the high prevalence of Alzheimer's disease (AD) in DS individuals, some interventions directed at treating AD have been tested. However, acetylcholinesterase inhibitors and the N-methyl-D-aspartate (NMDA) receptor antagonist memantine (common AD treatments; see chapter "Alzheimer's Disease") have been unsuccessful or inconclusive when applied to DS patients [46].

Hypothyroidism in general and in DS is easily controlled with T_4 replacement therapy. However, artificial elevation of T_4 levels reduces TSH (see above), which can lead to significant side effects. Because TSH stimulates production of thyroid hormones, reduction of TSH leads to decreased hormone secretion affecting hormones such as calcitonin. Calcitonin is important for preventing Ca^{2+} loss (see chapter "Teeth and Bones: Overview" under part "Teeth and Bones"), and synthetic replacement of T_4 has been associated with increased osteoporosis (see chapter "Osteoporosis") [47]. Nutritional therapies such as iodine, L-tyrosine, and zinc have also been suggested for treatment of hypothyroidism [48].

Perspectives

Given the genetic and phenotypic complexity underlying Down syndrome, therapeutic interventions have been limited. The significant increase in life expectancy that individuals with DS enjoy today requires to be complemented with effective treatments to enhance cognition and to prevent age- and AD-related cognitive decline. Dendritic spine pathology has been associated with intellectual disability in DS and other neurological disorders, yet no therapeutic approach exists to prevent or restore spine structure and function. This is a promising area of active research, which may pave the way for the discovery of new therapies to ameliorate functional connectivity at the cellular level in DS and other neurodevelopmental and neurodegenerative conditions.

Questions and Answers

Question 1 What is the genetic alteration associated with DS?

Answer 1 Partial or total triplication of chromosome 21 is the genetic cause of DS.

Question 2 What is the main source of free radical generation in cells?

Answer 2 The main source of free radical generation in cells are mitochondria.

Question 3 Why is hypothyroidism associated with delayed sexual maturity and cognitive deficits?

Answer 3 Hypothyroidism is associated with delayed sexual maturity and cognitive deficits because it alters gonadotropin-releasing hormone (GnRH) pulsatile secretion.

References

1. Roizen NJ, Patterson D (2003) Down's syndrome. Lancet 361:1281–1289. https://doi.org/10.1016/S0140-6736(03)12987-X
2. Shaw CK, Thapalial A, Nanda S, Shaw P (2006) Thyroid dysfunction in Down syndrome. Kathmandu Univ Med J 4:182–186
3. Patterson D (2009) Molecular genetic analysis of Down syndrome. Hum Genet 126:195–214. https://doi.org/10.1007/s00439-009-0696-8
4. Ferrari M, Stagi S (2021) Autoimmunity and genetic syndromes: a focus on Down syndrome. Genes (Basel) 12. https://doi.org/10.3390/genes12020268
5. Lott IT (2012) Neurological phenotypes for Down syndrome across the life span. Prog Brain Res 197:101–121. https://doi.org/10.1016/B978-0-444-54299-1.00006-6
6. Head E, Phelan MJ, Doran E, Kim RC, Poon WW, Schmitt FA, Lott IT (2017) Cerebrovascular pathology in Down syndrome and Alzheimer disease. Acta Neuropathol Commun 5:93. https://doi.org/10.1186/s40478-017-0499-4
7. Mann DM, Esiri MM (1989) The pattern of acquisition of plaques and tangles in the brains of patients under 50 years of age with Down's syndrome. J Neurol Sci 89:169–179. https://doi.org/10.1016/0022-510x(89)90019-1
8. Coskun PE, Busciglio J (2012) Oxidative stress and mitochondrial dysfunction in Down's syndrome: relevance to aging and dementia. Curr Gerontol Geriatr Res 2012:383170. https://doi.org/10.1155/2012/383170
9. Delabar JM, Allinquant B, Bianchi D, Blumenthal T, Dekker A, Edgin J, O'Bryan J, Dierssen M, Potier MC, Wiseman F, Guedj F, Creau N, Reeves R, Gardiner K, Busciglio J (2016) Changing paradigms in Down syndrome: the first international conference of the trisomy 21 research society. Mol Syndromol 7:251–261. https://doi.org/10.1159/000449049
10. Reeves RH, Delabar J, Potier MC, Bhattacharyya A, Head E, Lemere C, Dekker AD, De Deyn P, Caviedes P, Dierssen M, Busciglio J (2019) Paving the way for therapy: the second international conference of the trisomy 21 research society. Mol Syndromol 9:279–286. https://doi.org/10.1159/000494231
11. Dierssen M, Herault Y, Helguera P, Martinez de Lagran M, Vazquez A et al (2021) Building the future therapies for down syndrome: the third international conference of the T21 research society. Mol Syndromol 12:202–218. https://doi.org/10.1159/000514437
12. Hamlett ED, Flores-Aguilar L, Handen B, Potier MC, Granholm AC, Sherman S, Puig V, Santoro JD, Carmona-Iragui M, Rebillat AS, Head E, Strydom A, Busciglio J (2022) Innovating therapies for Down syndrome: an international virtual conference of the T21 research society (in English). Mol Syndromol 14:89. https://doi.org/10.1159/000526021
13. Lott IT, Head E, Doran E, Busciglio J (2006) Beta-amyloid, oxidative stress and down syndrome. Curr Alzheimer Res 3:521–528. https://doi.org/10.2174/156720506779025305
14. Brooksbank BW, Martinez M, Balazs R (1985) Altered composition of polyunsaturated fatty acyl groups in phosphoglycerides of Down's syndrome fetal brain. J Neurochem 44:869–874. https://doi.org/10.1111/j.1471-4159.1985.tb12896.x
15. Slonim DK, Koide K, Johnson KL, Tantravahi U, Cowan JM, Jarrah Z, Bianchi DW (2009) Functional genomic analysis of amniotic fluid cell-free mRNA suggests that oxidative stress is significant in Down syndrome fetuses. Proc Natl Acad Sci USA 106:9425–9429. https://doi.org/10.1073/pnas.0903909106
16. Busciglio J, Yankner BA (1995) Apoptosis and increased generation of reactive oxygen species in Down's syndrome neurons in vitro. Nature 378:776–779. https://doi.org/10.1038/378776a0
17. Busciglio J, Pelsman A, Wong C, Pigino G, Yuan M, Mori H, Yankner BA (2002) Altered metabolism of the amyloid beta precursor protein is associated with mitochondrial dysfunction in Down's syndrome. Neuron 33:677–688. https://doi.org/10.1016/s0896-6273(02)00604-9
18. Wang C, Youle RJ (2009) The role of mitochondria in apoptosis. Annu Rev Genet 43:95–118. https://doi.org/10.1146/annurev-genet-102108-134850
19. Helguera P, Seiglie J, Rodriguez J, Hanna M, Helguera G, Busciglio J (2013) Adaptive downregulation of mitochondrial function in Down syndrome. Cell Metab 17:132–140. https://doi.org/10.1016/j.cmet.2012.12.005
20. Pfeiffer M, Kayzer EB, Yang X, Abramson E, Kenaston MA, Lago CU, Lo HH, Sedensky MM, Lunceford A, Clarke CF, Wu SJ, McLeod C, Finkel T, Morgan PG, Mills EM (2011) Caenorhabditis elegans UCP4 protein controls complex II-mediated oxidative phosphorylation through succinate transport. J Biol Chem 286:37712–37720. https://doi.org/10.1074/jbc.M111.271452
21. Iulita MF, Do Carmo S, Ower AK, Fortress AM, Flores Aguilar L, Hanna M, Wisniewski T, Granholm AC, Buhusi M, Busciglio J, Cuello AC (2014) Nerve growth factor metabolic dysfunction in Down's syndrome brains. Brain 137:860–872. https://doi.org/10.1093/brain/awt372
22. Iulita MF, Cuello AC (2016) The NGF metabolic pathway in the CNS and its dysregulation in Down syndrome and Alzheimer's disease. Curr Alzheimer Res 13:53–67. https://doi.org/10.2174/1567205012666150921100030
23. Garcia O, Torres M, Helguera P, Coskun P, Busciglio J (2010) A role for thrombospondin-1 deficits in

astrocyte-mediated spine and synaptic pathology in Down's syndrome. PLoS One 5:e14200. https://doi.org/10.1371/journal.pone.0014200

24. Torres MD, Garcia O, Tang C, Busciglio J (2018) Dendritic spine pathology and thrombospondin-1 deficits in Down syndrome. Free Radic Biol Med 114:10–14. https://doi.org/10.1016/j.freeradbiomed.2017.09.025

25. Christopherson KS, Ullian EM, Stokes CC, Mullowney CE, Hell JW, Agah A, Lawler J, Mosher DF, Bornstein P, Barres BA (2005) Thrombospondins are astrocyte-secreted proteins that promote CNS synaptogenesis. Cell 120:421–433. https://doi.org/10.1016/j.cell.2004.12.020

26. Iwamoto T, Yamada A, Yuasa K, Fukumoto E, Nakamura T, Fujiwara T, Fukumoto S (2009) Influences of interferon-gamma on cell proliferation and interleukin-6 production in Down syndrome derived fibroblasts. Arch Oral Biol 54:963–969. https://doi.org/10.1016/j.archoralbio.2009.07.009

27. Sullivan KD, Lewis HC, Hill AA, Pandey A, Jackson LP, Cabral JM, Smith KP, Liggett LA, Gomez EB, Galbraith MD, DeGregori J, Espinosa JM (2016) Trisomy 21 consistently activates the interferon response. elife 5. https://doi.org/10.7554/eLife.16220

28. Nickoloff BJ, Riser BL, Mitra RS, Dixit VM, Varani J (1988) Inhibitory effect of gamma interferon on cultured human keratinocyte thrombospondin production, distribution, and biologic activities. J Invest Dermatol 91:213–218. https://doi.org/10.1111/1523-1747.ep12465005

29. Maheshwari RK, Banerjee DK, Waechter CJ, Olden K, Friedman RM (1980) Interferon treatment inhibits glycosylation of a viral protein. Nature 287:454–456. https://doi.org/10.1038/287454a0

30. Kang S, Byun J, Son SM, Mook-Jung I (2018) Thrombospondin-1 protects against Abeta-induced mitochondrial fragmentation and dysfunction in hippocampal cells. Cell Death Discov 4:31. https://doi.org/10.1038/s41420-017-0023-4

31. Flores-Aguilar L, Iulita MF, Kovecses O, Torres MD, Levi SM, Zhang Y, Askenazi M, Wisniewski T, Busciglio J, Cuello AC (2020) Evolution of neuroinflammation across the lifespan of individuals with Down syndrome. Brain 143:3653–3671. https://doi.org/10.1093/brain/awaa326

32. Pinto B, Morelli G, Rastogi M, Savardi A, Fumagalli A, Petretto A, Bartolucci M, Varea E, Catelani T, Contestabile A, Perlini LE, Cancedda L (2020) Rescuing over-activated microglia restores cognitive performance in juvenile animals of the Dp(16) mouse model of Down syndrome. Neuron 108:887–904.e12. https://doi.org/10.1016/j.neuron.2020.09.010

33. Jin M, Xu R, Wang L, Alam MM, Ma Z, Zhu S, Martini AC, Jadali A, Bernabucci M, Xie P, Kwan KY, Pang ZP, Head E, Liu Y, Hart RP, Jiang P (2022) Type-I-interferon signaling drives microglial dysfunction and senescence in human iPSC models of Down syndrome and Alzheimer's disease. Cell Stem Cell 29:1135–53.e8. https://doi.org/10.1016/j.stem.2022.06.007

34. Kanavin OJ, Aaseth J, Birketvedt GS (2000) Thyroid hypofunction in Down's syndrome: is it related to oxidative stress? Biol Trace Elem Res 78:35–42. https://doi.org/10.1385/BTER:78:1-3:35

35. Kennedy RL, Jones TH, Cuckle HS (1992) Down's syndrome and the thyroid. Clin Endocrinol 37:471–476. https://doi.org/10.1111/j.1365-2265.1992.tb01475.x

36. Tsou AY, Bulova P, Capone G, Chicoine B, Gelaro B, Harville TO, Martin BA, McGuire DE, McKelvey KD, Peterson M, Tyler C, Wells M, Whitten MS, Global Down Syndrome Foundation Medical Care Guidelines for Adults with Down Syndrome Workgroup (2020) Medical care of adults with Down syndrome: a clinical guideline. JAMA 324:1543–1556. https://doi.org/10.1001/jama.2020.17024

37. Dittrich R, Beckmann MW, Oppelt PG, Hoffmann I, Lotz L, Kuwert T, Mueller A (2011) Thyroid hormone receptors and reproduction. J Reprod Immunol 90:58–66. https://doi.org/10.1016/j.jri.2011.02.009

38. Shahid MA, Ashraf MA, Sharma S (2023) Physiology, thyroid hormone. StatPearls, Treasure Island

39. Manfredi-Lozano M, Leysen V, Adamo M, Paiva I, Rovera R et al (2022) GnRH replacement rescues cognition in Down syndrome. Science 377:eabq4515. https://doi.org/10.1126/science.abq4515

40. Carroll KN, Arbogast PG, Dudley JA, Cooper WO (2008) Increase in incidence of medically treated thyroid disease in children with Down syndrome after rerelease of American Academy of Pediatrics Health Supervision guidelines. Pediatrics 122:e493–e498. https://doi.org/10.1542/peds.2007-3252

41. Lott IT (2012) Antioxidants in Down syndrome. Biochim Biophys Acta 1822:657–663. https://doi.org/10.1016/j.bbadis.2011.12.010

42. Shichiri M, Yoshida Y, Ishida N, Hagihara Y, Iwahashi H, Tamai H, Niki E (2011) Alpha-tocopherol suppresses lipid peroxidation and behavioral and cognitive impairments in the Ts65Dn mouse model of Down syndrome. Free Radic Biol Med 50:1801–1811. https://doi.org/10.1016/j.freeradbiomed.2011.03.023

43. Tiano L, Busciglio J (2011) Mitochondrial dysfunction and Down's syndrome: is there a role for coenzyme Q(10)? Biofactors 37:386–392. https://doi.org/10.1002/biof.184

44. Tiano L, Padella L, Santoro L, Carnevali P, Principi F, Bruge F, Gabrielli O, Littarru GP (2012) Prolonged coenzyme Q10 treatment in Down syndrome patients: effect on DNA oxidation. Neurobiol Aging 33(626):e1–e8. https://doi.org/10.1016/j.neurobiolaging.2011.03.025

45. Miles MV, Patterson BJ, Chalfonte-Evans ML, Horn PS, Hickey FJ, Schapiro MB, Steele PE, Tang PH, Hotze SL (2007) Coenzyme Q10 (ubiquinol-10) supplementation improves oxidative imbalance in children with trisomy 21. Pediatr Neurol 37:398–403. https://doi.org/10.1016/j.pediatrneurol.2007.08.003

46. de la Torre R, Dierssen M (2012) Therapeutic approaches in the improvement of cognitive performance in Down syndrome: past, present, and future. Prog Brain Res 197:1–14. https://doi.org/10.1016/B978-0-444-54299-1.00001-7

47. Fowler PB, McIvor J, Sykes L, Macrae KD (1996) The effect of long-term thyroxine on bone mineral density and serum cholesterol. J R Coll Physicians Lond 30:527–532

48. Thiel R, Fowkes SW (2007) Down syndrome and thyroid dysfunction: should nutritional support be the first-line treatment? Med Hypotheses 69:809–815. https://doi.org/10.1016/j.mehy.2007.01.068

Eye

Eye: Overview

Anja Ehlers and Frank Müller

Introduction

Vision is our dominant sense, providing approximately 70–80% of our total sensory information. The human visual system is remarkable not only in quantity and quality of information it supplies about the world but also in the enormous range of stimulus intensity it covers from faint light of stars at night to bright sunlight. A single glance seems to be sufficient to analyze the location, shape, and color of an object as well as to predict the trajectory of a moving object. However, this outstanding performance comes at high costs involving parallel information processing in many cortical and subcortical areas of our brain. In this chapter, we will focus on the physiology of the eye, where the magic begins.

Anatomy and Physiology of the Eye

The eyes, specialized for the detection of light, are our most important sensory organs. The eyeball is well protected in a cavity of the skull called orbit. The frontal part of the human eye can be covered by the eyelids. The surface of the eye is moisturized by tears produced by the lacrimal gland (situated in the upper outer part of the orbit). Tears are drained by the lacrimal puncta in the inner corner of the eyelids via the lacrimal sac into the nose (see chapter "Glaucoma"). Most of the eyeball is covered by the tough white sclera (Fig. 1). In the front, the sclera changes over to the transparent cornea, the first and most refractive part of the optical apparatus. The iris is the colored part of the eye. It is a pigmented muscle tissue situated between the cornea and lens. Depending on the brightness of ambient light, the iris can contract or expand, thus changing the diameter of the pupil and controlling the amount of light falling into the eye. The lens is suspended by the zonula fibers attached to the ciliary muscles. Humans can adjust their focus to different viewing distances by contracting the ciliary muscles, which ultimately leads to changes in the shape of the lens. For their function, the cornea and lens must be transparent and, therefore, devoid of blood vessels. They are nourished by diffusion from the aqueous humor, a transparent fluid located in the anterior and posterior chamber of the eye. The compartment between the lens and retina is filled with the jellylike vitreous body. The pressure of both aqueous and vitreous humor, the so-called intraocular pressure, is slightly elevated, keeping the eyeball spherical.

The optical apparatus projects an image of our environment onto the back of the eye covered by the retina. The retina is a neuronal tissue

A. Ehlers · F. Müller (✉)
Institute of Biological Information Processing 1,
Molecular and Cellular Physiology (IBI-1),
Forschungszentrum Jülich GmbH, Jülich, Germany
e-mail: a.ehlers@fz-juelich.de;
f.mueller@fz-juelich.de

E. Lammert, M. Zeeb (eds.), *Metabolism of Human Diseases*,
https://doi.org/10.1007/978-3-031-96019-2_9

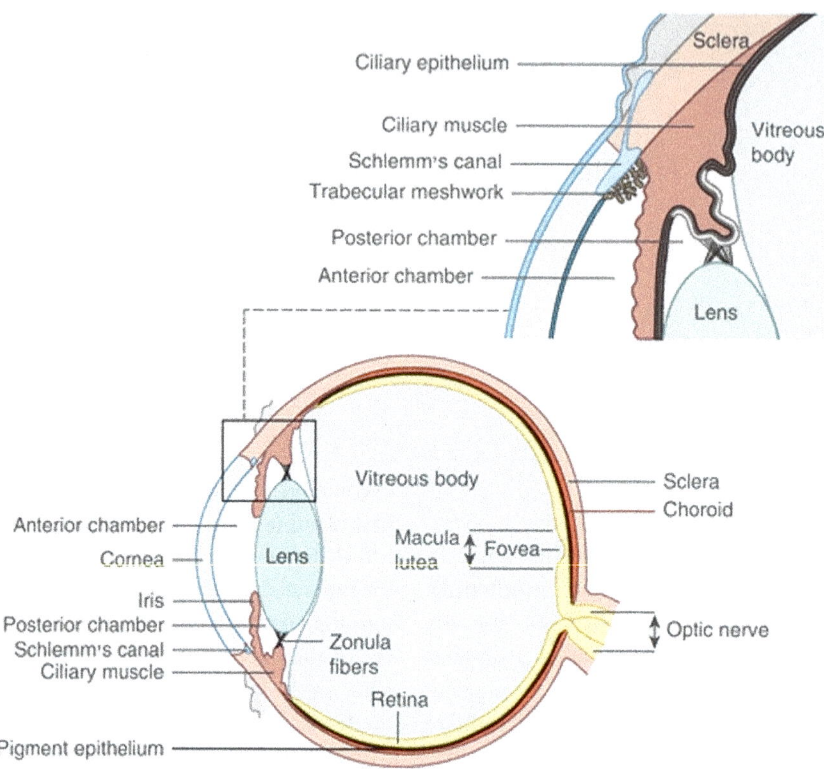

Fig. 1 Schematic cross-section of an eye. The different layers perform important tasks: the choroid is a layer of blood vessels important for photoreceptor supply. The retina is a neuronal tissue comprising the photoreceptor cells, a network of neurons, and the retinal ganglion cells as output neurons that relay the information to the brain; it also contains the macula lutea with the fovea, the central part of the retina with the highest visual acuity. The retinal pigment epithelium, a monolayer of pigmented cells that forms part of the blood–retinal barrier, is involved in the regeneration of the photopigment and the phagocytosis of the photoreceptors' outer segments. The vitreous body is a jellylike transparent substance that maintains the eye in its spherical shape. Inset: Structures responsible for the control of aqueous humor. The ciliary epithelium secrets aqueous humor into the posterior chamber. The trabecular meshwork is a spongy tissue, which drains the aqueous humor from the anterior chamber together with Schlemm's canal

originating from the brain during embryonic development and is, thus, a true part of the central nervous system. The retina is a well-layered, approximately 200 μm thick tissue and can be divided into an outer part that harbors the light-sensitive cells—the rod and cone photoreceptor cells—and an inner part that comprises a neuronal network [1]. This network performs the first steps of information processing before the signal is relayed by the retinal ganglion cells to the brain via the optic nerve. A human retina harbors around 120 million rod and 6 million cone photoreceptor cells. Rods are highly sensitive, can respond to single light quanta, and provide vision during night and at twilight. Cones are less sensitive and provide color vision during daylight.

Behind the retina, two more layers are located: the retinal pigment epithelium (RPE) and the choroid (see Fig. 1), a dense network of blood capillaries, which provides nutrients and oxygen to the photoreceptors. RPE and choroid are separated by Bruch's membrane, an elastin- and collagen-rich structure [2].

Due to its function, the eye has to cope with special problems. Retinal cells are the only neurons exposed to light. Bright illumination can result in the generation of free radicals that damage the retinal cells (see below). As photoreceptors (like most other central neurons) cannot be replaced, the eye has developed several mechanisms of protection. The optical apparatus absorbs high-energy ultraviolet light, which would otherwise damage the retina [3]. Moreover, in the center of the retina, protective pigments are embedded that absorb light of short wavelengths, giving this area a yellowish appearance, called *macula lutea* (Latin: yellow spot) or briefly macula (Fig. 1). Finally, the light-sensitive outer segments of the rod and cone photoreceptors are continuously renewed.

Metabolic and Molecular Pathways and Processes in the Eye

Production of Aqueous Humor

The aqueous humor is secreted into the posterior chamber of the eye by a part of the ciliary epithelium situated close to the region where the zonula fibers are attached (Fig. 1, inset). The composition of aqueous humor is relatively similar to blood plasma.

However, its protein concentration is low (less than 1%), whereas ascorbate is up to 50 times higher than in blood plasma. Oxygen is derived by diffusion from the cornea and from the vasculature of the iris [4]. The aqueous humor flows from the posterior chamber through the pupil into the anterior chamber (Fig. 1, inset) and drains away at the angle between the cornea and iris, where it passes through a porous tissue—the trabecular meshwork—into a collecting channel (Schlemm's canal), which empties into veins and thus into the bloodstream. In the healthy eye, the delicate balance between aqueous fluid production, circulation, and drainage must be maintained in order to keep the intraocular pressure at a constant level. Slow drainage of aqueous humor or overproduction may lead to an increase in intraocular pressure that can result in the death of retinal ganglion cells—a disease called glaucoma (see chapter "Glaucoma").

Retinal Metabolism

The retina is inverse, that is, before light reaches the photoreceptors, it has to pervade the different retinal layers: three layers of somata (termed gan-

glion cell layer, inner nuclear layer, and outer nuclear layer), separated by two synaptic layers (i.e., inner and outer plexiform layer). The ganglion cells are the output neurons of the retina. Their axons form the optic nerve (Fig. 1). As all neurons, retinal cells conduct information by generating electrical signals at their plasma membrane (see chapter "Brain: Overview"under part "Brain"). Ions are unequally distributed across the plasma membrane. For example, inside the cell, there is a high concentration of K^+ ions and large polyanions (such as proteins and nucleic acids), while outside the Na^+ ion and Cl^- ion concentration is high. Overall, this difference in electric charges results in a resting membrane potential of -70 mV in most neurons.

Photoreceptor cells can be roughly divided into two compartments (Fig. 2a). The inner compartment comprises the cell body, the axon with the synaptic region, as well as the biochemical machinery with the mitochondria and ribosomes for routine cell metabolism found in the inner segment. The outer segment harbors all proteins to absorb light, to amplify the signal, and to generate an electrical signal in response to light.

Rod outer segments are effective light catchers. The outer segment of a human rod contains a stack of up to 800 flat, hollow membrane compartments, called discs (Fig. 2b). The latter contain a high concentration of the photopigment rhodopsin in their membranes (Fig. 2c). Altogether 50–150 million rhodopsin molecules are found within a photoreceptor cell. Rhodopsin belongs to the family of G-protein-coupled receptors (see below). It consists of a protein part (the opsin) and a light-absorbing cofactor, the aldehyde form of vitamin A, retinal (Fig. 2c, d).

Retinal can exist in two different conformations. The folded 11-*cis*-retinal is covalently bound within the opsin molecule (Fig. 2c). Absorption of a light quantum causes 11-*cis*-retinal to switch to the elongated all-*trans*-retinal (Fig. 2d), inducing a conformational change in the rhodopsin molecule. In this state, rhodopsin activates the G-protein transducin, which in turn activates phosphodiesterase 6 (PDE6). The latter hydrolyzes cyclic guanosine monophosphate (cGMP), which acts as a signal transducer, ampli-

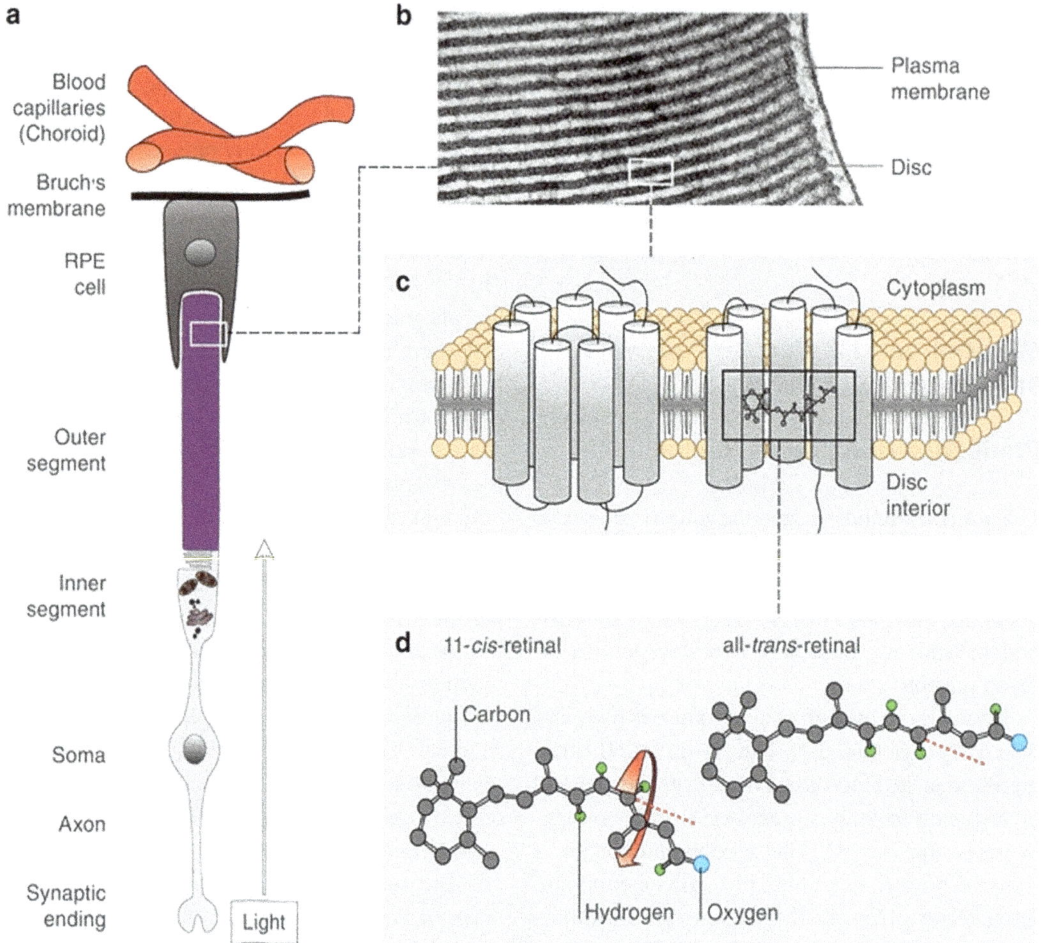

Fig. 2 Rod photoreceptor and rhodopsin. (**a**) Schematic overview of a rod photoreceptor and its surrounding layers. The blood capillaries of the choroid provide nourishment and oxygen to the photoreceptors. The retinal pigment epithelium (RPE) cells fulfill many functions (e.g., retinal metabolism and blood–retinal barrier). Note that the center of the eye is towards the bottom. The outer segments of the photoreceptor cells contain light-absorbing discs. The inner segments contain cell body and normal metabolic machinery. (**b**) An electron micrograph of the region indicated in (**a**), showing an outer segment of a photoreceptor cell with its membranous discs. (**c**) Schematic of a disc membrane harboring the light-sensitive pigment rhodopsin, a G-protein-coupled receptor with seven transmembrane helices (*left*). Two helices are removed to reveal the retinal-binding pocket (*right*). (**d**) Chemical structure of free 11-*cis*- and all-*trans*-retinal. Light converts 11-*cis*-retinal into all-*trans*-retinal

fier, and molecular switch. In the dark, due to an elevated cGMP concentration, cGMP-dependent ion channels (called cyclic nucleotide-gated ion channels) in the outer segments are open, leading to an influx of Na^+ and Ca^{2+}, depolarization of the membrane potential to approximately -30 to -40 mV, and transmitter release. Upon illumination and cGMP hydrolysis via PDE6, cGMP-dependent ion channels close. As fewer Na^+ and Ca^{2+} enter the cell, the membrane potential becomes more negative. Thus, during illumination, the cells are more hyperpolarized, and fewer transmitter molecules are released from the synapse. This kind of membrane potential modulation, also termed graded potentials, controls the activity of photoreceptor cells, like that of most retinal cells. Graded potentials are small variations in the magnitude of membrane potential in contrast to the "all-or-nothing" large amplitude action potentials. To shut off the signaling cascade, rhodopsin becomes phosphorylated [5] and then sealed by a protein called arrestin, which ultimately stops the interaction with transducin.

The photoreceptor information is relayed synaptically via the bipolar cells to the ganglion cells. Along this path, lateral neuronal interactions and feedback loops are provided by horizontal cells in the outer and by amacrine cells in the inner plexiform layer. As retinal output neurons, ganglion cells communicate via action potentials—short (1–2 ms) stereotyped changes in membrane potential that propagate in an all-or-none fashion along the axons of neurons (chapter "Brain: Overview" under part "Brain").

The physiology of photoreceptors is extraordinary in the sense that a large ionic current must be sustained in the dark that is switched off in the light. Therefore, the energy consumption of the retina is four times higher in the dark than during illumination [6]. In the dark, about 50% of the energy is used by the Na^+/K^+-ATPase to pump out excess Na^+ that enters the photoreceptors through open cGMP-dependent ion channels in the outer segment [7]. As during illumination the ion flux into the photoreceptor outer segment decreases, energy consumption by the Na^+/K^+-ATPase drops. However, the energy expenditure for the subsequent phosphorylation of rhodopsin and the regeneration of 11-*cis*-retinal increases.

Photoreceptor cells produce ATP from glucose via two pathways. Large mitochondria are found densely packed in the photoreceptor inner segments, yet only a fraction of glucose is metabolized via oxidative phosphorylation, the bigger part via glycolysis [8]. This reminds of the glucose metabolism found in rapidly growing cancer cells that have a high demand of anabolic intermediates (Warburg effect). While photoreceptors are postmitotic (meaning that they do not divide any longer), they also have high anabolic demand, as they shed approximately 10% of their outer segment every day to minimize photo-induced cell damage, and the need to replace the lost proteins and lipids is challenging. While glycolysis yields less energy in form of ATP, it produces just these intermediates (e.g., pyruvate) for the synthesis of amino acids and lipids. Glucose and oxygen are supplied from the capillary network

of the choroid that is separated from the photoreceptors by the RPE (Figs. 1 and 2a). The oxygen consumption of the retina is approximately 20% higher than the oxygen consumption reported for the brain [5].

The Role of the Retinal Pigment Epithelium

Together with the retina, the RPE is among the most metabolically active tissues in the body. The RPE serves many functions. First, melanin in RPE cells absorbs light that has passed through the retina in order to prevent light scattering and to reduce light-induced damage. Second, the tight junctions between RPE cells form a barrier between the choroid and photoreceptors [9]. Analogous to the blood–brain barrier, this blood–retinal barrier controls exchange between blood and retina and thus maintains the specialized environment of the photoreceptors. The barrier function of the RPE is physically supported by Bruch's membrane that acts as a semipermeable molecular sieve [2]. At the same time, RPE cells utilize glucose transporters that allow passive transport of glucose from the choroid to the photoreceptors [10]. Third, the RPE is involved in the regeneration of 11-*cis*-retinal. Upon illumination, 11-*cis*-retinal in the rhodopsin is converted to all-*trans*-retinal (see above). All-*trans*-retinal detaches from the opsin and is enzymatically reduced to all-*trans*-retinol. Retinoid binding proteins shuttle the retinol from the outer segments to the RPE, where the rest of the retinal metabolism takes place. Here, all-*trans*-retinol is first esterized with palmitate, then isomerized to 11-*cis*-retinol, and finally, oxidized to 11-*cis*-retinal. The latter is transported back to the outer segment, where it spontaneously reacts with opsin to regenerate rhodopsin, thus completing the visual cycle.

Finally, the RPE is of utmost importance for the regeneration of the photoreceptor outer segments. The strong illumination of the outer segments in a high-oxygen environment can lead to photochemical damage. If the energy from a photon is transferred from a light-absorbing molecule to oxygen, reactive oxygen species (e.g., singlet oxygen) can be created. Those can break molecular bonds or induce photooxidation. Accumulation of such events can ultimately lead to damage of the outer segments. Therefore, photoreceptor outer segments must be constantly renewed. Around sunrise, the tips of the outer segments are shed off and are phagocytosed by RPE cells, while new discs are added at the bases of the outer segments [11]. Complete renewal of the rod outer segment takes approximately 10 days. Due to the high amount of material, phagocytosis in RPE cells is metabolically demanding. Some of the breakdown products may be recycled, while others are discharged into the choriocapillaris. Some undigested proteins, lipids, and retinoids remain as an aggregation complex called lipofuscin [9]. Chemical reactions between these components lead to the formation of retinoid-lipid complexes in lipofuscin, the so-called bisretinoids [12]. Lipofuscin can absorb light, shows autofluorescence, and is susceptible to photochemical changes. Lipofuscin accumulation is thought to contribute to the development of age-related macular degeneration (see chapter "Age-Related Macular Degeneration").

Inside-Out: Signals from the Eye Affecting Other Organs and Tissues

The impact of the eye on other organs results mainly from the projection of the retinal output neurons, the ganglion cells, which relay the signals from the retina via their axons to the brain. Most of the ganglion cell output provides the basis for visual information processing that involves at least 30% of the cerebral cortex.

Recently, a special class of ganglion cells has been described that serves an entirely different function. These ganglion cells express their own photopigment called melanopsin, which is distantly related to rhodopsin. This cell class provides input to the suprachiasmatic nucleus (SCN) in the brain (together forming the retinohypothalamic tract). The SCN functions as pacemaker, responsible for the generation of the circadian clock. Retinal input from the melanopsin-containing ganglion cells resets the clock in the SCN every day ("entrainment") [13]. SCN cells project to the paraventricular nucleus of the hypothalamus—a site of hormone production. The projection from the SCN to the pineal gland controls the release of melatonin, the "hormone of the night" that is involved in the regulation of our sleep-wake cycle. Circadian rhythms also influence body temperature, blood pressure, and heart frequency (see chapter "Rheumatoid Arthritis").

Outside-In: Signals and Metabolites Affecting the Function of the Eye

Vitamin A, the precursor of retinal, is an essential vitamin [14]. It is stored in the liver (see chapter "Liver: Overview" under part "Liver") and transported in the blood (see chapter "Blood: Overview" under part "Blood") by retinoid binding proteins. RPE cells absorb vitamin A from the choroidal circulation and convert it to retinal to supply the photoreceptors. As retinal is pivotal for rhodopsin function, lack of vitamin A may lead to night blindness.

Diabetes (see chapter "Diabetes Mellitus") can affect the performance of the eye in two major ways. First, in the lens, glucose can be converted into sorbitol, which is later—but more slowly—converted into fructose [15]. Under normal conditions, only small amounts of sorbitol are synthesized. However, unphysiologically high blood glucose levels cause excess sorbitol production, which results in osmotic swelling of the lens (as sorbitol and fructose cannot leave the lens and thus are highly osmotic) that can ultimately lead to lens clouding, that is, cataract. Second, diabetes triggers pathological changes in retinal blood vessels. In early diabetic retinopathy (the so-called non-proliferative stage), retinal blood vessels become blocked, which results in reduced nourishment and oxygen supply. In response to the hypoxic state, in a second phase (called proliferative diabetic retinopathy), new blood vessels are formed. These vessels are thinner, mechanically less stable and their endothelial cells form a less effective blood–retinal barrier. Consequences are bleedings into the retina and vitreous as well as scar formation and detachment of the retina. Diabetic retinopathy may ultimately lead to blindness.

Perspectives

The main functions of the eye are to provide visual information about our environment and to synchronize our internal clock to the day/night cycle. Partial or total blindness can result from several pathological mechanisms, including the destruction of retinal ganglion cells and optic nerve due to an increase in intraocular pressure, as in glaucoma (see chapter "Glaucoma"), or the loss of photoreceptor cells triggered by photochemical damage or other factors, such as in age-related macular degeneration (see chapter "Age-Related Macular Degeneration"). In an aging society, it will become more and more important to provide effective therapeutic approaches for these diseases. The therapeutic bandwidth ranges from pharmacological treatment to gene transfer or retinal implants. In any case, success will rest on a thorough understanding of the physiological features of the eye and the retina.

Questions and Answers

Question 1 Try to give a rough estimate of the amount of rhodopsin (approximately molecular weight 40 kDa) that needs to be synthesized every day in a human eye. Assume that per eye there are 120 million rods, each with 100 million rhodopsin molecules of which 10% have to be replaced every day due to phagocytosis of the outer segment tip.

Answer 1 10^8 rods/retina $\times$ 10^8 rhodopsin molecules/rod $= 10^{16}$ rhodopsin molecules/retina. Ten percent need to be replaced per day $= 10^{15}$ molecules. 40 kDa means 40 kg/mol $= 40{,}000$ g/6.023×10^{23} molecules. $10^{15} \times 40{,}000$ g/6.023×10^{23} makes roughly 80 µg rhodopsin per eye and day.

Question 2 In 1877, Wilhelm Kühne observed that a living frog retina prepared in the dark is purple (this color is due to the presence of the visual pigment rhodopsin). Upon exposing the retina to light, the purple color vanished, which he interpreted as a light-induced chemical reaction of the visual pigment. The bleaching of the rhodopsin is triggered by the isomerization of 11-*cis*-retinal to all-*trans* retinal, followed by the detachment of the retinal from the opsin. He also found that the purple color was reconstituted in the dark but only, if the retina was still in contact with the RPE. If he detached the retina from the RPE before putting the eye back into the dark, the retina remained pale. Try to interpret this observation. Why is the contact between retina and RPE necessary?

Answer 2 The rhodopsin contains 11-*cis*-retinal. Upon light absorption, 11-*cis*-retinal is converted to all-*trans*-retinal that needs to be reisomerized to 11-*cis*-retinal. To this end, the retinoids must be shuttled between photoreceptors and RPE, as many steps in this "visual cycle" are performed by RPE cells. Detachment of the retina interrupts this exchange between RPE and photoreceptors and the rhodopsin cannot be reconstituted.

Question 3 In certain diseases, for example, retinitis pigmentosa in humans or in corresponding animal models, the rods and cones in the retina die. As they cannot be replaced, this degeneration results in blindness. What do you think: can an organism whose rods and cones have been completely lost still exhibit light entrainment of the circadian rhythm (i.e., day–night cycle)?

Answer 3 Yes. The circadian clock in the brain is reset by a special class of ganglion cells that express their own visual sensor protein (melanopsin). These cells can still detect light even if rods and cones are gone and relay this information to the pacemaker in the brain to achieve entrainment of the circadian clock.

References

1. Masland R (2012) The neuronal organization of the retina. Neuron 76:266–280
2. Booij JC, Baas DC, Beisekeeva J, Gorgels TG, Bergen AA (2010) The dynamic nature of Bruch's membrane. Prog Retin Eye Res 29:1–18
3. Hunter JJ, Morgan JI, Merigan WH, Sliney DH, Sparrow JR, Williams DR (2012) The susceptibility of the retina to photochemical damage from visible light. Prog Retin Eye Res 31:28–42
4. Shui YB, Fu JJ, Garcia C, Dattilo LK, Rajagopal R, McMillan S, Mak G, Holekamp NM, Lewis A, Beebe DC (2006) Oxygen distribution in the rabbit eye and oxygen consumption by the lens. Invest Ophthalmol Vis Sci 47:1571–1580
5. Yau KW, Hardie R (2009) Phototransduction motifs and variations. Cell 139:246–264
6. Okawa H, Sampath AP, Laughlin SB, Fain GL (2008) ATP consumption by mammalian rod photoreceptors in darkness and in light. Curr Biol 18:1917–1921
7. Ames A 3rd, Li YY, Heher EC, Kimble CR (1992) Energy metabolism of rabbit retina as related to function: high cost of Na$^+$ transport. J Neurosci 12:840–853
8. Narayan DS, Chidlow G, Wook JPM, Casson RJ (2017) Glucose metabolism in mammalian photoreceptor inner and outer segments. Clin Experiment Ophthalmol 45:730–741
9. Cunha-Vaz JG (1976) The blood-retinal barriers. Doc Ophthalmol 41:287–327
10. Senanayake P, Calabro A, Hu JG, Bonilha VL, Darr A, Bok D, Hollyfield JG (2006) Glucose utilization by the retinal pigment epithelium: evidence for rapid uptake and storage in glycogen, followed by glycogen utilization. Exp Eye Res 83:235–246
11. Kevany BM, Palczewski K (2010) Phagocytosis of retinal rod and cone retinal. Physiology (Bethesda) 25:8–15
12. Keller JN, Dimayuga E, Chen Q, Thorpe J, Gee J, Ding Q (2004) Autophagy, proteasomes, lipofuscin, and oxidative stress in the aging brain. Int J Biochem Cell Biol 36:2376–2391
13. Berson D (2003) Strange vision: ganglion cells as circadian photoreceptors. Trends Neurosci 26:314–320
14. D'Ambrosio DN, Clugston RD, Blaner WS (2011) Vitamin A metabolism: an update. Nutrients 3:63–103
15. Kinoshita JH (1965) Pathways of glucose metabolism in the lens. Investig Ophthalmol 4:619–628

Age-Related Macular Degeneration

Marie D. Just, Sam Mehr, Frank G. Holz, and Monika Fleckenstein

Introduction

Age-related macular degeneration (AMD) is the leading cause of irreversible blindness in the elderly population of industrialized countries and the fourth most common cause of blindness globally [1]. The prevalence of AMD in the world population ages 45–85 is estimated to be 8.69% [2]. Based on a meta-analysis, the prevalence of early-stage AMD among this population is estimated to be 8.01%, with advanced-stage AMD estimated at 0.37% [2]. Due to exponential population aging, the number of affected persons is likely to increase. The projected number of individuals worldwide with AMD in 2040 is 288 million [2].

M. D. Just · S. Mehr · M. Fleckenstein (✉)
Moran Eye Center, Ophthalmology, University of Utah, Salt Lake City, UT, USA
e-mail: marie@just-med.de;
monika.fleckenstein@hsc.utah.edu

F. G. Holz
Department of Ophthalmology, University Hospital Bonn, Bonn, Germany
e-mail: frank.holz@ukbonn.de

Pathogenesis and Therapeutical Strategies

As the name implies, aging processes play a major role in the pathogenesis of AMD. The age at which patients develop the disease can vary greatly, with incidence rates approximately quadrupling with each decade of age [3]. The additive genetic influence on AMD accounts for up to 71% heritability, depending on the stage of the disease [4]. Genome-wide association studies have indicated the evidence for strong genetic predisposition [5] and have associated the highest risk of AMD with variants in *CFH-CFHR5* on chromosome 1 and variants in *ARMS2/HTRA1* on chromosome 10 [6]. Most variants are localized in genetic loci encoding signaling pathways of lipid metabolism, the complement system, or extracellular matrix remodeling [5]. It has been suggested that certain genetic characteristics may lead to distinct alterations during the natural course of AMD, indicating the involvement of different pathological mechanisms [7].

While aging and genetic predisposition are nonmodifiable influences, a notable impact of modifiable risk factors has been reported as well. Cigarette smoking is consistently reported to have a strong positive correlation with all forms of AMD. This preventable exposure increases overall AMD risk over twofold in men and women [8, 9]. Furthermore, there are several lines of evidence regarding the impact of diet on

E. Lammert, M. Zeeb (eds.), *Metabolism of Human Diseases*,
https://doi.org/10.1007/978-3-031-96019-2_10

AMD progression. Intake of nutrient-rich foods such as vegetables and fruits, as well as increased consumption of omega-3 fatty acids, play a beneficial role in preventing or delaying the progression of AMD [10]. Moreover, abdominal obesity increases the risk of AMD progression, whereas physical activity seems to decrease the risk [11], once more reinforcing the importance of individual health.

The interaction of these demographic, genetic, systemic, and environmental factors is thought to induce pathophysiological changes. Most of them promote earlier onset and progression of AMD by acting on different biological pathways. The mechanisms involved include lipid deposition, oxidative stress, immune activation, and pathological tissue remodeling.

Various cell layers are implicated in the disease process: the choroid, Bruch's membrane (BrM), the retinal pigment epithelium (RPE), and the photoreceptors. An imbalance in the homeostasis of these layers may result in pathological alterations.

Histopathological studies revealed a loss of choriocapillaris density as a consequence of the natural aging process [12, 13]. These observations were underlined by recent in vivo studies with optical coherence tomography angiography (OCT-A) imaging, noting flow deficits in the 1-mm region around the macula in normal eyes at older ages [14]. The most prominent clinical and histopathologic lesions of AMD involve BrM, a thin connective tissue between the basal RPE and the choriocapillaris. Molecules essential for RPE and photoreceptor function traverse BrM, which is responsible for lipoprotein clearance. BrM increases in thickness due to the lifelong entrapment of cholesterol-rich and apo-B/E-containing lipoproteins and lipoprotein-like particles [15]. Structural changes and accumulation of material within BrM may reduce the free flow of molecules between the choriocapillaris and the RPE, reducing blood supply between the RPE and photoreceptors [16]. The implication of lipid transport pathways in AMD pathogenesis is supported by variants in several lipid-related genes, including *LIPC*, *CETP*, *ABCA1*, and *APOE* [6].

A characteristic of early- to advanced-stage AMD are drusen—extracellular accumulations in BrM that contain lipids, minerals, or proteins. Triggered by oxidized lipoproteins [17], drusen formation can progress, which in turn appears to stimulate inflammatory infiltration of molecules and cellular debris [18]. Depending on size and composition, drusen become visible on conventional ophthalmoscopy and may serve as a predictive factor for disease progression [19]. Of note, the appearance of drusen in terms of localization, number, and morphology of drusen and drusen-like deposits (e.g., basal laminar and linear deposits, or subretinal drusenoid deposits) can differ widely and call for further investigations and classifications [20, 21].

The RPE contributes to the lipid metabolism of the retina, recycling lipids via endolysosomal phagocytosis and autophagy from the photoreceptor outer segments byproducts [22]. The incomplete degradation of lipofuscin, an indigestible autofluorescent lipopigment, leads to accumulation in RPE cells with age [23]. Whether excessive lipofuscin accumulation possesses toxic properties is under current investigation [24]. The compromised metabolism is followed by the death of the photoreceptors with the degeneration of rods followed by cones [25], leading to pathologic alterations and clinical manifestation. Photoreceptor cell death can be considered the "final stage" of AMD, manifesting as irreversible vision loss.

Potential preceding stressors, such as demographic, systemic, and environmental risk factors, interact with genetic predisposition and initiate a cascade of pathological events, resulting in cell layer dysfunction, intraretinal flow impairment, injured metabolic clearance, and imbalanced protection mechanisms (Fig. 1).

As mentioned above, aging, smoking, and high-fat diets, all sources of oxidative stress, are risk factors for AMD. The retina provides an ideal environment for the generation of reactive oxygen species (ROS) due to its specific anatomical and metabolic characteristics, such as high energy demand and oxygen consumption by the outer retina-RPE complex (see chapter "Eye: Overview"), high levels of cumulative irradiation, and abundance of photosensitizers within the neurosensory retina and RPE [27]. Furthermore, photoreceptor outer segment membranes are rich in polyunsaturated fatty acids, which are readily oxidized and thus can initiate a cytotoxic chain reaction [28]. Alleviation of oxidative stress is thought to be crucial in preventing progression to advanced stages of AMD. Among patients with bilateral intermediate or unilateral advanced AMD, the Age-Related Eye Disease Study (AREDS) found that vitamins C and E, β-carotene, zinc, and cupric oxide can reduce the risk of progression to advanced AMD [29]. This formula was made available as a supplement for patients. In patients with intermediate AMD in at least one eye, these AREDS supplements reduced the risk of progression to neovascular AMD by ~25% in 5 years, but this result was not found in those with bilateral early AMD [29]. A link between β-carotene and increased risk of lung disease in current and former smokers resulted in the AREDS2 trial, which examined the efficacy of replacing β-carotene with lutein and zeaxanthin in an effort to achieve similar results to β-carotene without the drawbacks [30]. The study resulted in an altered AREDS2 supplement, consisting of vitamins C and E, cupric oxide, lutein, zeaxanthin, and zinc.

The immune system serves to maintain tissue homeostasis and preserve tissue functionality. Low-grade inflammation is part of a protective response that proves harmful when this fragile system becomes unbalanced. Oxidative stress triggers the release of inflammatory cytokines and chronic inflammation, called para-inflammation, which appears to lead to cell senescence and cell death [31]. The exact sequence of molecular damage, altered immune response, and subsequent systemic inflammation remains to be elucidated [32]. The local inflammatory process appears to contribute to drusen biogenesis. In addition, the damaged blood–retinal barrier provides a surface for retinal lesions [31]. The complement system belongs to the

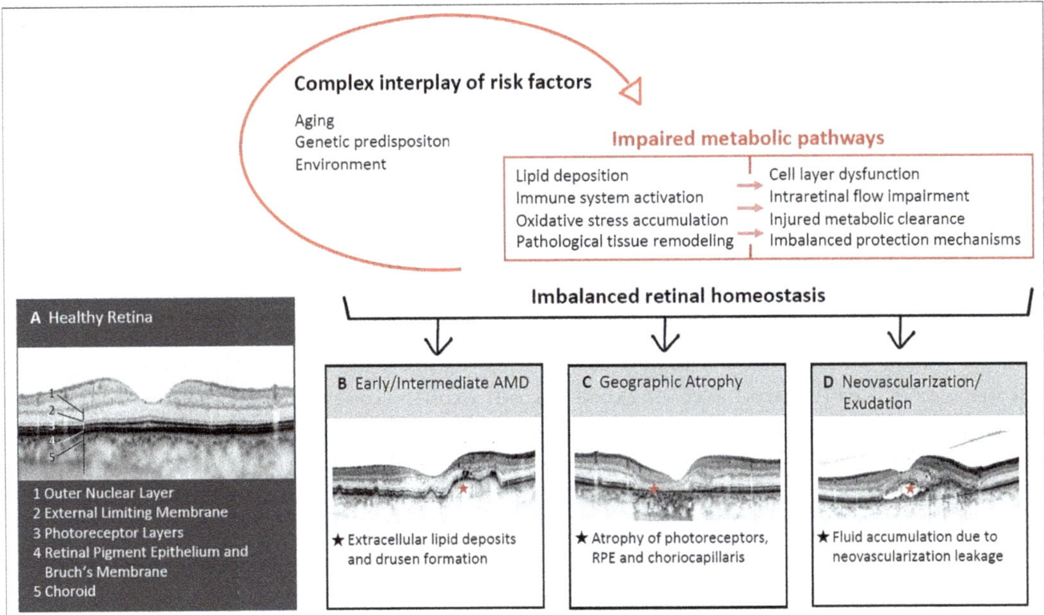

Fig. 1 Model of metabolic pathways contributing to AMD pathogenesis. Age-related macular degeneration (AMD) is known as a multifactorial disease affecting the central part of the macula. Complex interplay between risk factors leads to phenotypic alterations visible in OCT. Compared to the healthy retina with intact and distinguishable layers (**a**), extracellular lipid deposits and drusen—the hallmark lesions of AMD—appear in early and intermediate stages as a consequence of impaired metabolic pathways (**b**), possibly progressing to late stages such as geographic atrophy (**c**) or macular neovascularization (**d**). (Source OCT-images [26])

innate immune system. The role of the complement system pathway in AMD pathogenesis has been exemplified by the discovery of genes that encode complement system regulatory proteins such as Complement factor H [33], Factor B/Complement component 2 [34], and Complement component 3 [35]. Specific polymorphisms in corresponding genes are thought to be associated with abnormal complement activation [36], causing lower tolerance to oxidative and metabolic stress that result in RPE damage and photoreceptor cell atrophy [37, 38].

However, the pathophysiological processes causing drusen formation and subsequent neurodegeneration require further investigation.

Different forms of progression and phenotypic manifestations of AMD have been categorized. The early and intermediate forms of the disease are denoted by drusen and pigmentary alterations, often remaining asymptomatic. More advanced stages can take years to develop. The advanced stages of AMD are subdivided into a nonexudative and exudative form. The historical nomenclature referred to as "dry" or "wet," though still used, is evolving to characterize the heterogeneous subtypes more precisely.

The nonexudative, "dry," late-stage of AMD is called "geographic atrophy" (GA) and describes the enlarging areas of loss of choriocapillaris, RPE, and photoreceptors [39]. The progressive cell death of all affected layers leads to corresponding visual loss.

The exudative form is characterized by exudation from macular neovascularizations (MNV). The new vessels invade the outer retina, subretinal space, or sub-RPE space. "Wet" or exudative stage becomes apparent when those new pathological formations leak or rupture. Consequences can be observed in the form of extracellular fluid and/or hemorrhages in the above-mentioned retinal space [40]. While the accumulation of extracellular fluid is potentially reversible, persistent exudations may ultimately culminate in discoid retinal scarring and permanent vision loss.

Due to the heterogeneous clinical picture of AMD, different classification systems have been established and accepted over the years. In addition to classification according to drusen size of early, intermediate, and advanced AMD used in epidemiological population studies, clinical studies based on color fundus photography often refer to the AREDS classification system [41]. Efforts are underway to develop more precise classification systems incorporating state-of-the-art imaging, such as OCT, fluorescein-, indocyanine green-, or OCT-A, and fundus autofluorescence (FAF) imaging [42, 43].

The nonexudative and exudative late-stages of AMD are not mutually exclusive, as both forms can coexist in the same eye [44]. It has been hypothesized that GA is the natural end-stage of AMD if the incident of MNV does not arise. Specifications of different phenotypes of GA and MNV have become approachable through systematic in vivo retinal imaging, functional assessment, and clinical observations.

Neovascularization, or angiogenesis, has been target in prevention and treatment of exudative AMD [45]. Angiogenesis involves choroidal endothelial cells and RPE cells migrating into the neurosensory retina to proliferate and develop new vessels [46]. Some approaches consider MNV to be a nonspecific rescue mechanism which appears to assume the task of the choriocapillaris and slow down photoreceptor and RPE degeneration [47, 48].

Ocular angiogenesis is mainly induced by the hypoxia-driven vascular endothelial growth factor-A (VEGF-A, also abbreviated as VEGF) due to diminished choroidal blood circulation [49]. VEGF is a signaling protein that plays a key part in pathological angiogenesis and vascular permeability in the retina [50]. It is released in response to oxidative stress, complement activation, and regions of vascular abandonment. RPE-cells, retinal neurons (particularly amacrine and ganglion cells), and retinal glia cells (Müller cells) [51] release cytokines such as tumor necrosis factor alpha that stimulate retinal cells to express VEGF [52]. VEGF binds to the VEGFR2 receptor to activate multiple downstream pathways through signaling intermediates, which promote endothelial proliferation and angiogenesis [53]. Another key component of the angiogenesis cascade is the angiopoietin-2 molecule (Ang2). Ang2, a proangiogenic cytokine, may be critical in the progression of neovascular AMD by way of cooperation with VEGF to promote neovascularization. Ang2 serves the purpose of vessel destabilization by competing with Angiopoetin-1 (Ang1)1 and thereby restricting functionality of the Tie2 receptor expressed on endothelial cells of blood vessel walls [54]. In a Porcine retinal endothelial cell model, VEGF has been shown to be twice as potent as Ang2 in causing vessel hyperpermeability; though, both VEGF and Ang2 function together resulted in around three times more permeability than the effects of VEGF alone, which is why Ang2 is regarded to heighten sensitivity of blood vessels to the effects of VEGF [55]. Due to this, effective inhibition of Ang2 may be of significance in the prevention of AMD progression [56].

Most effective treatment for exudative AMD is through inhibition of VEGF. This approach— that includes delivery of anti-VEGF substances into the vitreous—has largely replaced angio-occlusive approaches such as laser coagulation or photodynamic therapy [57]. The introduction of biologics that target VEGF have been a milestone in the treatment of retinal disorders, including AMD [58]. Anti-VEGF drugs inhibit VEGF, mainly affecting permeability of neovascularizations. They have widely become the default choice for AMD treatment, with a reduction of approximately 50% in the rate of legal blindness in patients with AMD aged 50 years and older [59]. Multiple anti-VEGF drugs are already on the market and others are undergoing preclinical or clinical trials (Fig. 2).

The first intravitreally injected anti-VEGF medication to be approved for neovascular AMD treatment was **pegaptanib**, an oligonucleotide aptamer with a high binding specificity for the VEGF-A isoforms [60]. It inhibits amplification of VEGF receptor signaling. Pegaptanib was the first foray into anti-VEGF treatments, thus opening the door to further exploration of anti-VEGF drugs, but it has since been replaced for the most part by more effective drugs. **Ranibizumab**, a humanized monoclonal antibody neutralizing all VEGF-A isoforms, has a similar molecular structure and functionality to bevacizumab [61]. Monthly injected ranibizumab has been shown to

stabilize vision in over 90% of patients and significantly improve vision in over 30% of patients [62]. **Bevacizumab**, a medication approved as a cancer therapeutic, is a full-length humanized monoclonal antibody against VEGF-A. Though not approved for use in treating exudative AMD, it is not inferior to ranibizumab [63] and is widely used as an off-label alternative to more expensive anti-VEGF treatments. **Aflibercept** is a recombinant fusion protein, consisting of the key domains of the VEGF receptors 1 and 2, coupled to the Fc part of a human IgG molecule. It has a high affinity for all VEGF-A and VEGF-B isoforms as well as placental growth factor and has also been proven to be noninferior to ranibizumab [64].

Brolucizumab is a humanized single-chain antibody fragment against VEGF-A with comparable visual function results to aflibercept [65]. Given its low molecular weight, it can be applied in higher molar doses and injected at longer intervals after three monthly loading doses [66]. Adverse drug reactions such as intraocular inflammation, retinal vasculitis, and retinal vascular occlusion have been reported, so the indication for the use of brolucizumab should be made with caution [67].

Faricimab is the first bispecific antibody approved for the treatment of exudative AMD. It is composed of two antigen-binding fragments targeting both VEGF-A and Ang2, promising prolonged treatment intervals and improved patient compliance [68].

Though multiple proven treatments for exudative AMD are now widely available, effectively treating this disease continues to present a challenge in healthcare. Since these often lifelong injections must be administered sometimes as frequently as every month, the burden of continued visits to receive treatment can be difficult and costly for both patients and healthcare systems. Furthermore, consistent treatment cannot be guaranteed, and studies have shown that undertreatment continues to be a significant reason for vision loss regarding anti-VEGF injections [69]. Because of these difficulties, long-acting treatments, such as intraocular port delivery systems for anti-VEGF drugs, are under development. One of these long-acting treatments that has been

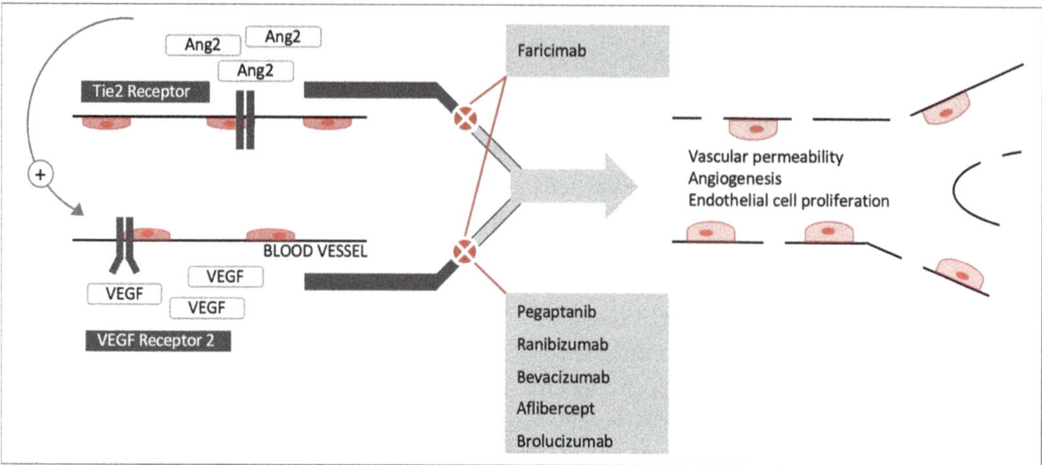

Fig. 2 Therapeutic strategies currently used in the treatment of exudative neovascular AMD. Hypoxia induces angiogenetic cascades mediated by VEGF and Ang2. VEGF receptor 2 and Tie2 receptor, both expressed on endothelial cells of blood vessel walls, are activated by aforementioned signal molecules and promote vascular permeability, angiogenesis, and proliferation of endothelial cells. Ang2 sensitizes vessels to the effect of VEGF. Anti-angiogenetic drugs for the treatment of exudative neovascular AMD that competitively bind VEGF are: pegaptanib, ranibizumab, bevacizumab, aflibercept, and brolucizumab. The bispecific antibody Faricimab targets both the VEGF- and Ang2-pathway

approved is a slow-release system involving a customized formulation of ranibizumab delivered by way of surgically placed permanent refillable intraocular implant [70]. However, the implant is associated with a threefold higher rate of endophthalmitis compared to monthly intravitreal ranibizumab injections in clinical trials [71].

In case of GA, various ongoing preclinical and clinical trials are investigating potential treatments for this type of AMD, particularly involving complement inhibition, neuroprotection, visual cycle modulators, cell-based therapies, and anti-inflammatory agents [72]. Pegcetacoplan and avacincaptad pegol have been shown to slow GA progression. In 2023, these two complement-inhibitors were approved as an intravitreal injection for the treatment of GA in the USA, but have not yet been approved in the EU.

Perspectives

Though our understanding of pathways and effective treatment of AMD has seen tremendous advances in recent years, significant challenges remain to be overcome. As the world population increases, the number of people with AMD continues to grow alongside it. Health systems face potentially overwhelming demand for treatment, of which the most effective options available are only capable of halting or slowing progression of select manifestations of the disease. Though AMD is referred to as one disease, it represents a wide spectrum of phenotypes, with varied symptoms and pathological pathways for each patient. As such, taking a more individualistic approach to treatment based on each patient's particular manifestation of AMD could help maintain efficacy of treatment while reducing treatment burden. Promising areas of interest are methods involving targeting of complement pathways, gene therapy, cellular regeneration, and retinal prostheses [6]. Though our knowledge of AMD remains limited by our incomplete understanding of the minutiae of metabolic interactions in the body, we expect to understand it in more detail as studies continue to elucidate the nuances of this disease.

Acknowledgments Not applicable.

Questions and Answers

Question 1 How can patients influence the progression of their disease?

Answer 1 At diagnosis, patients should be informed that smoking and an unhealthy diet are important influencing factors for onset, further and faster progression of AMD. In addition to a healthy lifestyle, eye medications such as AREDS may be beneficial and should be recommended. Regular examinations are essential to ensure therapeutic intervention at the earliest possible time. Patients should be aware that AMD may continue to progress despite the best efforts to halt it.

Question 2 Which retinal layers are involved in the pathogenesis of AMD?

Answer 2 AMD affects the central part of the retina, especially the macula. Pathological changes involve the complex of photoreceptors, retinal pigment epithelium (RPE), Bruch's membrane (BrM), and the choriocapillaris, the innermost layer of the choroid.

Question 3 What symptoms present AMD and which examinations are relevant for diagnosis?

Answer 3 Patients initially report difficulty in dim light and low-contrast situations, which can result in definitive visual field and vision loss. A basic screening method for metamorphopsia and distortion vision is the Amsler grid, which is commonly used in clinical examination. Retinal changes such as drusen, pigmentary changes, atrophy, or neovascularization are visible on fundoscopy. OCT, FAF, or angiography can detect manifestations of early to late stages of AMD.

Question 4 Which pathways are targeted by drug therapy, and which might be promising in the future?

Answer 4 Angiogenic processes play an important role in the progression of AMD to late stages.

Current treatments mainly target VEGF and Ang2. Promising targets for the treatment of other subtypes of the AMD spectrum are modulation of the complement system pathway and gene therapy.

References

1. Jonas JB, Cheung CMG, Panda-Jonas S (2017) Updates on the epidemiology of age-related macular degeneration. Asia Pac J Ophthalmol (Phila) 6(6):493–497
2. Wong WL et al (2014) Global prevalence of age-related macular degeneration and disease burden projection for 2020 and 2040: a systematic review and meta-analysis. Lancet Glob Health 2(2):e106–e116
3. Rudnicka AR et al (2015) Incidence of late-stage age-related macular degeneration in American Whites: systematic review and meta-analysis. Am J Ophthalmol 160(1):85–93.e3
4. Seddon JM et al (2005) The US twin study of age-related macular degeneration: relative roles of genetic and environmental influences. Arch Ophthalmol 123(3):321–327
5. Fritsche LG et al (2016) A large genome-wide association study of age-related macular degeneration highlights contributions of rare and common variants. Nat Genet 48(2):134–143
6. Fleckenstein M et al (2021) Age-related macular degeneration. Nat Rev Dis Primers 7(1):31
7. van Asten F et al (2018) A deep phenotype association study reveals specific phenotype associations with genetic variants in age-related macular degeneration: age-related eye disease study 2 (AREDS2) report no. 14. Ophthalmology 125(4):559–568
8. Seddon JM et al (1996) A prospective study of cigarette smoking and age-related macular degeneration in women. JAMA 276(14):1141–1146
9. Christen WG et al (1996) A prospective study of cigarette smoking and risk of age-related macular degeneration in men. JAMA 276(14):1147–1151
10. Merle BMJ et al (2019) Mediterranean diet and incidence of advanced age-related macular degeneration: the EYE-RISK consortium. Ophthalmology 126(3):381–390
11. McGuinness MB et al (2017) Physical activity and age-related macular degeneration: a systematic literature review and meta-analysis. Am J Ophthalmol 180:29–38
12. Ramrattan RS et al (1994) Morphometric analysis of Bruch's membrane, the choriocapillaris, and the choroid in aging. Invest Ophthalmol Vis Sci 35(6):2857–2864
13. Mullins RF et al (2014) The membrane attack complex in aging human choriocapillaris: relationship to

macular degeneration and choroidal thinning. Am J Pathol 184(11):3142–3153

14. Zheng F et al (2019) Age-dependent changes in the macular choriocapillaris of normal eyes imaged with swept-source optical coherence tomography angiography. Am J Ophthalmol 200:110–122

15. Curcio CA et al (2001) Accumulation of cholesterol with age in human Bruch's membrane. Invest Ophthalmol Vis Sci 42(1):265–274

16. Friedman E (1997) A hemodynamic model of the pathogenesis of age-related macular degeneration. Am J Ophthalmol 124(5):677–682

17. Yamada Y et al (2008) Oxidized low density lipoproteins induce a pathologic response by retinal pigmented epithelial cells. J Neurochem 105(4):1187–1197

18. Curcio CA et al (2011) The oil spill in ageing Bruch membrane. Br J Ophthalmol 95(12):1638–1645

19. Ouyang Y et al (2013) Optical coherence tomography-based observation of the natural history of drusenoid lesion in eyes with dry age-related macular degeneration. Ophthalmology 120(12):2656–2665

20. Schlanitz FG et al (2017) Drusen volume development over time and its relevance to the course of age-related macular degeneration. Br J Ophthalmol 101(2):198–203

21. Khan KN et al (2016) Differentiating drusen: drusen and drusen-like appearances associated with ageing, age-related macular degeneration, inherited eye disease and other pathological processes. Prog Retin Eye Res 53:70–106

22. Lakkaraju A et al (2020) The cell biology of the retinal pigment epithelium. Prog Retin Eye Res 78:100846

23. Sparrow JR et al (2003) A2E, a byproduct of the visual cycle. Vis Res 43(28):2983–2990

24. Zhang Q et al (2019) Highly differentiated human fetal RPE cultures are resistant to the accumulation and toxicity of lipofuscin-like material. Invest Ophthalmol Vis Sci 60(10):3468–3479

25. Curcio CA (2001) Photoreceptor topography in ageing and age-related maculopathy. Eye (Lond) 15(Pt 3):376–383

26. Holz FG, Schmitz-Valckenberg S, Fleckenstein M (2014) Recent developments in the treatment of age-related macular degeneration. J Clin Invest 124(4):1430–1438

27. Jabbehdari S, Handa JT (2021) Oxidative stress as a therapeutic target for the prevention and treatment of early age-related macular degeneration. Surv Ophthalmol 66(3):423–440

28. Seddon JM (2017) Macular degeneration epidemiology: nature-nurture, lifestyle factors, genetic risk, and gene-environment interactions – the Weisenfeld Award Lecture. Invest Ophthalmol Vis Sci 58(14):6513–6528

29. A randomized, placebo-controlled, clinical trial of high-dose supplementation with vitamins C and E, beta carotene, and zinc for age-related macular degeneration and vision loss: AREDS report no. 8. Arch Ophthalmol, 2001. 119(10): 1417–1436

30. Chew EY et al (2012) The age-related eye disease study 2 (AREDS2): study design and baseline characteristics (AREDS2 report number 1). Ophthalmology 119(11):2282–2289

31. Chen M, Xu H (2015) Parainflammation, chronic inflammation, and age-related macular degeneration. J Leukoc Biol 98(5):713–725

32. Rozing MP et al (2020) Age-related macular degeneration: a two-level model hypothesis. Prog Retin Eye Res 76:100825

33. Maller J et al (2006) Common variation in three genes, including a noncoding variant in CFH, strongly influences risk of age-related macular degeneration. Nat Genet 38(9):1055–1059

34. Gold B et al (2006) Variation in factor B (BF) and complement component 2 (C2) genes is associated with age-related macular degeneration. Nat Genet 38(4):458–462

35. Maller JB et al (2007) Variation in complement factor 3 is associated with risk of age-related macular degeneration. Nat Genet 39(10):1200–1201

36. Skerka C et al (2007) Defective complement control of factor H (Y402H) and FHL-1 in age-related macular degeneration. Mol Immunol 44(13):3398–3406

37. Borras C et al (2019) CFH exerts anti-oxidant effects on retinal pigment epithelial cells independently from protecting against membrane attack complex. Sci Rep 9(1):13873

38. Armento A et al (2020) Loss of complement factor H impairs antioxidant capacity and energy metabolism of human RPE cells. Sci Rep 10(1):10320

39. Fleckenstein M et al (2018) The progression of geographic atrophy secondary to age-related macular degeneration. Ophthalmology 125(3):369–390

40. Campochiaro PA et al (1999) The pathogenesis of choroidal neovascularization in patients with age-related macular degeneration. Mol Vis 5:34

41. Ferris FL et al (2005) A simplified severity scale for age-related macular degeneration: AREDS Report No. 18. Arch Ophthalmol 123(11):1570–1574

42. Sadda SR et al (2018) Consensus definition for atrophy associated with age-related macular degeneration on OCT: classification of atrophy report 3. Ophthalmology 125(4):537–548

43. Spaide RF et al (2020) Consensus nomenclature for reporting neovascular age-related macular degeneration data: consensus on neovascular age-related macular degeneration nomenclature study group. Ophthalmology 127(5):616–636

44. Sunness JS et al (1999) The development of choroidal neovascularization in eyes with the geographic atrophy form of age-related macular degeneration. Ophthalmology 106(5):910–919

45. Bressler SB (2009) Introduction: understanding the role of angiogenesis and antiangiogenic agents in age-related macular degeneration. Ophthalmology 116(10 Suppl):S1–S7

46. Wang H et al (2011) Upregulation of CCR3 by age-related stresses promotes choroidal endothelial cell

migration via VEGF-dependent and -independent signaling. Invest Ophthalmol Vis Sci 52(11):8271–8277

47. Pfau M et al (2020) Type 1 choroidal neovascularization is associated with reduced localized progression of atrophy in age-related macular degeneration. Ophthalmol Retina 4(3):238–248

48. Christenbury JG et al (2018) Progression of macular atrophy in eyes with type 1 neovascularization and age-related macular degeneration receiving long-term intravitreal anti-vascular endothelial growth factor therapy: an optical coherence tomographic angiography analysis. Retina 38(7):1276–1288

49. Metelitsina TI et al (2008) Foveolar choroidal circulation and choroidal neovascularization in age-related macular degeneration. Invest Ophthalmol Vis Sci 49(1):358–363

50. Witmer AN et al (2003) Vascular endothelial growth factors and angiogenesis in eye disease. Prog Retin Eye Res 22(1):1–29

51. Famiglietti EV et al (2003) Immunocytochemical localization of vascular endothelial growth factor in neurons and glial cells of human retina. Brain Res 969(1–2):195–204

52. Wang H, Hartnett ME (2016) Regulation of signaling events involved in the pathophysiology of neovascular AMD. Mol Vis 22:189–202

53. Abhinand CS et al (2016) VEGF-A/VEGFR2 signaling network in endothelial cells relevant to angiogenesis. J Cell Commun Signal 10(4):347–354

54. Ng DS et al (2017) Elevated angiopoietin 2 in aqueous of patients with neovascular age related macular degeneration correlates with disease severity at presentation. Sci Rep 7:45081

55. Peters S et al (2007) Angiopoietin modulation of vascular endothelial growth factor: effects on retinal endothelial cell permeability. Cytokine 40(2):144–150

56. Joussen AM et al (2021) Angiopoietin/Tie2 signalling and its role in retinal and choroidal vascular diseases: a review of preclinical data. Eye (Lond) 35(5):1305–1316

57. Brown DM et al (2009) Ranibizumab versus verteporfin photodynamic therapy for neovascular age-related macular degeneration: two-year results of the ANCHOR study. Ophthalmology 116(1):57–65.e5

58. Miller JW (2016) VEGF: from discovery to therapy: the Champalimaud Award Lecture. Transl Vis Sci Technol 5(2):9

59. Bloch SB, Larsen M, Munch IC (2012) Incidence of legal blindness from age-related macular degeneration in Denmark: year 2000 to 2010. Am J Ophthalmol 153(2):209–213.e2

60. Vinores SA (2006) Pegaptanib in the treatment of wet, age-related macular degeneration. Int J Nanomedicine 1(3):263–268

61. Rosenfeld PJ et al (2006) Ranibizumab for neovascular age-related macular degeneration. N Engl J Med 355(14):1419–1431

62. Brown DM et al (2006) Ranibizumab versus verteporfin for neovascular age-related macular degeneration. N Engl J Med 355(14):1432–1444

63. Martin DF et al (2011) Ranibizumab and bevacizumab for neovascular age-related macular degeneration. N Engl J Med 364(20):1897–1908

64. Heier JS et al (2012) Intravitreal aflibercept (VEGF trap-eye) in wet age-related macular degeneration. Ophthalmology 119(12):2537–2548

65. Dugel PU et al (2020) HAWK and HARRIER: phase 3, multicenter, randomized, double-masked trials of brolucizumab for neovascular age-related macular degeneration. Ophthalmology 127(1):72–84

66. Dugel PU et al (2021) HAWK and HARRIER: ninety-six-week outcomes from the phase 3 trials of brolucizumab for neovascular age-related macular degeneration. Ophthalmology 128(1):89–99

67. Monés J et al (2021) Risk of inflammation, retinal vasculitis, and retinal occlusion-related events with brolucizumab: post hoc review of HAWK and HARRIER. Ophthalmology 128(7):1050–1059

68. Nicolò M et al (2021) Faricimab: an investigational agent targeting the Tie-2/angiopoietin pathway and VEGF-A for the treatment of retinal diseases. Expert Opin Investig Drugs 30(3):193–200

69. Holz FG et al (2015) Multi-country real-life experience of anti-vascular endothelial growth factor therapy for wet age-related macular degeneration. Br J Ophthalmol 99(2):220–226

70. Holekamp NM et al (2022) Archway randomized phase 3 trial of the port delivery system with ranibizumab for neovascular age-related macular degeneration. Ophthalmology 129(3):295–307

71. Ranade SV et al (2022) The port delivery system with ranibizumab: a new paradigm for long-acting retinal drug delivery. Drug Deliv 29(1):1326–1334

72. Ammar MJ et al (2020) Age-related macular degeneration therapy: a review. Curr Opin Ophthalmol 31(3):215–221

Glaucoma

Samuel D. Crish and Gina N. Wilson

Introduction

The term "glaucoma" broadly categorizes several optic neuropathies. Together, this group of disorders make up the leading cause of blindness worldwide, affecting approximately 80 million people [1]. Glaucoma diagnoses are predicted to rise to almost 112 million people globally by the year 2040 [2]. Primarily, these neuropathies are characterized by a progressive vision loss due to dysfunction and degeneration of retinal ganglion cells (RGC) and their axons [3–12].

Glaucoma is usually associated with elevated intraocular pressure (IOP). This is for good reason, as elevated IOP is the main modifiable risk factor for glaucoma and the sole therapeutic target [13–15]. Treatments (i.e., medications or surgeries) that lower IOP slow the progression of vision loss in most, but not all patients. However, these treatments cannot reverse pathology or ultimately halt the underlying mechanisms driving further disease progression.

Elevated IOP selectively affects RGCs, the projection neurons of the eye. These neurons extend unmyelinated axons along the surface of the retina until they reach the optic disc, where these fibers exit the eye and form the optic nerve. After leaving the retina, RGC axons remain unmyelinated, for a distance of about 1 mm, until they exit the ocular cavity where nearly all of them become myelinated (>99%). This initial, unmyelinated portion is known as the optic nerve head (ONH) and it is a chief site of glaucoma pathogenesis.

Major mechanisms by which elevated IOP can damage RGCs include compression of RGC axons at the ONH, vascular insufficiency, and direct effects on neurons through mechanically gated channels [16]. However, the relationship between IOP and glaucomatous pathology is complicated. A significant proportion of glaucomatous vision loss occurs in the absence of elevated IOP ("normal-tension" glaucoma), and most people with high IOP never develop glaucoma [17]. Despite this, reducing IOP in patients with normal-tension glaucoma can still slow disease progression. Although there are numerous underlying factors that can interact and contribute to pathology, it is clear that IOP plays an important role in glaucoma, and systemic metabolic factors can have an influence here.

While glaucoma is characterized by the degeneration of the retina and its projection, problems often begin in the front of the eye,

S. D. Crish (✉)
Department of Pharmaceutical Sciences, Northeast Ohio Medical University, Rootstown, OH, USA
e-mail: scrish@neomed.edu

G. N. Wilson
Department of Neuroscience, Rockefeller Neuroscience Institute, West Virginia University, Morgantown, WV, USA
e-mail: gina.wilson@hsc.wvu.edu

E. Lammert, M. Zeeb (eds.), *Metabolism of Human Diseases*,
https://doi.org/10.1007/978-3-031-96019-2_11

93

where fluid balance influences IOP. Glaucoma can be divided into two main types: open-angle and closed-angle. The distinction between them is the morphology of the angle between the iris and cornea: the iridocorneal angle, where aqueous humor (AH) drains out of the anterior chamber into venous circulation or into other parts of the eye (see chapter "Eye: Overview" under part "Eye"). Primary open-angle glaucoma (POAG), where the angle appears normal, is the most common type, comprising 90% of diagnosed glaucomas [18]. Whether due to overproduction or impaired drainage, AH builds up in the eye, moderately raising intraocular pressure. Largely asymptomatic, the disease may progress for years until a glaucoma diagnosis is made. Closed-angle glaucoma, where the angle is substantially reduced, can share those features. However, in acute closed-angle glaucoma, drainage is suddenly blocked and produces a large spike in IOP. Acute closed-angle glaucoma is a medical emergency, producing intense ocular pain, headache, abnormal vision, and nausea. IOP must promptly be reduced through medication or surgery before irreversible damage occurs.

Glaucoma has a genetic component, with family history and ethnicity increasing its likelihood. For example, African-Americans are up to five times more likely to develop POAG—and at an earlier age—than their Caucasian counterparts and are 15 times more likely to suffer complete vision loss from the disease [18, 19]. Recent studies have also shown that males are more often diagnosed with glaucoma than females [19]. As with other age-related, chronic neurodegenerations, glaucoma genetics are multifactorial, involving interactions between multiple alleles associated with various risks and environmental factors [19–21].

Within the retina, interrelated mechanisms common to other neurodegenerative conditions such as oxidative stress, calcium dysregulation, hemodynamic changes, metabolic defects, intracellular transport deficits, gliosis, and neuroinflammation are implicated in the pathogenesis and progression of glaucomatous pathology [22]. These disturbances result in several defects in retinal ganglion cells: physiological deficits in neuronal signaling, abnormal axonal transport, and degeneration of axons and dendrites, all eventually leading to apoptotic cell death. To date, the mechanisms underlying these changes are not well understood.

Pathophysiology of Glaucoma and Metabolic Alterations

The retina and optic nerve are metabolically demanding, so an association between glaucoma and systemic metabolic conditions has a rich history [23; reviewed in 17]. Most notably, metabolic syndrome factors such as hypertension, insulin resistance, and obesity contribute to elevated intraocular pressure (IOP) and impaired energy supply of the retina and optic nerve. This may also explain the association between diabetes and increased IOP, a potential risk factor for glaucoma (Fig. 1).

Cardiovascular Factors

Metabolic dysfunction is often coupled to cardiovascular indications such as abnormal blood pressure. Mean arterial blood pressure outside of a normal range, whether hypertensive or hypotensive, is a risk factor for glaucoma [24, 25]. Systemic hypertension increases ciliary capillary pressure and can increase aqueous humor (AH) filtration through the ciliary body of the eye [38]. In a mechanism like that described for fat deposits and blood viscosity (see below), increased blood volume associated with elevated systemic blood pressure can increase episcleral and ciliary vein resistance [43]. This resistance opposes flow through the drainage canals and ultimately increases intraocular pressure (IOP) [44]. Further, hypertension-mediated vascular damage also impairs blood perfusion.

Other studies have associated low blood pressure with increased likelihood of developing glaucoma [44, 45]. Hypotension, whether due to systemic factors or antihypertensive therapy, can reduce ocular perfusion pressure (OPP; estimated as the difference between arterial pressure and

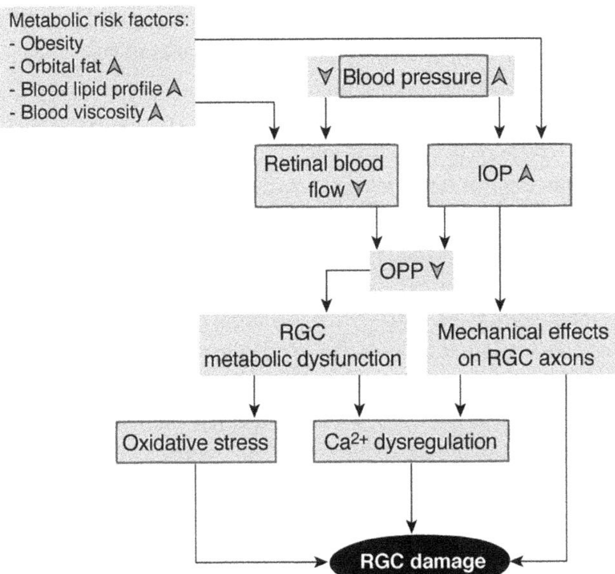

Fig. 1 Relationship between conditions associated with metabolic syndrome, diabetes, and glaucoma. Different factors act to (1) increase intraocular pressure (IOP) or (2) reduce blood flow to the retina (due to either low blood pressure or microvascular changes). Either of these effects will reduce ocular perfusion pressure (OPP), leading to a decrease in retinal blood supply. Ultimately, this impairs retinal function and leads to retinal ganglion cell (RGC) damage and degeneration through specific (i.e., Ca^{2+}-activated degenerative cascades) and nonspecific (i.e., oxidative stress-induced damage) mechanisms. Another major factor in glaucoma is age. Terms highlighted in framed boxes change with age in a way that age increases the prevalence and severity of the disease

IOP). Reductions of OPP, whether due to systemic hypotension or elevated IOP, reduces blood flow to the retina [26]. As the retina is one of the most metabolically demanding tissues in the body, even relatively subtle reductions in oxygen and nutrient supply may impair retinal function and produce increased oxidative stress, calcium dysregulation, abnormal neuronal signaling, and other damaging stressors.

Systemic Metabolic Factors

Obese individuals may have fat deposits around the eye and increased blood viscosity due to elevated levels of hematocrit, lipids, fibrinogen, and immunoglobulins, as well as changes in red blood cell rigidity. Both fat deposits and high blood viscosity can increase resistance in the episcleral vein, which in turn impedes aqueous humor (AH) outflow and increases IOP [23, 36].

There are conflicting reports on the relationship between diabetes and primary open-angle glaucoma (POAG) [27–31]. While diabetic retinopathies that affect retinal ganglion cells (RGCs) are well described, reasons for the increased incidence of glaucoma in diabetic patients is unclear [24]. Diabetes may act as a risk factor through effects on IOP and vascular changes [23, 27, 28, 30]. Further, abnormal blood vessel growth in the trabecular outflow pathways can lead to closed-angle neovascular glaucoma.

Insulin resistance, a key characteristic of type 2 diabetes (see chapter "Diabetes Mellitus"), and elevated blood glucose concentrations may alter the osmotic gradient to cause fluid accumulation in the intraocular space and lead to elevated IOP [30]. Notably, pharmacological blockade of insulin receptors in rodent models result in significantly increased IOP and loss of RGCs [37].

The association between aberrant insulin signaling and glaucoma is strong [37, 38]. While insulin is most associated with its systemic role in maintenance of blood glucose levels, recent work has shown that insulin is also synthesized within neurons and that central insulin signaling is critical for general neuronal survival. Research has suggested that preservation or restoration of insulin signaling within the central nervous system (CNS) may promote RGC survival and be therapeutically targeted to preserve vision [39, 40].

Insulin signaling, mediated by a G-protein coupled insulin receptor, is carried out primarily through the phospho-inositol 3 kinase (PI3K)/AKT pathway. AKT signaling is associated with anti-apoptotic effects and the downregulation or inactivation of other proteins such as Bad or glycogen synthase kinase 3β (GSK3β) [39–41]. GSK3β, if overactivated (i.e., due to insulin resistance), can lead to increases in hyperphosphorylated tau—a component of the hallmark neurofibrillary tangles observed in Alzheimer's disease (see chapter "Alzheimer's Disease"), but recently shown to be present in glaucomatous optic projections as well [9]. GSK3β may also activate calpain via the p25 pathway, which can lead to dysregulated calcium signaling and excitotoxicity [39, 40]. Insulin signaling via the PI3K/AKT pathway has been associated with stimulation of mitochondrial biogenesis and function, while mitochondrial function is reduced under insulin resistant conditions [39, 41].

While the breadth of evidence linking dysfunctional insulin signaling and/or diabetes with increased risk of glaucoma is striking, the interaction of insulin levels and activity between systemic circulation and the CNS is still unknown. Some studies report lower concentrations in the CNS compared with plasma, while other report higher. Additionally, studies have demonstrated that increases in systemic insulin concentrations do not correspond to proportional increases in insulin concentrations within the CNS [42]. More work is needed to parse these remaining questions. However, it is clear that aberrations in glucose and insulin signaling are not solely a systemic issue, but also an issue of the CNS that impacts retinal function and may partially explain the comorbidity between metabolic disorders (e.g., diabetes) and glaucoma.

Glaucoma and Dietary Considerations

As with many neurodegenerative disorders, recent research has indicated that diet may play a role in glaucoma risk, development, and progression. This is not surprising given the dietary risk factors for diabetes mellitus and the proposed comorbidity between diabetes and glaucoma. For reference, one study found the prevalence of diabetes in those diagnosed with primary open-angle glaucoma (POAG) to be approximately 20% of glaucoma patients 60 years of age or older [46]. While chronically high or uncontrolled blood glucose levels are associated with increased risk of diabetes and ocular dysfunction, sugar intake and carbohydrate metabolism is not the only dietary consideration that can impact risk, development, or progression of glaucoma. Many elements of our diet may be manipulated to positively or negatively impact disease outcomes.

Polyphenols have long been praised for their antioxidant properties and ability to modulate markers of neurodegenerative disease [46]. Resveratrol has been shown to reduce oxidative stress markers within the trabecular meshwork and lower intraocular pressure (IOP) [47, 48]. Additionally, in experimental models, resveratrol treatment was shown to promote retinal ganglion cell (RGC) survival and reduce inflammation [49, 50]. Not surprisingly, given the intricate relationship between metabolic dysfunction and glaucoma, polyphenols such as resveratrol have demonstrated positive efficacy in modulating blood glucose levels, improving blood lipid profiles, and reducing hypertension associated with diabetes [51]. However, mechanisms by which these benefits are conferred remain unclear.

Niacin is another important nutrient that investigators have observed is deficient in patients with POAG, specifically normotensive cases [52]. Niacin is a critical dietary component of nicotinamide adenine dinucleotide (NAD) and nicotinamide adenine dinucleotide phosphate (NADP)—coenzymes involved in hundreds of metabolic reactions [53]. The NAD redox couple participates in approximately 400 catabolic and

30 anabolic reactions of metabolism necessary for glycolysis, the Krebs cycle, and many more. Niacin and its NAD-related counterparts also aid in DNA repair, cell signaling, and antioxidant defenses [53, 54]. It is no surprise, therefore, that deficient niacin supply leads to multifarious health problems throughout the body. With diminished capacity of NAD and NADP to function in their role of converting nutrients to oxygen and adenosine triphosphate (ATP), metabolically active tissues such as the retina become energy deficient. Retinal ganglion cells become unable to efficiently carry out critical functions such as cellular respiration and axonal transport. Like other neurodegenerative diseases, an early feature of glaucoma is a reduction in axonal transport—an event that initiates prior to overt structural loss or visual deficits [55–57].

Other dietary factors have been studied in the context of glaucomatous neurodegeneration. High fat diets have been shown to improve outcome measures of RGC structure and function in rodent models of glaucoma within the laboratory, altering normal respiratory pathways by pushing cells into β-oxidation [57]. While there are reports on the benefits of ketone bodies in neurodegenerative disease, it has been posited that a ketogenic diet in humans may offer neuroprotection in glaucoma [58]. Although the long-term safety and efficacy is debated, ketogenic diets have also shown some success in improving metabolic control in patients with obesity or type 2 diabetes [59].

Treatment of Glaucoma

Although the dysfunction and degeneration of retinal ganglion cells and their projections are what cause blindness in glaucoma, existing therapies only target lowering intraocular pressure (IOP) through medication or surgery [60]. Currently, first-line treatment for glaucoma is the topical use of eyedrops to reduce IOP [61]. There are five main classes of IOP-lowering drugs, all of which either reduce the production of aqueous humor (AH), by reducing activity of the ciliary body, or increase AH outflow via action on the drainage canals in the iridocorneal angle (refer to Table 1). As noted above, lowering IOP often slows progression of vision loss even in normotensive patients [62].

AH production can be reduced with drugs targeting the adrenergic system, antagonizing β-adrenergic or agonizing α-adrenergic receptors (e.g., timolol or brimonidine, respectively), or with carbonic anhydrase inhibitors (e.g., dorzolamide), presumably by slowing the formation of bicarbonate ions with subsequent reduction in Na^+ and fluid transport. Alternatively, AH outflow can be increased by prostaglandin analogs (e.g., latanoprost, a prostaglandin F analog), α-adrenergic agonists, or parasympathomimetics that activate the cholinergic system (e.g., carbachol). The most recent drug approved by the Food and Drug Administration (FDA) for treatment of glaucoma was a prostaglandin EP2 receptor agonist, Omlonti [63].

Patients who are unresponsive to these medications or have other issues can undergo surgery to physically increase AH outflow by creating new drainage canals in the eye. While many of the surgical options are laser-based trabeculoplasties or iridotomies, incisional surgeries as well as implantation of drainage devices (e.g., shunts or cannulae) are additional options for those with more severe progression or who are resistant to less invasive techniques.

Influence of Treatment on Metabolism

Pharmacological treatment of glaucoma typically does not affect systemic metabolism to a large degree. The most common side effects impact the eye itself and may include irritation or discomfort due to corneal dryness, inflammation, or allergic reaction. However, some of the topical drugs used to treat glaucoma can cause systemic effects upon entering the circulation through the nasal mucosa by way of the lacrimal ducts (Fig. 2).

Drugs targeting β-adrenergic receptors can decrease systemic blood pressure and pulse rate (see chapter "Heart Failure"), as well as cause bronchospasm and negatively affect blood

Table 1 Classes and examples of the most important glaucoma treatments

Drug class	Examples	AH production	AH outflow
Prostoglandin analogs	Latanoprost, travoprost		↑
β-Blockers	Timolol, carteolol	↓	
α-Agonists	Brimonidine, apraclonidine	↓	↑
Carbonic anhydrase inhibitors	Dorzolamide, brinzolamide	↓	
Cholinergic agonists	Pilocarpine, carbachol		↑

Note: *AH* aqueous humor

glucose levels and lipid profile, the latter through impairing low-density lipoprotein (LDL) and triglyceride metabolism by inhibiting lipoprotein lipase (LPL, see chapter "Hyperlipidemia") [61]. Conflicting research on the safety of topical β-adrenergic blockers exist: several studies have found no increase in cardiovascular events [63] or respiratory issues in elderly populations [64], but others found an increased risk of developing airway obstruction [65]. Fortunately, multiple treatment options within this compound class have been developed that are selective for β1- or β2-adrenergic receptor subtypes, which differentially affect heart and lung tissue and can accordingly attenuate risks for those with cardiovascular or respiratory contraindications [66]. Unlike oral or injectable medications, the majority of glaucoma medications have the advantage of circumventing metabolic conversion and breakdown because they are applied directly to their target (i.e., the eye). Consequently, effective therapeutic doses can be very low, which minimizes the amount that may enter systemic circulation via lacrimal ducts and subsequently minimizes the risk of any off-target effects.

Perspectives

Until recently, most work has focused on the connection between metabolism and elevated intraocular pressure (IOP), a risk factor for glaucoma, rather than glaucoma itself. However, within the past decade, the specific relationship between metabolism and glaucoma has received more attention as the interactions between metabolic dysfunction and the pathogenesis of glaucoma, apart from IOP, have been recognized. Given the variety of glaucomas, the variability in disease progression, and the overlap of risk factors between glaucoma and systemic metabolic diseases, it is likely that several metabolic factors interact to impact the development or progression of glaucoma.

An area of great interest in glaucoma research is the development of a "complete therapy," where mechanisms of neurodegeneration are addressed in addition to lowering IOP [60]. At present, there are no dedicated neuroprotective agents used for the treatment of glaucoma (although some of the existing IOP-lowering drugs such as brimonidine are thought to have neuroprotective effects). Researchers are currently exploring several avenues—some of which could potentially have effects on metabolism—such as drugs affecting hemodynamics, mitochondrial function, or Ca^{2+} dynamics. Finally, further exploration of lifestyle factors and how they may provide neuroprotection against the disease or otherwise interact with underlying mechanisms to slow, halt, or reverse pathological progression offers an exciting opportunity to modify disease development and outcomes through changes in diet and exercise. With a projected increase in glaucoma prevalence, this multifaceted approach toward therapy represents one of the best hopes for reducing disease burden in the future.

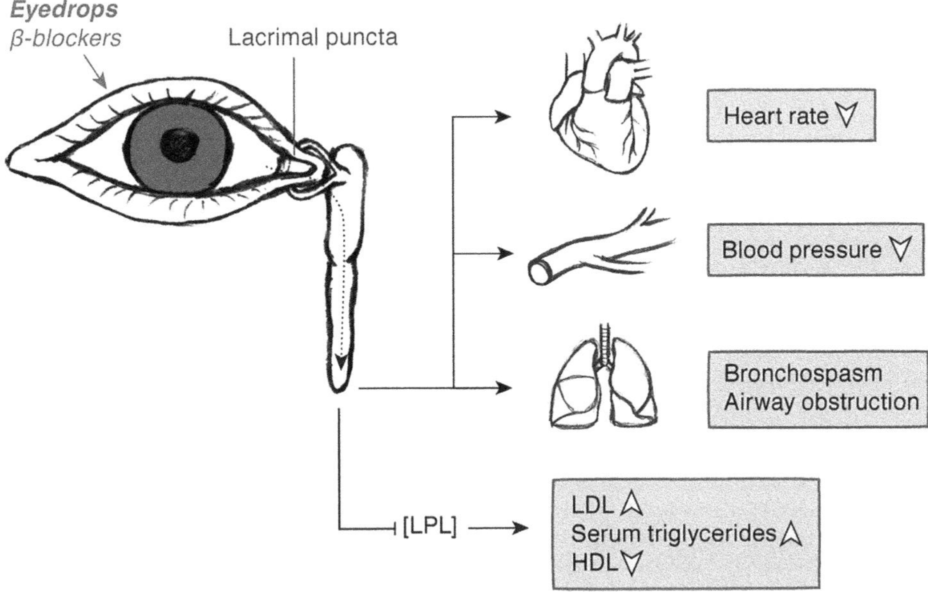

Fig. 2 Effects of topical β-blocker treatment (eyedrops) on systemic function. Eyedrops travel through the lacrimal puncta and enter systemic circulation through the nasolacrimal duct. Common intraocular pressure (IOP)-lowering drugs that inhibit the adrenergic system can decrease heart rate, lower blood pressure, impair respiratory function, and produce unfavorable blood lipid profiles through inhibition of lipoprotein lipase (LPL) in heart, muscle, and adipose tissue. *LDL* low-density lipoprotein, *HDL* high-density lipoprotein

Questions and Answers

Question 1 How can hypotension influence glaucomatous pathology?

Answer 1 Hypotension can reduce blood flow to the retina.

Question 2 Which physiological risk factor is targeted in glaucoma therapies?

Answer 2 Intraocular pressure (IOP).

Question 3 How can topical glaucoma medications enter systemic circulation?

Answer 3 Through the nasal mucosa via the lacrimal ducts.

References

1. Quigley HA, Broman AT (2006) The number of people with glaucoma worldwide in 2010 and 2020. Br J Ophthalmol 90:262–267
2. Tham YC, Li X, Wong TY, Quigley HA, Aung T, Cheng CY (2014) Global prevalence of glaucoma and projections of glaucoma burden through 2040: a systematic review and meta-analysis. Ophthalmology 121(11):2081–1090
3. Quigley HA, Hohman RM, Addicks EM, Massof RW, Green WR (1983) Morphologic changes in the lamina cribrosa correlated with neural loss in open-angle glaucoma. Am J Ophthalmol 95:673–691
4. Quigley HA, Dunkelberger GR, Green WR (1988) Chronic human glaucoma causes selectively greater loss of large optic nerve fibers. Ophthalmology 95:357–363
5. Quigley HA, Nickells RW, Kerrigan LA, Pease ME, Thibault DJ, Zack DJ (1995) Retinal ganglion cell death in experimental glaucoma and after axotomy occurs by apoptosis. Invest Ophthalmol Vis Sci 36:774–786
6. Kerrigan LA, Zack DJ, Quigley HA, Smith SD, Pease ME (1997) TUNEL-positive ganglion cells in human primary open angle glaucoma. Arch Ophthalmol (Chicago, Ill. 1960) 115:1031–1035

7. Howell GR, Libby RT, Jakobs TC, Smith RS, Phalan FC, Barter JW, Barbay JM, Marchant JK, Mahesh N, Porciatti V, Whitmore AV, Masland RH, John SWM (2007) Axons of retinal ganglion cells are insulted in the optic nerve early in DBA/2J glaucoma. J Cell Biol 179:1523–1537

8. Crish SD, Sappington RM, Inman DM, Horner PJ, Calkins DJ (2010) Distal axonopathy with structural persistence in glaucomatous neurodegeneration. Proc Natl Acad Sci 107(11):5196–5201

9. Wilson GN, Smith MA, Inman DM, Dengler-Crish CM, Crish SD (2016) Early cytoskeletal protein modifications precede overt structural degeneration in the DBA/2J mouse model of glaucoma. Front Neurosci 10:494. https://doi.org/10.3389/fnins.2016.00494

10. Risner ML, Pasini S, Cooper ML, Lambert WS, Calkins DJ (2018) Axogenic mechanism enhances retinal ganglion cell excitability during early progression in glaucoma. Proc Natl Acad Sci 115:E2393–E2402

11. Baltan S, Inman DM, Danilov CA, Morrison RS, Calkins DJ, Horner PJ (2010) Metabolic vulnerability disposes retinal ganglion cell axons to dysfunction in a model of glaucomatous degeneration. J Neurosci 30:5644–5652

12. Smith MA, Plyler ES, Dengler-Crish CM, Meier J, Crish SD (2018) Nodes of ranvier in glaucoma. J Neurosci 390:104–118

13. Gordon MO, Beiser JA, Brandt JD, Heuer DK, Higginbotham EJ, Johnson CA, Keltner JL, Miller JP, Parrish RK 2nd, Wilson MR, Kaas MA (2002) The ocular hypertension treatment study: baseline factors that predict the onset of primary open-angle glaucoma. Arch Ophthalmol 120:714–720

14. Guymer C, Wood JP, Chidlow G, Casson RJ (2019) Neuroprotection in glaucoma: recent advances and clinical translation. Clin Experiment Ophthalmol 1:88–105

15. Casson RJ (2022) Medical therapy for glaucoma: a review. Clin Experiment Ophthalmol 50(2):198–212

16. Flammer J, Orgül S, Costa VP, Orzalesi N, Krieglstein GK, Serra LM, Renard JP, Stefánsson E (2002) The impact of ocular blood flow in glaucoma. Prog Retin Eye Res 21(4):359–393

17. Killer H, Pircher A (2018) Normal tension glaucoma: review of current understanding and mechanisms of the pathogenesis. Eye 32:924–930

18. Shields MB (2008) Normal-tension glaucoma: is it different from primary open-angle glaucoma? Curr Opin Ophthalmol 19(2):85–88

19. Wiggs JL (2007) Genetic etiologies of glaucoma. Arch Ophthalmol 125(1):30–37

20. Gudiseva HV, Verma SS, Chavali VR, Salowe RJ, Lucas A, Collins DW, Rathi S, He J, Lee R, Merriam S, Bowman AS (2020) Genome wide-association study identifies novel loci in the primary open-angle African American glaucoma genetics (POAAGG) study. BioRxiv. https://doi.org/10.1101/2020.02.27.968156

21. Fuse N (2010) Genetic bases for glaucoma. Tohoku J Exp Med 221(1):1–10

22. Crish SD, Calkins DJ (2011) Neurodegeneration in glaucoma: progression and calcium-dependent intracellular mechanisms. Neuroscience 176:1–11

23. Marshall H, Mullany S, Qassim A, Siggs O, Hassall M, Ridge B, Nguyen T, Awadalla M, Andrew NH, Healey PR, Agar A, Galanopoulos A, Hewitt AW, MacGregor S, Graham SL, Mills R, Shulz A, Landers J, Casson RJ, Craig JE (2021) Cardiovascular disease predicts structural and functional progression in early glaucoma. Ophthalmology 1:58–69

24. Memarzadeh F, Ying-Lai M, Chung J, Azen SP, Varma R, Los Angeles Latino Eye Study Group (2010) Blood pressure, perfusion pressure, and open-angle glaucoma: the Los Angeles Latino eye study. Invest Ophthalmol Vis Sci 51:2872e2877

25. Cherecheanu AP, Garhofer G, Schmidl D, Werkmeister R, Schmetterer L (2013) Ocular perfusion pressure and ocular blood flow in glaucoma. Curr Opin Pharmacol 1:36–42

26. Ellis JD, Evans JM, Ruta DA, Baines PS, Leese G, MacDonald TM, Morris AD (2000) Glaucoma incidence in an unselected cohort of diabetic patients: is diabetes mellitus a risk factor for glaucoma? DARTS/MEMO collaboration. Diabetes audit and research in tayside study. Medicines monitoring unit. Br J Ophthalmol 84(11):1218–1224

27. Fang EN, Law SK, Walt JG, Chiang TH, Williams EN (2008 Feb) The prevalence of glaucomatous risk factors in patients from a managed care setting: a pilot evaluation. Am J Manag Care 14(1 Suppl):S28–S36

28. Marshall H, Mullany S, Qassim A, Siggs O, Hassall M, Ridge B, Nguyen T, Awadalla M, Andrew NH, Healey PR, Agar A, Galanopoulos A, Hewitt AW, MacGregor S, Graham SL, Mills R, Shulz A, Landers J, Casson RJ, Craig JE (2021) Cardiovascular disease predicts structural and functional progression in early glaucoma. Ophthalmology 128(1):58–69

29. Mapstone R, Clark CV (1985) Prevalence of diabetes in glaucoma. BMJ 291:93–95

30. Senthil S, Dada T, Das T, Kaushik S, Puthuran GV, Philip R, Rani PK, Rao H, Singla S, Vijaya L (2021) Neovascular glaucoma – a review. Indian J Ophthalmol 69(3):525–534

31. Zhao D, Cho J, Kim MH, Friedman DS, Guallar E (2015) Diabetes, fasting glucose, and the risk of glaucoma: a meta-analysis. Ophthalmology 122(1):72–78

32. Guiraudou M, Varlet-Marie E, Raynaud de Mauverger E, Brun JF (2013) Obesity-related increase in whole blood viscosity includes different profiles according to fat localization. Clin Hemorheol Microcirc 55:63–73

33. Newman-Casey PA, Talwar N, Nan B, Much DC, Stein JD (2011) The relationship between components of metabolic syndrome and open-angle glaucoma. Am Acad Ophthalmol 118:1318–1326

34. Bulpitt CJ, Hodes C, Everitt MG (1975) IOP and systemic blood pressure in the elderly. Br J Ophthalmol 59:717–720

35. Jonas JB, Grundler AE (1998) Prevalence of diabetes mellitus and arterial hypertension in primary and sec-

ondary open-angle glaucomas. Graefes Arch Clin Exp Ophthalmol 236:202–206

36. Al Hussein Al Awamlh S, Wareham LK, Risner ML, Calkins DJ (2021) Insulin signaling as a therapeutic target in glaucomatous neurodegeneration. Int J Mol Sci 22(9):4672

37. Dada T (2017) Is glaucoma a neurodegeneration caused by central insulin resistance: diabetes type 4? J Curr Glaucoma Pract 11(3):77–79

38. Blázquez E, Velázquez E, Hurtado-Carneiro V, Ruiz-Albusac JM (2014) Insulin in the brain: its pathophysiological implications for states related with central insulin resistance, type 2 diabetes and Alzheimer's disease. Front Endocrinol (Lausanne) 5:161

39. Jin N, Yin X, Yu D, Cao M, Gong CX, Iqbal K, Ding F, Gu X, Liu F (2015) Truncation and activation of GSK-3β by calpain I: a molecular mechanism links to tau hyperphosphorylation in Alzheimer's disease. Sci Rep 5:8187

40. Zhou H, Li XM, Meinkoth J, Pittman RN (2000) Akt regulates cell survival and apoptosis at a postmitochondrial level. J Cell Biol 151(3):483–494

41. Kleinridders A, Ferris HA, Cai W, Ronald Kahn C (2014) Insulin action in brain regulates systemic metabolism and brain function. Diabetes 63(7):2232–2243

42. Oh SW, Lee S, Park C, Kim DJ (2005) Elevated IOP is associated with insulin resistance and metabolic syndrome. Diabetes Metab Res Rev 21:434–440

43. Cherecheanu AP, Garhofer G, Schmidl D, Werkmeister R, Schmetterer L (2013) Ocular perfusion pressure and ocular blood flow in glaucoma. Curr Opin Pharmacol 13:36–42

44. Tielsch JM, Katz J, Sommer A, Quigley HA, Javitt JC (1995) Hypertension, perfusion pressure, and primary open-angle glaucoma. A population-based assessment. Arch Ophthalmol 113:216–221

45. Abu-Amero KK, Kondkar AA, Chalam KV (2016) Resveratrol and ophthalmic diseases. Nutrients 8(4):200

46. Luna C, Li G, Liton PB, Qiu J, Epstein DL, Challa P, Gonzalez P (2009) Resveratrol prevents the expression of glaucoma markers induced by chronic oxidative stress in trabecular meshwork cells. Food Chem Toxicol 47(1):198–204

47. Razali N, Agarwal R, Agarwal P, Kumar S, Tripathy M, Vasudevan S, Crowston JG, Ismail NM (2015) Role of adenosine receptors in resveratrol-induced intraocular pressure lowering in rats with steroid-induced ocular hypertension. Clin Experiment Ophthalmol 43:54–66

48. Nanjan MJ, Betz J (2014) Resveratrol for the management of diabetes and its downstream pathologies. Eur Endocrinol 10(1):31–35

49. Pirhan D, Yuksel N, Emre E, Cengiz A, Kursat Yildiz D (2015) Riluzole- and resveratrol-induced delay of retinal ganglion cell death in an experimental model of glaucoma. Curr Eye Res 41:59–69

50. Leske MC (2009) Ocular perfusion pressure and glaucoma: clinical trial and epidemiological findings. Curr Opin Ophthalmol 20:73–78

51. Jung KI, Kim YC, Park CK (2018) Dietary niacin and open-angle glaucoma: the Korean National Health and Nutrition Examination Survey. Nutrients 10(4):387

52. Kirkland JB (2012) Niacin requirements for genomic stability. Mutat Res 733(1–2):14–20

53. Meyer-Ficca M, Kirkland JB (2016) Niacin Adv Nutr 7(3):556–558

54. Crish SD, Balaram P, Inman DM, Horner PJ, Calkins DJ (2008) Early transport deficits and distal axonopathy in the DBA/2J mouse model of pigmentary glaucoma. Invest Ophthalmol Vis Sci 49(13):3664

55. Dengler-Crish CM, Smith MA, Inman DM, Wilson GN, Young JW, Crish SD (2014) Anterograde transport blockade precedes deficits in retrograde transport in the visual projection of the DBA/2J mouse model of glaucoma. Front Neurosci 17(8):290

56. Harun-Or-Rashid M, Pappenhagen N, Palmer PG, Smith MA, Gevorgyan V, Wilson GN, Crish SD, Inman DM (2018) Structural and functional rescue of chronic metabolically stressed optic nerves through respiration. J Neurosci 38(22):5122–5139

57. Żarnowski T, Chorągiewicz T, Tulidowicz-Bielak M, Thaler S, Rejdak R, Zarnowska I, Turski WA, Gasior M (2012) Ketogenic diet increases concentrations of kynurenic acid in discrete brain structures of young and adult rats. J Neural Transm 119:679–684

58. Bolla AM, Caretto A, Laurenzi A, Scavini M, Piemonti L (2019) Low-carb and ketogenic diets in type 1 and type 2 diabetes. Nutrients 11(5):962

59. McKinnon SJ, Goldberg LD, Peeples P, Walt JG, Bramley TJ (2008) Current management of glaucoma and the need for complete therapy. Am J Manag Care 14:S20–S27

60. Taniguchi T, Kitazawa Y (1997) The potential systemic effect of topically applied beta-blockers in glaucoma therapy. Curr Opin Ophthalmol 8:55–58

61. Sommer A, Tielsch J (2008) Blood pressure, perfusion pressure, and open-angle glaucoma. Arch Ophthalmol 126:741, author reply 741–742

62. Brown A (2022) FDA new drug approvals in Q3 2022. Nat Rev Drug Dis 21:788

63. Monane M, Bohn RL, Gurwitz JH, Glynn RJ, Choodnovskiy I, Avorn J (1994) Topical glaucoma medications and cardiovascular risk in the elderly. Clin Pharmacol Ther 55:76–83

64. Huerta C, García Rodríguez LA, Möller CS, Arellano FM (2001) The risk of obstructive airways disease in a glaucoma population. Pharmacoepidemiol Drug Saf 10:157–163

65. Kirwan JF, Nightingale JA, Bunce C, Wormald R (2002) Beta blockers for glaucoma and excess risk of airways obstruction: population based cohort study. BMJ 325:1396–1397

66. Brooks AM, Gillies WE (1992) Ocular beta-blockers in glaucoma management. Clinical pharmacological aspects. Drug Aging 2(3):208–221

Teeth and Bones

Teeth and Bones: Overview

Kazuhiko Kawasaki

Introduction

Bone and teeth are principal components of the mineralized skeleton in humans. Cartilage is also a vital skeletal component, though it is not necessarily mineralized. These mineralized or unmineralized skeletal components constitute the endoskeleton and the dermal skeleton, representing the internal and external skeletons, respectively. The endoskeleton was entirely cartilaginous when it arose in early vertebrates. The head and body of early vertebrates subsequently became covered with the dermal skeleton that was composed of basal bony plates and superficial tooth-like ornaments. The skeletal system of humans evolved from such an ancient plan. For example, our limb bones and vertebrae are part of the bony endoskeleton that arose from the cartilaginous endoskeleton and develops through ossification of cartilage primordia. By contrast, most bones of our skull and teeth are derived, respectively, from bony plates and superficial ornaments of the dermal skeleton [1].

Bone and teeth form in organic extracellular matrix that undergoes mineralization. The mineral component essentially consists of hydroxyapatite (calcium phosphate, $Ca_{10}(PO_4)_6(OH)_2$), with various potential substitutes, such as fluoride, replacing hydroxide. The organic constituents vary in different tissues and conditions [2]. Bone consists of ~65% (w/w) minerals, 25% organic matrix (~95% of which are collagen fibrils), and water. The tooth consists of a bulk of dentin covered with enamel on the crown and cementum on the root (Fig. 1a). Dentin is like bone in the proportion of minerals (~70% w/w) and collagen fibrils but includes different non-collagenous proteins. Cementum is also a bone-like tissue that facilitates the attachment of teeth to the jawbone. Unlike these collagenous mineralized tissues, enamel initially forms in a highly specialized collagen-free organic matrix, and its organic constituents are removed during formation. As a result, enamel matures into a highly mineralized (~97% w/w), virtually inorganic material [3].

The tooth is essential for mastication and also important for proper speech and aesthetics [3]. On the other hand, bone constitutes the structural framework, which supports and protects our body and enables movement. In addition, bone is critical to the homeostasis of calcium and phosphate [4]. Growth of these two organs is unique as explained below.

Bone Growth

Bone develops by two mechanisms: intramembranous ossification and endochondral ossification

K. Kawasaki (✉)
Department of Anthropology, Pennsylvania State University, University Park, PA, USA
e-mail: kuk2@psu.edu

© The Author(s), under exclusive license to Springer Nature Switzerland AG 2026
E. Lammert, M. Zeeb (eds.), *Metabolism of Human Diseases*,
https://doi.org/10.1007/978-3-031-96019-2_12

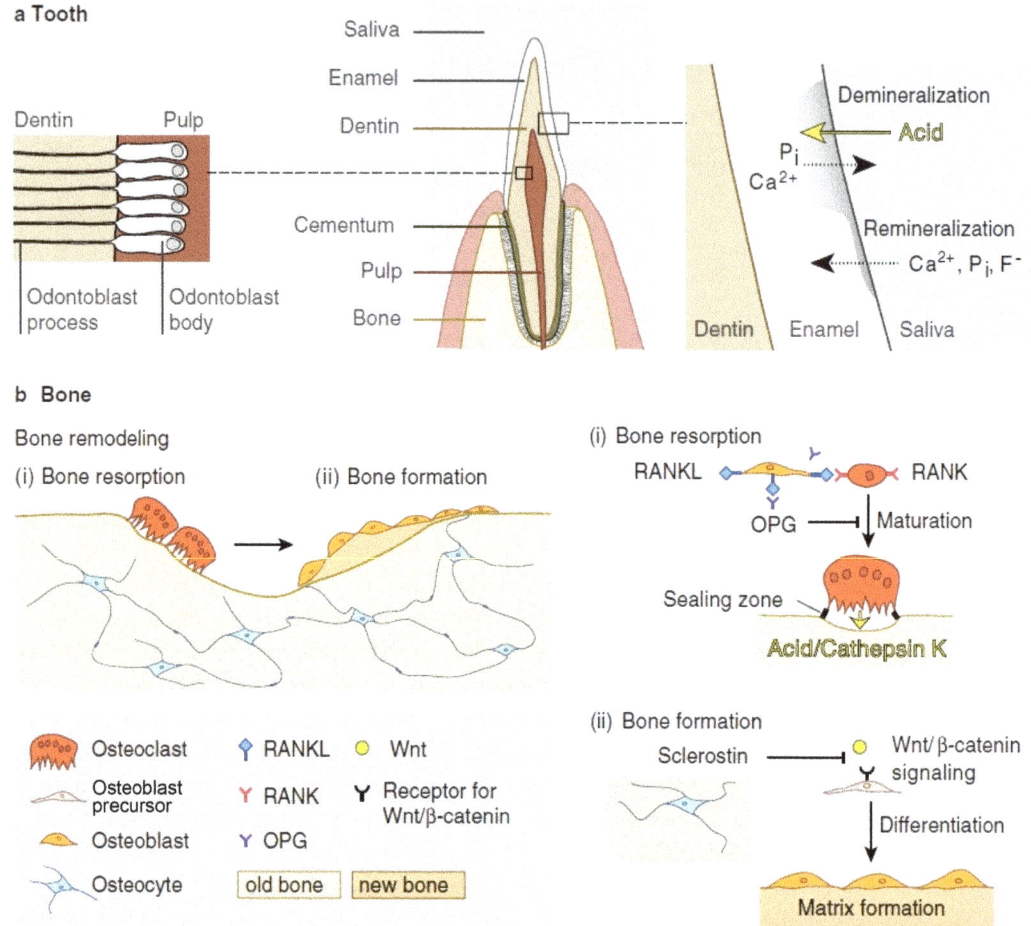

Fig. 1 Tooth-specific (**a**) and bone-specific (**b**) metabolic processes. (**a**) A sagittal section of a tooth is illustrated in the *middle* showing the different zones and structures in a human tooth. The pulp-dentin border is enlarged on the *left*. In the *right* enlargement, flows of calcium (Ca^{2+}), phosphate (P$_i$), and fluoride (F$^-$) during demineralization and remineralization are shown. The *yellow arrow* represents acid, for example, produced by cariogenic bacteria. (**b**) Bone remodeling progresses through (*i*) bone resorption and (*ii*) bone formation. Principal regulatory mechanisms in these processes are illustrated on the *right*. *RANK* receptor activator of nuclear factor-κB, *RANKL* RANK ligand, *OPG* osteoprotegerin

[5, 6]. In intramembranous ossification, mesenchymal cells proliferate and condense. Some of these cells differentiate into osteoblasts, which secrete organic matrix of bone. Intramembranous bones grow appositionally by adding new bone onto a preexisting bone surface.

Endochondral bone also develops from mesenchymal condensation, but, unlike intramembranous ossification, condensed cells differentiate into chondrocytes and form a cartilage model. Endochondral ossification progresses by replacing the cartilage model with bone initially through four distinct phases: (*i*) enlargement (hypertrophy) of chondrocytes, (*ii*) ossification of the cartilage surface (bone collar) and mineralization of the cartilage matrix around hypertrophic chondrocytes, (*iii*) invasion of vasculature along with osteoblast precursors and osteoclasts (see below) into cartilage, and (*iv*) resorption of mineralized cartilage by osteoclasts and secretion of the bone matrix onto resorbing cartilage by osteoblasts [7]. This endochondral bone forming site is referred to as the primary ossification center. As endochondral ossification progresses through cartilage and bone resorption and new bone formation, part of the remaining cartilage forms the growth plate, where chondrocytes continue to proliferate, differentiate, and secrete cartilage matrix [5]. In most long bones, the primary ossification center forms in the middle region of the shaft (diaphysis), and two growth plates develop at both ends of the primary ossification center. Subsequently, new bone forming sites, secondary ossification centers, develop within the cartilage at both ends of the growing endochondral bone (epiphysis) [6, 7]. Long bones grow longitudinally (interstitially) at the growth plates, while they increase in diameter by appositional growth, like intramembranous bones.

Throughout our lives, bone undergoes remodeling, the coupling of bone resorption by osteoclasts and bone formation by osteoblasts (Fig. 1b). By remodeling, bone can meet changing mechanical demands, prevent the accumulation of microdamage, and regulate calcium and phosphate homeostasis.

Tooth Growth

Teeth grow between two layers of cells, epithelial cells and mesenchymal cells. The organic matrix of dentin is secreted by mesenchymal cell-derived odontoblasts that align in the pulp cavity and extend the cell process to the dentin-enamel junction (Fig. 1a) [3]. The odontoblast process is thought to serve as the secretory conduit for particular matrix proteins. After primary dentin is completed in the root, odontoblasts continue to secrete the dentin matrix, but at reduced rates. This secondary dentin formation continues throughout life, and the pulp space reduces progressively with age.

Immediately after the dentin matrix begins mineralization, the enamel matrix is secreted onto the dentin surface by epithelial cells, known as ameloblasts. Whereas bone and dentin begin mineralization in a preformed collagen matrix, the enamel matrix is composed of highly specialized proteins that regulate mineralization begin-

ning immediately after secretion [3]. Enamel increases in thickness as ameloblasts retreat from the dentin-enamel junction. When enamel grows to the final thickness, its organic components are rapidly removed from the matrix, and mineral crystals grow into the vacant space. As a result, enamel matures into a hypermineralized inorganic tissue, the hardest tissue in our body.

As soon as the root begins to form, the tooth moves toward the occlusal surface. Eventually, the reduced enamel epithelium fuses with the oral epithelium. The tooth erupts through the core of the fused epithelia, which enables tooth eruption without a lesion of epithelium or bleeding [3]. Humans have two complete sets of dentition. The deciduous teeth are replaced by the permanent successors with additional permanent molars erupting behind the deciduous tooth row. The tooth replacement involves resorption of deciduous teeth and resorption and formation of the jawbone.

The Process of Calcium and Phosphate Metabolism and Physiological Reactions in Bone and Tooth

Bone Remodeling

Old bone is replaced by new bone throughout life by the bone remodeling process. Bone remodeling progresses through bone resorption and bone formation in a coordinated manner (Fig. 1b) [7]. It is in contrast with bone modeling that represents bone formation that is not coupled with bone resorption. For resorption, preosteoclasts, derived from the hematopoietic lineage, first mature into multinucleated osteoclasts, which adhere to the bone surface and form a closed compartment. Into this sealing zone, osteoclasts secrete acid (hydrogen ions) and proteolytic enzymes, including cathepsin K. The acid dissolves hydroxyapatite, whereas cathepsin K digests collagens and other bone matrix proteins. Following bone resorption, osteoblasts begin new bone formation, although the underlying molecular mechanisms remain unclear [8].

Osteoblasts produce the membrane-bound receptor activator of nuclear factor-κB ligand (RANKL), which plays a critical role in bone resorption (Fig. 1b) [9]. Binding of RANKL to its receptor, RANK, on osteoclast precursor cells induces their maturation (osteoclastogenesis). Furthermore, RANKL binding to RANK on mature osteoclasts mediates their activation and survival [10]. The RANKL–RANK interaction can be interrupted by antagonistic binding of osteoprotegerin (OPG) to RANKL (Fig. 1b). OPG is a soluble receptor secreted by osteoblasts and osteocytes (see below). Thus, the local RANKL/OPG ratio determines the activity of RANK signaling that induces bone remodeling [9–11].

For bone formation, osteocytes play an essential role as mechanosensory cells [12, 13]. Osteocytes represent osteoblast-derived cells that are embedded in the bone matrix (Fig. 1b). Osteoblasts that do not differentiate to osteocytes either line the surface of quiescent bone or undergo apoptosis. In bone, osteocytes lie in lacunae and extend their cell processes through canaliculi. These processes form a network with other osteocytes and osteoblasts present on the bone surface (Fig. 1b). It is thought that osteocytes are activated by sensing the flow of interstitial fluid filled in the lacuna-canalicular system [13, 14].

In response to mechanical loading, the expression of the sclerostin (*SOST*) gene is suppressed in osteocytes, whereas *SOST* expression in osteocytes is enhanced by mechanical unloading [9, 15]. Sclerostin antagonizes Wnt/β-catenin signaling that induces osteoblast differentiation (osteoblastogenesis). Suppressed *SOST* expression by mechanical loading thus activates bone formation (Fig. 1b). By contrast, bone formation is reduced by enhanced *SOST* expression in response to unloading. Mechanical loading may cause microdamage and apoptosis of osteocytes near the damage site [16]. Damage in bone, however, upregulates *RANKL* expression in surviving osteocytes, located adjacent to apoptotic osteocytes, and activates bone resorption (Fig. 1b). Subsequent bone formation eventually repairs the microdamage.

The bone remodeling cycle ends when resorbed bone is completely replaced by new bone. Imbalance of bone resorption and bone formation may lead to a loss of bone mass and strength, that is, osteoporosis (see chapter "Osteoporosis"). In addition to replacement or restoration, bone remodeling is important for homeostasis of calcium and phosphate, and bone serves as the reservoir of these ions in this process. Bone is thus highly vascularized, which is essential for the transport of calcium and phosphate, as well as oxygen and nutrients (common to teeth). A rich vasculature also facilitates the circulation of various hormones that regulate bone remodeling (see below). Another important function of bones is to accommodate the bone marrow, where multipotential stromal cells differentiate and give rise to osteoblasts, chondrocytes, hematopoiesis-supportive stromal cells, and marrow adipocytes [17]. The bone marrow also accommodates various hematopoietic lineage cells (see chapter "Blood: Overview"), including hematopoietic stem cells [18, 19].

Calcium Metabolism in Teeth

Unlike bone that undergoes dynamic remodeling, teeth are less metabolically active, resorbed only during replacement in normal physiological conditions. Yet the superficial layer of enamel repeats demineralization and remineralization (Fig. 1a), especially when challenged by acid [20]. Remineralization occurs by deposition of calcium, phosphate, and fluoride, supplied by saliva [21]. Fluoride-substituted hydroxyapatite is more acid-resistant, and fluoride selectively remains in enamel [3]. A similar process is known as post-eruptive maturation. Shortly after eruption, the enamel is more permeable and susceptible to acid dissolution. After cycles of demineralization and remineralization, net incorporation of calcium, phosphate, and especially fluoride into such enamel from saliva, drinking water (in some locations), and/or toothpaste increases surface hardness and resistance to caries [22].

If demineralization dominates remineralization in enamel, dental caries occur. When caries progress to reach dentin, odontoblasts respond by inductive mineralization in two principal ways, tubular sclerosis and tertiary dentin formation [3]. Tubular sclerosis represents mineralization of dentinal tubules that enclose odontoblast processes (Fig. 1a). Tertiary dentin forms in the pulp by preexisting odontoblasts or newly differentiated odontoblast-like cells.

The Role of Bone and Teeth in Calcium Phosphate Homeostasis

In adult humans, 5–10% of bone is remodeled each year. Such high rates of bone turnover are essential for the homeostasis of calcium and phosphate in the extracellular fluid (ECF; mostly blood plasma and interstitial fluid) to ensure physiological activities [4]. Approximately 99% of calcium and 85% of phosphate in our body reside in bone and teeth as hydroxyapatite crystals, primarily providing the rigidity of these structures. Thus, metabolically more active bone is considered the reservoir of calcium and phosphate. The homeostasis of calcium and phosphate is regulated in a dynamic equilibrium accomplished primarily by intestinal absorption, renal excretion, and bone remodeling through hormonal controls, largely by parathyroid hormone (PTH), vitamin D, and fibroblast growth factor 23 (FGF23; Fig. 2) [4, 23–26].

PTH is a peptide hormone synthesized in the parathyroid grands and is secreted after activation by proteolytic processing. Vitamin D is, by contrast, a steroid hormone, either synthesized in the skin in response to ultraviolet (UV) radiation or ingested with the food, and is activated by sequential hydroxylation, initially to $25(OH)D_3$ (calcidiol) in the liver and then to $1,25(OH)_2D_3$ (calcitriol) in the kidney. FGF23 is a peptide hormone secreted primarily from osteocytes and is responsible mainly for phosphate homeostasis. Calcium and phosphate concentrations in the ECF are restored through actions of PTH, $1,25(OH)_2D_3$, and FGF23, which are coupled with multiple negative and positive feedback loops [4].

In normal conditions, changes in the calcium concentration are detected by the calcium sensing receptor (CaSR) in the parathyroid glands [4]. Small decreases in calcium concentration enhance PTH secretion (Fig. 2). PTH binds its receptor (parathyroid hormone 1 receptor; PTH1R) in the kidney, which leads to calcium reabsorption and hydroxylation of $25(OH)D_3$ to $1,25(OH)_2D_3$. In the small intestine, $1,25(OH)_2D_3$ increases calcium absorption to restore the calcium concentration, which in turn suppresses PTH secretion.

Sufficient phosphate is usually supplied from the diet and absorbed in the small intestine [4]. In response to increases in phosphate concentration, PTH secretion is induced. PTH suppresses phosphate reabsorption and increases the $1,25(OH)_2D_3$ level in the kidney. Though $1,25(OH)_2D_3$ enhances intestinal phosphate absorption, the net effect of PTH secretion is phosphate excretion in the kidney [4] (Fig. 2). Whereas PTH and $1,25(OH)_2D_3$ are primarily responsible for calcium homeostasis, FGF23 is more important for the regulation of phosphate homeostasis. The secretion of FGF23 from osteocytes is triggered by increases in the phosphate concentration and suppresses phosphate reabsorption in the kidney, like PTH secretion (Fig. 2). In contrast to PTH, however, FGF23 downregulates the $1,25(OH)_2D_3$ level, which leads to the suppression of intestinal phosphate absorption. FGF23 thus effectively enhances renal phosphate excretion (Fig. 2) [26].

Bone remodeling is integral to the calcium and phosphate homeostasis. As described above, the primary effect of PTH is to increase the ECF calcium concentration, which involves bone resorption (Fig. 2) [27]. However, the effects of PTH on bone remodeling are highly complicated and depend on the temporal profile of its levels. This is because PTH affects both bone formation and resorption through several different mechanisms, including downregulation of *SOST* expression and upregulation of *RANKL* expression in osteocytes (Fig. 1b) [28]. Thus, whereas intermittent administration of PTH (e.g., osteoporosis treatment) enhances bone formation, a continuous elevation of PTH (e.g., hyperparathyroidism) increases bone resorption [27–30].

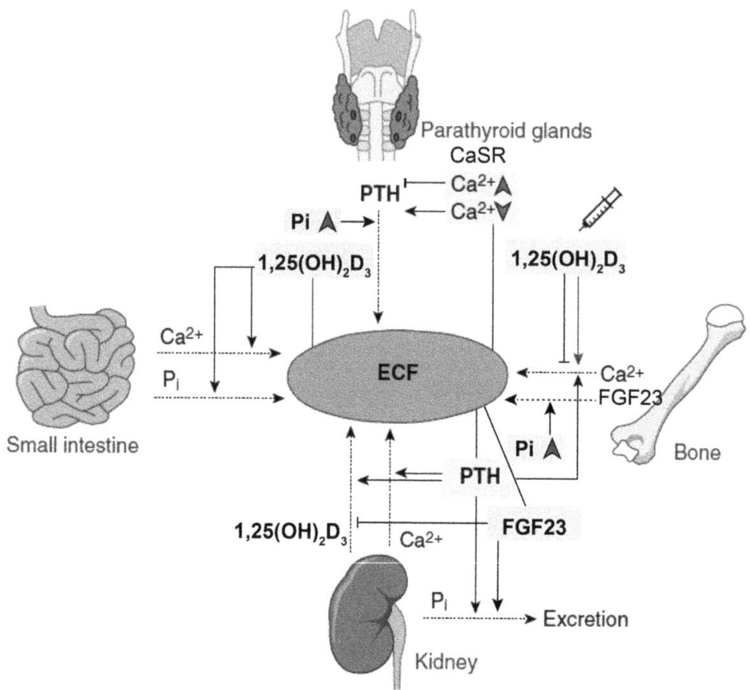

Fig. 2 Metabolic pathways of calcium (*Ca²⁺*) and phosphate (*Pi*) involving bone, kidney, small intestine, and parathyroid glands. An increase and a decrease of calcium concentration in the extracellular fluid (*ECF*) are detected by the calcium sensing receptor (*CaSR*). Although physiological doses of active vitamin D (1,25(OH)₂D₃) do not stimulate bone resorption, a pharmacological or toxic dose of 1,25(OH)₂D₃ induces bone resorption (dashed arrow). However, long-term administration of a pharmacological dose of 1,25(OH)₂D₃ suppresses bone resorption

Vitamin D signaling through its receptor on osteoblasts or their precursor cells does not appear to be essential for bone metabolism in physiological conditions but is important for pharmacological uses, like PTH. When administrated at pharmacological doses, 1,25(OH)₂D₃ increases bone resorption through upregulation of *RANKL* expression and downregulation of *OPG* expression in osteoblasts (Fig. 1b) [24, 31]. Paradoxically, long-term administration of 1,25(OH)₂D₃ at pharmacological doses was shown to suppress bone resorption [31].

Perspectives

As vertebrates invaded terrestrial environments, where calcium and phosphate may not be as freely available as they were in aquatic environ-ments, bone became increasingly important for the storage of calcium and phosphate that ensures the organism's metabolism. The tooth is less metabolically active. Yet, even though the tooth surface is acellular, it repeats demineralization and remineralization. An imbalance of calcium phosphate metabolism in bone and teeth may result in life-threatening conditions including osteoporosis and caries, respectively. It is notable that these disorders became widely prevalent health issues only recently in modern humans after the long successful history of hard tissues. These topics are described in following chapters.

Questions and Answers

Question 1 How does bone remodeling work?

Answer 1 Bone remodeling is a process including the resorption of old bone by osteoclasts and the formation of new bone by osteoblasts in a coupled manner.

Question 2 What happens when dental enamel is challenged by acid?

Answer 2 The enamel surface can be remineralized by the deposition of calcium, phosphate, and fluoride supplied by saliva. If demineralization dominates remineralization in enamel, dental caries occur.

Question 3 What is the physiological response to a small decrease in ECF calcium concentration?

Answer 3 Decreases in calcium concentrations are detected in the parathyroid glands and enhance PTH secretion. PTH binds its receptor in the kidney and increases calcium reabsorption and vitamin D activation. Active vitamin D ($1,25(OH)_2D_3$) increases calcium absorption in the small intestine.

Question 4 What is the function of FGF23 in response to increases in the phosphate concentration?

Answer 4 In response to increases in the concentration of phosphate, FGF23 is secreted from osteocytes and enhances phosphate extraction (suppresses phosphate reabsorption) in the kidney. In addition, FGF23 downregulates the $1,25(OH)_2D_3$ level, which leads to the suppression of intestinal phosphate absorption.

References

1. Kawasaki K, Richtsmeier JT (2017) Association of the chondrocranium and dermatocranium in early skul formation. In: Percival CJ, Richtsmeier JT (eds) Building bones: bone formation and development in anthropolgy. Cambridge University Press, Cambridge, pp 52–78. https://doi.org/10.1017/9781316388907

2. Kawasaki K, Buchanan AV, Weiss KM (2009) Biomineralization in humans: making the hard choices in life. Annu Rev Genet 43:119–142. https://doi.org/10.1146/annurev-genet-102108-134242

3. Nanci A (2016) Ten Cate's oral histology: development, structure, and function, 9th edn. Elsevier, St. Louis

4. DiMeglio LA, Imel EA (2019) Calcium and phosphate: hormonal regulation and metabolism. In: Burr DB, Allen MR (eds) Basic and applied bone biology, 2nd edn. Academic, pp 257–282. https://doi.org/10.1016/B978-0-12-813259-3.00013-0

5. Pawlina W, Ross MH (2018) Histology: a text and atlas, 8th edn. Wolters Kluwer, Philadelphia

6. Blumer MJF (2021) Bone tissue and histological and molecular events during development of the long bones. Ann Anat 235:151704. https://doi.org/10.1016/j.aanat.2021.151704

7. Allen MR, Burr DB (2019) Bone growth, modeling, and remodeling. In: Burr DB, Allen MR (eds) Basic and applied bone biology, 2nd edn. Academic, pp 85–100. https://doi.org/10.1016/B978-0-12-813259-3.00005-1

8. Sims NA, Martin TJ (2019) Coupling of bone formation and resorption. In: Bilezikian J, Martin TJ, Clemens T, Rosen C (eds) Principles of bone biology, vol 1, 4th edn. Academic, pp 219–243. https://doi.org/10.1016/B978-0-12-814841-9.00010-5

9. Komori T (2013) Functions of the osteocyte network in the regulation of bone mass. Cell Tissue Res 352:191–198. https://doi.org/10.1007/s00441-012-1546-x

10. Feng X, Teitelbaum SL (2013) Osteoclasts: new insights. Bone Res 1:11–26. https://doi.org/10.4248/BR201301003

11. Takegahara N, Kim H, Choi Y (2022) RANKL biology. Bone 159:116353. https://doi.org/10.1016/j.bone.2022.116353

12. Robling AG, Daly R, Fuchs RK, Burr DB (2019) Mechanical adaptation. In: Burr DB, Allen MR (eds) Basic and applied bone biology, 2nd edn. Academic, pp 203–233. https://doi.org/10.1016/B978-0-12-813259-3.00011-7

13. Klein-Nulend J, Bonewald LF (2019) The osteocyte. In: Bilezikian J, Martin TJ, Clemens T, Rosen C (eds) Principles of bone biology, vol 1, 4th edn. Academic, pp 133–162. https://doi.org/10.1016/B978-0-12-814841-9.00006-3

14. Wittkowske C, Reilly GC, Lacroix D, Perrault CM (2016) *In vitro* bone cell models: impact of fluid shear stress on bone formation. Front Bioeng Biotechnol 4:87. https://doi.org/10.3389/fbioe.2016.00087

15. Moriishi T, Komori T (2022) Osteocytes: their lacunocanalicular structure and mechanoresponses. Int J Mol Sci 23. https://doi.org/10.3390/ijms23084373

16. Burr DB (2014) Repair mechanisms for microdamage in bone. J Bone Miner Res 29:2534–2536. https://doi.org/10.1002/jbmr.2366

17. Bianco P, Robey PG (2015) Skeletal stem cells. Development 142:1023–1027. https://doi.org/10.1242/dev.102210

18. Kim MJ, Valderrabano RJ, Wu JY (2022) Osteoblast lineage support of hematopoiesis in health and disease. J Bone Miner Res 37:1823–1842. https://doi.org/10.1002/jbmr.4678

19. Calvi LM (2019) Bone marrow and the hematopoietic stem cell niche. In: Bilezikian J, Martin TJ, Clemens T, Rosen C (eds) Principles of bone biology, vol 1, 4th edn. Academic, pp 73–87. https://doi.org/10.1016/B978-0-12-814841-9.00003-8

20. Dawes C (2003) What is the critical pH and why does a tooth dissolve in acid? J Can Dent Assoc 69:722–724

21. Garcia-Godoy F, Hicks MJ (2008) Maintaining the integrity of the enamel surface: the role of dental biofilm, saliva and preventive agents in enamel demineralization and remineralization. J Am Dent Assoc 139(Suppl):25S–34S. https://doi.org/10.14219/jada.archive.2008.0352

22. Lynch RJM (2013) The primary and mixed dentition, post-eruptive enamel maturation and dental caries: a review. Int Dent J 63(Suppl 2):3–13. https://doi.org/10.1111/idj.12076

23. Bisello A, Friedman PA (2019) Parathyroid hormone and parathyroid hormone – related protein actions on bone and kidney. In: Bilezikian J, Martin TJ, Clemens T, Rosen C (eds) Principles of bone biology, vol 1, 4th edn. Academic, pp 645–689. https://doi.org/10.1016/B978-0-12-814841-9.00027-0

24. Christakos S, Pike JW (2019) Vitamin D gene regulation. In: Bilezikian J, Martin TJ, Clemens T, Rosen C (eds) Principles of bone biology, vol 1, 4th edn. Academic, pp 739–756. https://doi.org/10.1016/B978-0-12-814841-9.00030-0

25. Carpenter TO, Bergwitz C, Insogna KL (2019) Phosphorus homeostasis and related disorders. In: Bilezikian J, Martin TJ, Clemens T, Rosen C (eds) Principles of bone biology, vol 1, 4th edn. Academic, pp 469–507. https://doi.org/10.1016/B978-0-12-814841-9.00020-8

26. Fukumoto S (2019) Fibroblast growth factor 23. In: Bilezikian J, Martin TJ, Clemens T, Rosen C (eds) Principles of bone biology, vol 2, 4th edn. Academic, pp 1529–1538. https://doi.org/10.1016/B978-0-12-814841-9.00063-4

27. Bellido T, Gallant KMH (2019) Hormonal effects on bone cells. In: Burr DB, Allen MR (eds) Basic and applied bone biology, 2nd edn. Academic, pp 299–313. https://doi.org/10.1016/B978-0-12-813259-3.00015-4

28. Wein MN, Kronenberg HM (2018) Regulation of bone remodeling by parathyroid hormone. Cold Spring Harb Perspect Med 8:a031237. https://doi.org/10.1101/cshperspect.a031237

29. Jilka RL (2007) Molecular and cellular mechanisms of the anabolic effect of intermittent PTH. Bone 40:1434–1446. https://doi.org/10.1016/j.bone.2007.03.017

30. Rendina-Ruedy E, Rosen CJ (2022) Parathyroid hormone (PTH) regulation of metabolic homeostasis: an old dog teaches us new tricks. Mol Metab 60:101480. https://doi.org/10.1016/j.molmet.2022.101480

31. Nakamichi Y, Udagawa N, Suda T, Takahashi N (2018) Mechanisms involved in bone resorption regulated by vitamin D. J Steroid Biochem Mol Biol 177:70–76. https://doi.org/10.1016/j.jsbmb.2017.11.005

Osteoporosis

Emmanuel Biver

Introduction to Osteoporosis

Osteoporosis is a skeletal disorder characterized by low bone mass and microarchitectural deterioration of bone tissue, leading to an increase in bone fragility and susceptibility to fracture. An operational definition of osteoporosis has also been defined, based on a value for bone mineral density (BMD) 2.5 standard deviations or more below the young adult mean [1]. The most widely validated technique for the quantitative assessment of BMD is dual-energy X-ray absorptiometry (DXA). Bone loss is due to an imbalance between bone resorption by osteoclasts and bone formation by osteoblasts, which are part of the bone remodeling process. This results in a decrease in BMD and alterations of bone geometry and micro-architecture at the cortical and trabecular sites, leading to bone fragility and fractures. Osteoporosis has the potential to alter quality of life and to increase mortality. From the age of 50 years, 50% of women and 20% of men will be concerned by fracture during their remaining lifetime. This important public health issue will tend to increase due to the aging of the world population. Beyond aging, many pathological conditions (e.g., endocrine diseases, chronic inflammatory diseases) or treatments (e.g., corticosteroids) interfere with bone metabolism and induce osteoporosis.

Many factors regulating bone remodeling contribute to osteoporosis (Fig. 1). Increased bone resorption and decreased bone formation can contribute to its progression. Significant bone loss occurs in both women and men because of sex steroid deficiency and environmental risk factors, including nutritional intake and lifestyle habits.

- *Genetic and Lifestyle Habit Risk Factors and Comorbidities*

Genetic factors account for 50–80% of the variation among individuals in bone mass and structure. Peak bone density is achieved at the beginning of the third decade of life. Lower peak bone mass contributes to the risk of osteoporosis and fractures in later life [2]. Lifestyle habits involve excess alcohol intake or tobacco use (because of toxic effects on bone cells but also indirect effects due to endocrine changes and muscle weakness) or physical inactivity (which promotes bone resorption via RANKL activation and inhibits bone formation via sclerostin secretion, a negative regulator of bone formation produced by osteocytes). Sporadic causes of bone loss can also be identified in about 20% of women and 50% of men and include glucocorticoid therapy (which inhibits bone formation and

E. Biver (✉)
Division of Bone Diseases, Department of Medicine, Geneva University Hospital and Faculty of Medicine, Geneva, Switzerland
e-mail: emmanuel.biver@hug.ch

E. Lammert, M. Zeeb (eds.), *Metabolism of Human Diseases*,
https://doi.org/10.1007/978-3-031-96019-2_13

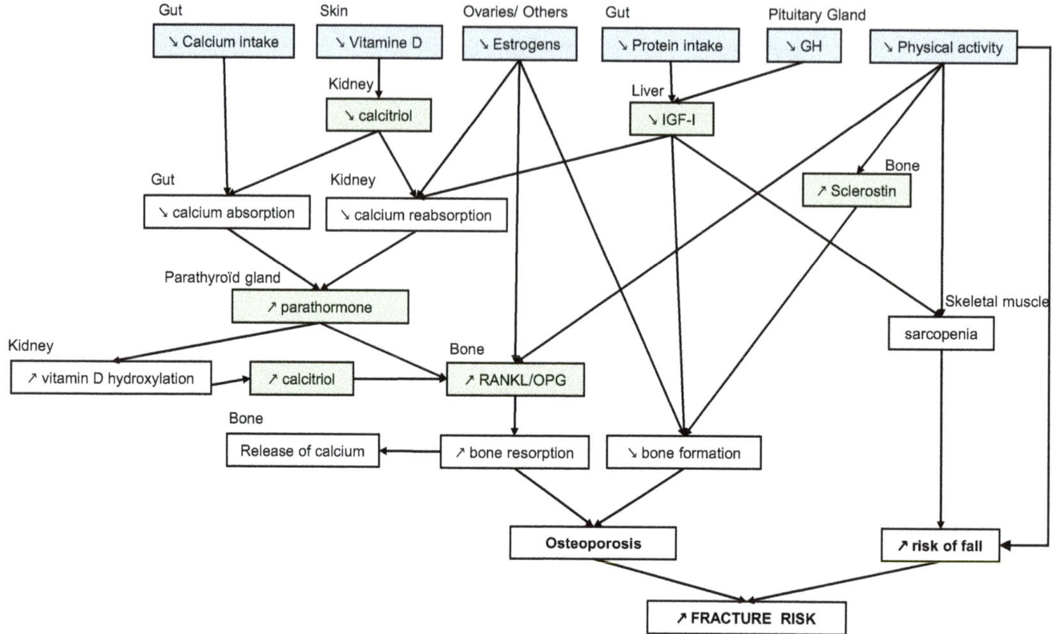

Fig. 1 Metabolic changes in age-related osteoporosis. Multiple metabolic processes in various organs are involved in the imbalance between bone resorption and formation leading to osteoporosis and contributing to the incidence of fractures. *GH* growth hormone, *IGF-1* insulin-like growth factor-1, *RANKL* receptor activator of nuclear factor-κB ligand, *OPG* osteoprotegerin

suppresses calcium (Ca^{2+}) absorption), malabsorption, idiopathic hypercalciuria (excess of urine calcium excretion without an apparent underlying etiology, due to genetic causes), or endocrine diseases, such as primary hyperparathyroidism or hyperthyroidism (all related to an imbalance of hormones affecting bone metabolism).

- *Endocrine Factors*

Ca^{2+} and phosphate (PO_4^{3-}, P_i) are essential components for bone mineral formation and their availability is mainly regulated by parathyroid hormone (PTH, increasing bone resorption) and vitamin D (increasing Ca^{2+} and P_i absorption, see chapter "Teeth and Bones: Overview" under part "Teeth and Bones").

One major cause of osteoporosis is the decline in estrogens after menopause in women, but also in men, resulting in a negative balance in bone remodeling. Estrogens control bone remodeling by estrogen receptors on osteoblasts and osteoclasts. Estrogen deficiency promotes osteoclastogenesis by upregulating the receptor activator of nuclear factor-κB ligand (RANKL, see chapter "Teeth and Bones: Overview" under part "Teeth and Bones") in bone marrow cells, whereas estrogens stimulate the production of its soluble decoy receptor osteoprotegerin (OPG) in osteoblasts. Estrogen deficiency also stimulates bone resorption by indirect effects, in particular via cells of the immune system and proinflammatory cytokines (see below). Moreover, estrogens modulate Ca^{2+} absorption and excretion.

These effects are often exacerbated by vitamin D insufficiency (calcidiol <50 nmol/L), which induces secondary hyperparathyroidism (meaning increased PTH). Impaired vitamin D synthesis in skin and reduced renal hydroxylation in old age lead to low calcitriol levels.

Renal synthesis of calcitriol is regulated not only by PTH but also by insulin-like growth factor-1 (IGF-1). IGF-1 is produced in the liver in response to growth hormone (somatotropin) from the pituitary (see chapter "Brain: Overview" under part "Brain") and dietary animal protein intake, as a diet rich in amino acids is a common inducer of growth. IGF-1 is an important regulator of muscle and bone growth, acting as an autocrine/paracrine growth factor in multiple tissues through IGF-1 receptor, a classical tyrosine kinase receptor. IGF-1 directly regulates renal tubular reabsorption of P_i and stimulates its transport into osteoblastic cells, resulting in activation of bone mineralization. Furthermore, age-related decrease in IGF-1 levels contributes to sarcopenia by decreasing muscle size and strength, protein synthesis, and increasing muscle cell apoptosis. Sarcopenia reduces muscle loading on the skeleton and is associated with osteoporosis [3].

Circulating serotonin from the gut is inversely associated with BMD. It inhibits bone mass accrual by decreasing osteoblast proliferation and bone formation [4]. Several gut-associated metabolites, such as short-chain fatty acids, secondary bile acids, or polyamines also regulate bone health by modulating calcium absorption and/or bone remodeling [5]. Finally, leptin secreted by adipocytes stimulates bone remodeling directly through increased osteoblast proliferation and differentiation, but indirectly decreases axial skeleton bone formation through a hypothalamic relay via the sympathetic nervous system, independently of its regulation of energy metabolism.

- *Nutritional Factors*

Dietary factors influence bone health, particularly calcium, and protein intakes. Ca^{2+} metabolism is altered in the elderly due to impaired vitamin D levels, reduced intestinal Ca^{2+} absorption, and a lower dietary Ca^{2+} intake. In addition, relative hypocalcemia causes secondary hyperparathyroidism, which stimulates bone resorption to maintain homeostasis in extracellular Ca^{2+} concentration (see chapter "Teeth and Bones: Overview" under part "Teeth and Bones"). In addition, a positive relationship has been found between protein intake in both men and women

and bone mass at various skeletal sites [6]. A Mediterranean-type diet is associated with better bone health, while vegan diets, and to a lesser extent, vegetarian diets, are associated with lower BMD and higher fracture risk [7].

- *Immune Factors and Gut Microbiota*

Many cytokines, such as interleukin 1, interleukin 6, and tumor necrosis factor-α, are involved in the pathogenesis of osteoporosis. In addition, estrogen deficiency has been associated with an increased production of pro-inflammatory cytokines leading to T cell activation. These T cells produce RANKL and tumor necrosis factor-α that stimulate osteoclastogenic activity. The gut microbiota might also increase the frequency of T cells in the bone marrow, thereby promoting the expression of inflammatory cytokines in bone and activating osteoclastogenesis [8]. Interactions between gut microbiota, immune system, and bone metabolism influence musculoskeletal health and osteoporosis treatments [9].

Osteoporosis Treatment and Influence on Bone Metabolism

The aim of osteoporosis treatment is the prevention of fragility fractures. This includes non-pharmacological interventions, Ca^{2+} and vitamin D supplements, and drugs with proven anti-fracture efficacy. The choice of first-line treatment greatly depends on the patient and his/her personal and disease characteristics, including age, comorbidities, fracture risk, expected adherence and potential adverse effects, and others.

Reduced intake of Ca^{2+}, proteins, and vitamin D contributes to osteoporosis. Thus, an adequate Ca^{2+} supply is needed to prevent secondary hyperparathyroidism and optimize bone mineralization, if necessary with dietary and Ca^{2+} supplements. Vitamin D deficiency is very frequent and hampers the clinical benefits of anti-resorbing therapies. Vitamin D supplements reduce the risk of fracture and propensity to falls by their effect on muscles and prevent secondary hyperparathyroidism. Daily dietary sources (even if rich in oily fish) and sun exposure fail to provide enough vitamin D in patients with vitamin D insufficiency, and supplements are needed to achieve an optimal circulating level of calcidiol (>75 nmol/L, target in osteoporotic patients). To improve vitamin D absorption, it should be taken with a fat-rich diet, such as milk, or at the end of the meal [10].

Correction of poor protein intake can restore an altered growth hormone–IGF-1 axis and improve BMD, as well as muscle mass and strength [6].

Promotion of physical activity has a positive impact on bone loss and prevention of falls, mechanical stimuli targeting osteocytes, and muscle strength. Mechanical loading also prevents the expression of sclerostin, inducing new bone formation [11].

Discontinuation of toxic habits should also be encouraged.

The primary objective of all anti-osteoporotic drugs is to reduce fracture risk, by improving bone strength. Except for menopause hormone therapy (MHT), the efficacy of all drugs was tested in patients receiving a combination of Ca^{2+} and vitamin D (Table 1). Depending on their mechanism of action and regular fracture risk assessment, sequential treatments of osteoporosis should be considered [12].

Most of the drugs (Table 1) decrease bone resorption: bisphosphonates, denosumab and SERMs (selective estrogen-receptor modulators). In osteoporotic women using MHT for treatment of menopausal symptoms, MHT is also effective in raising BMD and decreasing fracture risk.

Bisphosphonates were the first of these and are still the most common treatment. They inhibit osteoclastic resorption. Selective estrogen-receptor modulators (SERMs), nonsteroidal agents that bind to estrogen receptors (see chapter "Breast Cancer"), act as estrogen agonists on bone. Denosumab is a humanized monoclonal antibody that selectively inhibits the RANKL.

Two classes of bone anabolic agents (promoting bone formation) are used for the treatment of osteoporosis. PTH analogs, teriparatide and abaloparatide, increase bone formation based on the effects of PTH on bone turnover that depend on the pattern and duration of its elevation. While hyperparathyroidism is associated with increased bone resorption, daily administration of teriparatide results in upregulation of bone formation on the skeleton. Romosozumab, a monoclonal antibody that inhibits sclerostin, promotes bone formation while inhibiting bone resorption.

Perspectives

Osteoporosis is a common disorder associated with high morbidity and increased mortality; however, a significant proportion of women and men at high fracture risk do not receive therapy for osteoporosis (treatment gap) [13]. Its prevention is feasible by cost-effective strategies targeting risk factors or drugs increasing BMD and preventing fractures. With population aging and increased life expectancy, long-term treatment of osteoporosis needs to consider sequential therapeutic strategies.

Questions and Answers

Question 1 Which factors influence bone remodeling?

Answer 1 Multiple factors may influence bone remodeling. These include genetic factors, health related factors (comorbidity and drugs), and environmental factors. Bone remodeling is accelerated by estrogen deficiency. Dietary factors, particularly calcium and protein intakes, contribute to calcium and phosphate balances, regulated by parathyroid hormone and vitamin D. Interactions between gut microbiota and immune system also influence bone metabolism. Pro-inflammatory cytokines increase bone resorption. Lifestyle factors such as alcohol intake, tobacco use, or physical inactivity also negatively affect bone metabolism.

Table 1 Pharmacological agents preventing fracture risk used in postmenopausal osteoporosis

Drug class	Drug molecules	Bone targets and mechanism of action	Effects on bone remodeling		Side/off-target effects and adverse effects
			Bone resorption	Bone formation	
Bisphosphonates	Alendronate, risedronate, ibandronate, zoledronate	*Inhibitors of bone resorption:* inhibit osteoclast activity; induce osteoclast apoptosis	↓↓↓	↓↓↓	Oral: low bioavailability due to low absorption in gastrointestinal (<1%) Long-lasting remanent protection of bone Esophageal irritation Flu-like symptoms after intravenous injection Osteonecrosis of the jaw in patients receiving high doses Atypical fractures of the femur with long-term use
SERMs	Raloxifene, bazedoxifene	*Inhibitors of bone resorption:* Nonsteroidal agents that bind to the estrogen receptor and act as estrogen agonists on bone	↓↓	↓↓	Flushes ↓ Deep venous thromboembolism ↑
Anti-RANKL monoclonal antibodies	Denosumab	*Inhibitors of bone resorption:* Humanized monoclonal antibodies binding to RANKL. Prevent the effect of RANKL on osteoclasts differentiation, activation, and survival	↓↓↓	↓	Hypocalcemia Possibly associated with osteonecrosis of the jaw and atypical fractures of the femur for long term treatment
PTH	Teriparatide/ abaloparatide	*Activators of bone formation:* Increase bone remodeling in favor of osteoblasts ↑	↑↑	↑↑↑	Contraindicated in conditions with abnormally increased bone turnover, in patients with prior radiation therapy to the skeleton, skeletal malignancies, or bone metastases Limited to 24 months (risk of osteosarcoma)
Anti-sclerostin monoclonal antibodies		*Mixed effect on bone resorption and formation:* Inhibits bone resorption and stimulates modeling based-bone formation	↓	↑↑↑	Contraindicated in patients with a history of myocardial infarction or stroke

SERM selective estrogen-receptor modulator, *RANKL* receptor activator of nuclear factor-κB ligand

Question 2 What non-pharmacological interventions should patients with osteoporosis apply?

Answer 2 Non-pharmacological interventions in the management of osteoporosis include optimization of calcium and protein intake, correction of vitamin D deficiency or insufficiency, promotion of physical activity, prevention of falls, and discontinuation of toxic habits.

Question 3 How do anti-osteoporotic drugs affect bone metabolism?

Answer 3 Anti-osteoporotic drugs increase bone mineral density (BMD) and therefore increase bone strength. Bisphosphonates, denosumab, and selective estrogen-receptor modulators (SERMs) decrease bone resorption by targeting mainly osteoclasts. Teriparatide and abaloparatide increases bone remodeling in favor of bone formation. Romosozumab inhibits sclerostin and thereby promotes modeling-based bone formation and inhibits bone resorption.

References

1. Kanis JA (1994) Assessment of fracture risk and its application to screening for postmenopausal osteoporosis: synopsis of a WHO report. WHO Study Group. Osteoporosis 4(6):368–381
2. Chevalley T, Rizzoli R (2022) Acquisition of peak bone mass. Best Pract Res Clin Endocrinol Metab 36(2):101616
3. Alajlouni DA et al (2023) Muscle strength and physical performance contribute to and improve fracture risk prediction in older people: a narrative review. Bone 172:116755
4. Karsenty G, Gershon MD (2011) The importance of the gastrointestinal tract in the control of bone mass accrual. Gastroenterology 141(2):439–442
5. Bhardwaj A et al (2021) "Osteomicrobiology": the nexus between bone and bugs. Front Microbiol 12:812466
6. Rizzoli R, Biver E, Brennan-Speranza TC (2021) Nutritional intake and bone health. Lancet Diabetes Endocrinol 9(9):606–621
7. Biver E et al (2023) Dietary recommendations in the prevention and treatment of osteoporosis. Joint Bone Spine 90(3):105521
8. Ohlsson C, Sjögren K (2018) Osteomicrobiology: a new cross-disciplinary research field. Calcif Tissue Int 102(4):426–432
9. Papageorgiou M, Biver E (2021) Interactions of the microbiome with pharmacological and non-pharmacological approaches for the management of ageing-related musculoskeletal diseases. Ther Adv Musculoskelet Dis 13:1759720x211009018
10. Chevalley T et al (2022) Role of vitamin D supplementation in the management of musculoskeletal diseases: update from an European Society of Clinical and Economical Aspects of Osteoporosis, Osteoarthritis and Musculoskeletal Diseases (ESCEO) working group. Aging Clin Exp Res 34(11):2603–2623
11. Zhang L et al (2022) Exercise for osteoporosis: a literature review of pathology and mechanism. Front Immunol 13:1005665
12. Reid IR, Billington EO (2022) Drug therapy for osteoporosis in older adults. Lancet 399(10329):1080–1092
13. Willers C et al (2022) Osteoporosis in Europe: a compendium of country-specific reports. Arch Osteoporos 17(1):23

Joints

Joints: Overview

Jessica Bertrand, Jan Hubert,
and Miriam Bollmann

Anatomy and Physiology of Joints

Joint Classification

Joints are locations at which two or more bones come together. They are often constructed to allow movement and provide mechanical support. Joints can be classified structurally, describing how the bones connect to each other, as fibrous, cartilaginous, and synovial. Joints can also be classified functionally based on the amount of movement allowed: "synarthroses" are immovable, "amphiarthroses" are slightly movable, and "diarthroses" are freely movable. Synarthroses are fibrous joints that connect bones without allowing any movement and are found, for example, in the skull and pelvis and at the union of the spinous processes and vertebrae. In cartilaginous joints (amphiarthroses), the bones are attached by cartilage, and they allow for only a little movement, such as in the spine or ribs. Synovial joints, also called diarthroses, are of crucial importance for the skeletal function, as they permit a wider range of movement. The ends of the opposing skeletal elements in synovial joints are covered with articular cartilage. The spaces between the bones in synovial joints are filled with synovial fluid, which helps to lubricate and protect the cartilage and nourishes the tissues. The joint is surrounded by the synovial membrane and held together by the fibrous joint capsule (that insulates the joints from surrounding tissues) and ligaments (that hold the skeletal elements in place) [1].

Synovial joints can be distinguished again into six different types depending on the mobility and type of movement they allow for: gliding, hinge, pivot, condyloid, saddle, "ball and socket," and compound joints.

The following passages will describe the different structures of synovial joints, with special emphasis on biochemical processes that play a role during joint homeostasis.

Joint Formation

A process called endochondral ossification mediates the formation of long bones and thereby the formation of articular joints in arms and legs (see chapter "Teeth and Bones: Overview" under part "Teeth and Bones"). During the differentiation of mesenchymal cells into chondrocytes in the bone anlagen, the expression pattern of extracellular

J. Bertrand (✉) · M. Bollmann
Department of Orthopaedic Surgery, Otto- von- Guericke University, Magdeburg, Germany
e-mail: Jessica.bertrand@med.ovgu.de;
miriam.bollmann@gu.se

J. Hubert
Division of Orthopaedics, Department of Trauma and Orthopaedic Surgery, University Medical Center Hamburg-Eppendorf, Hamburg, Germany
e-mail: j.hubert@uke.de

matrix proteins changes. While the expression of collagen I decreases, chondrocytes start producing collagens II, IX, and XI as well as proteoglycans like aggrecan, link protein, and matrix Gla protein [2]. At both ends of the bones, this composition of cartilage matrix is retained, and the chondrocytes do not differentiate any further and form the future articular cartilage.

In the rest of the bone anlagen, the chondrocytes differentiate further and become hypertrophic. As part of their hypertrophic differentiation, they start to express collagen X. With beginning bone formation, cartilage undergoes vascularization [2, 3], and the extracellular matrix gets calcified. Subsequently, a continuous cycle of bone remodeling starts, driven by resorbing osteoclasts and bone-forming osteoblasts (see chapter "Teeth and Bones: Overview" under part "Teeth and Bones").

During osteoarthritis (see chapter "Osteoarthritis"), however, articular chondrocytes also undergo prehypertrophic to hypertrophic differentiation in the articular cartilage. This differentiation is accompanied by an increase in expression of certain marker genes, such as alkaline phosphatase [4] and collagen X [5], with subsequent calcification of the diseased cartilage [5].

Joint-Specific Pathways and Processes

Bone

The aforementioned balance between bone formation and resorption is important for normal bone function and is a continuous process. This balance is regulated by different factors, including macrophage colony-stimulating factor (M-CSF) and receptor activator of nuclear factor-κB ligand (RANKL) [6, 7]. M-CSF is constitutively expressed by osteoblasts and binds to receptors on monocytes, macrophages, and osteoclasts, thereby inducing osteoclast maturation and differentiation. RANKL is mainly expressed by osteoblasts in response to factors that stimulate bone resorption, such as parathy-roid hormone and calcitriol. RANKL binds to its receptor on osteoclast precursors, initiating their maturation to mature osteoclasts. Bone resorption induced by RANKL can be blocked by osteoprotegerin (OPG), which is also secreted by osteoblasts and osteogenic stromal stem cells (see chapter "Teeth and Bones: Overview" under part "Teeth and Bones").

The Wnt signaling pathways (canonical and noncanonical) play a central role in bone remodeling in both physiological and pathological conditions. Canonical Wnt signaling promotes differentiation of osteoblast precursor cells (Fig. 1) [8, 9]. Additionally, this pathway suppresses bone resorption by shifting the RANKL/OPG ratio toward OPG in mature osteoblasts [10]. However, the activation of the noncanonical pathway enhances the RANKL-induced osteoclast formation [11].

Cartilage

Cartilage is a flexible form of connective tissue. The predominant form in the human body is the hyaline cartilage, named after its glassy, translucent appearance. It is commonly associated with the skeletal system, as it covers the bones and represents the articular cartilage in joints. It is also found between the ribs and the sternum or breastplate, in the trachea and bronchi of the lungs, in the ear and nose, and in the larynx or voice box. Macroscopically, articular cartilage can be divided into the superficial zone, the transitional zone, the radial zone, and the calcified cartilage zone, where the cartilage interfaces with the bone (Fig. 1) [12]. These zones are characterized by a distinct organization of the collagen network, as well as by differences in the amounts and types of proteoglycans. Type II collagen is the principal molecular component in healthy articular cartilage, but collagens III, VI, IX–XII, and XIV all contribute in smaller amounts to the mature cartilage matrix [13–15], whereas the collagen X is specific for hypertrophic chondroytces in, e.g., osteoarthritic cartilage tissue. The main proteoglycan is aggrecan, but also other smaller proteoglycans are present and of important

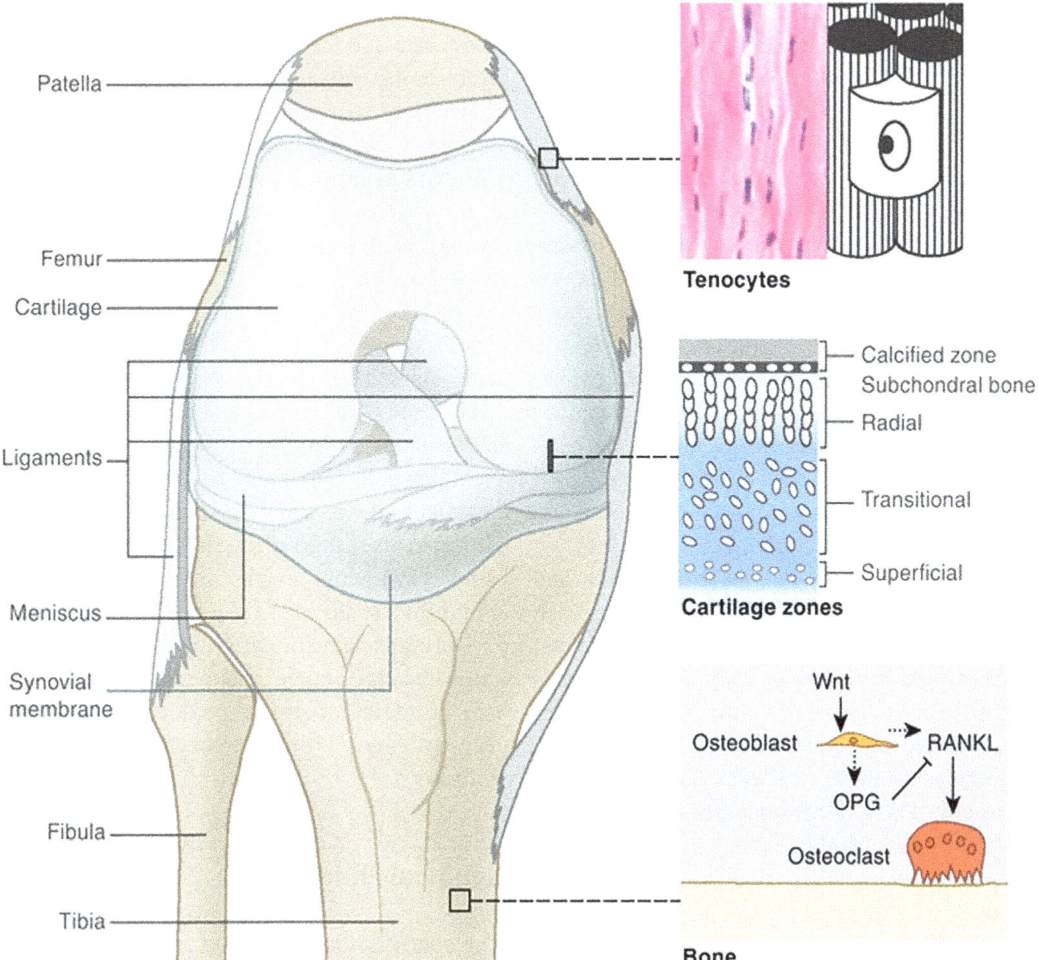

Fig. 1 Macroscopic and microscopic diagram of the anatomy of the human knee. The knee is a synovial joint; its anatomical structures, bones (patella, femur, fibula, and tibia), ligaments, and meniscus are shown (to the *left*), including the different cell compartments. The inserts to the *right* depict magnifications of important structures and tissue-specific pathways showing, from *top to bottom*, tendon, cartilage, and bone. Please note that the cartilage zones do not extend along the cartilage but into it, toward the femur, as indicated by the *small gray box*

function for the cartilage homeostasis. Aggrecan is important for the biomechanical properties of articular cartilage because it builds a hydrated gel structure due to its interaction with hyaluronan and link protein, thereby retaining water in the tissue. Due to these properties of aggrecan, the cartilage is a pressure-resistant and reversibly deformable tissue that distributes the loads in joints and transfers the compressive load to the underlying bone.

The only cells found in cartilage are chondrocytes. In adult cartilage, the chondrocytes remain resting in a non-proliferating state, but display moderate metabolic activity and the ability to maintain the surrounding matrix, which is called cartilage homeostasis.

Ligaments and Tendons

Tendons and ligaments are tough, fiber-rich connective tissues characterized by their excellent tensile strength. Both consist of fibroblasts (in tendons called tenocytes), which have long extensions that are located in rows in between the collagen fibrils, and the proteoglycan matrix that is synthesized by the fibroblasts into the intercellular space (Fig. 1). The collagen fibrils consist primarily of collagen type I and very small amounts of elastin and other collagens (types II–V, IX, and X).

Tendons are responsible for the power transmission as well as for the stabilization of joints and skeletal elements. In tendons the collagenous fibers, which are parallel and oriented in tensile direction, are divided by septa of loose connective tissue (peritendineum) to separate bundles. On the outside, the tendon is enveloped by a white fibrous sheath called epitendineum, which merges into the perimysium, the connective tissue surrounding the muscle.

Ligaments mediate the guidance of joints and skeletal elements. They are coarse, fiber-rich connective tissues that connect different skeletal elements and have mainly stabilizing functions. The histological structure of the ligament is very similar to the structure of tendons.

Damage of these structures due to a trauma commonly leads to an impairment of the joint function, is frequently associated with calcification, and may possibly result in changes of its biomechanical properties and subsequent osteoarthritis (see chapter "Osteoarthritis") [16]. Inflammatory processes, like in rheumatoid arthritis (see chapter "Rheumatoid Arthritis"), can cause a chronic damage to the tendons, tendon sheaths, the articular capsule, and the ligaments, thus deforming the joints [17].

Synovium

The synovial membrane (or synovium, Fig. 2) is specific to synovial joints and seals the synovial fluid from the surrounding tissue. It is only about four to five cell layers thick and has no basement membrane, which makes it differ significantly from regular epithelium. The composition of the synovium is very variable, but mostly has two layers: the superficial layer of the synovium is called (1) lining layer and can be separated histologically from the more loose network of fibroblasts underneath, which is called (2) sub-lining layer (Fig. 2). It consists of two cell types, fibroblasts and macrophages, which both differ from similar cells in other tissues. The sub-lining is intensely vascularized, providing nutrients to the synovium and the avascular cartilage. The (resident) macrophages are called type A synoviocytes and are rare in healthy synovium. Their main function is phagocytosis of undesirable substances from the synovial fluid, such as cell debris and dead cell tissues. The fibroblast-like type B synoviocytes provide the synovial cavity with lubricating factors, such as lubricin and hyaluronan, and produce components of the extracellular matrix of the synovial membrane, including collagens and fibronectin. As the lining layer lacks the basement membrane, it facilitates the rapid exchange of nutrients between blood and the synovium, making the synovial fluid an ultrafiltrate of the blood plasma.

During rheumatoid arthritis, type B synoviocytes undergo stable activation, meaning that they proliferate more and produce proinflammatory cytokines [18] (see chapter "Rheumatoid Arthritis").

Synovial Fluid

The synovial fluid ensures a smooth movement of the joint. It is of a straw color and its high viscosity is predominantly due to the high amount of the main lubricant proteins hyaluronic acid and lubricin. These lubricant proteins result in a thixotrophic behavior of the synovial fluid, ensuring a low friction movement of the articular surfaces.

Additionally, the synovial fluid fulfills metabolic and regulatory functions. The synovial fluid supplies the less or non-vascularized structures in the joint (e.g., cartilage, meniscus) with nutrients, which can, for example, be secreted by the synovium into the synovial fluid [19]. The synovial fluid also facilitates the cross- talk between the different cells in the joint [20].

Pathological conditions such as trauma, inflammation, as well as viral, bacterial, and fungal penetration can lead to a change in the composition or the volume of the synovial fluid. During joint inflammation, for example, the total

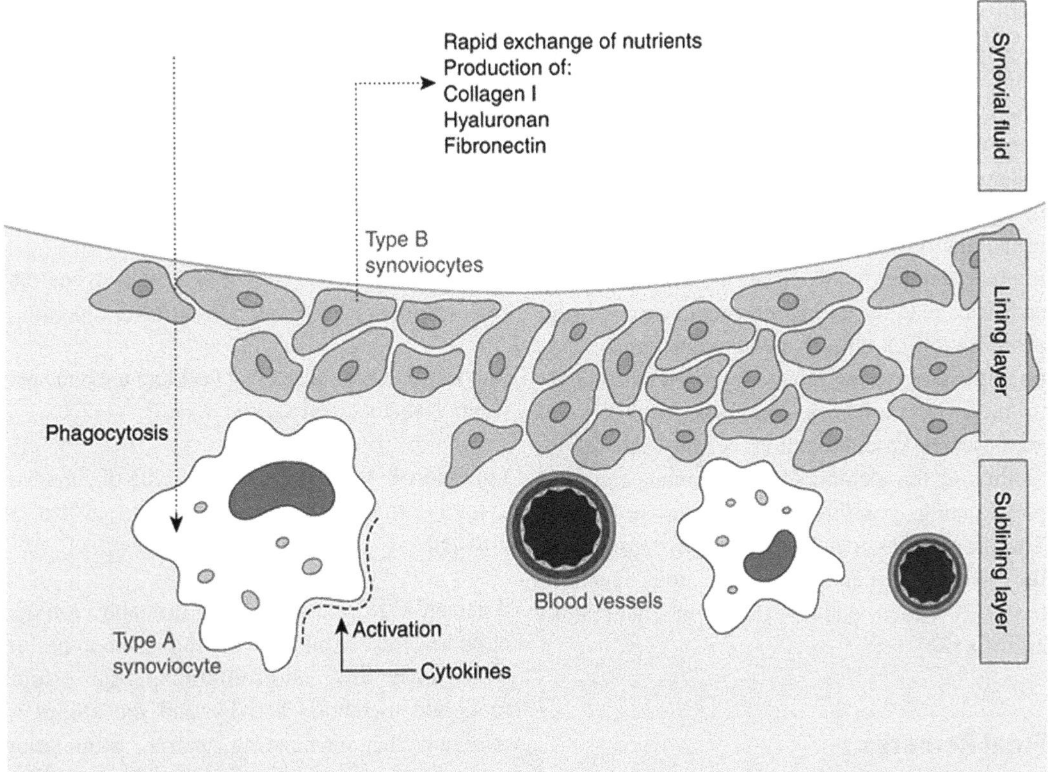

Fig. 2 Schematic view of the synovial membrane (or synovium) and its cell types. Type A synoviocytes are resident macrophages, which are rare in healthy synovium, and type B synoviocytes are fibroblast-like cells. Basal functions of these cell types are shown in black. The different layers of the synovium are labeled on the *right side*. As depicted, the sub-lining is extensively vascularized

protein content is increased as the synovium loses its ability to selectively filter proteins. Meanwhile the lubricant proteins hyaluronic acid and lubricin are decreased, impairing the good lubricating characteristics of the synovial fluid [21]. Therefore, the synovial fluid is a source to monitor changes of the metabolism of the joint or disease activity such as osteoarthritis or rheumatoid arthritis.

Inside-In and Outside-In Signaling: Metabolites Affecting the Joints

Several morphogens and growth factors have been implicated in regulating the sensitive homeostasis of resting chondrocytes in healthy cartilage. Among others, three groups of soluble proteins regulate chondrocyte differentiation in endochondral ossification: bone morphogenetic proteins (BMPs), growth factors, and Wnts [22]. The stimulation of resting chondrocytes with these factors leads to a loss of the resting phenotype and induces hypertrophic differentiation or proliferation of chondrocytes, such as in osteoarthritis. As cartilage tissue is avascular and the chondrocytes are isolated inside their lacunae, the communication between chondrocytes in the superficial zone and chondrocytes in the middle and deeper layers occurs through diffusion. Chondrocytes located in the superficial zone of adult cartilage communicate via gap junction channels consisting of connexin 43 and 45 [23].

The synovium plays an important role in the repair process induced by proinflammatory cytokines that are released in response to intra-articular damage [24]. Exposure to pro-inflammatory cytokines activates type A synoviocytes, which then function as tissue macrophages. The type B synoviocytes are also proposed to play a critical role in the switch from acute inflammation to adaptive immunity, tissue repair, as well as chronic inflammation. During chronic inflammation they can get stably activated, meaning that they fail to switch off their inflammatory program, leading to inappropriate survival and proliferation of type B synoviocytes and retention of leukocytes within the inflamed tissue. Due to these inflammatory conditions, there is an increase in vascularization of the sub-lining, facilitating the accumulation of immune cells and perpetuating possible autoimmune processes. These processes together lead to hyperplasia of the synovial membrane, and the resulting tissue is called pannus tissue in rheumatoid arthritis [25].

Final Remarks

The joint does not only make movement possible, but functions as a highly specialized organ. The different compartments work together with an extensive crosstalk. If one part of the joint is damaged or impaired in function, it affects the whole joint, as seen, for example, during osteoarthritis (see chapter "Osteoarthritis") or rheumatoid arthritis (see chapter "Rheumatoid Arthritis").

Questions and Answers

Question 1 Why is aggrecan such an important component in articular cartilage?

Answer 1 Aggrecan is important for the biomechanical properties of articular cartilage as it is allows, in interaction with other proteins, to retain water. This results in a pressure and reversible deformable tissue that distributes the loads in joints and transfers the compressive load to the underlying bone.

Question 2 In how many groups can joints be divided based on their amount of movement?

Answer 2 Based on the functionality, joints can be divided into three groups; "synarthroses" are immovable, "amphiarthroses" are slightly movable, and "diarthroses" are freely movable.

Question 3 Which are the two cells types that play an important role in normal bone function.

Answer 3 Osteoblasts that build up the bone and osteoclasts that degrade the bone.

Question 4 What is the normal state of chondrocytes in articular cartilage and how can that be affected?

Answer 4 Normally, in adult articular cartilage chondrocytes exhibit a resting phenotype, in which they are non-proliferating, but exhibit moderate metabolic activity and the ability to maintain the surrounding matrix. Stimulation with bone morphogenetic proteins (BMPs), growth factors, and Wnts induces a loss of the resting phenotype and hypertrophic differentiation or proliferation of chondrocytes, such as in osteoarthritis.

References

1. Pacifici M, Koyama E, Iwamoto M (2005) Mechanisms of synovial joint and articular cartilage formation: recent advances, but many lingering mysteries. Birth Defects Res C Embryo Today 75(3):237–248
2. DeLise AM, Fischer L, Tuan RS (2000) Cellular interactions and signaling in cartilage development. Osteoarthritis Cartilage 8(5):309–334
3. Olsen BR, Reginato AM, Wang W (2000) Bone development. Annu Rev Cell Dev Biol 16:191–220
4. Pfander D, Swoboda B, Kirsch T (2001) Expression of early and late differentiation markers (proliferating cell nuclear antigen, syndecan-3, annexin VI, and alkaline phosphatase) by human osteoarthritic chondrocytes. Am J Pathol 159(5):1777–1783
5. Fuerst M, Bertrand J, Lammers L, Dreier R, Echtermeyer F, Nitschke Y, Rutsch F, Schafer FK, Niggemeyer O, Steinhagen J et al (2009) Calcification of articular cartilage in human osteoarthritis. Arthritis Rheum 60(9):2694–2703

6. Suda T, Takahashi N, Udagawa N, Jimi E, Gillespie MT, Martin TJ (1999) Modulation of osteoclast differentiation and function by the new members of the tumor necrosis factor receptor and ligand families. Endocr Rev 20(3):345–357

7. Takahashi M, Hong YM, Yasuda S, Takano M, Kawai K, Nakai S, Hirai Y (1988) Macrophage colony-stimulating factor is produced by human T lymphoblastoid cell line, CEM-ON: identification by amino-terminal amino acid sequence analysis. Biochem Biophys Res Commun 152(3):1401–1409

8. Clement-Lacroix P, Ai M, Morvan F, Roman-Roman S, Vayssiere B, Belleville C, Estrera K, Warman ML, Baron R, Rawadi G (2005) Lrp5-independent activation of Wnt signaling by lithium chloride increases bone formation and bone mass in mice. Proc Natl Acad Sci USA 102(48):17406–17411

9. Kato M, Patel MS, Levasseur R, Lobov I, Chang BH, Glass DA 2nd, Hartmann C, Li L, Hwang TH, Brayton CF et al (2002) Cbfa1-independent decrease in osteoblast proliferation, osteopenia, and persistent embryonic eye vascularization in mice deficient in Lrp5, a Wnt coreceptor. J Cell Biol 157(2):303–314

10. Glass DA 2nd, Bialek P, Ahn JD, Starbuck M, Patel MS, Clevers H, Taketo MM, Long F, McMahon AP, Lang RA et al (2005) Canonical Wnt signaling in differentiated osteoblasts controls osteoclast differentiation. Dev Cell 8(5):751–764

11. Tu X, Joeng KS, Nakayama KI, Nakayama K, Rajagopal J, Carroll TJ, McMahon AP, Long F (2007) Noncanonical Wnt signaling through G protein-linked PKCdelta activation promotes bone formation. Dev Cell 12(1):113–127

12. Wong M, Carter DR (2003) Articular cartilage functional histomorphology and mechanobiology: a research perspective. Bone 33(1):1–13

13. Burgeson RE, Hebda PA, Morris NP, Hollister DW (1982) Human cartilage collagens. Comparison of cartilage collagens with human type V collagen. J Biol Chem 257(13):7852–7856

14. Eyre DR, Muir H (1975) The distribution of different molecular species of collagen in fibrous, elastic and hyaline cartilages of the pig. Biochem J 151(3):595–602

15. Poole AR, Kojima T, Yasuda T, Mwale F, Kobayashi M, Laverty S (2001) Composition and structure of articular cartilage: a template for tissue repair. Clin Orthop Relat Res 2001(391 Suppl):S26–S33

16. Fleming BC, Hulstyn MJ, Oksendahl HL, Fadale PD (2005) Ligament injury, reconstruction and osteoarthritis. Curr Opin Orthop 16(5):354–362

17. Ash Z, McGonagle D (2011) Joint appendages: the structures which have historically been overlooked in arthritis research and therapy development. Best Pract Res Clin Rheumatol 25(6):779–784

18. Korb-Pap A, Bertrand J, Sherwood J, Pap T (2016) Stable activation of fibroblasts in rheumatic arthritis-causes and consequences. Rheumatology (Oxford) 55(suppl 2):ii64–ii67

19. Pascual E, Jovani V (2005) Synovial fluid analysis. Best Pract Res Clin Rheumatol 19(3):371–386

20. Chou CH, Jain V, Gibson J, Attarian DE, Haraden CA, Yohn CB, Laberge RM, Gregory S, Kraus VB (2020) Synovial cell cross-talk with cartilage plays a major role in the pathogenesis of osteoarthritis. Sci Rep 10(1):10868

21. Brannan SR, Jerrard DA (2006) Synovial fluid analysis. J Emerg Med 30(3):331–339

22. Dreier R (2010) Hypertrophic differentiation of chondrocytes in osteoarthritis: the developmental aspect of degenerative joint disorders. Arthritis Res Ther 12(5):216

23. Mayan MD, Carpintero-Fernandez P, Gago-Fuentes R, Martinez-de-Ilarduya O, Wang HZ, Valiunas V, Brink P, Blanco FJ (2013) Human articular chondrocytes express multiple gap junction proteins: differential expression of connexins in normal and osteoarthritic cartilage. Am J Pathol 182(4):1337–1346

24. Smith MD, Barg E, Weedon H, Papengelis V, Smeets T, Tak PP, Kraan M, Coleman M, Ahern MJ (2003) Microarchitecture and protective mechanisms in synovial tissue from clinically and arthroscopically normal knee joints. Ann Rheum Dis 62(4):303–307

25. Neumann E, Lefevre S, Zimmermann B, Gay S, Muller-Ladner U (2010) Rheumatoid arthritis progression mediated by activated synovial fibroblasts. Trends Mol Med 16(10):458–468

Osteoarthritis

Johanne Martel-Pelletier, Camille Roubille, and Jean-Pierre Pelletier

Introduction to Osteoarthritis

Osteoarthritis (OA) is the most common form of arthritis, affecting about 31 million adults in the United States, and between 242 and 300 million individuals worldwide [1–3]. Its onset and development are strongly associated with catabolic alterations and inflammation, leading to joint tissue alterations. This disease is characterized by the degradation and loss of articular cartilage, hypertrophic bone changes with osteophyte formation (bony projections along the joint margins), subchondral bone remodeling, and inflammation of the synovial membrane. Other tissues of the joint, including the muscles and ligaments, are also altered during the OA process, making it a disease of the whole joint [4].

OA can be triggered by endogenous predisposing factors including age, high body mass index and genetics, and external factors such as trauma. This disease could progress slowly and silently, which hinders early diagnosis, and is typically noticed in the moderate-late stages, limiting the effectiveness of preventive measures. It

J. Martel-Pelletier (✉) · J.-P. Pelletier
Osteoarthritis Research Unit, University of Montreal Hospital Research Centre (CRCHUM),
Montreal, QC, Canada
e-mail: jm@martelpelletier.ca; dr@jppelletier.ca

C. Roubille
Internal Medicine Department, Montpellier Hospital Center (CHU), Montpellier, France

engenders pain and reduces quality of life and mobility, which could result in joint replacement.

The current therapies available to treat OA only relieve the symptoms, not the structural alterations of the joint. Although, at present, there are no regulatory agency-approved disease-modifying OA drugs (DMOADs), advancements have been made toward understanding its pathological processes, and promising DMOADs are being developed. These should slow the progression of the disease and improve its outcome.

Pathophysiology of Osteoarthritis

OA is a multifactorial disease resulting in the failure of the articular tissues to maintain a homeostatic balance between matrix synthesis and degradation. An initial phase of cartilage remodeling is characterized by edema, followed by degradation and loss of this tissue. These alterations are associated with synovial inflammation and subchondral bone remodeling (Fig. 1).

In cartilage, there is only one type of cell, the chondrocyte, which is responsible for the maintenance of this tissue's extracellular matrix (ECM, see chapter "Joints: Overview" under part "Joints"). Early during the OA process, increased biomechanical stress and/or biochemical stimuli can activate the anabolic function of chondrocytes

© The Author(s), under exclusive license to Springer Nature Switzerland AG 2026
E. Lammert, M. Zeeb (eds.), *Metabolism of Human Diseases*,
https://doi.org/10.1007/978-3-031-96019-2_15

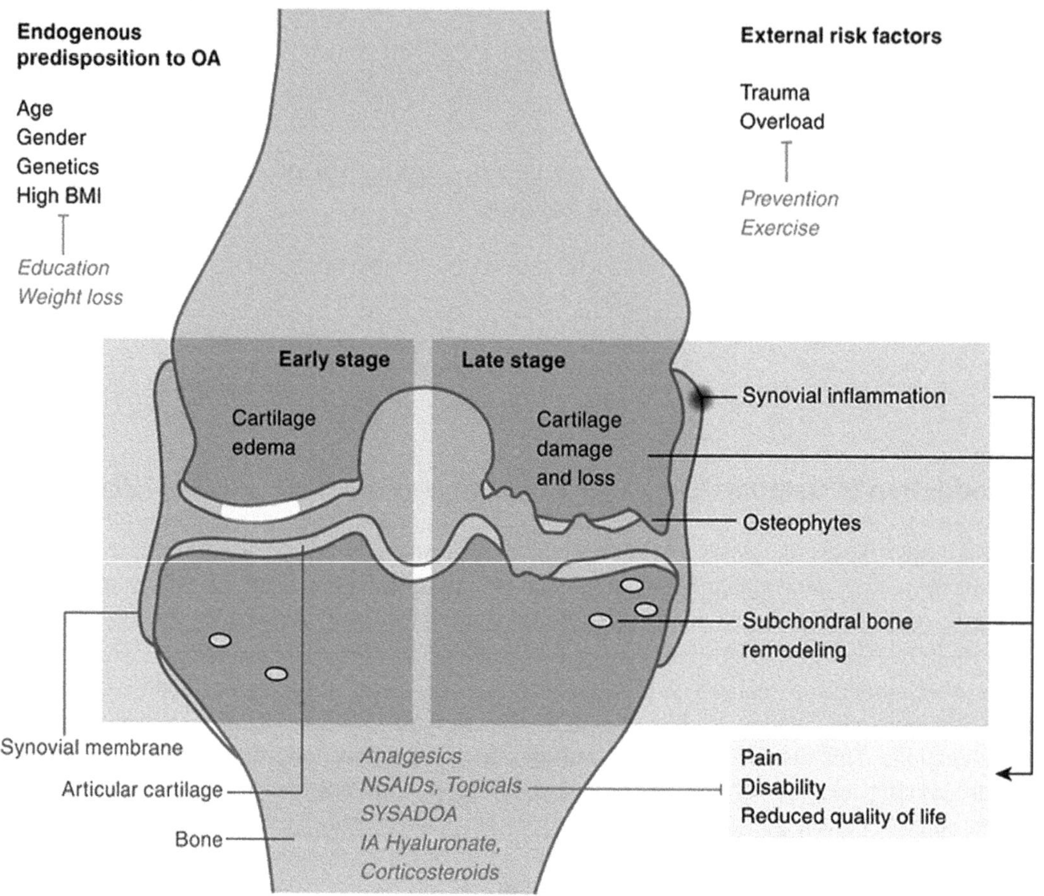

Fig. 1 A model of global pathophysiology of knee osteoarthritis and current non-pharmacological management. Osteoarthritis can be triggered by external factors, including endogenous predisposition factors such as age, genetics, high body mass index (BMI) and trauma. After an initial phase of cartilage edema, cartilage damage and loss occur, which is associated with synovial inflammation, osteophyte formation, and subchondral bone remodeling. These structural changes generate pain, disability, and reduced quality of life. Current osteoarthritis management consists of education (on osteoarthritis, its progression, and its risk factors), weight loss (if necessary), prevention of injury, exercise, and pharmacological symptomatic treatment. Common treatments include acetaminophen (paracetamol), nonsteroidal anti-inflammatory drugs (NSAIDs), opioids, intra-articular (IA) administration of hyaluronate and corticosteroids, and symptomatic slow-acting drugs for OA

to repair early cartilage damage. Over time, this anabolic attempt fails and leads to an imbalance favoring degradation. This degradation will induce a vicious circle in which the degradative fragments of ECM proteins (e.g., fibronectin and collagen) induce synovial membrane inflammation, which in turn will produce catabolic and inflammatory factors [5, 6], thus aggravating the OA process. Increasing evidence suggests that in OA there is a cross-talk between the cartilage, synovial membrane, and subchondral bone, which sustains the catabolic process [7, 8].

Synovial inflammation and subchondral bone remodeling, by their release of soluble mediators, affect the cartilage thus sustaining its degradation. The chondrocytes (autocrine pathway) and synoviocytes (see chapter "Joints: Overview" under part "Joints") in the synovial membrane (paracrine pathway) release catabolic and proinflammatory substances. These include proteinases, for example, matrix metalloproteinases (MMPs) and aggrecanases, and inflammatory cytokines, including interleukin (IL)-1β and tumor necrosis factor-α (TNF-α), which enhance the synthesis of proteinases and other catabolic factors to degrade the ECM of the articular tissues. Soon, these factors overwhelm endogenous inhibitors including, among others, tissue inhibitors of MMPs (TIMPs) and proinflammatory cytokine inhibitors including the IL-1β receptor antagonist (IL-1Ra) (Fig. 2). Additionally, in OA, the loss of integrity of the osteochondral junction is associated with microcracks and the invasion of articular cartilage by vascular channels originating from the subchondral bone. Continued ECM degradation in the articular tissues prevents the cartilage from withstanding normal mechanical factors.

Biomarkers

It is believed that OA patients may not all respond to the same intervention. To accurately manage the course of the disease, it is important to identify early predictors of the disease process. This will lead to precision medicine, ensuring the proper treatment at the right time. Biomarkers offer an interesting avenue and can assist in discriminating individuals who may respond to a given therapy. In OA, biomarkers can be in biological fluids (e.g., urine, blood) or imaging biomarkers (i.e., magnetic resonance imaging [MRI]). Studies of OA biomarkers have primarily focused on molecules, factors, and tissues associated with cartilage, bone metabolism, and inflammation.

As collagen type II and aggrecan are the most abundant proteins in cartilage, their synthesis and degradation products have been the focus of several OA biomarkers studies. Other proteins have also been evaluated, and those extensively studied include, among others, hyaluronic acid, cartilage oligomeric matrix protein (COMP), aggrecanases, MMP-1, -3, and -9, inflammatory factors including IL-1β, IL-6, and TNF-α, as well some bone-related Wnt signaling antagonists, sclerostin, and Dickkopf (DKK)-1.

The image-based methodology MRI has become an important tool for assessing OA, and data have demonstrated that it can detect articular tissue alterations before radiographic evidence of OA. Over the past two decades, MRI for assessing knee structure have been developed, including semi-quantitative (scoring) methods and both semi- and fully automated quantitative systems. These have enabled the quantification of alterations in many joint tissues involved in the pathology of OA. In addition to cartilage alterations changes in the menisci, ligaments, infrapatellar fat pad, bone curvature/shape, bone marrow, and joint effusion/swelling have also been studied.

Linked with the development of OA is genetic variability. Genome-wide association studies and quantitative trait analyses have identified many candidate genes within the loci associated with the increased risk of OA [9, 10]. Recently, three genes associated with cartilage formation and repair as well as structural disease progression, namely fibroblast growth factor 18 (FGF-18), growth differentiation factor 5 (GDF-5), and transforming growth factor beta (TGF-β) [10], have been studied for their DMOAD properties.

Data are emerging about OA disease-associated epigenetic regulatory mechanisms in the blood and articular tissues [11]. Suggested as pharmacological candidates for DMOADs are three enzymes including DNA methyltransferase (DNMT)3B, ten-eleven-translocation (TET)1, and sirtuin (SIRT)1 [12]. However, many challenges must be addressed for OA therapy; a major one is ensuring the specificity of modulating epigenetic regulators and preventing off-target effects.

Osteoarthritis Management

A multimodal approach combining non-pharmacological and pharmacological treatment (Fig. 1) [13] is currently the primary option for OA management. Presently, treatments primarily involve those aimed at reducing the dominant symptom of OA—joint pain. However, there is still a significant group of patients for whom these treatments do not provide adequate pain relief.

Pharmacological treatments are classified into rapid or slow-acting symptomatic agents

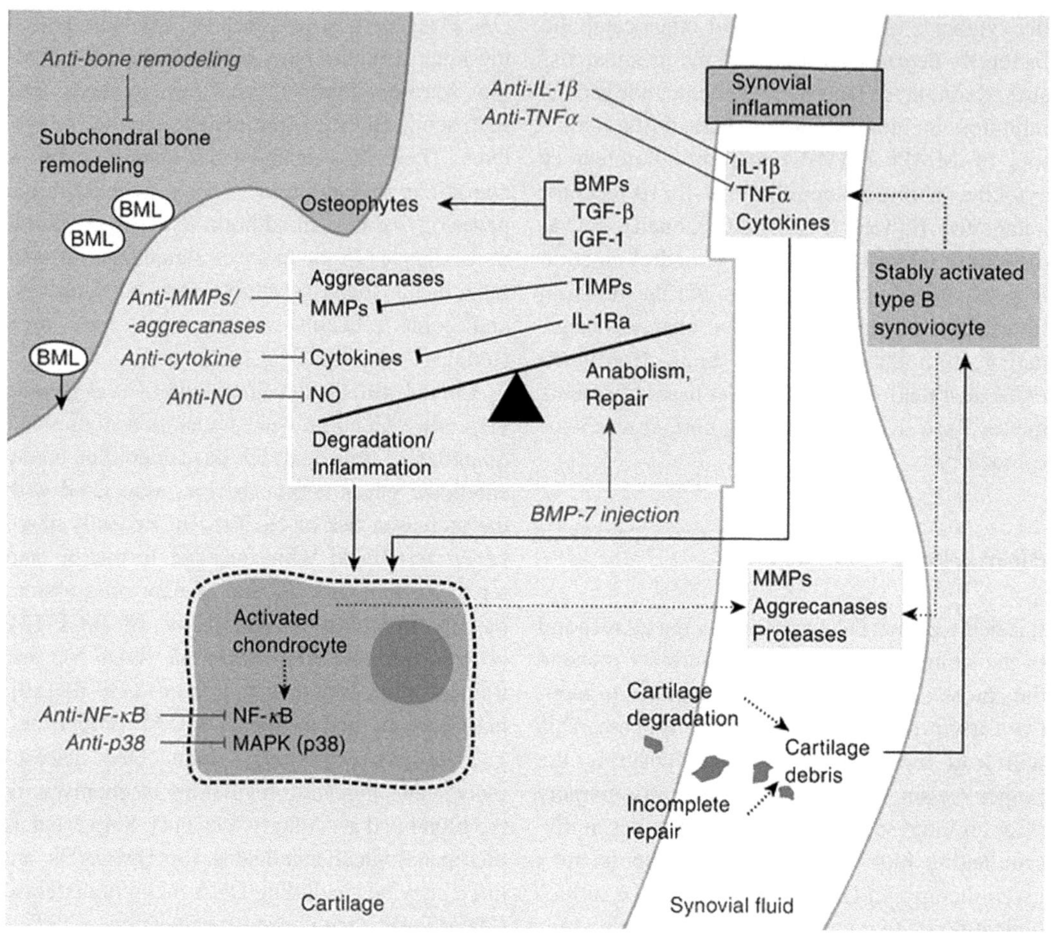

Fig. 2 A model of cross-talk between cartilage, synovial membrane, and subchondral bone leading to an imbalance favoring articular tissue degradation in knee osteoarthritis. Excessive production of proteinases (matrix metalloproteinases (MMPs) and aggrecanases), nitric oxide (NO), and inflammatory cytokines (interleukin-1β [IL-1β] and tumor necrosis factor α [TNF-α]) by the chondrocyte contribute to the degradation of articular tissue including the cartilage. This induces a vicious circle in which the cartilage fragments activate synoviocytes (see chapter "Joints: Overview" under part "Joints"), resulting in enhanced cartilage degradation and synovial inflammation. The attempt to repair, which could occur via growth factors such as bone morphogenetic proteins (BMPs) and transforming growth factor β (TGF-β), fails to achieve a complete repair of the extracellular matrix. Disease-modifying osteoarthritis drugs, such as inhibitors of MMPs and aggrecanases, anti-cytokine therapy, anti-bone remodeling, anti-nuclear factor-κB (NF-κB), and anti-mitogen-activated protein kinase (MAPK), have been developed, and aim to inhibit these factors and pathways. Bone marrow lesion (BML), insulin-like growth factor 1 (IGF-1), interleukin-1β receptor antagonist (IL-1Ra), tissue inhibitors of metalloproteinases (TIMPs)

(SYSADOAs), and some of the slow-acting may also contribute to reducing the natural progression of joint structural damage. When combined approaches are unsuccessful, surgical treatments may be considered. There is also hope for a regulatory agency-approved DMOAD.

Non-pharmacological Treatment

The combination of education, exercise, improvement in muscle strength, and weight loss (if overweight) is reported to be joint protective and is recommended by most guidelines [14]. In addition, orthotic (joint-stabilizing) devices and the prevention of injury are also recognized as beneficial interventions.

Pharmacological Treatment: Rapid-Acting Symptomatic Agents

The rapid-acting symptomatic treatments for OA consist mainly of analgesics, oral and topical nonsteroidal anti-inflammatory drugs (NSAIDs), and opioids.

At present, acetaminophen is no longer the first-line analgesic recommended in guidelines due to limitations in effect and a range of unwanted side effects [15, 16]. Opioids have become more widely prescribed (often in combination with acetaminophen), particularly for OA patients who experience a lack of efficacy, contraindications, or intolerance to NSAIDs [13] and those who cannot undergo total joint arthroplasty because of comorbidities contraindicating surgery and anesthesia [14]. However, opioids present several, sometimes severe adverse events. These include analgesia, sedation, vomiting, and respiratory depression. Other harms that outweigh the benefits of this drug include overdoses and deaths.

Duloxetine, another analgesic, has shown significant moderate benefits for knee pain, function, and quality of life [17]. It is a selective serotonin and norepinephrine reuptake inhibitor. These neurotransmitters are involved in the mediation of endogenous descending inhibitory pain pathways and central sensitization. In chronic pain states, their inhibitory effect is reduced or lost, and duloxetine increases their availability and activity [18]. The main adverse events include nausea, constipation, and hyperhidrosis (increased sweating).

A second class of rapid symptomatic treatments that aim to block or reduce joint inflammation is NSAIDs. NSAIDs can be used orally or topically with similar efficacy [19]; however, the topical application shows fewer gastrointestinal complications but possible local skin reactions. NSAIDs can be nonselective (inhibiting both cyclooxygenase [COX]-1 and -2; e.g., diclofenac) or selectively inhibit only COX-2, also

known as coxibs (e.g., celecoxib). These are recommended for patients who are unresponsive to acetaminophen, preferentially during inflammatory flares [13]. The use of NSAIDs is limited by gastrointestinal, renal, and cardiovascular side effects, which increase with age due to comorbidities. Coxibs demonstrate fewer gastrointestinal complications than nonselective NSAIDs but pose a potential cardiovascular risk [20]. Coxibs inhibit prostacyclin production (by COX-2) but do not inhibit thromboxane A2 release (from platelets, formed by COX-1). This potentially explains the coxib-related cardiovascular risk, as such an imbalance could create continued thromboxane A2 production, thus increasing the risk of thrombosis. Another explanation could be that some coxibs have been found to increase blood pressure.

Corticosteroids are potent anti-inflammatory drugs that inhibit, among other factors, phospholipase A2, reducing the release of proinflammatory phospholipids. Intra-articular corticosteroid injection is recommended for OA inflammatory flares [13].

Hyaluronic acid is a glycosaminoglycan component of ECM and synovial fluid. It is involved in the maintenance of joint homeostasis, and its concentration decreases in OA patients. Intra-articular hyaluronic acid injection (viscosupplementation) is recommended for knee OA patients with an inadequate response to initial therapy [14], despite possible induced transient pain and swelling at the injection site [21]. Compared with intra-articular corticosteroids, viscosupplementation has a delayed but prolonged effect [21].

Nerve growth factor (NGF) modulates signals that control the expression of peripheral and central pain substances and sensitized adjacent nociceptive neurons in response to stimulation. Pain relief and improvement in functional capacity are observed with intravenous administration of the NGF inhibitors fasinumab and tanezumab, as well as with subcutaneous administration of fulranumab [22–25]. However, side effects may include osteonecrosis.

Other novel treatments have been proposed for reducing clinical symptoms and include platelet-rich plasma and mesenchymal stem cells

(MSCs). However, recent studies do not support the use of platelet-rich plasma for managing knee OA symptoms [26, 27]. Additionally, there is a lack of high-quality clinical studies evaluating the efficacy of MSCs on symptoms and joint structure.

Pharmacological Treatment: Slow-Acting Symptomatic Drugs

Among the SYSADOAs for OA are glucosamine and chondroitin sulfate. Several clinical trials demonstrated that when using as outcomes radiography or cartilage volume loss assessed by MRI, glucosamine sulfate or chondroitin sulfate has moderate knee structural protective effects in patients with OA [28–31]. Moreover, a combination of both products reduced knee cartilage volume loss as assessed by MRI over two years [32].

Diacerein, an inhibitor of IL-1β and some proteinases, is effective in knee [33] and hip OA [34]. Diarrhea is the most frequent adverse event, which likely occurs due to prostaglandin synthesis induced by the active metabolite of diacerein, rhein, leading to an increase in gut motility.

Treatment with avocado-soybean unsaponifiables reduces pain in knee and hip OA, and the effect is prolonged even after treatment discontinuation [35, 36]. Inhibition of IL-1β and MMPs has been proposed as a potential mechanism of action.

Disease-Modifying Osteoarthritis Drugs

Currently, several classes of DMOADs are in development or being tested in clinical trials (Fig. 2).

Targeting Cartilage Catabolism and Anabolism

MMP inhibitors aim to block ECM degradation. One such drug, doxycycline, showed only a minimal structural benefit and no effect on pain in a clinical trial [37]. An inhibitor of inducible nitric oxide (NO) synthase (cindunistat) revealed a reduction in joint space narrowing in mild OA in the first year [38].

A randomized, double-blind, placebo-controlled trial of a recombinant human Fibroblast growth factor 18 (FGF-18) (sprifermin) showed no reduction in the primary outcome, that is, central medial femorotibial compartment cartilage thickness, by using MRI in symptomatic knee OA patients. However, there was a dose-dependent reduction in loss of total and lateral femorotibial cartilage thickness and volume, as evaluated by MRI, as well as in the radiographic joint space width narrowing in the lateral femorotibial compartment. This drug also appeared to reduce symptoms at 12 months [39]. Concurring with the first trial, a recent dose-finding multicenter randomized trial also reported an improvement in total femorotibial joint cartilage thickness after two years of treatment [40].

Targeting Synovial Inflammation by Anti-cytokine Therapy

A blockade of inflammatory cytokines focuses on IL-1β and TNF-α. Other potential targets include IL-6, nuclear factor-κB (NF-κB), and mitogen-activated protein kinase (MAPK), as they are part of the inflammation cascade in OA. Intra-articular administration of the IL-1β receptor antagonist anakinra shows controversial results with regard to symptoms [41], probably due to the short half-life of the drug and the protocol used for the clinical studies. Moreover, two antibodies against TNF-α (adalimumab, infliximab) were shown to alleviate symptoms of hand [42] and knee OA [43], yet the results remain controversial [44].

The conventional disease-modifying drugs used in inflammatory arthritides (DMARDs), such as hydroxychloroquine and methotrexate, have also been investigated for OA. A systematic review and meta-analysis concluded that hydroxychloroquine has no benefit in reducing pain and improving physical function in hand or knee OA patients [45]. Methotrexate is not routinely used due to insufficient evidence for its efficacy in OA.

Targeting Subchondral Bone Remodeling

Strontium ranelate inhibits bone resorption in subchondral bone [46]. This results from decreased differentiation and resorptive activity of osteoclasts, along with increased osteoclast apoptosis [47]. Strontium ranelate reduces the progression of spinal [48] and knee OA, as assessed in a Phase III trial using both X-rays [49] and MRI [50]. A meta-analysis also confirmed an improvement in joint space narrowing with strontium ranelate [51]. However, the original manufacturer ceased strontium ranelate production in 2017 due to concerns regarding cardiovascular safety, specifically thromboembolic side effects.

Bisphosphonates can inhibit bone resorption and, therefore, are the mainstream medications for osteoporosis. Although there are controversies in the literature about bisphosphonates in the treatment of OA, the most recent meta-analysis recorded on this subject concluded that such treatment did not provide symptomatic relief nor delay radiographic progression in knee OA, but it could have beneficial effects in certain subsets of patients who exhibit high rates of subchondral bone turnover [52].

In vitro, calcitonin reduces collagen degradation by inhibiting the expression and activity of MMPs in chondrocytes. A pivotal Phase III clinical trial evaluating oral salmon calcitonin in patients with knee OA demonstrated no reproducible clinical effects on pain and function [53].

Vitamin D deficiency was found to be associated with the development and progression of knee OA. However, whether vitamin D supplements can reduce OA pain and structural changes remains controversial.

Transforming growth factor beta (TGF-β) has been identified as an important mediator of subchondral bone development, and inhibition of its activity in subchondral bone appears to attenuate degeneration of articular cartilage [54]. Results from two Phase II [55, 56] and one Phase III [57] trials on the Tissue Gene-C delivery of human allogeneic chondrocytes expressing Transforming growth factor beta (TGF-β)1 directly to the injured joint reported no statistically significant improvement in the knee OA structure. However, it showed improvement in pain and function.

The Wnt/β-catenin signaling may also prove to be a promising target. Lorecivivint is a small-molecule Wnt pathway inhibitor. Results from a Phase IIa trial with intra-articular administration of Lorecivivint in patients with moderately to severely symptomatic knee OA did not show improvement in pain [58].

Perspectives

The management of OA is based on a wide spectrum of therapeutic options to relieve pain, but drugs aimed at delaying structural progression are lacking. The currently available therapies do not meet all the needs of OA patients, particularly for those with rapid structural progression of the disease. The focus is primarily on the development of DMOADs that could be combined with conventional therapy to provide a more effective treatment. Given the multiple pathways involved in this disease, a multidisciplinary approach to treatment is likely the future therapeutic strategy for OA.

Questions and Answers

Question 1 What is the name (acronym) of a drug that inhibits the structural OA disease progression?

Answer 1 Disease-modifying osteoarthritis drug (DMOAD).

Question 2 What is the dominant unmet symptom of OA?

Answer 2 Pain.

Question 3 What are most guidelines advocating for with regard to the treatment of OA?

Answer 3 Most guidelines are consistent in advocating for non-pharmacologic treatments including exercise, weight loss, and education as the initial interventions.

Question 4 What are the targeting molecules for the NSAIDs?

Answer 4 Both cyclooxygenase (COX)-1 and 2.

References

1. OARSI (2016) Osteoarthritis: a serious disease. In: Submitted to the U.S. Food and Drug Administration. https://www.oarsi.org/sites/default/files/library/2018/pdf/oarsi_white_paper_oa_serious_disease121416_1.pdf. Accessed 12 Dec 2019.
2. Vina ER, Kwoh CK (2018) Epidemiology of osteoarthritis: literature update. Curr Opin Rheumatol 30:160–167
3. Abramoff B, Caldera FE (2020) Osteoarthritis: pathology, diagnosis, and treatment options. Med Clin North Am 104:293–311
4. Loeser RF, Goldring SR, Scanzello CR, Goldring MB (2012) Osteoarthritis: a disease of the joint as an organ. Arthritis Rheum 64:1697–1707
5. Martel-Pelletier J, Lajeunesse D, Pelletier JP (2005) Chapter 109: etiopathogenesis of osteoarthritis. In: Koopman WJ, Moreland LW (eds) Arthritis & allied conditions. a textbook of rheumatology, 15th edn. Lippincott, Williams & Wilkins, Baltimore, pp 2199–2226
6. Martel-Pelletier J, Boileau C, Pelletier J-P, Roughley P (2008) Cartilage in normal and osteoarthritis conditions. In: Pap T (ed) Best practice & research clinical rheumatology, vol 22. Rapid Medical Media, East Sussex, pp 351–384
7. Martel-Pelletier J, Lajeunesse D, Pelletier JP (2007) Chapter 13: subchondral bone and osteoarthritis progression: a very significant role. In: Buckwalter JA, Lotz M, Stoltz JF (eds) Osteoarthritis, inflammation and degradation: a continuum. IOS Press, Amsterdam, pp 206–218
8. Martel-Pelletier J, Pelletier JP (2010) Is osteoarthritis a disease involving only cartilage or other articular tissues? Eklem Hastalik Cerrahisi 21:2–14
9. Loughlin J (2015) Genetic contribution to osteoarthritis development: current state of evidence. Curr Opin Rheumatol 27:284–288
10. Tachmazidou I, Hatzikotoulas K, Southam L, Esparza-Gordillo J, Haberland V, Zheng J, Johnson T, Koprulu M, Zengini E, Steinberg J, Wilkinson JM, Bhatnagar S, Hoffman JD, Buchan N, Suveges D, Arc OC, Yerges-Armstrong L, Smith GD, Gaunt TR, Scott RA, McCarthy LC, Zeggini E (2019) Identification of new therapeutic targets for osteoarthritis through genome-wide analyses of UK Biobank data. Nat Genet 51:230–236
11. Barter MJ, Bui C, Young DA (2012) Epigenetic mechanisms in cartilage and osteoarthritis: DNA methylation, histone modifications and microRNAs. Osteoarthr Cartil 20:339–349
12. Grandi FC, Bhutani N (2020) Epigenetic therapies for osteoarthritis. Trends Pharmacol Sci 41:557–569
13. Zhang W, Moskowitz RW, Nuki G, Abramson S, Altman RD, Arden N, Bierma-Zeinstra S, Brandt KD, Croft P, Doherty M, Dougados M, Hochberg M, Hunter DJ, Kwoh K, Lohmander LS, Tugwell P (2008) OARSI recommendations for the management of hip and knee osteoarthritis, part II: OARSI evidence-based, expert consensus guidelines. Osteoarthr Cartil 16:137–162
14. Kolasinski SL, Neogi T, Hochberg MC, Oatis C, Guyatt G, Block J, Callahan L, Copenhaver C, Dodge C, Felson D, Gellar K, Harvey WF, Hawker G, Herzig E, Kwoh CK, Nelson AE, Samuels J, Scanzello C, White D, Wise B, Altman RD, DiRenzo D, Fontanarosa J, Giradi G, Ishimori M, Misra D, Shah AA, Shmagel AK, Thoma LM, Turgunbaev M, Turner AS, Reston J (2020) 2019 American College of Rheumatology/Arthritis Foundation guideline for the management of osteoarthritis of the hand, hip, and knee. Arthritis Care Res (Hoboken) 72:149–162
15. Roberts E, Delgado Nunes V, Buckner S, Latchem S, Constanti M, Miller P, Doherty M, Zhang W, Birrell F, Porcheret M, Dziedzic K, Bernstein I, Wise E, Conaghan PG (2016) Paracetamol: not as safe as we thought? A systematic literature review of observational studies. Ann Rheum Dis 75:552–559
16. Leopoldino AO, Machado GC, Ferreira PH, Pinheiro MB, Day R, McLachlan AJ, Hunter DJ, Ferreira ML (2019) Paracetamol versus placebo for knee and hip osteoarthritis. Cochrane Database Syst Rev 2:CD013273
17. Osani MC, Bannuru RR (2019) Efficacy and safety of duloxetine in osteoarthritis: a systematic review and meta-analysis. Korean J Intern Med 34:966–973
18. Skljarevski V, Zhang S, Iyengar S, D'Souza D, Alaka K, Chappell A, Wernicke J (2011) Efficacy of duloxetine in patients with chronic pain conditions. Curr Drug ther 6:296–303
19. Underwood M, Ashby D, Cross P, Hennessy E, Letley L, Martin J, Mt-Isa S, Parsons S, Vickers M, Whyte K, team Ts. (2008) Advice to use topical or oral ibuprofen for chronic knee pain in older people: randomised controlled trial and patient preference study. BMJ 336:138–142

20. Trelle S, Reichenbach S, Wandel S, Hildebrand P, Tschannen B, Villiger PM, Egger M, Juni P (2011) Cardiovascular safety of non-steroidal anti-inflammatory drugs: network meta-analysis. BMJ 342:c7086

21. Bellamy N, Campbell J, Robinson V, Gee T, Bourne R, Wells G (2006) Viscosupplementation for the treatment of osteoarthritis of the knee. Cochrane Database Syst Rev:CD005321

22. Sanga P, Katz N, Polverejan E, Wang S, Kelly KM, Haeussler J, Thipphawong J (2013) Efficacy, safety, and tolerability of fulranumab, an anti-nerve growth factor antibody, in the treatment of patients with moderate to severe osteoarthritis pain. Pain 154:1910–1919

23. Brown MT, Herrmann DN, Goldstein M, Burr AM, Smith MD, West CR, Verburg KM, Dyck PJ (2014) Nerve safety of tanezumab, a nerve growth factor inhibitor for pain treatment. J Neurol Sci 345:139–147

24. Tiseo PJ, Kivitz AJ, Ervin JE, Ren H, Mellis SJ (2014) Fasinumab (REGN475), an antibody against nerve growth factor for the treatment of pain: results from a double-blind, placebo-controlled exploratory study in osteoarthritis of the knee. Pain 155:1245–1252

25. Schnitzer TJ, Ekman EF, Spierings EL, Greenberg HS, Smith MD, Brown MT, West CR, Verburg KM (2015) Efficacy and safety of tanezumab monotherapy or combined with non-steroidal anti-inflammatory drugs in the treatment of knee or hip osteoarthritis pain. Ann Rheum Dis 74:1202–1211

26. Bennell KL, Paterson KL, Metcalf BR, Duong V, Eyles J, Kasza J, Wang Y, Cicuttini F, Buchbinder R, Forbes A, Harris A, Yu SP, Connell D, Linklater J, Wang BH, Oo WM, Hunter DJ (2021) Effect of intra-articular platelet-rich plasma vs placebo injection on pain and medial tibial cartilage volume in patients with knee osteoarthritis: The RESTORE randomized clinical trial. JAMA 326:2021–2030

27. Glinkowski WM, Gut G, Śladowski D (2025) Platelet-Rich Plasma for Knee Osteoarthritis: A Comprehensive Narrative Review of the Mechanisms Preparation Protocols and Clinical Evidence. J Clin Med 14(11):3983. https://doi.org/10.3390/jcm14113983

28. Pavelka K, Gatterova J, Olejarova M, Machacek S, Giacovelli G, Rovati LC (2002) Glucosamine sulfate use and delay of progression of knee osteoarthritis: a 3-year, randomized, placebo-controlled, double-blind study. Arch Intern Med 162:2113–2123

29. Kahan A, Uebelhart D, De Vathaire F, Delmas PD, Reginster JY (2009) Long-term effects of chondroitins 4 and 6 sulfate on knee osteoarthritis: the study on osteoarthritis progression prevention, a two-year, randomized, double-blind, placebo-controlled trial. Arthritis Rheum 60:524–533

30. Lee YH, Woo JH, Choi SJ, Ji JD, Song GG (2010) Effect of glucosamine or chondroitin sulfate on the osteoarthritis progression: a meta-analysis. Rheumatol Int 30:357–363

31. Wildi LM, Raynauld JP, Martel-Pelletier J, Beaulieu A, Bessette L, Morin F, Abram F, Dorais M, Pelletier JP (2011) Chondroitin sulphate reduces both cartilage volume loss and bone marrow lesions in knee osteoarthritis patients starting as early as 6 months after initiation of therapy: a randomised, double-blind, placebo-controlled pilot study using MRI. Ann Rheum Dis 70:982–989

32. Martel-Pelletier J, Roubille C, Abram F, Hochberg MC, Dorais M, Delorme P, Raynauld JP, Pelletier JP (2015) First-line analysis of the effects of treatment on progression of structural changes in knee osteoarthritis over 24 months: data from the osteoarthritis initiative progression cohort. Ann Rheum Dis 74:547–556

33. Pelletier JP, Yaron M, Haraoui B, Cohen P, Nahir MA, Choquette D, Wigler I, Rosner IA, Beaulieu AD (2000) Efficacy and safety of diacerein in osteoarthritis of the knee: a double-blind, placebo-controlled trial. The Diacerein Study Group. Arthritis Rheum 43:2339–2348

34. Dougados M, Nguyen M, Berdah L, Mazieres B, Vignon E, Lequesne M (2001) Evaluation of the structure-modifying effects of diacerein in hip osteoarthritis: ECHODIAH, a three-year, placebo-controlled trial. Evaluation of the chondromodulating effect of diacerein in OA of the hip. Arthritis Rheum 44:2539–2547

35. Maheu E, Mazieres B, Valat JP, Loyau G, Le Loet X, Bourgeois P, Grouin JM, Rozenberg S (1998) Symptomatic efficacy of avocado/soybean unsaponifiables in the treatment of osteoarthritis of the knee and hip: a prospective, randomized, double-blind, placebo-controlled, multicenter clinical trial with a six-month treatment period and a two-month followup demonstrating a persistent effect. Arthritis Rheum 41:81–91

36. Maheu E, Cadet C, Marty M, Moyse D, Kerloch I, Coste P, Dougados M, Mazieres B, Spector TD, Halhol H, Grouin JM, Lequesne M (2014) Randomised, controlled trial of avocado-soybean unsaponifiable (Piascledine) effect on structure modification in hip osteoarthritis: the ERADIAS study. Ann Rheum Dis 73:376–384

37. Nuesch E, Rutjes AW, Trelle S, Reichenbach S, Juni P (2009) Doxycycline for osteoarthritis of the knee or hip. Cochrane Database Syst Rev:CD007323

38. Hellio le Graverand MP, Clemmer RS, Redifer P, Brunell RM, Hayes CW, Brandt KD, Abramson SB, Manning PT, Miller CG, Vignon E (2013) A 2-year randomised, double-blind, placebo-controlled, multicentre study of oral selective iNOS inhibitor, cindunistat (SD-6010), in patients with symptomatic osteoarthritis of the knee. Ann Rheum Dis 72:187–195

39. Lohmander LS, Hellot S, Dreher D, Krantz EF, Kruger DS, Guermazi A, Eckstein F (2014) Intraarticular sprifermin (recombinant human fibroblast growth factor 18) in knee osteoarthritis: a randomized, double-blind, placebo-controlled trial. Arthritis Rheumatol 66:1820–1831

40. Hochberg MC, Guermazi A, Guehring H, Aydemir A, Wax S, Fleuranceau-Morel P, Reinstrup Bihlet A, Byrjalsen I, Ragnar Andersen J, Eckstein F (2019)

Effect of intra-articular sprifermin vs placebo on femorotibial joint cartilage thickness in patients with osteoarthritis: the FORWARD randomized clinical trial. JAMA 322:1360–1370

41. Chevalier X, Goupille P, Beaulieu AD, Burch FX, Bensen WG, Conrozier T, Loeuille D, Kivitz AJ, Silver D, Appleton BE (2009) Intraarticular injection of anakinra in osteoarthritis of the knee: a multicenter, randomized, double-blind, placebo-controlled study. Arthritis Rheum 61:344–352

42. Fioravanti A, Fabbroni M, Cerase A, Galeazzi M (2009) Treatment of erosive osteoarthritis of the hands by intra-articular infliximab injections: a pilot study. Rheumatol Int 29:961–965

43. Maksymowych WP, Russell AS, Chiu P, Yan A, Jones N, Clare T, Lambert RG (2012) Targeting tumour necrosis factor alleviates signs and symptoms of inflammatory osteoarthritis of the knee. Arthritis Res Ther 14:R206

44. Verbruggen G, Wittoek R, Vander Cruyssen B, Elewaut D (2012) Tumour necrosis factor blockade for the treatment of erosive osteoarthritis of the interphalangeal finger joints: a double blind, randomised trial on structure modification. Ann Rheum Dis 71:891–898

45. Singh A, Kotlo A, Wang Z, Dissanayaka T, Das S, Antony B (2021) Efficacy and safety of hydroxychloroquine in osteoarthritis: a systematic review and meta-analysis of randomized controlled trials. Korean J Intern Med

46. Tat SK, Pelletier JP, Mineau F, Caron J, Martel-Pelletier J (2011) Strontium ranelate inhibits key factors affecting bone remodeling in human osteoarthritic subchondral bone osteoblasts. Bone 49:559–567

47. Martel-Pelletier J, Wildi LM, Pelletier JP (2012) Future therapeutics for osteoarthritis. Bone 51:297–311

48. Bruyere O, Delferriere D, Roux C, Wark JD, Spector T, Devogelaer JP, Brixen K, Adami S, Fechtenbaum J, Kolta S, Reginster JY (2008) Effects of strontium ranelate on spinal osteoarthritis progression. Ann Rheum Dis 67:335–339

49. Reginster JY, Badurski J, Bellamy N, Bensen W, Chapurlat R, Chevalier X, Christiansen C, Genant H, Navarro F, Nasonov E, Sambrook PN, Spector TD, Cooper C (2013) Efficacy and safety of strontium ranelate in the treatment of knee osteoarthritis: results of a double-blind, randomised placebo-controlled trial. Ann Rheum Dis 72:179–186

50. Pelletier JP, Roubille C, Raynauld JP, Abram F, Dorais M, Delorme P, Martel-Pelletier J (2015) Disease-modifying effect of strontium ranelate in a subset of patients from the Phase III knee osteoarthritis study SEKOIA using quantitative MRI: reduction in bone marrow lesions protects against cartilage loss. Ann Rheum Dis 74:422–429

51. Gregori D, Giacovelli G, Minto C, Barbetta B, Gualtieri F, Azzolina D, Vaghi P, Rovati LC (2018) Association of pharmacological treatments with long-term pain control in patients with knee osteoarthritis: a systematic review and meta-analysis. JAMA 320:2564–2579

52. Vaysbrot EE, Osani MC, Musetti MC, McAlindon TE, Bannuru RR (2018) Are bisphosphonates efficacious in knee osteoarthritis? A meta-analysis of randomized controlled trials. Osteoarthr Cartil 26:154–164

53. Karsdal MA, Byrjalsen I, Alexandersen P, Bihlet A, Andersen JR, Riis BJ, Bay-Jensen AC, Christiansen C, CSMC021C2301/2 investigators (2015) Treatment of symptomatic knee osteoarthritis with oral salmon calcitonin: results from two phase 3 trials. Osteoarthr Cartil 23:532–543

54. Zhen G, Wen C, Jia X, Li Y, Crane JL, Mears SC, Askin FB, Frassica FJ, Chang W, Yao J, Carrino JA, Cosgarea A, Artemov D, Chen Q, Zhao Z, Zhou X, Riley L, Sponseller P, Wan M, Lu WW, Cao X (2013) Inhibition of TGF-beta signaling in mesenchymal stem cells of subchondral bone attenuates osteoarthritis. Nat Med 19:704–712

55. Ha CW, Cho JJ, Elmallah RK, Cherian JJ, Kim TW, Lee MC, Mont MA (2015) A multicenter, single-blind, phase IIa clinical trial to evaluate the efficacy and safety of a cell-mediated gene therapy in degenerative knee arthritis patients. Hum Gene Ther Clin Dev 26:125–130

56. Lee MC, Ha CW, Elmallah RK, Cherian JJ, Cho JJ, Kim TW, Bin SI, Mont MA (2015) A placebo-controlled randomised trial to assess the effect of TGF-ss1-expressing chondrocytes in patients with arthritis of the knee. Bone Joint J 97-B:924–932

57. Kim MK, Ha CW, In Y, Cho SD, Choi ES, Ha JK, Lee JH, Yoo JD, Bin SI, Choi CH, Kyung HS, Lee MC (2018) A multicenter, double-blind, phase III clinical trial to evaluate the efficacy and safety of a cell and gene therapy in knee osteoarthritis patients. Hum Gene Ther Clin Dev 29:48–59

58. Yazici Y, McAlindon TE, Gibofsky A, Lane NE, Clauw D, Jones M, Bergfeld J, Swearingen CJ, DiFrancesco A, Simsek I, Tambiah J, Hochberg MC (2020) Lorecivivint, a novel intraarticular CDC-like kinase 2 and dual-specificity tyrosine phosphorylation-regulated kinase 1A inhibitor and Wnt pathway modulator for the treatment of knee osteoarthritis: a phase II randomized trial. Arthritis Rheumatol 72:1694–1706

Rheumatoid Arthritis

Emanuele Gotelli, Stefano Soldano,
Vanessa Smith, and Maurizio Cutolo

Introduction to Rheumatoid Arthritis

Rheumatoid arthritis (RA) is a chronic, autoimmune, and inflammatory systemic disease that mainly involves diarthrodial joints (see chapter "Joints: Overview"), with cartilage degradation, bone erosion, and progressive loss of articular function [1].

The prevalence of RA ranges from 0.5 to 1% of the whole population and women are more frequently affected than men (estimated ratio F:M = 3:1) [2].

The cause of RA is unknown, but a cross-talk between genetic, non-genetic, and epigenetic factors has been identified. The consequence is the pathological modification of synovial antigens that activate both innate (neutrophils, monocytes/macrophages, natural killer cells) and adaptive (B and T lymphocytes) immune response. The release of several proinflammatory cytokines, such as tumor necrosis factor-α (TNF-α), interleukin (IL)-1, IL-6, IL-17 and interferon (IFN)-γ, causes joint damage. Moreover, B lymphocytes produce auto-antibodies (rheumatoid factor—RF and anti-citrullinated protein antibodies—ACPA) that strengthen innate response and perpetuate articular injury [1].

This intense activation of immune cells and pro-inflammatory cytokines produces systemic symptoms (fever, morning stiffness, weight loss, asthenia, decreased appetite), involving also extra-articular tissues (lungs, eyes, heart, kidneys and small vessels) [1].

E. Gotelli · S. Soldano · M. Cutolo (✉)
Laboratory of Experimental Rheumatology and Academic Division of Clinical Rheumatology, Department of Internal Medicine and Medical Specialties (Di.M.I.), University of Genova, Genova, Italy
e-mail: emanuele.gotelli@unige.it;
stefano.soldano@unige.it; mcutolo@unige.it

V. Smith
Department of Rheumatology, Ghent University Hospital, Ghent, Belgium

Department of Internal Medicine, Ghent University Hospital, Ghent, Belgium

Unit for Molecular Immunology and Inflammation, Vlaams Institute voor Biotechnologie (VIB), Inflammation Research Center (IRC), Ghent, Belgium
e-mail: vanessa.smith@ugent.be

Main Text

The Neuroendocrine Immune System in Course of Rheumatoid Arthritis

The alterations of the endocrine system have been the object of the first studies concerning RA etiopathogenesis. In 1950, Edward Calvin Kendall, Tadeus Reichstein, and Philip Showalter Hench won the Nobel Prize in Physiology or Medicine "for their discoveries relating to the hormones of the adrenal cortex, their structure and biological effects" [3].

In pioneering research, they identified cortisol for the first time, demonstrating pleiotropic immune properties with downregulation of RA inflammatory response [4]. Moreover, the observation that pregnancy improved clinical manifestations of RA in women drew attention on the biological effects of sex hormones (estrogens, progesterone and androgens) [5]. All these hormones undergo daily fluctuations of their serum concentrations. This is due to the light/dark environmental stimulus that activates the hypothalamic suprachiasmatic nucleus (SCN), through light-sensitive retinal ganglion cells. SCN is activated by darkness and promotes the release of melatonin from the pineal gland that enhances nocturnal immune-inflammatory response and synchronizes the circadian release of peripheral neuropeptides and hormones [6].

This cross-talk between central nervous, endocrine, and immune networks is named neuroendocrine immune (NEI) system. It is mainly constituted by three axes that are perturbated in course of RA:

- The hypothalamic–pituitary–adrenal axis (HPA).
- The hypothalamic–pituitary–gonadal axis (HPG).
- The endocrine system of vitamin D (D hormone).

The Hypothalamic-Pituitary-Adrenal Axis

The HPA axis (see also chapters "Brain: Overview" under paragraphs "Brain" and "Major Depression and Metabolic Syndrome") deeply influences the activation of the immune system [7].

In response to a stressful event, the hypothalamus secretes the corticotropin-releasing hormone (CRH) that stimulates the pituitary synthesis of adrenocorticotropic hormone (ACTH). ACTH increases the production of cortisol in the zona

fasciculata and dehydroepiandrosterone (DHEA) and androstenedione (ASD) in the zona reticularis of adrenal glands. DHEA and ASD are both precursors of androgens and estrogens. Cortisol shows anti-inflammatory properties by downregulating stimulation, activation, proliferation, differentiation, recruitment, and survival of several immune cells, including neutrophils, lymphocytes, monocytes-macrophages, basophils, and eosinophils [8].

At RA onset, endogenous cortisol allows a quick control of pro-inflammatory cytokines release, but chronic inflammation directly impairs HPA axis both at hypothalamic-pituitary and adrenal levels (peripheral steroidogenesis). As a consequence, RA patients have an insufficient production of cortisol in response to nocturnal release of melatonin (pro-inflammatory) that favors the progression of the disease (so called "disproportion principle") (Fig. 1) [9].

The Hypothalamic-Pituitary-Gonadal Axis

With the onset of puberty, HPG axis is constitutively activated and the hypothalamus secretes gonadotropin-releasing hormone (GnRH) with a pulsatile rhythm with subsequent release of follicle-stimulating hormone (FSH) and luteinizing hormone (LH) by the pituitary gland. FSH and LH stimulate the gonadal production of testosterone in men and estrogens and progesterone in women (see chapter "Reproductive System: Overview" under part "Reproductive System") [10].

All sex hormones derive from cholesterol and are structurally related to steroid hormones (i.e. glucocorticoids), thus deeply influencing immune cells. Testosterone shows anti-inflammatory effects, decreasing the secretion of TNF-α, IL-1, and IL-6 by immune cells [10]. Androgens' serum levels are reduced in men with active RA, and men suffering from primary (i.e., Klinefelter syndrome) or secondary (i.e., androgen deprivation therapy for prostate cancer) hypogonadism are at high risk to develop autoimmune diseases, including RA [11–13].

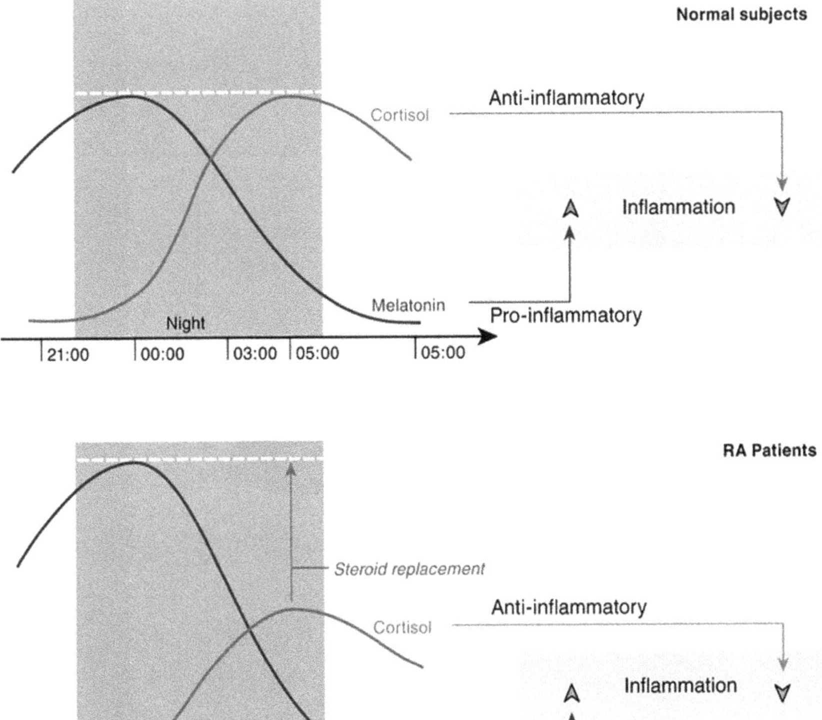

Fig. 1 Relatively impaired cortisol secretion in rheumatoid arthritis (RA) patients compared to healthy subjects. The cortisol nadir at midnight and the parallel increase of the immune-stimulatory hormone melatonin drive the increase of nightly proinflammatory mediators (not shown). This triggers a delayed expression/release of cortisol, which suppresses the surge in inflammatory tumor necrosis factor-α (TNF-α), interleukin (IL)-1, and IL-6 mediators at around 5 a.m. In healthy subjects, the cortisol secretion is proportional to melatonin levels, whereas in RA patient it is impaired in relation to melatonin (immune enhancer hormone), thus favoring the immune-inflammatory stimulation. This is the reason why steroid replacement is mandatory as etiological treatment in RA patients and in other chronic inflammatory diseases

Estrogens are pro-inflammatory hormones at physiological low serum concentrations (0.05–1.00 nmol/L), while at higher serum concentrations (above 3–5 nmol/L, such as during pregnancy) they exert anti-inflammatory properties, suppressing macrophages, T helper 1 (Th1), and Th17 cells activation. However, high serum concentrations of estrogens enhance B cell response, thus worsening B cell-mediated connective tissue diseases (i.e., systemic lupus erythematosus) [14].

As a consequence, women in reproductive age are at higher risk of developing RA (and other autoimmune conditions) than men; RA symptoms fluctuate during period and often improve during pregnancy. After menopause, estrogens' gonadal production ends with a significant decrease of B cells activation. In this phase, RA in women is mainly driven by pro-inflammatory T cells, and their role becomes more relevant than that of B cells [15].

At last, peripheral aromatases are expressed by RA synovial macrophages to convert gonadal testosterone and adrenal DHEA/ASD into pro-inflammatory estrogens (intracrinological mechanism) (Fig. 2) [16]. For this reason, men suffering from RA have peripheral low serum concentrations of androgens, with

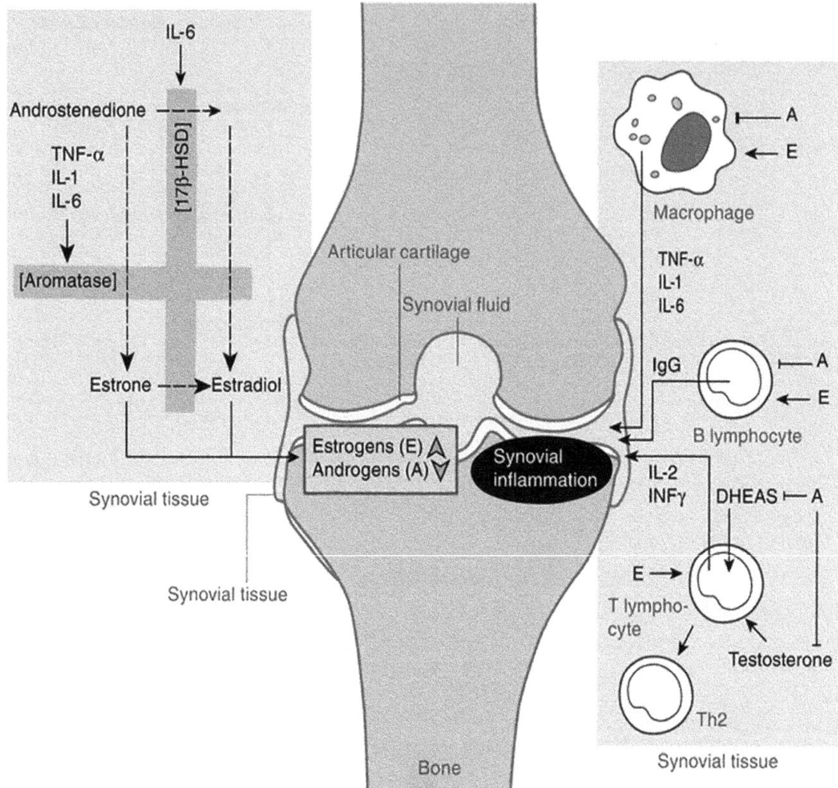

Fig. 2 Sex hormones' synthesis pathways and their influence on cytokine expression by synovial cells. *Left side*: Androgens (such as testosterone) and estrogens (such as estrone and estradiol) originate from common precursors (see chapter "Reproductive System: Overview" under part "Reproductive System"). Proinflammatory cytokines promote peripheral (incl. synovial) conversion of androgens to estrogens by acting on the respective enzymes (*light gray boxes*). *Right side*: Effects of estrogens and androgens on the immune cells and their cytokine production. Estrogens (*E*) may exert stimulatory effects on some activities of macrophages and T-helper 2 (*T h 2*) cells in producing inflammatory cytokines and induce B lymphocytes to secrete IgG (immunoglobulins G). Androgens (*A*) have an inhibitory effect on the same cells. *TNF-α* tumor necrosis factor-α, *IL* Interleukin, *IFN-γ* Interferon-γ, *17β-HSD* 17β-Hydroxysteroid dehydrogenase

higher concentrations of estrogens metabolites in the inflamed tissues (so called "androgen drain") [17].

The Endocrine System of Vitamin D (D Hormone)

Vitamin D is a steroid hormone, whose dietary intake satisfies only 20% of the human daily need. Cholecalciferol (pre vitamin D_3) is synthesized for the most part in the skin from cutaneous 7-dehydro cholesterol, under the effects of ultraviolet-B (UV-B) exposition. Cholecalciferol is subsequently hydroxylated first into calcidiol ($25OHD_3$) in the liver and then into calcitriol [$1,25(OH)_2D_3$] in the kidney (Fig. 3) [18].

For its steroidal structure, calcitriol is a true hormone (D hormone) that exerts its biological effect through the nuclear vitamin D receptor (VDR), widely expressed on immune cells. Calcitriol regulates both innate and adaptive immunity, downregulating neutrophils, monocytes-macrophages, T and B lymphocytes activation, so reducing the release of pro-inflammatory cytokines. Moreover, calcitriol interacts with HPG axis functions, for example, by reducing aromatase activation in RA synovial macrophages (Fig. 3) [19, 20].

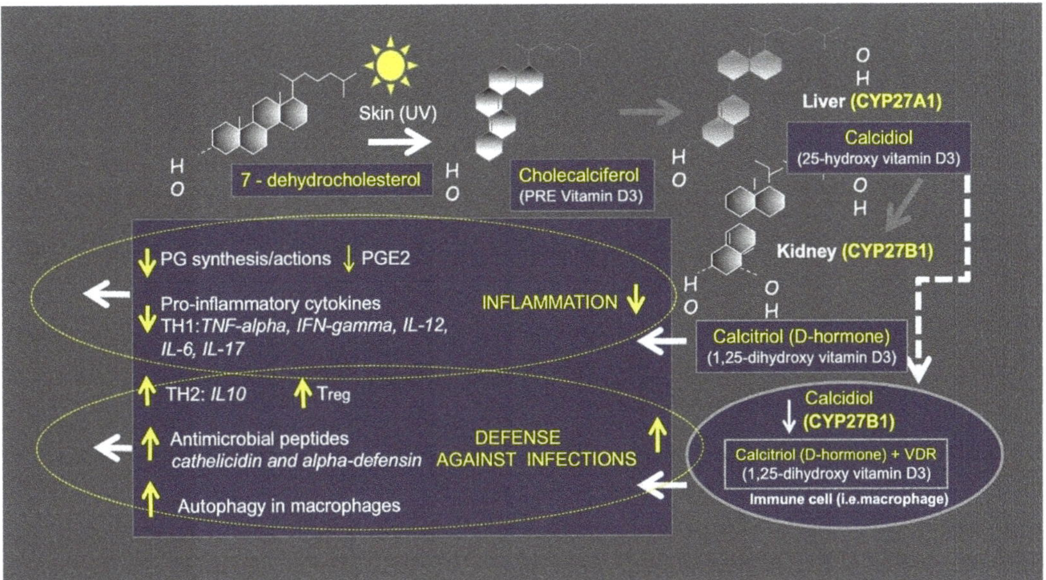

Fig. 3 Endogenous hormone D synthesis pathways and its influence on immune system. The final biological effects are the downregulation of inflammatory response and the increased defense against infections. *CYP27A1* cytochrome P 450 family 27 subfamily A member 1, *CYP27B1* cytochrome P 450 family 27 subfamily B member 1, *IFN-gamma* interferon gamma, *IL* interleukin, *PG* prostaglandin, *PGE2* prostaglandin E2, *TH1* T helper 1, *TH2* T helper 2, *TNF-α* tumor necrosis factor-alpha, *T reg* T regulatory cells, *UV* ultraviolet, *VDR* vitamin D receptor

Low serum concentrations of vitamin D can be present in RA patients, correlate negatively with disease severity (clinical and radiographic), and are influenced by the seasonality of exposure to the sun [21].

Circadian and Circannual Rhythms to Tailor Rheumatoid Arthritis Management

In 2017, Jeffrey Hall, Michael Rosbash, and Michael Young won the Nobel Prize in Physiology or Medicine "for their discoveries of molecular mechanisms controlling the circadian rhythm" [22].

As previously reported, a large network of genes and proteins is controlled by a light/dark 24-hour cycle [6]. In particular, the immune system is upregulated during the night as a consequence of the redistribution of energy-rich substrates from cardiovascular, musculo-skeletal, and nervous systems to inflammatory/prolifera-

tive processes. Serum concentrations of pro-inflammatory cytokines, such as IL-1, IL-6, and TNF-α rise at 2–3 a.m. following melatonin stimulation, reaching a peak around 5 a.m., while cortisol rises 60–120 minutes later and decreases during the morning up to the evening.

As a consequence, several RA clinical symptoms are exacerbated in the early morning (particularly joint stiffness), due to the impairment of cortisol production in chronic inflammation (above discussed "disproportion principle"). This fact explains why exogenous glucocorticoids still represent the first line of RA therapy, in particular when administered at a low dose (less than 5 mg of prednisone-equivalent daily), long term and with nocturnal formulation (Fig. 1) [23]. The administration of modified tablets of prednisone at 10 p.m. that release the active drug in the stomach at 2–3 a.m., allows a more marked reduction in RA symptoms and serum inflammatory markers, replacing the endogenous production of cortisol (concept of "replacement therapy") [24]. This tailored administration has the advantage of minimizing exogenous glucocorticoids-related side effects, such as arterial hypertension, hyperglycemia, insomnia, osteopenia/osteoporosis, and, specifically, progressive HPA inhibition due to a negative feedback [25, 26].

At last, vitamin D serum concentrations follow circannual fluctuations, with the lowest values in winter and springtime. The supplementation with vitamin D analogues should ensure a serum concentration > 40 ng/mL for all the year and cooperate with standard drugs to reduce RA activity [21]. Rheumatologists should use specific Patient Reported Outcome questionnaires to estimate the patients' risk of vitamin D insufficiency in order to promptly suggest exogenous supplementation [27].

Perspectives

Taking into account circadian and circannual rhythms for RA treatment also seems promising for the treatment of other chronic inflammatory conditions.

Further investigations are needed to identify the best daily dosages of vitamin D supplementation and to verify the efficacy of night-release formulation of other common drugs in RA treatment, such as non-steroidal anti-inflammatory drugs (NSAIDs) and disease-modifying antirheumatic drugs (DMARDs), in particular methotrexate and leflunomide.

Acknowledgements The previous version of this chapter was co-authored by Marianna Meroni and Elena Bernero under the supervision of Maurizio Cutolo.

All the authors are members of the European Alliance of Associations for Rheumatology (EULAR) Study Group on Neuroendocrine Immunology of Rheumatic Diseases (NEIRD).

Questions and Answers

Question 1 Which of the following scenarios is an example of "androgen drain" in course of RA?

Answers 1 Adrenal production of dehydroepiandrosterone (DHEA) and androstenedione (ASD).

- Gonadal production of testosterone.
- Peripheral conversion of androgens into estrogens.
- Peripheral conversion of estrogens into androgens.

(Correct answer: 3rd)

Question 2 Which of the following axes shows a clear inadequate hormonal production in response to chronic inflammation?

Answers 2 Hypothalamic–pituitary–adrenal axis.

- Hypothalamic–pituitary–gonadal axis.
- Endocrine system of vitamin D (hormone D).
- None of the above.

(Correct answer: 1st)

Question 3 Which of these is the best pharmacological use of exogenous glucocorticoids in course of RA?

Answers 3 High dosage (i.e., 50 mg) of prednisone equivalent per day for long term in order to definitively suppress inflammatory response.

- High dosage (i.e., 50 mg) of prednisone equivalent per day for 3 weeks, because long-term assumption has an unacceptable profile of risk at any dosage.
- Low dosage (i.e., less than 5 mg) of prednisone equivalent per day after lunch, when morning stiffness is resolved.
- Low dosage (i.e., less than 5 mg) of modified release prednisone equivalent per day at 10 p.m. for long term in order to replace cortisol impaired production.

(Correct answer: 4th)

References

1. Smolen JS, Aletaha D, Barton A et al (2018) Rheumatoid arthritis. Nat Rev Dis Primers 4:18001
2. van der Woude D, van der Helm-van Mil AHM (2018) Update on the epidemiology, risk factors, and disease outcomes of rheumatoid arthritis. Best Pract Res Clin Rheumatol 32(2):174–187
3. Raju TN (1999) The Nobel chronicles. 1950: Edward Calvin Kendall (1886-1972); Philip Showalter Hench (1896-1965); and Tadeus Reichstein (1897-1996). Lancet 353(9161):1370
4. Hench PS, Slocumb CH, Holley HF, Kendall EC (1950) Effect of cortisone and pituitary adrenocorticotropic hormone (ACTH) on rheumatic diseases. JAMA J Am Med Assoc 144(16):1327–1335
5. Hench PS (1938) The ameliorating effect of pregnancy on chronic atrophic (infectious, rheumatoid) arthritis, fibrositis, and intermittent hydrarthrosis. Proc Staff Meet Mayo Clin 13:161–167
6. Cutolo M (2019) Circadian rhythms and rheumatoid arthritis. Joint Bone Spine 86(3):327–333
7. Straub R, Buttgereit F, Cutolo M (2011) Alterations of the hypothalamic-pituitary-adrenal axis in systemic immune diseases – a role for misguided energy regulation. Clin Exp Rheumatol 29(5 Suppl 68):S15–S23
8. Ronchetti S, Migliorati G, Bruscoli S, Riccardi C (2018) Defining the role of glucocorticoids in inflammation. Clin Sci (Lond) 132(14):1529–1543
9. Straub RH, Cutolo M (2016) Glucocorticoids and chronic inflammation. Rheumatology (Oxford) 55(suppl 2):ii6–ii14
10. Cutolo M, Straub RH (2020) Sex steroids and autoimmune rheumatic diseases: state of the art. Nat Rev Rheumatol 16:628–644
11. Cutolo M (2009) Androgens in rheumatoid arthritis: when are they effectors? Arthritis Res Ther 11(5):126
12. Baillargeon J, Al Snih S, Raji MA et al (2016) Hypogonadism and the risk of rheumatic autoimmune disease. Clin Rheumatol 35(12):2983–2987
13. Yang DD, Krasnova A, Nead KT et al (2018) Androgen deprivation therapy and risk of rheumatoid arthritis in patients with localized prostate cancer. Ann Oncol 29(2):386–391
14. Pacini G, Paolino S, Andreoli L et al (2020) Epigenetics, pregnancy and autoimmune rheumatic diseases. Autoimmun Rev 19:102685
15. Straub RH, Bijlsma JW, Masi A, Cutolo M (2013) Role of neuroendocrine and neuroimmune mechanisms in chronic inflammatory rheumatic diseases–the 10-year update. Semin Arthritis Rheum 43(3):392–404
16. Rubinow KB (2018) An intracrine view of sex steroids, immunity, and metabolic regulation. Mol Metab 15:92–103
17. Capellino S, Straub RH, Cutolo M (2014) Aromatase and regulation of the estrogen-to-androgen ratio in synovial tissue inflammation: common pathway in both sexes. Ann N Y Acad Sci 1317:24–31
18. Trombetta AC, Meroni M, Cutolo M (2017) Steroids and autoimmunity. Front Horm Res 48:121–132
19. Cutolo M, Paolino S, Sulli A et al (2014) Vitamin D, steroid hormones, and autoimmunity. Ann N Y Acad Sci 1317:39–46
20. Villaggio B, Soldano S, Cutolo M (2012) 1,25-dihydroxyvitamin D3 downregulates aromatase expression and inflammatory cytokines in human macrophages. Clin Exp Rheumatol 30(6):934–938
21. Cutolo M, Soldano S, Sulli A et al (2021) Influence of seasonal vitamin D changes on clinical manifestations of rheumatoid arthritis and systemic sclerosis. Front Immunol 12:683665
22. Huang RC (2018) The discoveries of molecular mechanisms for the circadian rhythm: the 2017 Nobel prize in physiology or medicine. Biom J 41(1):5–8
23. Cutolo M, Paolino S, Gotelli E (2021) Glucocorticoids in rheumatoid arthritis still on first line: the reasons. Expert Rev Clin Immunol 17(5):417–420
24. Buttgereit F, Doering G, Schaeffler A et al (2008) Efficacy of modified-release versus standard prednisone to reduce duration of morning stiffness of the joints in rheumatoid arthritis (CAPRA-1): a double-blind, randomised controlled trial. Lancet 371(9608):205–214
25. Cutolo M (2016) Glucocorticoids and chronotherapy in rheumatoid arthritis. RMD Open 2(1):e000203
26. Paolino S, Cutolo M, Pizzorni C (2017) Glucocorticoid management in rheumatoid arthritis: morning or night low dose? Reumatologia 55(4):189–197
27. Vojinovic J, Tincani A, Sulli A et al (2017) European multicentre pilot survey to assess vitamin D status in rheumatoid arthritis patients and early development of a new patient reported outcome questionnaire (D-PRO). Autoimmun Rev 16(5):548–554

Gastrointestinal Tract: Overview

Satish Keshav and Philip Allan

Anatomy and Physiology of the Gastrointestinal Tract

The gastrointestinal tract (GIT) is divided into distinct subunits, with morphological and functional differences: mouth including teeth, esophagus, stomach, pancreas and biliary tract, liver, duodenum, jejunum, ileum, colon, and anus. The different functions are reflected in macroscopic and microscopic anatomy. Macroscopically, the stomach, for instance, has longitudinal, oblique, and circular muscle layers to mix and churn food and so aid digestion. Microscopically, the cells lining the stomach include specialized H cells that produce hydrochloric acid (see below).

In contrast, the ileum is an approximately 3 m long tube-like structure, with circular and smooth muscle optimized for peristalsis, and microscopically, the predominant cell type is the enterocyte, specialized for absorption of nutrients. The surface area of the ileum is increased by internal folds covered with projecting villi. These in turn are microscopically covered with microvilli, which project from the luminal surface of the epithelial cells.

The colon is specialized for extracting water, solidifying waste into feces before excretion. Bacteria in the colon aid metabolism by produc-

ing short chain fatty acids (SCFAs) such as butyrate, which is a source of energy for colonic epithelial cells. In this chapter, we review important functions of the gastrointestinal tract, particularly the stomach, ileum, and colon, emphasizing nutritional and metabolic implications.

Gastrointestinal Tract-Specific Metabolic/Molecular Pathways and Processes

The Stomach

The stomach expands to accommodate food after a meal, initiates the release of hormones that coordinate subsequent events, and secretes hydrochloric acid (HCl), which converts pepsinogen into functional pepsin. Pepsin digests dietary proteins and peptides into smaller peptides that are further digested and absorbed as oligo- and dipeptides in the duodenum, jejunum, and ileum.

HCl creates an environment that is inhospitable to most microorganisms. Epithelial cells are protected from the effects of HCl by a number of mechanisms including gastric mucus. Additionally, gastric acid favors the reduction in iron from ferrous to ferric (Fe^{2+} to Fe^{3+}), which promotes its absorption in the proximal duodenum. In addition, intrinsic factor (IF) secreted by the gastric mucosa binds to dietary cobalamin,

S. Keshav · P. Allan (✉)
P. Allan Experimental Medicine Division,
John Radcliffe Hospital, Oxford, UK
e-mail: p.allan@doctors.org.uk

E. Lammert, M. Zeeb (eds.), *Metabolism of Human Diseases*,
https://doi.org/10.1007/978-3-031-96019-2_17

protecting it from digestion and enabling absorption in the terminal ileum (Fig. 1).

The Small Intestine: Duodenum, Jejunum, and Ileum

The small intestine is differentiated along its length, with distinct functions occurring in duodenum, jejunum, and ileum. Acidic chyme from stomach is released gradually into the duodenum, where it is mixed with alkaline bile and pancreatic secretions rich in HCO_3^-. The duodenal mucosa is specialized for absorption.

The pancreas produces the bulk of digestive enzymes for carbohydrates, lipids, and proteins, and bile acids contribute to the formation of mixed micelles of lipid, aiding their digestion.

Intestinal epithelial cells contribute to digestion by secreting enzymes such as lactase (see chapter "Lactose and Food Intolerance") that are bound to their surface and express in their cell membranes protein transporters that enable nutrients to pass from the lumen of the intestine into the epithelial cell and then out of the cell and into the lymphatic and vascular circulation (lipophilic and hydrophilic substances, respectively). In addition, epithelial cells absorb electrolytes and metallic trace elements such as Na^+, K^+, Ca^{2+}, Mg^{2+}, Zn^{2+}, Cu^{2+}, and water. Paracellular movement of water and electrolytes also occurs, and the direction, either into or out of the lumen, depends on osmotic and ionic gradients. Tight junctions between epithelial cells are more effective distally, and paracellular transport in the healthy intestine plays only a minor role.

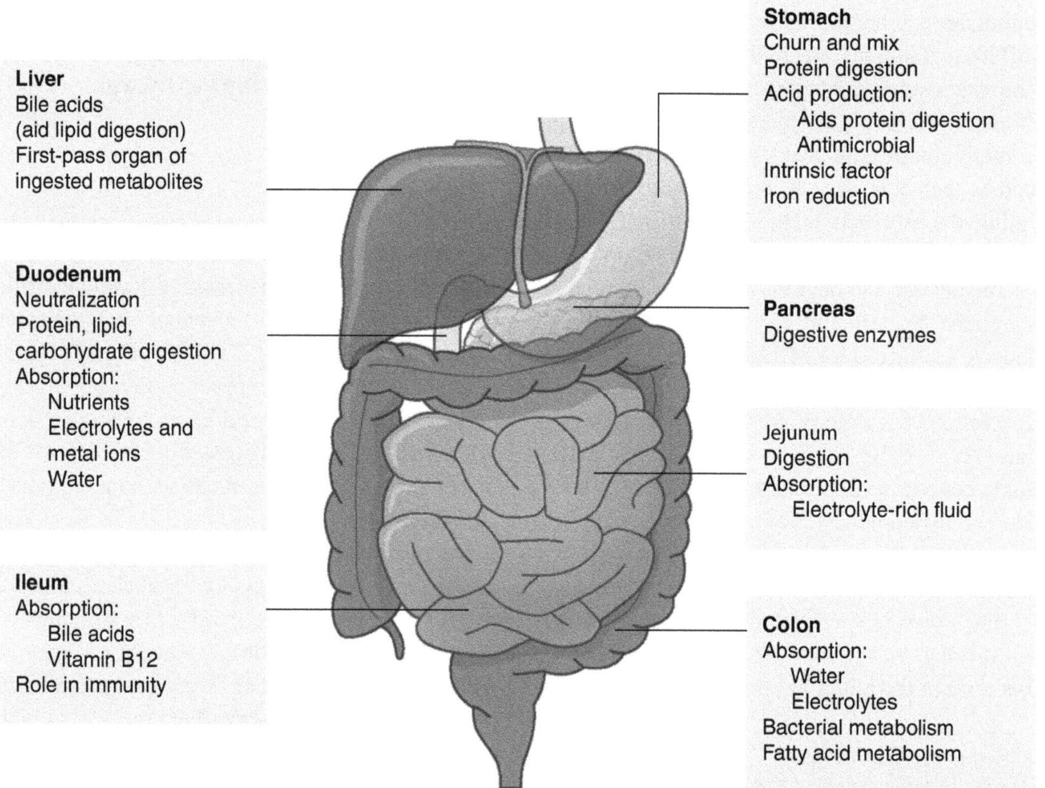

Fig. 1 Overview of the gut with summary of main functions demonstrating absorption of water and electrolytes, carbohydrate, lipid, proteins, vitamins, and minerals. Gut physiology is determined by the specialized cells lining the epithelium with absorption of different molecules in different parts of the gut

Digestion and absorption continue in the jejunum, where much of the electrolyte-rich fluid that is secreted by the stomach, pancreas, and duodenum is reabsorbed. Without this reabsorption, total body water, Na^+, K^+, and Mg^{2+} are rapidly depleted, presenting a clinical problem for patients who undergo intestinal resection leaving them with a jejunostomy or proximal ileostomy.

The terminal ileum is particularly important for the absorption of bile acids and vitamin B_{12}. As a consequence, disease of the ileum can cause vitamin B_{12} deficiency and diarrhea caused by loss of the enterohepatic circulation of bile salts (see chapter "Liver: Overview" under part "Liver"), with consequent excess of these in the colon, where they cause secretion of water and electrolytes [1]. The ileum also acts as a barrier between the large and small intestine, with effects on bacterial populations and intestinal motility. Finally, the ileum plays a prominent role for the immune system (see below) [2].

Absorption of carbohydrates requires digestion of complex carbohydrates into monosaccharides, as polysaccharides and disaccharides cannot be absorbed. Lactose, a disaccharide of galactose and glucose, is the main sugar in bovine and human milk and is digested by lactase, which is produced by intestinal epithelial cells. Lack of lactase causes lactose intolerance, in which colonic bacteria digest lactose to produce fermentation by-products, which, with the lactose, increases the osmotic gradient and results in diarrhea (see chapter "Lactose and Food Intolerance"). Furthermore, excess lactose may alter colonic bacteria. Avoiding dietary dairy can lead to Ca^+ deficiency and potentially osteomalacia (Fig. 2).

The Large Intestine: Cecum, Colon, and Rectum

The main function of the specialized colonic epithelium is to reabsorb H_2O from 3 to 6 L/day of fluid that enters the cecum, so that approximately 200 mL is excreted as fecal waste. Reabsorbed water is accompanied by electrolytes, particularly Na^+, K^+, and Mg^{2+}. Colonic bacteria metabolize fats and contribute to the provision of essential fatty acids. Tight junctions between colonic epithelial cells prevent salt and water loss from the interstitial tissue space and limit the translocation of bacteria and toxins from the colonic lumen into the circulation.

The practical importance of reabsorption is illustrated by patients who have undergone extensive resection of the small intestine. When they have a proximal jejunostomy or ileostomy, loss of salt and water means that they require regular intravenous supplementation, even when absorption of dietary carbohydrate, fat, and protein is sufficient. In such cases, anastomosis of the small intestine to the colon typically reverses dependence on intravenous fluids.

Colonic bacteria contribute metabolically to the host. Food residue in the colon is fermented, yielding gases such as methane (CH_4), H_2, and CO_2 and short chain fatty acids (SCFAs), such as acetate, butyrate, and propionate. However, fermentation can also produce toxic metabolites (see below). Butyrate is a major energy source for colonic epithelial cells and its absence can result in inflammation (see also chapter "Colorectal Cancer").

Moreover, the mix of colonic bacterial species, termed the microbiome, varies between individuals. This variation may be associated with diseases including obesity and autoimmunity [3].

Inside-In: Metabolites of the Gastrointestinal Tract Affecting Itself

Regulation of digestive processes involves structures in the hypothalamus (see chapters "Brain: Overview" under part "Brain", "Diabetes Mellitus", and "Brain: Overview" under part "Brain"), yet it also occurs to a major part within the gastrointestinal tract, starting with the amount of food intake sensed by chewing and stomach filling. Most importantly, the stomach releases hormones, such as gastrin, cholecystokinin (CCK), and secretin, which in turn promote the release of HCl, stimulate contraction of the gallbladder, and promote secretion from the

Fig. 2 Diagram of the microscopic appearance of a stylized crypt-villus unit and colonic crypt, showing specialized cells, enzymes, absorption, secretion, and immune functions

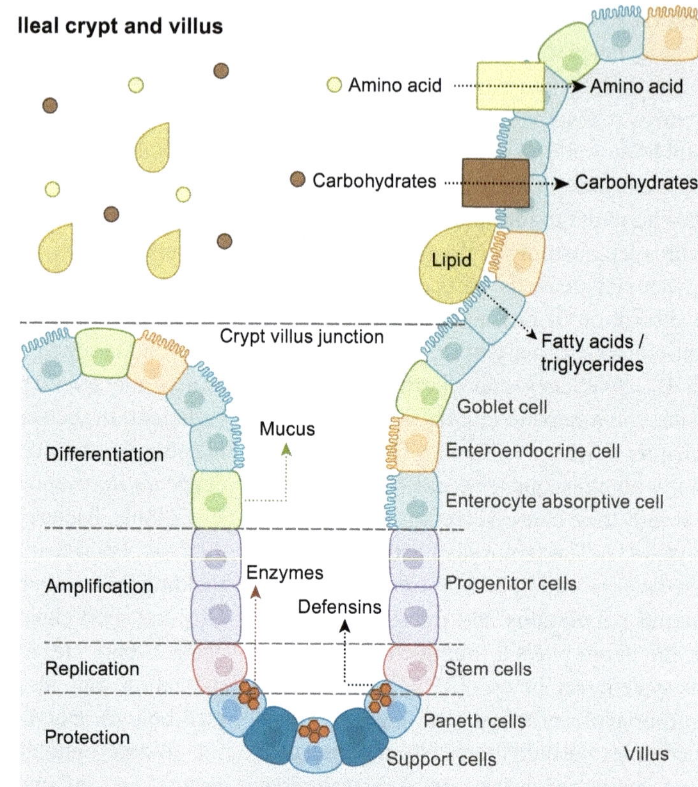

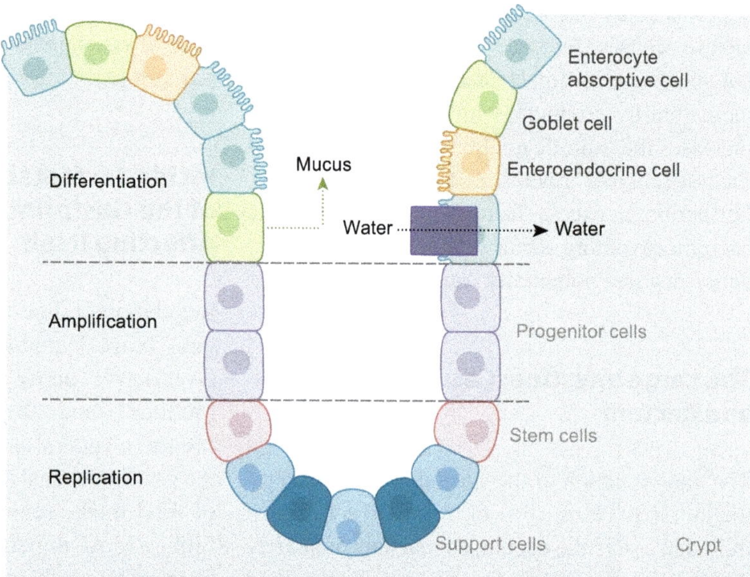

pancreatobiliary system, while enhancing intestinal propulsion.

Outside-In: Metabolites of Other Tissues and External Factors Affecting the Gastrointestinal Tract

Stomach

Helicobacter pylori (*H. pylori*) is an infectious bacterial strain that has adapted to gastric acid by secreting a urease that generates NH_4^+ from urea, thus raising the local pH.

As a consequence, *H. pylori* infection is widespread and is associated with a number of diseases including gastritis, peptic ulcer (see chapter "Peptic Ulcer Disease"), and gastric cancer.

Consequently, *H. pylori* can cause metabolic and nutritional problems, with hypochlorhydria, reduced IF production, and gastritis leading to bleeding.

Small Intestine

There are several examples how circulating hormones and neural signals can regulate intestinal absorption [4]. However, a detailed discussion is not within the scope of this chapter.

Iron transporters particularly are concentrated proximally, where the majority of iron is absorbed. As an example, iron absorption is inhibited by the circulating hormone hepcidin, levels of which are increased by infection and inflammation contributing to the anemia seen in many chronic diseases.

The small intestine is also highly relevant to the interaction with endogenous material and antigens. Thus, intestinal lymphoid tissue is concentrated in the ileum, with many prominent Peyer's patches, and specialized cells that form part of the innate immune system (see chapter "Anatomy and Physiology of the Immune System" under part "Immune System"), such as Paneth cells, are most abundant in the ileum [2]. Paneth cells and goblet cells, among others, pro-

duce many antibacterial proteins such as α and β defensins, regenerating (REG) proteins, and cathelicidins. Innate and adaptive immune mechanisms, as well the effects of constant propulsion along the intestinal tract, and chemical and mechanical barrier functions of glycosaminoglycan-rich mucus maintain near sterility of the small intestine, with approximately 10^5–10^7 bacteria/ml of luminal content, compared to 10^{11}–10^{12}/ml in the colon. Commensal bacteria in the intestine probably have a symbiotic relationship with the host, preventing colonization by pathogens by competing for the ecological niche and contributing partly to the production of nutrients such as vitamin K.

Inside-Out: Metabolites of the Gastrointestinal Tract Affecting Other Tissues

Obviously, the main function of the gastrointestinal tract is to supply the whole body with nutrients, electrolytes, and trace elements required for regular metabolism and homeostasis.

Small Intestine

Additionally, the gastrointestinal tract, more specifically the small intestine, can act as an endocrine organ secreting incretins, which act on pancreatic β-cells causing an increase in insulin secretion even before blood glucose is elevated or delay gastric emptying (see chapters "Pancreas: Overview" under part "Pancreas" and "Diabetes Mellitus").

Colon

Fermentation by colonic bacteria can produce toxic metabolites that can be absorbed into the circulation to exert systemic effects (see chapter "Cirrhosis"). There is therefore now a concerted research effort aimed at understanding how the microbiome interacts with the intestine and the whole body in health and disease.

Typical Pathology of the Gastrointestinal Tract

Stomach and Small Intestine

Vomiting causes loss of acid. This loss results in concomitant K^+ loss in the urine to favor retention of hydrogen (H^+) ions (see chapter "Kidney: Overview" under part "Kidney"). The loss of K^+ can be exacerbated by diarrhea, which also causes loss of Mg^{2+} and Ca^{2+}. The absorption of vitamins and minerals can also be affected by intestinal transit time and absorptive capacity. Acute illnesses, such as gastroenteritis, do not typically cause nutritional deficiencies. However, in chronic disease, water-soluble vitamins, e.g., thiamine (B1), riboflavin (B2), pyridoxine (B6), niacin, folate, B12, and ascorbic acid (C), initially, and later fat-soluble vitamins (vitamins A, D, E, and K) may become deficient. Essential minerals, such as Zn^{2+}, Cu^{2+}, selenium, and phosphate, are also typically depleted in chronic malabsorptive states [5].

Colon

The capacity of the colon to reabsorb water and electrolytes is reduced by inflammation, including infection, ulcerative colitis, and microscopic colitis. Bile acid diarrhea results from colonic secretion stimulated by bile acids that are not reabsorbed in the ileum, when they enter the colon. Laxatives and microbial toxins can also stimulate secretion and cause diarrhea. An example is the heat stable enterotoxin a (STa) of *Escherichia coli*, which binds to a guanylate cyclase-linked receptor expressed on the surface of colonic epithelial cells, causing secretion of Cl^- and water secondary to increased intracellular cGMP.

Colorectal cancer (CRC, see chapter "Colorectal Cancer") is a major cause of morbid-ity and mortality. The interaction of diet and the microbiome may provide a mechanistic link for these observations. Chronic inflammation of the colon, including longstanding ulcerative colitis increases the risk of CRC [5].

Final Remarks

The intestine is critically important for nutrition and metabolism and has to meet major challenges posed by the ingestion of food containing potential toxins and pathogens. Specialized structures at the macroscopic and microscopic level contribute to the function of the gastrointestinal tract and are altered in disease. A symbiotic relationship with commensal intestinal microorganisms is only now being understood in more detail, and alterations in this relationship may be a major determinant of health and disease. This introductory chapter is followed by more detailed chapters on important and common intestinal diseases, i.e., *H. pylori* infection (see chapter "Peptic Ulcer Disease"), gastroenteritis, lactose intolerance (see chapter "Lactose and Food Intolerance"), and colorectal cancer (see chapter "Colorectal Cancer"). For further information, the reader is referred to more extensive textbooks of gastroenterology [5, 6].

References

1. Chiang JY (2013) Bile acid metabolism and signaling. Compr Physiol 3:1191–1212
2. Keshav S (2006) Paneth cells: leukocyte-like mediators of innate immunity in the intestine. J Leukoc Biol 80:500–508
3. Karlsson F, Tremaroli V, Nielsen J, Bäckhed F (2013) Assessing the human gut microbiota in metabolic diseases. Diabetes 62:3341–3349
4. Mourad FH, Saadé NE (2011) Neural regulation of intestinal nutrient absorption. Prog Neurobiol 95:149–162
5. Keshav S, Culver E (2011) Gastroenterology: clinical cases uncovered, 1st edn. Wiley-Blackwell, New York
6. Keshav S, Bailey E (2012) The gastrointestinal system at a glance, 2nd edn. Wiley-Blackwell, New York

Peptic Ulcer Disease

Peter C. Konturek and Stanislaw J. Konturek

Introduction to Peptic Ulcer Disease

Peptic ulcer disease (PUD) is defined as a defect in the lining of the gastrointestinal (GI) mucosa, with appreciable depth or involvement of submucosa [1]. The development of peptic ulceration is a result of an imbalance between factors potentially damaging the gastric mucosa (aggressive) and protective (defensive) factors. The former include endogenous factors (e.g., gastric acid and pepsin) and exogenous factors, including chronic *Helicobacter pylori* (Hp) infection, use of non-steroidal anti-inflammatory drugs (NSAIDs), consumption of alcohol, smoking, and exposure to stress [2]. The latter comprise pre-epithelial defense (secretion of mucus and bicarbonates), epithelial defense (epithelial restitution and replication, extracellular buffers including bicarbonates), and post-epithelial-defense (mucosal microcirculation, tissue acid-base balance).

Sometimes the definitive cause of the ulcer cannot be established ("idiopathic ulcer"). The possible pathogenetic factors associated with this rare peptic ulceration are older age, multimorbidity, recent surgery, underlying sepsis, and medication (other than NSAIDs ulcerogenic drugs).

Peptic ulcer is typically localized in stomach (gastric ulcer) and duodenum (duodenal ulcer), most commonly in the duodenal bulb (the part of the duodenum closest to the stomach) due to the high exposure to gastric acid. Small areas of the duodenum colonized by Hp, called duodenal gastric metaplasia, may additionally contribute to the pathogenesis of PUD in duodenum [3].

The most common symptoms of PUD are abdominal pain, vomiting, hematemesis (vomiting of blood), and melena (the passage of dark stools stained with altered blood). In contrast, older patients (> 80 years) often lack abdominal pain. Pain occurs typically during the night, when gastric acidity is increased (due to circadian changes) and buffering food intake is minimal [5].

Complications of PUD include bleeding, perforation of stomach or duodenum, and gastric outlet obstruction (i.e., an obstructed pylorus) that could be fatal if untreated [6]. Severe or persistent peptic ulcers increase the risk of gastric cancer.

Duodenoscopy is the current standard for diagnosis of PUD revealing its localization and possible complications (especially bleeding). In addition, endoscopy allows obtaining biopsies from gastric mucosa for detection of Hp infection [7].

P. C. Konturek (✉)
Second Department of Internal Medicine, Thuringia Clinic Saalfeld, Saalfeld/Saale, Germany
e-mail: pkonturek@thueringen-kliniken.de

S. J. Konturek
Institute of Physiology, Jagiellonian University Cracow, Krakow, Poland

© The Author(s), under exclusive license to Springer Nature Switzerland AG 2026
E. Lammert, M. Zeeb (eds.), *Metabolism of Human Diseases*,
https://doi.org/10.1007/978-3-031-96019-2_18

Pathophysiology of Peptic Ulcer Disease and Metabolic Alterations

Among the endogenous aggressive factors, gastric acid and pepsin play a crucial role, as shown by Schwartz' dictum: "no acid, no ulcer" [8]. Both are required for digestion of nutrients but also pose a threat to the gastric mucosa due to acidity and protein degradation capabilities.

Secretion of acid and pepsin in the stomach is stimulated by gastrin and inhibited by prostaglandins, mainly prostaglandin E2 (PGE_2). PGE_2 also reinforces the mucosal defense by increased production of bicarbonate and mucus in the stomach (see chapter "Colorectal Cancer"). Other defensive factors include mucosal blood flow, nitric oxide (NO), hydrogen sulfide (H_2S), locally produced mucosal growth factors such as epidermal growth factor (EGF) and vascular endothelial growth factor-A (VEGF-A), and others mediators (Fig. 1) [2, 3].

The mucosal microvasculature is of critical importance as it maintains a relatively high pH in the superficial epithelium and releases prostaglandins, NO, pro-coagulant factors, and protective growth factors [9]. Chronic Hp infection of gastric mucosa is the most important risk factor for PUD [10]. Hp induces a chronic inflammatory immune response, the intensity of which depends on strain-specific virulence factors, host genetic factors, duration of infection, and the location (antrum predominant gastritis or gastritis of the whole gastric mucosa, called pangastritis, see chapter "Colorectal Cancer") [11].

In antrum predominant gastritis, the concentration of somatostatin decreases due to the reduction in the number of somatostatin producing cells in the antrum causing hypergastrinemia. This, in turn, increases gastric acid secretion and causes the appearance of gastric metaplasia, that is, small surface areas resembling gastric mucosa in the duodenal bulb. Metaplastic epithelium in duodenal bulb can further become infected by Hp causing an inflammatory response (duodenitis) followed by peptic ulcer formation [12, 13].

In pangastritis, gastric acid output is decreased due to pro-inflammatory cytokines (interleukin 1β, tumor necrosis factor α) and gastric atrophy. However, the chronic inflammatory response and the increased release of pro-inflammatory cytokines impair mucus secretion and epithelial defense mechanisms (cell restitution and mucosal reparative mechanisms) resulting in ulceration in the stomach [14].

The second most important cause of PUD is use of NSAIDs (including aspirin and ibuprofen), which may profoundly inhibit the mucosal defense mechanisms and cause peptic ulcer and bleeding, synergistic to Hp. 5–20% of frequent NSAID users develop PUD. The ulcerogenic effect of many NSAIDs is mainly caused by inhibition of the cyclooxygenases 1 and 2 (COX-1, COX-2) in the gastric mucosa, key enzymes in the synthesis of prostaglandins. This decreases mucus and bicarbonate production, mucosal blood flow, epithelial cell proliferation and restitution, and increases gastric acid secretion. Other pathogenetic factors are also involved in NSAID-induced ulcerogenesis, such as direct topical injury [15]. NSAIDs might also inhibit mucosal generation of growth factors, decrease synthesis of protective polyamines and H_2S, and increase nitric oxide (NO) release due to upregulation of inducible NO synthase [16].

Other causes of Hp-negative (Hp^-) ulcers are smoking, alcohol consumption, and the presence of different comorbidities, such as liver cirrhosis (see chapter "Cirrhosis") and chronic renal failure (see chapter "Kidney: Overview") [17]. Cigarette smoking stimulates basal acid output, increases generation of reactive oxygen species (ROS) in gastrointestinal mucosa, and reduces secretion of EGF and PGE_2 [18]. Alcohol consumption damages the mucosal barrier by inducing a range of effects from subtle mucosal erythema (i.e., redness due to increased blood flow, for example, as a result of vasodilation) to hemorrhagic (bleeding) gastritis. The damaging effect of ethanol appears as early as 30 min after ingestion of ethanol and reaches a peak at about 60 min. The damaging effect of ethanol is caused by impairment of mucosal microcirculation, declined level of prostaglandins, and increased generation of ROS and NO [19]. Interestingly, coffee consumption shows no association with increased risk for the development of PUD [20].

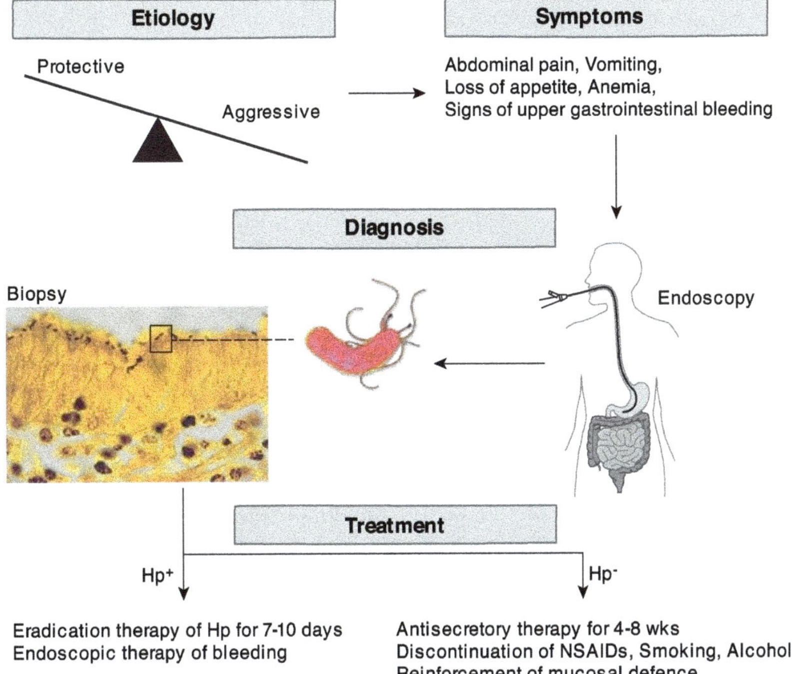

Fig. 1 Pathogenesis of peptic ulcer disease (PUD). PUD develops due to an imbalance between aggressive and protective factors within the gastric mucosa and commonly causes vomiting, loss of appetite, anemia, and upper gastrointestinal bleeding. It is diagnosed by esophagogastroduodenoscopy (upper endoscopy), which allows localization and treatment (endoscopic hemostasis) of the ulcer. Additionally, biopsies are taken to determine the presence (Hp+) or absence (Hp-) of Helicobacter pylori infection resulting in eradication therapy or treatment with antisecretory drugs, respectively. Drugs reinforcing mucosal defense can be used instead of anti-secretory drugs. Alcohol and smoking should be avoided

Treatment

The treatment strategy of PUD depends on the presence or absence of Hp infection (Fig. 1).

In Hp+ ulcers, eradication of Hp is usually achieved by a combination of acid-inhibiting therapy and two antibiotics (standard triple therapy). Whereas antibiotics specifically kill Hp, acid inhibitors prevent progression and complications of PUD. Acid inhibitors, especially proton pump inhibitors (PPI), synergize with the antibiotics by effectively increasing gastric pH. However, antibiotic resistance of Hp is rising, hampering the eradication rates. Alternative, second-line therapeutic strategies include quadruple therapy (adding treatment with bismuth salts, which shows strong antimicrobial activity), sequential therapy (adding different antibiotics stepwise), or fluoroquinolone-based (antibiotic) regimens [21]. Eradication of Hp has become the standard prophylaxis treatment to prevent recurrence of PUD [22] and even reduces the incidence of PUD and the risk of peptic ulcer bleeding in NSAID/aspirin users [23].

Treatment of Hp ulcers is governed by reduction in factors that attack the gastroduodenal mucosa (such as gastric acid) or by reinforcement of the mucosal defense mechanisms. Inhibition of aggressive factors includes (i) compounds neutralizing acid, called antacids; (ii) inhibitors of acid and pepsin secretion, such as anticholinergics, histamine H2 receptor antagonists (H2RA), and proton pump inhibitors (PPIs); and (iii) compounds eliminating potentially cytotoxic elements in the gastroduodenal secretions like bile acids or lecithin (Fig. 2).

H2RAs block the stimulatory effect on acid secretion of histamine, released from enterochro-

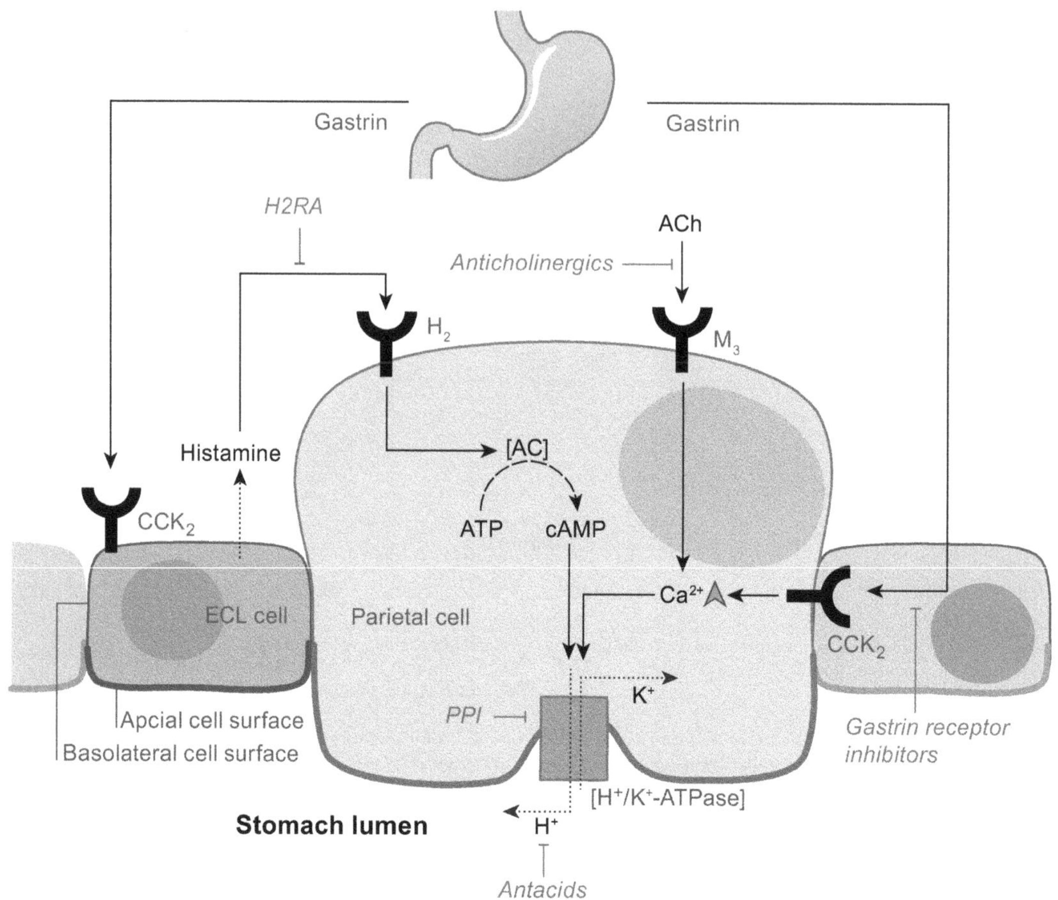

Fig. 2 Working mechanism of anti-secretory drugs. Anti-secretory drugs prevent the release of gastric acid by acting on the H^+/K^+-ATPase. Histamine H_2 receptor (H_2R) antagonists (H2RA) competitively and reversibly block H_2R, inhibiting signaling via adenylate cyclase (AC) to reduce cAMP levels and thus H^+ secretion. PPIs (proton pump inhibitors) directly inhibit the ATPase. Anticholinergics block acetylcholine (ACh) effects via the M_3 receptors, decreasing intracellular Ca^{2+}. Similarly, blockade of gastrin receptors (CCK_2) reduces intracellular Ca^{2+} but also decreases histamine release from enterochromaffin (ECL) cells. However, these drugs are currently not used in PUD. Antacids directly neutralize acid within the gastrointestinal tract

maffin cells (ECL). PPIs covalently bind to the H^+/K^+-ATPase on parietal cells within the stomach epithelium (see chapter "Colorectal Cancer") to inhibit this transporter and thus acid secretion. PPIs show improved peptic ulcer healing and faster pain relief compared to H2RAs [24]. Pharmacological strategies to reinforce the mucosal defense mechanisms increase protective factors (PGE_2, EGF), gastric mucus and alkaline secretion, and gastric mucosal blood flow, trigger or facilitate reparative mechanisms, and decrease ROS [25].

Bleeding remains an important complication of PUD with a high mortality rate (5–15%), especially in the elderly due to increased use of low dose-aspirin for the prevention of cardiovascular diseases. Management of bleeding includes multimodal endoscopic techniques to stop the gastrointestinal bleeding (endoscopic hemostasis) followed by pharmacological acid suppression with PPI or H2RA. Moreover, high doses of intravenously administered PPIs are recommended to reduce risk of re-bleeding after endoscopic hemostasis [26]. If repeated endoscopies fail or perforation (of the gastrointestinal wall) is present, surgical therapy is indicated, the exact type of which is controversial, ranging from oversewing to partial gastrectomy [27].

Influence of Treatment on Metabolism and Consequences for Patients

Therapy of Hp⁺ PUD

The commonly used first-line standard triple therapy includes two antibiotics, such as clarithromycin or metronidazole, and amoxicillin along with a PPI, such as omeprazole, lansoprazole, pantoprazole, rabeprazole, or esomeprazole and is given for 7–14 days. Sequential therapy, which shows significantly higher eradication rates [28], generally consists of PPI and amoxicillin for the first 5 days, followed by PPI along with clarithromycin and metronidazole for 5 days.

Second-line treatment adds bismuth salts or a fluoroquinolone-based antibiotic (e.g., levofloxacin or moxifloxacin) to PPI and the other antibiotics. If second-line therapy fails, further eradication depends on the antibiotic susceptibility [29].

In patients with antrum-predominant gastritis, eradication of Hp restores impaired inhibition of gastrin release, decreases basal and stimulated acid secretion in the stomach, and acid load in the duodenum. However, the atrophy of gastric mucosa does not reverse after successful eradication of Hp [30].

Therapy of Hp⁻ PUD

Most antacids are inorganic salts (magnesium and/or aluminum hydroxide and/or bicarbonate, and calcium carbonate), which neutralize gastric acid directly. Aluminum additionally increases synthesis of PGE_2 and gastric mucosal microcirculation, yet it may cause constipation. Magnesium can cause diarrhea. Due to their adsorptive capacity, these antacids may hinder intestinal absorption of concurrent medication. Moreover, antacids containing sodium bicarbonate, magnesium hydroxide, and calcium carbonate are contraindicated in patients with renal insufficiency [31].

Anticholinergics (e.g., pirenzepine or telenzepine) speed up the healing of peptic ulcers due to their inhibitory effect on gastric acid secretion. However, side effects include visual disturbances, photophobia, and dryness of the mouth, limiting their use, especially in patients with glaucoma (see chapter "Osteoporosis"), enlarged prostate, or stenosis of the pylorus sphincter [32].

H2RAs (e.g., cimetidine, ranitidine, famotidine, nizatidine, roxatidine) competitively and reversibly block the H2R on the basolateral membrane of the parietal cells inhibiting mainly basal and partially meal-stimulated gastric acid secretion. Possible side effects of H2RAs include the development of tolerance and rebound acid hypersecretion (a temporary increase in gastric acid secretion after the abrupt withdrawal of H2RA) [33].

PPIs (e.g., omeprazole, lansoprazole, pantoprazole, and rabeprazole) inhibit acid secretion by the parietal cell [34]. Although gastric acid inhibition is very effective, prolonged suppression favors small intestinal bacterial overgrowth and concomitant malabsorption and facilitates enteric infections (especially with *Clostridium difficile*), as bacteria are no longer effectively eliminated in the stomach [35]. Additionally, osteoporosis (due to parathyroid hyperplasia, see chapter "Rheumatoid Arthritis") and hyperplasia of enterochromaffin cells (due to chronic hypergastrinemia) can develop (Fig. 2) [36].

To enhance mucosal defense, initially, misoprostol, a prostaglandin E1 analogue, was used. Unfortunately, its use was limited by abdominal pain and diarrhea due to its stimulatory effect on intestinal motility [37]. Sucralfate exerts protective action by increasing the synthesis of mucosal growth factors, mucosal microcirculation, and by angiogenic actions [38]. Colloidal bismuth subcitrate accelerates ulcer healing by the formation of occlusive complexes and the accumulation of epidermal growth factor (EGF) in the ulcer crater [39].

Rebamipide is cytoprotective and enhances gastric mucus production, stimulates the release of endogenous prostaglandins, and inhibits the generation of ROS [40].

Perspectives

Thanks to a declining prevalence of Hp infection and introduction of effective anti-Hp-therapies, PUD frequency is decreasing [4] and NSAIDs-induced ulcers now emerge as the most important cause of PUD.

With regard to the increasing resistance rate of Hp and rising frequency of Hp⁻ ulcers, novel therapies to strengthen the mucosal defense are urgently needed. New candidates to reinforce the mucosal defense include melatonin, probiotics, H_2S, NO, and CO [41]. Melatonin enhances gastric microcirculation and exerts anti-oxidative effects on the gastric mucosa, independently of prostaglandin synthesis [42]. Some probiotics increase production of mucosal growth factors and reduce proinflammatory cytokines [43]. Finally, H_2S, NO, and CO were shown to increase protective prostaglandins in gastroduodenal mucosa and inhibit the generation of proinflammatory cytokines [44].

Questions and Answers

Question 1 Where does a peptic ulcer normally occur within the GI tract?

Answer 1 Peptic ulcer is typically localized in stomach and duodenum, most commonly in the duodenal bulb due to the high exposure to gastric acid.

Question 2 What kind of endogenous and exogenous factors exist that trigger PUD?

Answer 2 Endogenous factors include gastric acid and pepsin, while exogenous factors include chronic *Helicobacter pylori* (Hp) infection, use of non-steroidal anti-inflammatory drugs (NSAIDs), consumption of alcohol, smoking, and exposure to stress.

Question 3 How can you treat Hp⁺ peptic ulcer disease?

Answer 3 Eradication of Hp is usually achieved by a combination of acid-inhibiting therapy and two antibiotics (standard triple therapy).

References

1. Malfertheiner P, Chan FKL, McColll KEL (2009) Peptic ulcer disease. Lancet 374:1449–1461
2. Konturek PC (1997) Physiological, immunohistochemical and molecular aspects of gastric adaptation to stress, aspirin and to H. pylori-derived gastrotoxins. J Physiol Pharmacol 48:3–42
3. Tytgat GNJ (2011) Ethiopathogenetic principles and peptic ulcer disease classification. Dig Dis 29:454–458
4. Konturek PC, Bielanski W, Konturek SJ, Hahn EG (1999) Helicobacter pylori associated gastric pathology. J Physiol Pharmacol 50(5):695–710
5. Choung RS, Talley NJ (2008) Epidemiology and clinical presentation of stress-related peptic damage and chronic peptic ulcer. Curr Mol Med 8(4):253–257
6. Milosavljevic T, Kostić-Milosavljević M, Jovanović I, Krstić M (2011) Complications of peptic ulcer disease. Dig Dis 29:491–493
7. Graham DY, Kato M, Asaka M (2008 Jul) Gastric endoscopy in the 21st century: appropriate use of an invasive procedure in the era of non-invasive testing. Dig Liver Dis 40(7):497–503
8. Schwarz K (1910) Über penetrierende Magenund Jejunalgeschwüre. Beitr Klin Chir 67:96–128
9. Tarnawski AS, Ahluwalia A, Jones MK (2012) The mechanisms of gastric mucosal injury: focus on microvascular endothelium as a key target
10. Graham DY, Lew GM, Klein PD, Evans DG, Evans DJ Jr, Saeed ZA, Malaty HM (1992) Effect of treatment of helicobacter pylori infection on the long-term recurrence of gastric or duodenal ulcer. A randomized, controlled study. Ann Intern Med 116(9):705–708
11. Marshall BJ (1995) The 1995 Albert Lasker Medical Research Award. Helicobacter pylori. The etiologic agent for peptic ulcer. JAMA 274(13):1064–1066
12. Konturek PC, Konturek SJ, Bobrzynski A, Kwiecien N, Obtulowicz W, Stachura J, Hahn EG, Rembiarz K (1997) Helicobacter pylori and impaired gastric secretory functions associated with duodenal ulcer and atrophic gastritis. J Physiol Pharmacol 48(3):365–373

13. McColl KE, El-Omar E, Gillen D (2000) Helicobacter pylori gastritis and gastric physiology. Gastroenterol Clin N Am 29(3):687–703

14. Malfertheiner P, Leodolter A, Peitz U (2000) Cure of helicobacter pylori-associated ulcer disease through eradication. Bailliers Best Pract Res Clin Gastroenterol 14(1):119–132

15. Takeuchi K (2012) Pathogenesis of NSAID-induced gastric damage: importance of cyclooxygenase inhibition and gastric hypermotility. World J Gastroenterol 18(18):2147–2160

16. Chan FKL, Leung WK (2002) Peptic ulcer disease. Lancet 360:933–941

17. Gisbert JP, Calvet X (2009) Review article: helicobacter pylori-negative duodenal ulcer. Aliment Pharmacol Ther 30:791–815

18. Maity P, Biswas K, Roy S, Banerjee RK, Bandyopadhyay U (2003) Smoking and the pathogenesis of gastroduodenal ulcer-recent mechanistic update. Mol Cell Biochem 253(1–2):329–338

19. Franke A, Teyssen S, Singer MV (2005) Alcohol-related diseases of the esophagus and stomach. Dig Dis 23(3–4):204–213

20. Shimamoto T, Yamamichi N, Kodashima S, Takahashi Y, Fujishiro M, Oka M, Mitsushima T, Koike K (2013) No association of coffee consumption with gastric ulcer, duodenal ulcer, reflux esophagitis and non-erosive reflux disease: a crpss-sectional study of 8,013 healthy subjects in Japan. PLoS One 8(6):e65996

21. Selgrad M, Malfertheiner P (2011) Treatment of helicobacter pylori. Curr Opin Gastroenterol 27(6):565–570

22. Chan FK, Ching JY, Suen BY, Tse YK, Wu JC, Sung JJ (2013) Effects of Helicobacter pylori infection on long-term risk of peptic ulcer bleeding in low-dose-aspirin users. Gastroenterology 144(34):528–535

23. Malfertheiner P, Megraud F, O'Morain C, Bazzoli F, El-Omar E, Graham D, Hunt R, Rokkas T, Vakil N, Kuipers EJ (2007) Current concepts in the management of Helicobacter pylori infection: the Maastricht III consensus report. Gut 56(6):772–781

24. Sachs G, Shin JM, Hunt R (2010) Novel approaches to inhibition of gastric acid secretion. Curr Gastroenterol Rep 12(6):437–447

25. Holle GE (2010) Pathophysiology and modern treatment of ulcer disease. Int J Mol Med 25(4):483–491

26. Holster IL, Kuipers EJ (2011) Update on the endoscopic management of peptic ulcer bleeding. Curr Gastroenterol Rep 13:525–531

27. Lee CW, Sarosi GA Jr (2011) Emergency ulcer surgery. Surg Clin North Am 91(5):1001–1013

28. Ford AC, Delaney BC, Forman D, Moayyedi P (2006) Eradication therapy for peptic ulcer disease in helicobacter pylori positive patients. Cochrane Database Syst Rev 2:CD003840

29. Graham DY, Calvet X (2012) Guide regarding choice of second-line therapy to obtain a high cumulative cure rate. Helicobacter 17:243–245

30. McColl KE (1997) Helicobacter pylori and acid secretion: where are we now? Eur J Gastroenterol Hepatol 9(4):333–335

31. Konturek SJ (1993) New aspects of clinical pharmacology of antacids. J Physiol Pharmacol 44(3 Suppl 1):5–21

32. Warner CW, McIsaac RL (1991) The evolution of peptic ulcer therapy. A role for temporal control of drug delivery. Ann N Y Acad Sci 618:504–516

33. Huang JQ, Hunt RH (2001) Pharmacological and pharmacodynamic essentials of H2-receptor antagonists and proton pump inhibitors for the practising physician. Best Pract Res Clin Gastroenterol 15(3):355–370

34. Mössner J, Caca K (2005) Developments in the inhibition of gastric acid secretion. Eur J Clin Investig 35(8):469–475

35. Williams C, McColl KE (2006) Review article: proton pump inhibitors and bacterial overgrowth. Aliment Pharmacol Ther 23(1):3–10

36. Abraham NS (2012) Proton pump inhibitors: potential adverse effects. Curr Opin Gastroenterol 28(6):615–620

37. Reskin JB, White RH, Jackson JE, Weaver AL, Tindall EA, Lies RB, Stanton DS (1995) Misoprostol dosage in the prevention of non-steroidal anti-inflammatory drug-induced gastric and duodenal ulcers: a comparision of three regimens. Ann Intern Med 123:344–350

38. Konturek SJ, Brzozowski T, Majka J, Szlachcic A, Bielanski W, Stachura J, Otto W (1993) Fibroblast growth factor in the gastroprotection and ulcer healing: interaction with sucralfate. Gut 34(7):881–887

39. Hall DW (1989) Review of the modes of action of colloidal bismuth subcitrate. Scand J Gastroenterol Suppl 157:3–6

40. Song KH, Lee YC, Fan DM, Ge ZZ, Ji F, Chen MH, Jung HC, Bo J, Lee SW, Kim JH (2011) Healing effects of rebamipide and omeprazole in helicobacter pylori-positive gastric ulcer patients after eradication therapy: a randomized double-blind, multinational, multi-institutional comparative study. Digestion 84:221–229

41. Al-Jiboury H, Kaunitz JD (2012) Gastroduodenal mucosal defense. Curr Opin Gastroenterol 28(6):594–601

42. Celinski K, Konturek SJ, Konturek PC, Brzozowski T, Cichoz-Lach H, Slomka M, Bielanski W, Reiter RJ (2011) Melatonin or L-tryptophan accelerates healing of gastroduodenal ulcers in patients treated with omeprazole. J Pineal Res 50(4):389–394

43. Konturek PC, Sliwowski Z, Koziel J, Ptak-Belowska A, Burnat G, Brzozowski T, Konturek SJ (2009) Probiotic bacteria E.Coli Nissle 1917 attenuates acute gastric lesion induced by stress. J Physiol Pharmacol 6:41–48

44. Wallace JL (2012) Hydrogen sulphide: a rescue molecule for mucosal defence and repair. Dig Dis Sci 57(6):1432–1434

Lactose and Food Intolerance

Anthony K. Campbell and Stephanie B. Matthews

Introduction to Lactose and Food Intolerance

Lactose and food intolerance are biochemical conditions caused by the inability to digest fully the sugar in milk, that is, lactose [1], other sugars, and sugar alcohols (Fig. 1). The condition of lactose intolerance is better described as "lactose sensitivity," as everyone can tolerate some lactose, although this amount varies considerably between individuals [2, 3]. It is essential to distinguish lactose intolerance from an allergy to milk proteins as there are major differences in the management of these two conditions. Milk allergy occurs in 3–5% of infants. An intolerance is caused by a biochemical defect, such as the loss of a particular enzyme. Whereas an allergy involves the immune system, often the immunoglobulin IgE or IgG. IgE binds to mast cells, stimulating a rise in intracellular calcium to activate secretion of histamine. An allergy to milk typically involves an immune response to one of the milk proteins, for example, the major milk protein casein. Lactose intolerance, on the other hand, is caused by loss or inhibition of the enzyme lactase or disruption of lactose digestion. Throughout this article, the term "lactose intolerance" will be used, since this is the one everyone is familiar with in medicine and elsewhere. The issue is what level of lactose the individual can tolerate without causing symptoms. Many Northern Europeans can drink one or two glasses of milk, up to 500 mL, without exhibiting any symptoms. However, other ethnic groups can be so sensitive that just 10 mL of milk can cause adverse symptoms. Some 4 billion people around the world express low lactase levels (see below) and are thus potentially sensitive to lactose, suffering a range of gut and systemic symptoms (Fig. 2), unless diagnosed and then managed correctly. Lactose sensitivity is associated with two common gut conditions [3], irritable bowel syndrome (IBS, a disease characterized by abdominal pain, diarrhea, or constipation, or both alternating), and inflammatory bowel disease (IBD, a group of inflammatory conditions affecting the small intestine and colon, including Crohn's disease and ulcerative colitis) [4]. Thus, the possibility of IBS or IBD should be taken into account when diagnosing and treating these patients for lactose and food intolerance [5]. The genetics of lactose sensitivity are complex [6, 7].

Lactose, galactose-1,4-β-glucose, is only found naturally in significant amounts in mammalian milk (cow's milk 49 g/L; human milk 70 g/L). Lactose is one sixth as sweet as sucrose, and provides some 40% of the energy

A. K. Campbell (✉)
School of Pharmacy and Pharmaceutical Sciences, Cardiff University, Cardiff, UK
e-mail: campbellak@cardiff.ac.uk

S. B. Matthews
Welston Court Science Centre, Welston Court, Milton, Pembrokeshire, UK

E. Lammert, M. Zeeb (eds.), *Metabolism of Human Diseases*,
https://doi.org/10.1007/978-3-031-96019-2_19

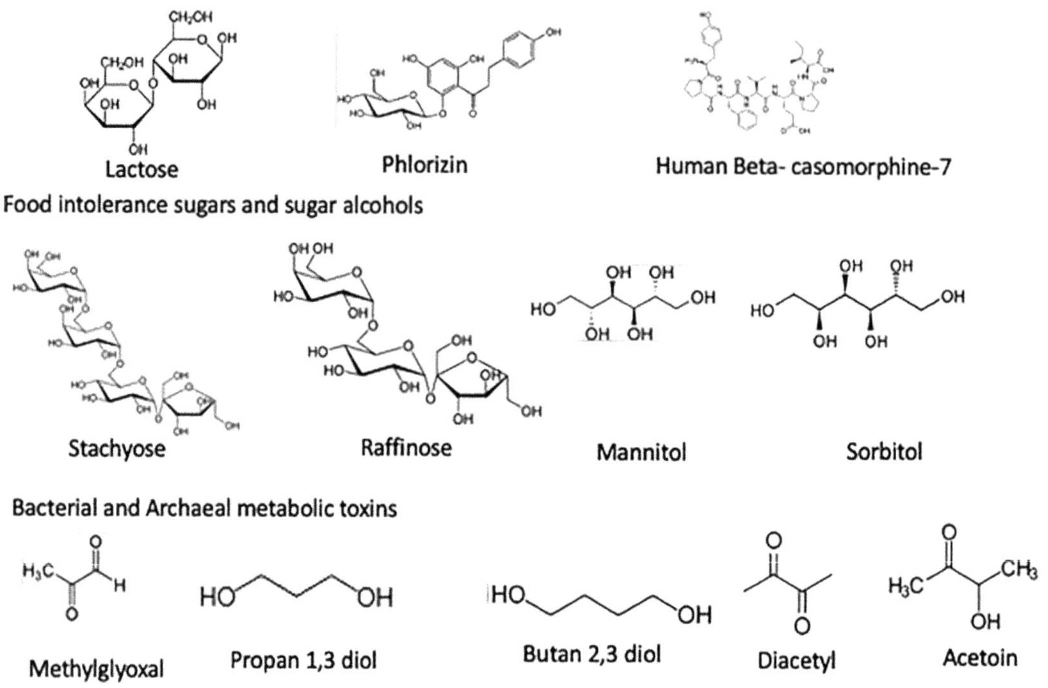

Lactose   **Phlorizin**   **Human Beta- casomorphine-7**

Food intolerance sugars and sugar alcohols

Stachyose   **Raffinose**   **Mannitol**   **Sorbitol**

Bacterial and Archaeal metabolic toxins

Methylglyoxal   **Propan 1,3 diol**   **Butan 2,3 diol**   **Diacetyl**   **Acetoin**

Fig. 1 Key sugars and bacterial toxins in lactose and food intolerance

requirements of a suckling infant. However, adults do not need lactose. Disaccharides (e.g., lactose, sucrose, and isomaltose) cannot be absorbed directly in the intestine, requiring an enzyme to cleave them into monosaccharides. Lactose is cleaved by lactase (lactase-phlorizin hydrolase), which occurs in the small intestinal microvilli. Galactose and glucose are subsequently absorbed into the enterocyte via the sodium-activated glucose transporter 1 (SGLUT1, also called SGLT1). Glucose and galactose are then taken up into the blood by the transporter GLUT2. Galactose is converted to glucose mainly by the Leloir pathway, where galactose is converted to uridine diphosphate (UDP)-glucose, particularly in the liver via galactose 1-phosphate (Gal-1-P) and UDP-galactose. Alternatively, humans can use the De Ley Doudoroff pathway. Inherited mutations of enzymes in the Leloir pathway cause galactosemias, affecting 1 in 55,000 individuals. For example, galactosemia type 1 (galactose-1-phosphate uridylyltransferase deficiency) is seen in suckling infants causing severe illness, including not wanting to drink, vomiting, jaundice, hypoglycemia, enlarged liver (hepatomegaly), enlarged spleen (splenomegaly), proximal tubulous kidney damage, cataract, mental retardation, and failure to thrive.

Galactose synthesis occurs mainly in the mammary gland, via the reversible conversion of UDP-glucose to UDP-galactose. UDP-galactose then reacts with glucose to form lactose, though 65% of galactose for lactose synthesis comes from the diet. Galactose can also be synthesized de novo from glycerol. Galactose has an important role in cerebrosides which consist of galactose or glucose and a ceramide.

SGLT1 is inhibited by tri- and tetra-saccharides (e.g., raffinose and stachyose) found in root vegetable, pulses, and soya. People who eat a lot of these can exhibit symptoms similar to hypolactasia (lack or reduced amounts of lactase).

A further problem is the increasing prevalence of sensitivity to fructose. There is an increased use of fructose as a sweetener in many foods and drinks, and sensitivity to fructose can lead to similar symptoms to those of lactose intolerance.

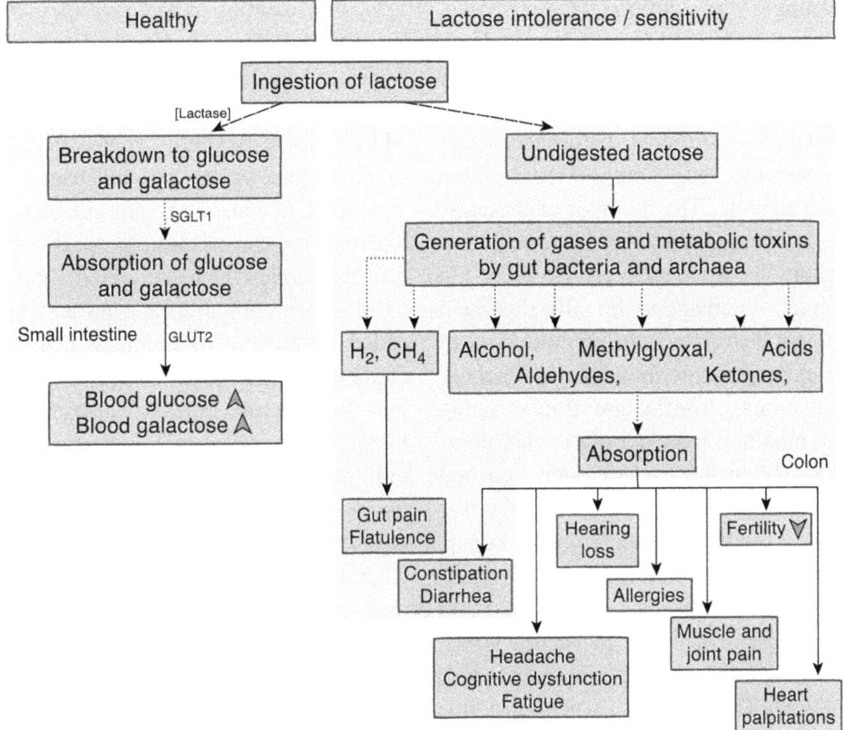

Fig. 2 Main metabolic changes in lactose intolerance. Under healthy conditions, lactose is digested by lactase and incorporated causing an increase in blood glucose and galactose levels (left side). In patients who cannot hydrolyze lactose fully in the small intestine (lactose intolerant / sensitive, right side) lactose is transferred to the large intestine and metabolized to metabolic toxins. The metabolic toxin groups are shown, with the respective gut and systemic symptoms. *SGLT1* sodium-activated glucose transporter 1, *GLUT2* glucose transporter 2, *H₂* hydrogen, *CH₄* methane

Fructose is absorbed into the enterocyte by GLUT5. Like glucose and galactose, fructose is transferred into the blood by GLUT2.

In diagnosis, hypolactasia is often overlooked, and not tested for if the patient only complains of systemic symptoms, as opposed to gut symptoms. Yet, in many hypolactasia/lactose intolerant patients, systemic symptoms can be more significant than those in the gut.

Interestingly, the symptoms that affected Charles Darwin for 50 years match exactly those of lactose and food intolerance [8]. Yet, he was never diagnosed, because his doctors did not understand the link between the gut and the peripheral tissues, even though lactose itself was recognized in the nineteenth century.

Lactose intolerance is often confused in babies with allergy to milk proteins, particularly αS1-casein, as symptoms are similar [1]. Protein allergy usually disappears by 3 years, whereas lactose intolerance can get worse with age. The allergy severity ranges from mild to anaphylactic. Depending on the allergen, changing the origin of the milk, for example, to goats milk, can be effective, though this is not always the case.

Additionally, many patients suffering from lactose intolerance are also sensitive to other substances in foods, the most common being wheat.

Pathophysiology of Lactose Intolerance and Metabolic Alterations

There are six causes of hypolactasia, where the lactase enzyme is considerably lower than in people who have retained their enzyme into adulthood: (i) congenital loss (very rare), (ii) inherited loss on weaning—very common, (iii) gut infections, such as rotavirus and *Giardia* (a monocellular parasite), (iv) damage to the villi in the small intestine caused by radiotherapy or bacterial overgrowth, (v) hormonal disturbance (e.g., thyroid), menopause, and ageing, (vi) inhibition of the lactase enzyme. The first two of these are irreversible. In contrast, the other four are potentially reversible [1], after recovery from the gut infection, repair of damage to the villi, treatment of the hormonal disturbance, or identification of the substances responsible for inhibiting lactase. Almost all mammals lose lactase after weaning [1, 2]. Thus, inherited hypolactasia is very common, being as high as 90% in Chinese people and > 80% in Asians. In evolution, apes would never see milk after coming off the mother's breast. Lactase persistency in adult humans is only present in some ethnic groups, such as Bedouins and Northern Europeans (in which it reaches 90%). Sensitivity to lactose generally increases with southern origins. The loss after weaning correlates with two polymorphisms—C/T_{13910} and G/A_{22018}—occurring within introns of a helicase, upstream from the lactase gene on the long arm of chromosome 2 [5, 7]. There is close correlation between the level of lactase and the C and G genotypes, those with CC and GG having the lowest levels. The molecular basis of this correlation is unknown. But the key is the number of cells expressing lactase rather than the level of lactase in each cell [1, 2]. In Chinese people, the loss of lactase can be >90% by the age of 5 years, but in other ethnic groups it may take until teenage or later before the nadir of lactase is reached.

Hypolactasia causes a maldigestion of lactose in the small intestine. Lack of lactose uptake does not cause severe symptoms as the human body does not require lactose as such and can synthesize galactose in some non-mammary cells if required. However, undigested lactose is transported to the colon where it is subject to degradation by bacteria or archaeans forming gases such as hydrogen and methane as well as metabolic toxins such as methylglyoxal and other alcohols, diols, aldehydes, ketones, and acids [9]. Due to the variable tolerance of lactose between individuals [3, 4] and the different degradation products, the resulting symptoms vary in type and severity (Fig. 2) [3, 5, 10].

Gas in the large intestine causes gut symptoms, like distension, borborygmi, and flatulence. Moreover, the metabolic toxins affect the balance of microflora in the gut, and cause diarrhea or constipation. This surprising difference between diarrhea or constipation in particular patients reflects whether the bacterial metabolic toxins act to block smooth muscle contraction, and thus cause constipation, or act in an analogous way to cholera or enterotoxin to signal fluid secretion into the large intestine. After absorption into the blood, the bacterial metabolic toxins result in a multitude of systemic symptoms in peripheral tissues, ranging from fatigue, headache, cognitive dysfunction, muscle and joint pain to heart palpitations, exacerbation of allergies [1, 9–11], and liver and kidney problems (see Chapter "Liver: Overview" and "Kidney: Overview"). These symptoms are the result of direct effects of the toxins on ionic signaling in peripheral tissues [9]. Most prominently, methylglyoxal reacts with hormones (e.g., insulin) and neurotransmitters (e.g., serotonin and dopamine), inactivating them [9], a biochemical process called the Pictet-Spengler reaction. Methylglyoxal also acts on neurons directly, causing pain.

Lactose intolerance is associated with IBS [5] and IBD [4]. The symptoms in IBS are a result of poor lactose digestion and sugar absorption in the small intestine. In IBD, it is not clear whether lactose intolerance is a cause or consequence of the intestinal inflammation. Although there are endogenous mechanisms capable of inactivating bacterial toxins, their role in lactose intolerance, IBS, and IBD, is as yet unknown.

Treatment

Treatment depends critically on diagnosis and on the reason for the hypolactasia. Measurement of breath hydrogen and methane after a food challenge provides a useful way of identifying a patient's sensitivity to lactose and other foods. A handheld device linked to your mobile phone is now available. Failure to treat correctly may

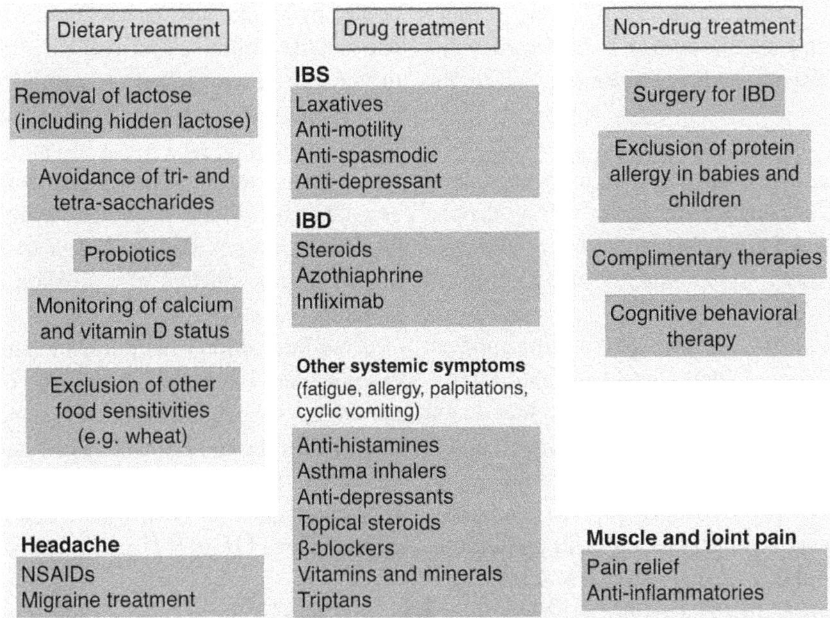

Fig. 3 Influence of disease treatments on disease. Three general types of treatment are available for the various symptoms in lactose sensitive patients: (i) dietary manipulation, particularly lactose exclusion, (ii) drugs, which are often not very effective, if at all, and (iii) non-drug therapy. The most effective and common treatment is removal of lactose from the diet. Other dietary restrictions might be helpful as well. Comorbidities and symptoms are often targeted using specific drugs. Cognitive behavioral therapy involves discussions with a psychologist, using standard psychological methods. Complimentary therapies involve acupuncture, reflexology, and exercise. *IBS* Irritable bowel syndrome, *IBD* inflammatory bowel disease, *NSAIDs* non-steroidal anti-inflammatory drugs

leave patients with long-term damage to the intestine or other tissues and also other conditions, such as IBD or celiac disease. As congenital loss and loss after weaning cannot be "reactivated," avoidance of lactose in the diet is, in general, inevitable (Fig. 3) [1, 2]. However, this is difficult because lactose is "hidden" in many foods and drinks, not normally associated with milk (e.g., chocolate, bread, biscuits, processed meat) [1, 3]. Lactose is used in processed foods because it improves shelf life, browning, mouth feel, and flavor. As a filler, lactose bulks up expensive ingredients and often replaces fat removed from products in low-fat alternatives.

Labeling of added lactose is poor as it is commonly part of "flavorings" or "added sugar," being described as "natural." It is possible to take the equivalent of a liter of milk per day in "hidden" lactose, for example, after starting a weight loss regime using meal substitute drinks high in lactose. Symptoms of lactose intolerance are often aggravated by tri- and tetra-saccharides inhibiting SGLT1 (see above) and leading to unabsorbed sugars reaching the bacteria in the large intestine.

Natural yoghurt is often tolerated, since the lactose content is much reduced. Substitutes, such as lactose-free cow's milk or soya milk, are generally useful, if soya intolerance is excluded.

Symptoms caused only by lactose can disappear within a few days after removing lactose from the diet. Alternatively, the enzyme marketed as "lactase," actually β-galactosidase, taken before eating lactose, may alleviate symptoms, as it is supposed to digest lactose similar to the endogenous enzyme. β-galactosidase hydrolyses lactose to galactose and glucose and is expressed in many bacteria, as well as several molds. Yet, it has no sequence similarity to mammalian lactase, contains only one active center for lactose, and cannot hydrolyze cerebrosides (also known as monoglycosylceramides) as lactase does. The use of β-galactosidase is not routinely recommended, as most of the enzyme is degraded in the stomach. Timing and dose are critical to be effective.

Despite the best therapeutic approach of a lactose-free diet being drug free, many drugs are prescribed, or bought over the counter by people with lactose sensitivity, in an attempt to alleviate gut and systemic symptoms, the most common being IBS.

Treatment of IBS includes antispasmodics (e.g., mebeverine), bulk laxatives, antimotility drugs (e.g., loperamide), antidepressants (e.g., the tricyclic amitriptyline or a serotonin re-uptake inhibitor), which reduce pain and cramps, and analgesics. Probiotics can also reduce symptoms [12]. Patients often also take antacids and proton pump inhibitors to reduce stomach acidity and reflux. However, these have little or no consistent effect on the gut and systemic symptoms caused by lactose. Non-drug treatments of IBS include cognitive behavioral therapy, stress management, complementary therapies, including acupuncture and reflexology, and exercise.

Lactose sensitivity is commonly seen in IBD [4]. IBD treatment involves aminosalicylates, corticosteroids, and immunosuppressants. These are used to reduce the effect of inflammation in the intestine, but do not alleviate the symptoms caused by lactose.

Influence of Treatment on Metabolism and Consequences for Patients

Removal of lactose from the diet or uptake of digesting enzymes does not have adverse effects on the patient's metabolism. It is vital to ensure calcium and vitamin D levels when milk, their major dietary source, is avoided. Advice on the possible use of probiotics is also required. However, dietary removal of lactose may not solve all the problems, if the patient is intolerant to other foods. These need to be identified and treated, most commonly by dietary means.

The systemic symptoms of lactose intolerance lead to unnecessary drug usage. Headaches are treated by regular analgesics and the risk of overuse is high. Migraine and cyclical vomiting syndrome can be controlled by regular β-blockers and triptans (see chapter "Down Syndrome"). Other commonly used drugs are antihistamines, anti-inflammatories, steroids, and asthma inhalers, taken to alleviate allergic symptoms (see chapter "Stroke"). Yet, unknown to most asthmatics, many inhalers also contain significant amounts of lactose.

Drug interaction from polypharmacy is a serious potential risk, especially as many of the drugs are self-prescribed. There is also a danger of drug overuse. Furthermore, there can be an iatrogenic problem of taking too much lactose in drugs, as it is often included in pills, which then contributes to morbidity [13].

FODMAPS, Gluten, and Food Intolerance

Lactose is not the only food that causes this array of gut and systemic symptoms [14]. These conditions include IBS, IBD, celiac and non-celiac gluten sensitivity, and FODMAPS syndrome standing for "fermentable oligosaccharides, disaccharides, monosaccharides, and polyols" found in a range of foods [15–17]. They include sugars and other molecules that are not fully absorbed in the small intestine or block the uptake of glucose and galactose. They include FOS, an acronym for Fructans (Fructo-oligosaccharide), Inulins, and GOS, an anacronym for Galacto-oligosaccharides such as raffinose and stachyose. Fructose is found in many fruits and is particularly high in apples, mangos, pears, watermelon, certain corn-syrups, honey, and agave. Fructans are high in rye, wheat, asparagus, broccoli, onions, cabbage, and garlic. Inulins are fructan polysaccharides found in thousands of plants and consist of glucosyl residues linked to repetitive fructosyl residues through a beta-link. Inulins are high in onions and wheat. Raffinose and stachyose are found in certain root vegetables such as parsnips. Sorbitol and mannitol have also been identified as culprits of food intolerance. Sorbitol is high in several fruits such as plums, peaches, cherries, and dried fruit, but is also used as a sugar substitute in baked foods. Similarly, mannitol is found in most fruits and vegetables, and is particularly high in mushrooms. Mannitol is used in confectionary products such as chocolate coatings and chewing gum. People who have a sensitivity to FODMAPS suffer from the same gut and systemic symptoms as lactose intolerance.

Gluten is a structural group of proteins found in cereal grains such as wheat, barley, and rye, and is an important component of wheat flour giving dough elasticity, helping bread to rise with the right final shape. Gluten proteins in wheat include prolamins and glutelins, as well as inulins, can cause inflammatory, immunological or autoimmune problems in susceptible people. There are two main types of disorder—celiac disease, which is an autoimmune reaction causing antibodies to gluten proteins, and non-celiac gluten sensitivity which does not involve the immune system directly [18]. Both conditions cause gut and systemic symptoms, similar to those caused by lactose intolerance. However, celiac disease tends to result in more severe abdominal pain and other gut symptoms, as well as diarrhea or constipation, bloating, nausea, and acid reflux. Also, children with untreated celiac disease often do not grow normally. As with lactose intolerance, many patients can suffer from depression. The only treatment for celiac disease or non-celiac gluten intolerance is to follow a rigorous gluten free diet by avoiding all breads and other foods such as pizzas, tortillas, pasta, and soy source containing wheat flour that has gluten. Many restaurants and hotels now offer not only vegetarian and vegan menus, but also menus free of gluten. A major problem is that as many as 80% of people sensitive to gluten remain undiagnosed.

COVID-19 Related to Lactose and Food Intolerance

In January 2020 the highly contagious viral disease coronavirus disease 2019 (COVID-19) spread worldwide, resulting in a pandemic. There are two interactions of COVID-19 that have been associated with lactose and food intolerance [19]. First, if you are lactose intolerant, then you can be more susceptible to a serious COVID-19 infection. Second, the SARS-CoV-2 virus that causes COVID-19 can infect the gut in addition to infection of the respiratory tract. As with other gut viral infections this leads to damage of the villi carrying lactase, and thus reduce a person's threshold to lactose, causing gut and systemic symptoms. In addition to symptoms associated with infection of the respiratory tract, many patients with COVID-19 have a high prevalence of diarrhea, nausea, vomiting, and abdominal pain, as well as headache, muscle and joint pain, and fatigue. All of these are associated with the biochemical mechanism causing the symptoms of lactose and food intolerance. What is unclear is how reversible this effect is. Long COVID-19 may result in long-term damage to the small intestine and thus induce long-term sensitivity to lactose. This requires further investigation.

Perspectives

In diagnosis, the gold standard [5] for a patient presenting with unexplained gut and systemic symptoms should first be to test for the polymorphism C/T_{13910}. All CC should immediately change to a lactose-free diet, whereas CT or TT patients should undergo a hydrogen and methane breath test after lactose ingestion, together with a record of both gut and systemic symptoms. A few patients can show no positive hydrogen or methane breath test, but still exhibit symptoms after lactose ingestion. Hypolactasia caused by infections, for example, *Giardia* or rotavirus, which damage the lactase-containing intestinal villi (see chapter "Colorectal Cancer"), or hormonal imbalance, which can reduce lactase levels, should be investigated, if there is no evidence of family history. If the breath hydrogen or methane is increased, together with induction of symptoms by lactose, the patient should change to a lactose-free diet. All patients with a significant increase in symptoms after the lactose load should undergo a supervised trial to determine their lactose threshold. Every patient should be followed up in 12 weeks for a definitive diagnosis, based on clinical improvement, following lactose removal from the diet. Similar challenges can be performed and interpreted after ingestion of fructose, inulin, sorbitol, and mannitol.

There are two types of milk from either A1 or A2 cows, the latter being descendants of the ancient cow [20]. The main difference in the milk is a mutation in beta-casein at amino-acid 67, from proline in A2 milk to histidine in A1. When casein in A1 milk is broken down in the gut, it generates the heptapeptide beta-casomorphin (Fig. 1) which binds to opioid receptors. This can activate nerves, increasing the pleasant sensation when this milk is drunk by calves and humans. There is a high incidence of type 1 diabetes in countries where A1 cows are the main source of milk. A number of active peptides can be generated during the digestion of milk proteins.

Lactose sensitivity and food intolerance via metabolic toxins produced by gut bacteria and archaeans (Fig. 1) provide a new mechanism linking the gut to other illnesses [9, 10, 21–24], such as type 2 diabetes (see chapter "Metabolic Syndrome" and "Diabetes Mellitus"), Alzheimer's disease (see chapter "Alzheimer's Disease"), Parkinson's disease, some cancers, and periodontal disease. An important toxin produced by gut bacteria and Archaea is methylglyoxal which can covalently modify and thus inactivate 5-hydroxytryptamine and insulin [9, 25, 26]. These toxins induce calcium signals in bacteria [27–29]. The molecular mechanisms responsible for the production of bacterial metabolic toxins are a good target for drug discovery in developing a new treatment for these conditions.

A crucial issue is that many people who have lactose intolerance also suffer from intolerance to other foods, particularly those containing tri- and tetra-saccharides and sugar alcohols.

Questions and Answers

Question 1 Name four of the systemic symptoms in lactose and food intolerance.

Answer 1 Heart palpitations, headache, severe fatigue, cognitive dysfunction, muscle and joint pain, exacerbation of allergies, liver and kidney problems.

Question 2 Name four causes of loss of lactase—hypolactasia.

Answer 2 Congenital loss, inherited loss in weaning, gut infections, damage to the intestinal villi, hormonal disturbance, inhibition of lactase, COVID-19.

Question 3 Name two sugars in vegetables that can cause gut symptoms.

Answer 3 Raffinose, stachyose.

Question 4 What does FODMAP stand for?

Answer 4 Fermentable oligosaccharides, disaccharides, monosaccharides, and polyols.

References

1. Campbell A, Jenkins-Waud J, Matthews S (2005) The molecular basis of lactose intolerance. Sci Prog 92:241–287
2. Campbell AK, Matthews SB (2005) Tony's lactose free cookbook: the science of lactose intolerance and how to live without lactose. The Welston Press, Pembrokeshire, p 204
3. Matthews SB, Waud JP, Roberts AG, Campbell AK (2005) Systemic lactose intolerance: a new perspective on an old problem. Postgrad Med J 81:167–173
4. Eadala P, Matthews SB, Waud JP, Green JT, Campbell AK (2011) Association of lactose sensitivity with inflammatory bowel disease – demonstrated by analysis of genetic polymorphism, breath gases and symptoms. Aliment Pharmacol Ther 34:735–746
5. Waud JP, Matthews SB, Campbell AK (2008) Measurement of breath hydrogen and methane, together with lactase genotype, defines the current best practice for investigation of lactose sensitivity. Ann Clin Biochem 45:50–58
6. Flatz G (1987) Genetics of lactose digestion in humans. Adv Hum Genet 16:1–77
7. Enattah NS, Sahi T, Savilahti E, Terwilliger JD, Peltonen L, Jarvela I (2002 Feb) Identification of a variant associated with adult-type hypolactasia. Nature Genet 30(2):233–237

8. Campbell AK, Matthews SB (2015) Darwin diagnosed? Biol J Linn Soc 116:964–984
9. Campbell AK, Matthews SB, Vassell N, Cox C, Naseem R, Chaichi MJ, Holland IB, Wann KT (2010) Bacterial metabolic 'toxins': a new mechanism for lactose and food intolerance, and irritable bowel syndrome. Toxicology 278(3):268–276
10. Matthews SB, Campbell AK (2000) When sugar is not so sweet. Lancet 355(9212):1330
11. Grimbacher B, Peters T, Peter HH (1997) Lactose-intolerance may induce severe chronic eczema. Int Arch Allergy Immunol 113(4):516–518
12. Zuccotti G (2008) Probiotics in clinical practice: an overview. J Int Med Res 36(1):1–53
13. Eadala P, Waud JP, Matthews SB, Green JT, Campbell AK (2009) Quantifying the 'hidden' lactose in drugs used for the treatment of gastrointestinal conditions. Aliment Pharmacol Ther 29(6):677–687
14. Bischoff SC (2012) Food intolerance. Laryngo-Rhino-Otologie 91:720–738
15. Barrett JS, Gibson PR (2012) Fermentable oligosaccharides, disaccharides, monosaccharides and polyols (FODMAPs) and nonallergic food intolerance: FODMAPs or food chemicals? Ther Adv Gastroenterol 5:261–268
16. Gibson PR (2017) History of the low FODMAP diet. J Gastroenterol Hepatol 32:5–7
17. Gibson PR, Halmos EP, Muir JG (2020) Review article: FODMAPS, prebiotics and gut health. The FODMAP hypothesis revisited. Aliment Pharmacol Ther 52:233–246. 2012;5:261–8
18. Leonard MM, Sapone A, Catassi C, Fasano A (2017) Celiac disease and nonceliac gluten sensitivity a review. JAMA 318:647–656
19. Elbeltagi R, Al Beltagi M, Kamal Saeed N, Bediwy AS (2023) COVID-19-induced gastrointestinal dysfunction: a systematic review. World J Clin Cases 11(22):5252–5272
20. Clemens RA (2011) Milk A1 and A2 peptides and diabetes. Milk Milk Prod Hum Nutr 67:187–195
21. Gomez-Martinez S, Redondo N, Marcos A (2020) The gut microbiota in relation to food allergies and intolerances. Ann Nutr Metab 76:37–38
22. Kugler C, Ring J, Schnopp C (2009) Skin diseases from food intolerance. Ernahrungs-Umschau 56:682–687
23. Felice VD, Quigley EM, Sullivan AM, O'Keeffe GW, O'Mahony SM (2016) Microbiota-gut-brain signalling in Parkinson's disease: implications for non-motor symptoms. Parkinsonism Relat Disord 27:1–8
24. Elbeltagi R, AL-Beltagi M, Kamal Saeed N, Bediwy AS (2023) COVID-19-induced gastrointestinal dysfunction: a systematic review. World J Clin Cases 11(22):5252–5272
25. Lee KM, Lee CY, Zhang G, Lyu AP, Yue KKM (2019) Methylglyoxal activates osteoclasts through JNK pathway leading to osteoporosis. Chem Biol Interact 308:147–154
26. Simoes CD, Maganinho M, Sousa AS (2022) FODMAPs, inflammatory bowel disease and gut microbiota: updated overview on the current evidence. Eur J Nutr 61:1187–1198
27. Zhang S, Jiao TW, Chen YS, Gao N, Zhang LL, Jiang M (2014) Methylglyoxal induces systemic symptoms of irritable bowel syndrome. PLoS One:9
28. Campbell AK, Naseem R, Wann K, Holland IB, Matthews SB (2007) Fermentation product butane 2,3-diol induces Ca2+ transients in E. Coli through activation of lanthanum-sensitive Ca2+ channels. Cell Calcium 41:97–106
29. Campbell AK, Naseern R, Holland IB, Matthews SB, Wann KT (2007) Methylglyoxal and other carbohydrate metabolites induce lanthanum-sensitive Ca2+ transients and inhibit growth in E. coli. Arch Biochem Biophys 468:107–113

Colorectal Cancer

Soeren Ocvirk, Anika Sander,
and Stephen J. D. O'Keefe

Introduction to Colorectal Cancer

Colorectal cancer (CRC) is the most common gastrointestinal cancer. It is the second most common cancer in women and third most common in men by incidence, and represents the second most common cause of cancer mortality in both sexes worldwide [1]. CRC is mainly a disease of Western populations with more than 50% of cases recorded in high-income countries [2]. The vast majority of CRC cases are sporadic and caused by a complex interplay between genetic, host, and environmental factors, in particular diet [3]. There is convincing evidence for Western diets, rich in red and processed meat, animal fat, and low in fiber, being a major risk factor for CRC [4]. Recent research has further highlighted the key role of the gut microbiota in mediating the dietary risk of colon cancer [5]. In addition, other environmental factors such as alcohol, smoking, as well as diseases such as inflammatory bowel diseases and obesity contribute to increased CRC risk [6].

Pathophysiology of Colorectal Cancer

CRC results from a stepwise accumulation of genetic defects and clonal proliferation of mutated colonic epithelial cells in an adenoma–carcinoma transformation sequence of normal colonic mucosa (Fig. 1), a protuberant growth known as polyp, adenoma, and ultimately adenocarcinoma [7]. Mutations of the adenomatous polyposis colon (APC) tumor suppressor gene are the most common (~80%) genetic defects observed in sporadic CRC. The non-mutated protein product of the APC gene prevents the accumulation of β-catenin protein, its nuclear translocation, and inappropriate activation of gene transcription via the canonical Wnt pathway that promotes cell proliferation [8]. A plethora of carcinogens, for example, present in tobacco smoke or heated meat, reach the colonic mucosal epithelium and cause genetic mutations. Poor

S. Ocvirk
Division of Gastroenterology, Hepatology and Nutrition, Department of Medicine, University of Pittsburgh, Pittsburgh, PA, USA

ZIEL – Institute for Food and Health, Technical University of Munich, Freising, Germany

A. Sander
ZIEL – Institute for Food and Health, Technical University of Munich, Freising, Germany

S. J. D. O'Keefe (✉)
Division of Gastroenterology, Hepatology and Nutrition, Department of Medicine, University of Pittsburgh, Pittsburgh, PA, USA

African Microbiome Institute, Department of Biomedical Sciences, Faculty of Medicine and Health Sciences, Stellenbosch University, Cape Town, South Africa
e-mail: sjokeefe@pitt.edu

© The Author(s), under exclusive license to Springer Nature Switzerland AG 2026
E. Lammert, M. Zeeb (eds.), *Metabolism of Human Diseases*,
https://doi.org/10.1007/978-3-031-96019-2_20

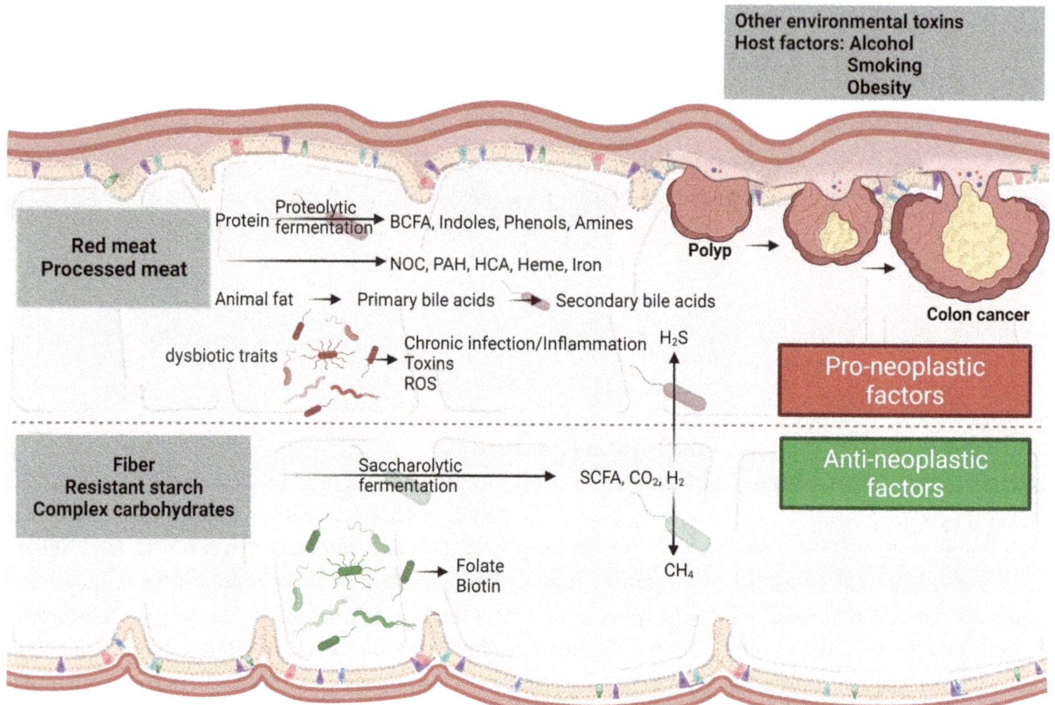

Fig. 1 Overview on microbial metabolites and dietary factors affecting the risk of colorectal cancer (CRC). The balance of dietary and microbial metabolites with health-promoting or detrimental effects on the colon determines the risk of CRC. High fiber and resistant starch diets promote saccharolytic fermentation by bacteria and enhance the production of short-chain fatty acids (SCFA) that have anti-inflammatory and anti-tumorigenic activity. High intake of red meat and fat promote the production of pro-inflammatory end products of proteolytic fermentation by bacteria and tumor-promoting secondary bile acids. *BCFA* branched-chain fatty acids, *NOC* N-Nitroso compounds, *PAH* polyaromatic hydrocarbons, *HCA* heterocyclic amines, *ROS* reactive oxygen species, *SCFA* short-chain fatty acids. (Created with BioRender.com)

folate intake among heavy alcoholics and interference of its absorption by alcohol can result in genetic defects from impaired folate-mediated DNA synthesis, DNA methylation, and repair processes [9]. There is evidence of a positive association between consumption of red and processed meat and CRC, whereas intake of dietary fibers shows linear dose-dependent inverse relationship with CRC risk (Fig. 1) [4, 10]. Evidence supporting the cancer-protective effect of dietary components such as vitamin D, folate, fish, fruits, vegetables, and selenium is suggestive but limited. Interestingly, the colonic microbiota plays a crucial role in mediating the influence of diet on CRC.

Pathophysiologic Role of the Colonic Microbiota and Its Metabolites

Besides its role in regulation of fluid conservation, electrolyte balance, and terminal conduit of undigested human excreta, the colon is inhabited by approximately 10^{13} to 10^{14} microbial cells belonging to a diverse group of microorganisms, termed the gut microbiome, which holds a rich repertoire of metabolic functions [11, 12]. In a symbiotic relationship, the microbiota depends on undigested food residues and in turn produces metabolites that are essential for host health [13]. Recent advances in molecular identification and characterization techniques have led to a better understanding of the microbiota composition and appreciation of their metabolic potential [11, 12]. New analyses identified the gut microbiota to cover more than 3,500 species genomes that belong to more than 2,300 different bacterial genera [14]. While ~90% of protein and carbohydrate are digested and absorbed along the small intestine, residual food is metabolized by the colonic microbiota through fermentation, producing protective and vital metabolites, such as short-chain fatty acids (SCFAs), or vitamins such as folate and biotin that are essential for DNA synthesis and repair. However, the gut microbiota can also produce toxins and detrimental metabolites such as hydrogen sulfide (H_2S), reactive oxygen species (ROS), and secondary bile acids that may promote inflammation and neoplastic progression under specific conditions (see below). Chronic intestinal inflammation is triggered and perpetuated by some gut bacteria through signaling pathways such as induction of toll-like receptors, upregulation of cyclooxygenase-2 (COX-2), and activation of mitogen-activated protein kinases (MAPKs) that promote proinflammatory cytokine release and epithelial cell proliferation, which may ultimately result in genetic mutation and neoplastic transformation [15]. Thus, intestinal health and epithelial homeostasis arise from the fine balance between mutualistic and detrimental traits of the microbiota and their metabolites (Fig. 1). In this context, commensal gut bacteria may also express factors with pathogenic potential (e.g., virulence factors, toxins, ROS) that contribute to disease pathogenesis under specific conditions (e.g., genetic susceptibility or Western dietary patterns)—these are also termed "pathobionts," since they belong to the normal gut microbiota of healthy individuals [16]. Several bacteria were shown to express virulence factors such or toxins with DNA damaging and tumor-promoting effects or were identified to be more abundant in CRC [17–21]. A disturbed microbial composition and function result in a state termed "dysbiosis," which is characterized by a lower microbial diversity compared with a healthy gut and altered gut microbial metabolism (Fig. 1). Dysbiotic traits of

the gut microbiota were associated with many chronic diseases such as diabetes, obesity, inflammatory bowel diseases, and CRC. However, dysbiosis of the gut microbiota is predominantly detected in patients and in association with disease. Thus, it often remains unclear, if the dysbiotic state of the gut microbiota is causally linked to disease initiation or progression, or if it is a secondary observation following disease or perturbed intestinal conditions.

Indeed, diet is the main driver of CRC risk [3]. Native African people consume a diet rich in indigestible fiber and resistant starch and low in animal products, whereas African Americans consume more animal protein, red meat, and saturated fat and lower amounts of complex carbohydrates, resulting in a higher CRC incidence [22]. Alaska Native people, which consume a traditional diet rich in fat and low in dietary fiber, have the world's highest CRC risk and a gut microbiota that is characterized by lower diversity and adaptation to bile acid metabolism [23]. Indigestible fiber, resistant starch, and complex carbohydrates undergo saccharolytic fermentation predominantly in the proximal colon yielding SCFAs (e.g., acetate, propionate, and butyrate), ethanol, and gases such as carbon dioxide (CO_2) and hydrogen (H_2) (Fig. 1). Acetate and propionate are the major (~85%) fraction of SCFAs but are absorbed mostly into the systemic circulation, with acetate being taken up by the liver for cholesterol synthesis and propionate participating in gluconeogenesis. Butyrate, on the other hand, exerts its principal actions locally in the colon serving as the chief energy source for colonic epithelial cells, acts as regulator of epithelial cell growth and differentiation along the crypt-villus axis, and has anti-inflammatory and antineoplastic properties [4, 24, 25]. It causes hyperacetylation of histones by inhibiting histone deacetylases and modulates transcription factors to regulate gene expression and cell function [26]. Its actions are also mediated by signaling pathways involving upregulation of peroxisome proliferator- activated receptor-γ (PPARγ), G-protein coupled receptor signalling (e.g., GPR43, GPR41, GPR109), and suppression of nuclear factor-κB (NF-κB) activation [27].

Finally, butyrate plays a critical role in regulating the intestinal barrier by enhancing mucus production, and low-fiber diets were associated with higher numbers of mucus-degrading bacteria, which aggravated experimental colitis [28]. A lack of butyrate production by gut bacteria results in higher colonic tumor numbers in experimental CRC [25]. On the contrary, undigested protein residues reaching the distal colon undergo proteolytic fermentation by bacteria producing branched-chain fatty acids and inflammatory nitrogenous metabolites such as phenolic and indolic compounds that have been shown to cause colonocyte DNA damage in experimental models (Fig. 1) [4]. High consumption of red meat promotes a shift to proteolytic fermentation capacity of the gut microbiota by providing large amounts of undigested protein residues [29]. Red meat is also responsible for increasing CRC risk in several other ways. Hydrogen (H_2), produced during fermentation, is generally excreted in the breath (directly or as methane). However, it can also be converted to H_2S by sulfide-reducing bacteria using methionine and cysteine from animal protein [30]. H_2S induces mucosal hyperproliferation and free radical-mediated genotoxicity, effects that can be reversed by butyrate [31]. Aromatic amino acids, which are abundant in red meat, undergo bacterial decarboxylation and N-nitrosation resulting in formation of N-nitroso compounds [32]. In addition, meat processing and cooking practices that expose meat to very high temperatures also result in formation of several mutagens (such as polycyclic aromatic hydrocarbons, and heterocyclic amines) that cause DNA base alkylation and formation of DNA adducts that are markers of chemical carcinogenesis [32, 33]. Finally, a high intake of fat promotes hepatic synthesis of bile acids that escape ileal reabsorption and enter the colonic lumen, where primary bile acids are deconjugated and converted to secondary bile acids by gut bacteria. The main secondary BAs are deoxycholic acid and lithocholic acid, which have tumor-promoting and genotoxic activity [34]. Bacteria producing these secondary bile acids are more abundant in fecal samples of high-risk cohorts and CRC patients [23, 29]. High-fat diet also stimulates delivery of sulfur-rich taurine conjugates to the colon promoting certain detrimental bacterial strains and inducing colitis by proinflammatory Th1-mediated immune responses and bacterial by-products such as H_2S and secondary bile acids [35]. In addition, high-fat diet induces altered gut microbial metabolism towards higher levels of lysophosphatidic acid, shown to promote proliferation of CRC epithelial cell lines [36].

Prevention of Colorectal Cancer

In general, adopting a healthy lifestyle with increased physical activity, limited consumption of alcohol, avoidance of tobacco use, and, most importantly, dietary modifications (see below) can mitigate CRC risk.

Dietary Modifications for Minimizing Risk of Colorectal Cancer

In light of the influence of diet on CRC risk, it is prudent to consume a balanced diet that can maintain or modulate the microbial composition to produce tumor-suppressive metabolites, such as butyrate. Enhancing butyrate production can limit the tumor-promoting effects of secondary bile acids, proliferative effects of H_2S, and DNA damage induced by red meat [4, 10, 37]. Increasing intake of dietary fiber, resistant starch, and complex carbohydrates and moderating red meat and animal fat portions seem to be a simple step for promoting colonic mucosal health. A reduction in red meat does not disturb general metabolism and has several beneficial effects for human health. Our demand for dietary protein, important for our structural and metabolic needs, can quite easily be met by consumption of other protein-rich diets such as white meat, fish, and legumes. Recently, also ketogenic diets demonstrated tumor-suppressive effects in experimental CRC [38].

Cancer Screening and Chemoprevention

The adenoma–carcinoma sequence usually takes 7–10 years offering an adequate window period to screen and intervene. At least 60% of deaths from CRC can be prevented by early detection of precancerous polyps through diligent screening of people who are 50 years or older (or earlier in case of higher CRC risk) using high-sensitivity fecal occult blood testing (meaning detection of blood in the stool), flexible sigmoidoscopy (i.e., an investigation of the rectum and last third of the colon by insertion of a camera mounted on a flexible scope into the anus and its guidance through the rectum into the colon), and/or colonoscopy (or coloscopy, that is, an endoscopic examination of the large bowel with a flexible tube inserted via the anus) with the latter allowing the removal of polyps (surgical prevention) [39]. New non-invasive fecal screening tests targeting DNA mutations are promising tools for early detection of polyps and CRC [40]. Several pharmacological agents such as aspirin and other nonsteroidal anti-inflammatory agents (NSAIDs), statins, calcium, vitamin D, selenium, and post-menopausal hormone replacement therapy potentially reduce the incidence or recurrence of adenoma (chemoprevention) [6]. Significant associated risks (e.g., gastrointestinal bleeding with NSAIDs) render chemoprevention less attractive for the general population, yet it can be considered when the potential benefits outweigh the risks especially in those with high CRC risk.

Treatment of Colorectal Cancer

Surgery, chemotherapy (including targeted monoclonal antibody therapies), and/or radiotherapy are the common therapeutic modalities, utilized either alone or in combination. Tumor location and stage at diagnosis (local extent, spread to lymph nodes, or distant metastasis) determine the treatment strategy [41]. Surgery is the cornerstone of CRC treatment, and resection can be curative when the tumor is localized to the colon or rectum and sometimes even when isolated metastatic foci in the liver or lung are amenable to resection. Systemic chemotherapy is used in combination with surgery either postoperatively (adjuvant therapy) when regional lymph nodes are involved or preoperatively (neoadjuvant chemoradiotherapy) to shrink metastatic foci before resection. Chemotherapy alone is used to prolong survival and for palliation in non-resectable advanced or metastatic CRC patients. 5-Fluorouracil (5-FU) in combination with leucovorin, capecitabine, oxaliplatin, and irinotecan and monoclonal antibody therapy targeting vascular endothelial growth factor-A (bevacizumab) or epidermal growth factor receptor (cetuximab, panitumumab) are the traditional chemo-immunotherapeutic agents. Better screening and treatment options have helped to improve the 5-year survival rates for CRC to >90% (local cancer), 70% (regional spread), and 15% (distant metastasis) based on the staging at diagnosis [42]. Clearly, prevention is the way to go.

Perspectives

Our ability to manipulate the gut microbiota and their metabolic profiles in order to minimize CRC risk through dietary intervention or administration of probiotics needs to be validated further through rigorous and interdisciplinary research. In this context, studies need to focus on causal relationships between diet, the gut microbiota and CRC risk, and provide mechanistic insight. Moreover, better understanding of the molecular characteristics of CRC and their variations that can affect prognosis and response to treatment and development of novel-targeted antibody therapies in the present era of "personalized medicine" could enhance therapeutic success rates immensely. Advances in minimally invasive surgical techniques can help to improve tumor resectability and minimize surgical complications. With the pandemic of obesity, the incidence rate of CRC is expected to rise further throughout the world. However, a preemptive strategy of addressing and mitigating the risk factors complemented by a diligent screening strategy can help to decrease the incidence of CRC.

Questions and Answers

Question 1 Which dietary pattern is associated with a high colorectal cancer (CRC) risk?

Answer 1 There is convincing evidence for Western diets, rich in red and processed meat, animal fat and low in fiber, being a major risk factor for CRC.

Question 2 How is a dysbiotic gut microbiota defined that may be implicated in CRC?

Answer 2 A dysbiotic gut microbiota is characterized by a lower microbial diversity, changes in composition, and altered microbial metabolism compared with a healthy gut.

Question 3 What microbial or diet-derived metabolites may have tumor-promoting effects in the colon?

Answer 3 Metabolites that were shown to promote CRC risk are, for example, secondary bile acids, hydrogen sulfide, reactive oxygen species (ROS), polycyclic aromatic hydrocarbons, and heterocyclic amines.

References

1. Sung H, Ferlay J, Siegel RL, Laversanne M, Soerjomataram I, Jemal A, Bray F (2021) Global cancer statistics 2020: GLOBOCAN estimates of incidence and mortality worldwide for 36 cancers in 185 countries. CA Cancer J Clin 71:209–249. https://doi.org/10.3322/caac.21660
2. Arnold M, Sierra MS, Laversanne M, Soerjomataram I, Jemal A, Bray F (2017) Global patterns and trends in colorectal cancer incidence and mortality. Gut 66:683–691. https://doi.org/10.1136/gutjnl-2015-310912
3. Lichtenstein P, Holm NV, Verkasalo PK, Iliadou A, Kaprio J, Koskenvuo M, Pukkala E, Skytthe A, Hemminki K (2000) Environmental and heritable factors in the causation of cancer–analyses of cohorts of twins from Sweden, Denmark, and Finland. N Engl J Med 343:78–85. https://doi.org/10.1056/NEJM200007133430201
4. O'Keefe SJD (2016) Diet, microorganisms and their metabolites, and colon cancer. Nat Rev Gastroenterol Hepatol 13:691–706. https://doi.org/10.1038/nrgastro.2016.165
5. Wong SH, Yu J (2019) Gut microbiota in colorectal cancer: mechanisms of action and clinical applications. Nat Rev Gastroenterol Hepatol 16:690–704. https://doi.org/10.1038/s41575-019-0209-8
6. Chan AT, Giovannucci EL (2010) Primary prevention of colorectal cancer. Gastroenterology 138:2029–2043.e10. https://doi.org/10.1053/j.gastro.2010.01.057
7. Nguyen LH, Goel A, Chung DC (2020) Pathways of colorectal carcinogenesis. Gastroenterology 158:291–302. https://doi.org/10.1053/j.gastro.2019.08.059
8. Schneikert J, Behrens J (2007) The canonical Wnt signalling pathway and its APC partner in colon cancer development. Gut 56:417–425. https://doi.org/10.1136/gut.2006.093310
9. Kherbek H, Daoud R, Soueycatt T, Soueycatt Y, Ali Z, Ehsan J, Alshehabi Z, Georgeos M (2022) The relationship between folic acid and colorectal cancer; a literature review. Ann Med Surg 80:104170. https://doi.org/10.1016/j.amsu.2022.104170
10. Reynolds A, Mann J, Cummings J, Winter N, Mete E, Morenga LT (2019) Carbohydrate quality and human health: a series of systematic reviews and meta-analyses. Lancet 393. https://doi.org/10.1016/S0140-6736(18)31809-9
11. Sonnenburg JL, Bäckhed F (2016) Diet–microbiota interactions as moderators of human metabolism. Nature 535:56–64. https://doi.org/10.1038/nature18846
12. Tremaroli V, Bäckhed F (2012) Functional interactions between the gut microbiota and host metabolism. Nature 489:242–249. https://doi.org/10.1038/nature11552
13. Chen L, Zhernakova DV, Kurilshikov A, Andreu-Sánchez S, Wang D, Augustijn HE, Vich Vila A, Weersma RK, Medema MH, Netea MG, Kuipers F, Wijmenga C, Zhernakova A, Fu J (2022) Influence of the microbiome, diet and genetics on inter-individual variation in the human plasma metabolome. Nat Med 28:2333–2343. https://doi.org/10.1038/s41591-022-02014-8
14. Leviatan S, Shoer S, Rothschild D, Gorodetski M, Segal E (2022) An expanded reference map of the human gut microbiome reveals hundreds of previously unknown species. Nat Commun 13:3863. https://doi.org/10.1038/s41467-022-31502-1
15. Schmitt M, Greten FR (2021) The inflammatory pathogenesis of colorectal cancer. Nat Rev Immunol 21:653–667. https://doi.org/10.1038/s41577-021-00534-x
16. Jochum L, Stecher B (2020) Label or concept – what is a pathobiont? Trends Microbiol 28:789–792. https://doi.org/10.1016/j.tim.2020.04.011

17. Wang N, Fang J-Y (2023) Fusobacterium nucleatum, a key pathogenic factor and microbial biomarker for colorectal cancer. Trends Microbiol 31:159–172. https://doi.org/10.1016/j.tim.2022.08.010

18. Pleguezuelos-Manzano C, Puschhof J, Rosendahl Huber A, van Hoeck A, Wood HM, Nomburg J, Gurjao C, Manders F, Dalmasso G, Stege PB, Paganelli FL, Geurts MH, Beumer J, Mizutani T, Miao Y, van der Linden R, van der Elst S, Genomics England Research Consortium, Garcia KC, Top J, Willems RJL, Giannakis M, Bonnet R, Quirke P, Meyerson M, Cuppen E, van Boxtel R, Clevers H (2020) Mutational signature in colorectal cancer caused by genotoxic pks+ E. coli. Nature 580:269–273. https://doi.org/10.1038/s41586-020-2080-8

19. Haghi F, Goli E, Mirzaei B, Zeighami H (2019) The association between fecal enterotoxigenic B. Fragilis with colorectal cancer. BMC Cancer 19:879. https://doi.org/10.1186/s12885-019-6115-1

20. Arthur JC, Perez-Chanona E, Mühlbauer M, Tomkovich S, Uronis JM, Fan T-J, Campbell BJ, Abujamel T, Dogan B, Rogers AB, Rhodes JM, Stintzi A, Simpson KW, Hansen JJ, Kekuk TO, Fodor AA, Jobin C (2012) Intestinal inflammation targets cancer-inducing activity of the microbiota. Science 338:120–123. https://doi.org/10.1126/science.1224820

21. Wang X, Huycke MM (2007) Extracellular superoxide production by enterococcus faecalis promotes chromosomal instability in mammalian cells. Gastroenterology 132:551–561. https://doi.org/10.1053/j.gastro.2006.11.040

22. O'Keefe SJD, Li JV, Lahti L, Ou J, Carbonero F, Mohammed K, Posma JM, Kinross J, Wahl E, Ruder E, Vipperla K, Naidoo V, Mtshali L, Tims S, Puylaert PGB, DeLany J, Krasinskas A, Benefiel AC, Kaseb HO, Newton K, Nicholson JK, de Vos WM, Gaskins HR, Zoetendal EG (2015) Fat, fibre and cancer risk in African Americans and rural Africans. Nat Commun 6:6342. https://doi.org/10.1038/ncomms7342

23. Ocvirk S, Wilson AS, Posma JM, Li JV, Koller KR, Day GM, Flanagan CA, Otto JE, Sacco PE, Sacco FD, Sapp FR, Wilson AS, Newton K, Brouard F, DeLany JP, Behnning M, Appolonia CN, Soni D, Bhatti F, Methé B, Fitch A, Morris A, Gaskins HR, Kinross J, Nicholson JK, Thomas TK, O'Keefe SJD (2020) A prospective cohort analysis of gut microbial co-metabolism in Alaska native and rural African people at high and low risk of colorectal cancer. Am J Clin Nutr 111:406–419. https://doi.org/10.1093/ajcn/nqz301

24. Kaiko GE, Ryu SH, Koues OI, Collins PL, Solnica-Krezel L, Pearce EJ, Pearce EL, Oltz EM, Stappenbeck TS (2016) The colonic crypt protects stem cells from microbiota-derived metabolites. Cell 165:1708–1720. https://doi.org/10.1016/j.cell.2016.05.018

25. Donohoe DR, Holley D, Collins LB, Montgomery SA, Whitmore AC, Hillhouse A, Curry KP, Renner SW, Greenwalt A, Ryan EP, Godfrey V, Heise MT, Threadgill DS, Han A, Swenberg JA, Threadgill DW, Bultman SJ (2014) A gnotobiotic mouse model demonstrates that dietary fiber protects against colorectal tumorigenesis in a microbiota- and butyrate-dependent manner. Cancer Discov 4:1387–1397. https://doi.org/10.1158/2159-8290.CD-14-0501

26. Donohoe DR, Collins LB, Wali A, Bigler R, Sun W, Bultman SJ (2012) The Warburg effect dictates the mechanism of butyrate-mediated histone acetylation and cell proliferation. Mol Cell 48:612–626. https://doi.org/10.1016/j.molcel.2012.08.033

27. Bultman SJ (2014) Molecular pathways: gene-environment interactions regulating dietary fiber induction of proliferation and apoptosis via butyrate for cancer prevention. Clin Cancer Res 20:799–803. https://doi.org/10.1158/1078-0432.CCR-13-2483

28. Desai MS, Seekatz AM, Koropatkin NM, Kamada N, Hickey CA, Wolter M, Pudlo NA, Kitamoto S, Terrapon N, Muller A, Young VB, Henrissat B, Wilmes P, Stappenbeck TS, Núñez G, Martens EC (2016) A dietary fiber-deprived gut microbiota degrades the colonic mucus barrier and enhances pathogen susceptibility. Cell 167:1339–1353.e21. https://doi.org/10.1016/j.cell.2016.10.043

29. Wirbel J, Pyl PT, Kartal E, Zych K, Kashani A, Milanese A, Fleck JS, Voigt AY, Palleja A, Ponnudurai R, Sunagawa S, Coelho LP, Schrotz-King P, Vogtmann E, Habermann N, Niméus E, Thomas AM, Manghi P, Gandini S, Serrano D, Mizutani S, Shiroma H, Shiba S, Shibata T, Yachida S, Yamada T, Waldron L, Naccarati A, Segata N, Sinha R, Ulrich CM, Brenner H, Arumugam M, Bork P, Zeller G (2019) Meta-analysis of fecal metagenomes reveals global microbial signatures that are specific for colorectal cancer. Nat Med 25:679. https://doi.org/10.1038/s41591-019-0406-6

30. Abu-Ghazaleh N, Chua WJ, Gopalan V (2021) Intestinal microbiota and its association with colon cancer and red/processed meat consumption. J Gastroenterol Hepatol 36:75–88. https://doi.org/10.1111/jgh.15042

31. Guo F-F, Yu T-C, Hong J, Fang J-Y (2016) Emerging roles of hydrogen sulfide in inflammatory and neoplastic colonic diseases. Front Physiol 7

32. Zhu Y, Wang PP, Zhao J, Green R, Sun Z, Roebothan B, Squires J, Buehler S, Dicks E, Zhao J, Cotterchio M, Campbell PT, Jain M, Parfrey PS, Mclaughlin JR (2014) Dietary N-nitroso compounds and risk of colorectal cancer: a case-control study in Newfoundland and Labrador and Ontario, Canada. Br J Nutr 111:1109–1117. https://doi.org/10.1017/S0007114513003462

33. Cross AJ, Ferrucci LM, Risch A, Graubard BI, Ward MH, Park Y, Hollenbeck AR, Schatzkin A, Sinha R (2010) A large prospective study of meat consumption and colorectal cancer risk: an investigation of potential mechanisms underlying this association. Cancer Res 70:2406–2414. https://doi.org/10.1158/0008-5472.CAN-09-3929

34. Ocvirk S, O'Keefe SJD (2021) Dietary fat, bile acid metabolism and colorectal cancer. Semin

Cancer Biol 73:347–355. https://doi.org/10.1016/j. semcancer.2020.10.003

35. Devkota S, Wang Y, Musch MW, Leone V, Fehlner-Peach H, Nadimpalli A, Antonopoulos DA, Jabri B, Chang EB (2012) Dietary-fat-induced taurocholic acid promotes pathobiont expansion and colitis in Il10−/− mice. Nature 487:104–108. https://doi.org/10.1038/nature11225

36. Yang J, Wei H, Zhou Y, Szeto C-H, Li C, Lin Y, Coker OO, Lau HCH, Chan AWH, Sung JJY, Yu J (2022) High-fat diet promotes colorectal tumorigenesis through modulating gut microbiota and metabolites. Gastroenterology 162:135–149.e2. https://doi.org/10.1053/j.gastro.2021.08.041

37. Sheng L, Jena PK, Hu Y, Liu H-X, Nagar N, Kalanetra KM, French SW, French SW, Mills DA, Wan Y-JY (2017) Hepatic inflammation caused by dysregulated bile acid synthesis is reversible by butyrate supplementation. J Pathol 243:431–441. https://doi.org/10.1002/path.4983

38. Dmitrieva-Posocco O, Wong AC, Lundgren P, Golos AM, Descamps HC, Dohnalová L, Cramer Z, Tian Y, Yueh B, Eskiocak O, Egervari G, Lan Y, Liu J, Fan J, Kim J, Madhu B, Schneider KM, Khoziainova S, Andreeva N, Wang Q, Li N, Furth EE, Bailis W, Kelsen JR, Hamilton KE, Kaestner KH, Berger SL, Epstein JA, Jain R, Li M, Beyaz S, Lengner CJ, Katona BW, Grivennikov SI, Thaiss CA, Levy M (2022) B-Hydroxybutyrate suppresses colorectal cancer. Nature 605:160–165. https://doi.org/10.1038/s41586-022-04649-6

39. Levin B, Lieberman DA, McFarland B, Smith RA, Brooks D, Andrews KS, Dash C, Giardiello FM, Glick S, Levin TR, Pickhardt P, Rex DK, Thorson A, Winawer SJ (2008) Screening and surveillance for the early detection of colorectal cancer and adenomatous polyps, 2008: a joint guideline from the American Cancer Society, the US Multi-Society Task Force on colorectal cancer, and the American College of Radiology*†. CA Cancer J Clin 58:130–160. https://doi.org/10.3322/CA.2007.0018

40. Imperiale TF, Ransohoff DF, Itzkowitz SH, Levin TR, Lavin P, Lidgard GP, Ahlquist DA, Berger BM (2014) Multitarget stool DNA testing for colorectal-cancer screening. N Engl J Med 370:1287–1297. https://doi.org/10.1056/NEJMoa1311194

41. Cunningham D, Atkin W, Lenz H-J, Lynch HT, Minsky B, Nordlinger B, Starling N (2010) Colorectal cancer. Lancet 375:1030–1047. https://doi.org/10.1016/S0140-6736(10)60353-4

42. Miller KD, Nogueira L, Devasia T, Mariotto AB, Yabroff KR, Jemal A, Kramer J, Siegel RL (2022) Cancer treatment and survivorship statistics, 2022. CA Cancer J Clin 72:409–436. https://doi.org/10.3322/caac.21731

Pancreas

Pancreas: Overview

Laura Leonhardt, Alexandra E. Folias,
and Matthias Hebrok

Anatomy and Physiology of the Pancreas

The pancreas is a complex organ integral for controlling energy homeostasis through the digestion, absorption, and dynamic supply of nutrients. Located in the abdominal region, the head of the pancreas is nestled in the curve of the duodenal

L. Leonhardt
TUM School of Medicine, Technical University Munich, Munich, Germany

Center for Organoid Systems, Technical University Munich, Garching, Germany

Institute for Diabetes and Organoid Technology, Helmholtz Diabetes Center, Helmholtz Zentrum München, Neuherberg, Germany
e-mail: laura.leonhardt@tum.de

A. E. Folias
Diabetes Center, Department of Medicine, University of California San Francisco, San Francisco, CA, USA

M. Hebrok (✉)
TUM School of Medicine, Technical University Munich, Munich, Germany

Center for Organoid Systems, Technical University Munich, Garching, Germany

Institute for Diabetes and Organoid Technology, Helmholtz Diabetes Center, Helmholtz Zentrum München, Neuherberg, Germany

Diabetes Center, Department of Medicine, University of California San Francisco, San Francisco, CA, USA
e-mail: matthias.hebrok@tum.de;
matthias.hebrok@helmholtz-munich.de

loop of the small intestine (Fig. 1a) with the tail situated next to the spleen. The pancreas is organized in lobes surrounding a branching network of ducts that merge into the main pancreatic duct and open into the duodenum via the papilla of Vater [1, 2]. Two functionally and morphologically distinct cellular compartments can be distinguished: the exocrine (acinar, centroacinar and duct cells) and endocrine (islets) pancreas. The majority of the tissue is comprised of acinar cells (~95%) forming grape-like structures termed acini at the tip of intercalated, or terminal, ducts [3] (Fig. 1b, c). Centroacinar cells are the terminal end duct cells interfacing with the acini. The ductal system transports bicarbonate-rich fluid containing digestive enzymes synthesized and secreted by acinar cells to the duodenum following the intake of food. In contrast, the endocrine compartment of the pancreas is known as Islets of Langerhans (~1–2% of pancreatic cells). Given their distinctive morphology and function that sets them apart from the surrounding exocrine tissue, islets are often referred to as "micro-organs" (Fig. 1b, d). Islet clusters are mainly composed of four hormone-secreting cell types, α-cells (glucagon, ~30% of cells within islets), β-cells (insulin, ~60%), δ-cells (somatostatin, ~10%), and PP-cells (pancreatic polypeptide, <3–5%) [4, 5]. A fifth type, ghrelin-expressing ϵ-cells, is rare in the adult pancreas and found primarily during gestational development [6]. An extensive capillary network and innervation by

© The Author(s), under exclusive license to Springer Nature Switzerland AG 2026
E. Lammert, M. Zeeb (eds.), *Metabolism of Human Diseases*,
https://doi.org/10.1007/978-3-031-96019-2_21

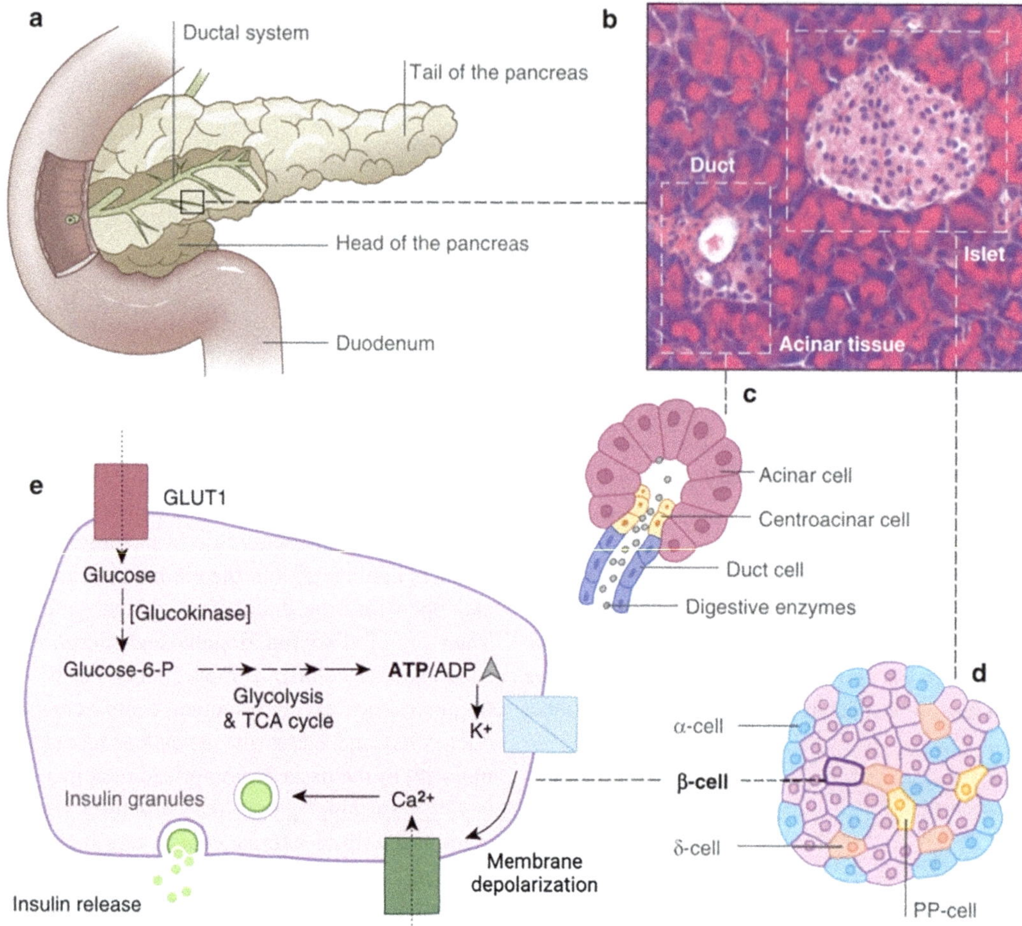

Fig. 1 Pancreas anatomy. (**a**) Diagram showing the head of the pancreas located adjacent to the duodenal loop and the pancreatic ductal system. (**b**) Microscopic image of the pancreas (stained with hematoxylin and eosin) showing the exocrine (acinar, centroacinar and duct cells) and endocrine (islet cells) compartments. (**c**) Schematic of the architecture of an acinus connected to a terminal duct. (**d**) Schematic of the hormone-secreting endocrine cell types found in the islets of Langerhans: α-cells (glucagon), β-cells (insulin), δ-cells (somatostatin), and PP cells (pancreatic polypeptide). (**e**) Schematic of glucose stimulated insulin secretion in a β-cell. *GLUT1* glucose transporter 1, *TCA* tricarboxylic acid. (Image modified from Metabolism of Human Diseases, 2014)

autonomic neurons within the islets allow the efficient monitoring of blood glucose levels and dynamic release of hormones into the bloodstream. It is important to note that while the endocrine cell types are similarly present across mammals, there are species-specific differences in islet architecture and composition [5, 7]. Most cells in the adult pancreas are postmitotic with little cellular turnover. In the event of pancreatic damage such as pancreatitis, mitotic activity of the exocrine compartment is increased and the tissue can regenerate to some extent. In contrast, endocrine cells show very limited regenerative capacity leading to severe insulin deficiency in the context of β-cell destruction or exhaustion, as seen in diabetes [8, 9].

Metabolic Pathways and Processes in the Pancreas

In adults, the exocrine pancreas exhibits the highest level of protein synthesis of all organs. Acinar cells are almost exclusively dedicated to producing and secreting large amounts of digestive enzymes, including lipases, proteinases (e.g., trypsin, chymotrypsin, carboxypeptidase), and amylases. This becomes evident from electron micrographs showing acinar cells containing an extensive rough endoplasmic reticulum (ER) network and abundant apically located zymogen granules [2, 3]. Zymogen granules store the inactive proenzymes which are released into the acinar lumen via exocytosis upon a proper stimulus. Duct cells secrete bicarbonate (HCO_3^-) to neutralize stomach acidity as well as mucins and fluids which transport the digestive enzymes. HCO_3^- ion release from ductal cells is coupled to the transport of chloride via the cystic fibrosis transmembrane conductance regulator (CFTR). A defect in this gene leads to cystic fibrosis which affects the pancreas and several other secretory epithelia [10]. Other pathologies from the exocrine pancreas include acute or chronic pancreatitis, a state of inflammation or acinar cell destruction with multiple underlying causes, and pancreatic cancer, mainly ductal adenocarcinomas [11].

Glucose is one of the three major metabolic fuels and is particularly important for satisfying quick energy demands. To utilize glucose, it must first be taken up by the cells from the bloodstream. The peptide hormone insulin, synthesized and secreted by β-cells, is the major stimulator of glucose uptake by peripheral tissues. Disruption in insulin secretion or its peripheral effects can lead to the manifestation of diabetes. Both hyperglycemia (elevated blood glucose levels) and hypoglycemia (low blood glucose levels) can lead to serious long- and short-term complications (e.g., coma, heart disease, blindness) highlighting the importance of tightly regulating blood glucose levels. β-cells continually monitor blood glucose and respond with the release of appropriate levels of insulin through the internal metabolization of glucose, a process known as glucose-stimulated insulin secretion (GSIS).

To act as "glucose sensors," β-cells can efficiently sample and metabolize a wide range of glucose levels. Glucose is transported into the β-cell at levels proportional to the bloodstream by glucose transport protein 1 (GLUT1; GLUT2 in mice) with low-affinity/high-capacity characteristics [4, 12]. Upon uptake, glucose is phosphorylated by glucokinase in a rate-limiting step that retains it within the cell. The very high K_m of glucokinase (4–10 mmol/l) allows the β-cell to metabolize and "sense" a range of glucose levels without becoming saturated [12]. Further metabolization of glucose through glycolysis and mitochondrial oxidative phosphorylation leads to an accumulation of ATP and an increase in intracellular ATP/ADP ratio. As a consequence, ATP-sensitive potassium (K^+) channels are closed, triggering the depolarization of the plasma membrane. This change in membrane potential opens voltage-gated calcium (Ca^{2+}) channels. Influx of Ca^{2+} then induces the exocytosis of secretory vesicles storing insulin [8] (Fig. 1e). The released, active form of insulin is derived from the proteolytic cleavage of proinsulin that removes the C-peptide portion. This process of GSIS is characterized by two phases, a first phase of an insulin spike followed by a second phase of sustained insulin exocytosis.

Inside-In: Metabolites of the Pancreas Affecting Itself

A number of external and internal signals contribute to the fine-tuning of hormone secretion from islets in response to glucose (Fig. 2a). The anatomical organization of the islet is considered important for its efficient function, allowing the first layer of regulation to occur directly within these micro-organs. β-cells communicate with each other through direct cell–cell contact mediated by ephrin receptors to inhibit basal insulin release and enhance GSIS [13]. Gap junctions between neighboring β-cells are important for the synchronization of Ca^{2+} oscillations during GSIS [14]. Other islet cell types regulate insulin secretion from β-cells via paracrine interactions. Glucagon released from α-cells amplifies insulin secretion through binding to glucagon or glucagon-like peptide 1 (GLP-1) receptors on β-cells, suggesting a negative feedback loop to prevent hyperglycemia. In contrast, somatostatin from δ-cells both inhibits and is stimulated by insulin and glucagon secretion [8]. In turn, factors secreted from β-cells (insulin, γ-amino butyric acid (GABA), zinc) can directly inhibit α-cells to secrete glucagon, while β-cell derived urocortin 3 (UCN3) is co-released with insulin during GSIS to potentiate somatostatin release in response to glucose [15]. These intricate regulatory mechanisms are critical to tailor hormone output to restoring energy homeostasis.

Most of the regulation of exocrine function is believed to occur via indirect mechanisms ("Outside-In"), described below. Yet, there is evidence that islet hormones can induce (insulin) or inhibit (somatostatin, pancreatic polypeptide) the secretion of digestive enzymes, thus representing an intrapancreatic islet-acinar axis [16].

Outside-In: Metabolites of Other Tissues Affecting the Pancreas

Both the synthesis and secretion of digestive enzymes are highly regulated by external factors (Fig. 2a). The main secretagogues are the peptide hormone cholecystokinin (CCK) and the neurotransmitter acetylcholine (ACh) [17, 18]. CCK, released from enteroendocrine cells (I-cells) in

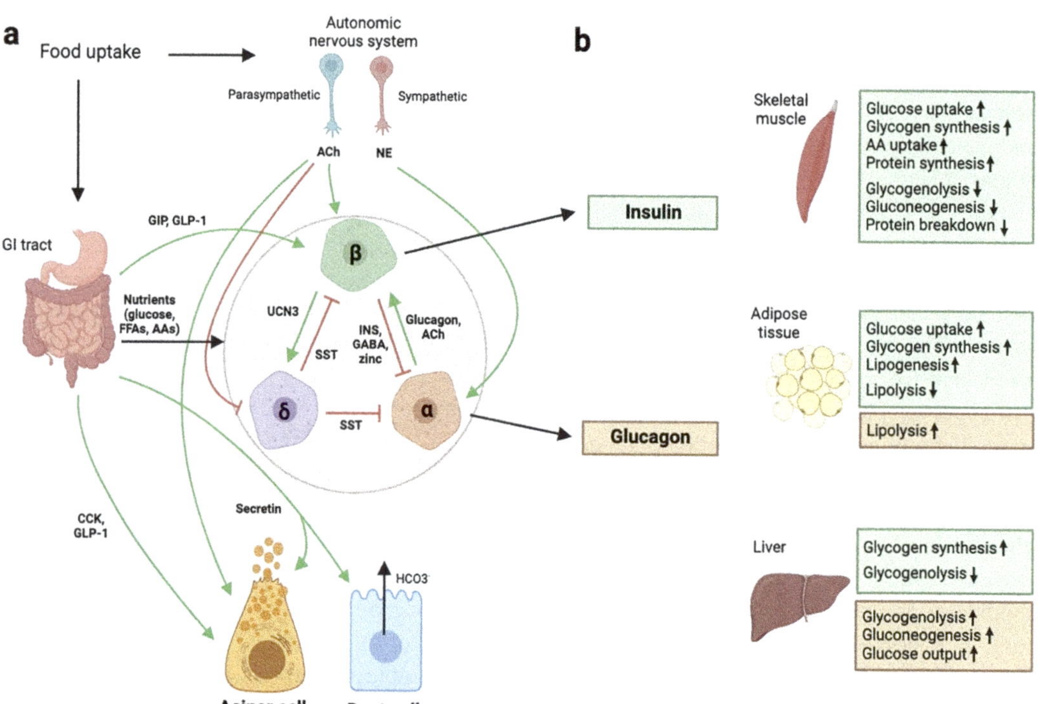

Fig. 2 Regulation of secretion and effect of pancreatic hormones. (**a**) Schematic of the multilayered regulation of secretion by endocrine islet (α-, β- and δ-cells) and exocrine (acinar, duct) cells following the uptake of food. (**b**) Effect of insulin and glucagon on peripheral tissues. Insulin promotes the uptake and storage of energy sources while glucagon mobilizes fuels during states of low fuel availability (fasting/exercise). *ACh* Acetylcholine, *NE* norepinephrine, *GIP* glucose-dependent insulinotropic polypeptide, *GLP-1* glucagon-like peptide 1, *CCK* cholecystokinin, *INS* insulin, *SST* somatostatin, *UCN3* urocortin 3, *GABA* γ-amino butyric acid, *FFAs* free fatty acids, *AA* amino acids. (Image adapted from Noguchi & Huising, 2019 and created with BioRender.com)

the small intestine mainly in response to free fatty acids, activates CCK receptors on acinar cells to promote secretion of digestive enzymes [18, 19]. The intake of food further stimulates parasympathetic (cholinergic) neurons innervating the pancreas to release ACh. A number of other factors also positively regulate acinar secretion including secretin, the incretin hormone GLP-1, and vasoactive intestinal peptide (VIP) [16, 18]. After receptor activation, an increase in intracellular free Ca^{2+} leads to zymogen exocytosis. Bicarbonate secretion from duct cells occurs on a low basal level and is induced by the gastrointestinal hormone secretin in response to a postprandial dropping of the duodenal pH [20].

Nutrients, neuronal pathways and enteroendocrine hormones regulate islet hormone secretion (Fig. 2a). Glucose is the primary trigger of insulin secretion when levels rise above 5 mM, while hypoglycemic conditions cause α-cells to release glucagon. Like β-cells, δ-cells are similarly stimulated by glucose in a process called glucose-stimulated somatostatin secretion (GSSS) leading to synchronous insulin and somatostatin release. Besides glucose, other circulating nutrients such as amino acids and fatty acids stimulate insulin secretion from β-cells via different mechanisms, including direct membrane depolarization or coupling to metabolic processes. α-cells are highly sensitive to amino acids resulting in a spike of glucagon secretion after protein ingestion [7]. Following food intake, incretins such as GLP-1 and glucose-dependent insulinotropic polypeptide (GIP) are released and potentiate GSIS from β-cells, which highly express the GLP-1 receptor. Studies suggest that this so-called incretin effect accounts for as much as half of the insulin response to ingested carbohydrates [21]. Collectively, incretin hormones are attractive candidates for clinical use in patients with impaired blood glucose regulation. Food intake further activates the autonomic nervous system, even before food has reached the gastrointestinal tract. A net activation of intraislet sympathetic neurons increases glucagon release while simultaneously inhibiting insulin and somatostatin secretion. Conversely, net parasympathetic signaling (via ACh) activates insulin and glucagon secretion while decreasing that of somatostatin [7, 22].

Inside-Out: Metabolites of the Pancreas Affecting Other Tissues

Food intake stimulates the release of digestive fluids from the exocrine pancreas to facilitate the intestinal breakdown and absorption of macronutrients (i.e., carbohydrates, lipids, proteins). Following a meal, enzyme delivery to the duodenum occurs rapidly with peak output as early as as 20–30 min postprandially [19]. The secreted inactive proenzymes (zymogens) become activated in a cascade of proteolytic cleavage steps initiated by enteropeptidase localized in the duodenum [11]. The active enzymes catalyze the hydrolysis of macronutrients to yield monomers and oligomers that can be taken up by enterocytes in the small intestine (see chapter "Gastrointestinal Tract: Overview" under the part "Gastrointestinal Tract"). Disrupted exocrine function for example due to pancreatitis can lead to exocrine pancreatic insufficiency characterized by malabsorption of nutrients, which can be fatal if enzymatic substitution is not provided.

Insulin is the major anabolic hormone that promotes the uptake and storage of glucose, lipids and amino acids in peripheral tissues (Fig. 2b). While stimulating glycogen synthesis (storage of glucose in the polymeric form of glycogen) in the liver and skeletal muscle, lipogenesis (synthesis of fatty acids and triglycerides) in adipose tissue and protein synthesis in skeletal muscle, it simultaneously decreases glycogenolysis, lipolysis and protein breakdown in these organs. Collectively, these insulin effects lead to the restoration of normoglycemia after a meal [7, 23]. Conversely, the catabolic hormone glucagon mobilizes fuels, particularly glucose, during states of fasting or exercising. In response to glucagon, hepatic glycogenolysis and gluconeogenesis (conversion of other carbon sources such as pyruvate, lactate and glycerol into glucose) are increased to elevate blood glucose levels. Unlike in the liver, glucose derived by glycogenolysis in the skeletal muscle is used directly and not released into the bloodstream. Additionally, glucagon stimulates lipolysis and protein breakdown to provide fatty acids and amino acids as substrates as well as hepatic beta oxidation to supply energy for gluconeogenesis [24].

Final Remarks

Both the endocrine and the exocrine pancreas are vital for the homeostatic control of energy balance in the body. Exocrine acinar and duct cells release digestive enzymes and bicarbonate-rich fluid into the intestine to aid with the breakdown and uptake of macronutrients. As these nutrients are absorbed, the endocrine islet cells release hormones to stimulate their proper use and storage in peripheral tissues. In the fasted state, islet hormones promote the mobilization of energy stores. The importance of proper pancreas function is exemplified by the various pathological conditions that can arise, such as exocrine insufficiency and diabetes (see chapter "Diabetes Mellitus").

Questions and Answers

Question 1 What are the main functions of the pancreatic ductal system?

Answer 1 Secretion of bicarbonate, mucins, and fluids; transport of digestive enzymes to the duodenum.

Question 2 Why do the glucose transporters of the beta cells have a high K_m?

Answer 2 To allow β-cells to rapidly take up glucose without becoming saturated thus being able to "sense" a wide range of glucose levels in the bloodstream.

Question 3 What is the so-called incretin effect?

Answer 3 The potentiation of glucose-stimulated insulin secretion (GSIS) by incretin hormones such as GLP-1 and GIP in response to food uptake, particularly carbohydrates.

Question 4 Are digestive enzymes active directly upon secretion from acinar cells?

Answer 4 No, acinar enzymes are secreted as inactive proenzymes (zymogens) that become activated by a series of proteolytic cleavage steps in the duodenum.

References

1. Reichert M, Rustgi AK (2011) Pancreatic ductal cells in development, regeneration, and neoplasia. J Clin Invest 121:4572–4578
2. Kern HF (1993) Fine structure of the human exocrine pancreas. In: Go VLW (ed) The pancreas: biology, pathobiology, and disease. Raven Press, pp 9–19
3. Motta PM, Macchiarelli G, Nottola SA, Correr S (1997) Histology of the exocrine pancreas. Microsc Res Tech 37:384–398
4. da Silva Xavier G (2018) The cells of the islets of Langerhans. J Clin Med 7:54
5. Cabrera O et al (2006) The unique cytoarchitecture of human pancreatic islets has implications for islet cell function. Proc Natl Acad Sci USA 103:2334–2339
6. Andralojc KM et al (2009) Ghrelin-producing epsilon cells in the developing and adult human pancreas. Diabetologia 52:486–493
7. Noguchi GM, Huising MO (2019) Integrating the inputs that shape pancreatic islet hormone release. Nat Metab 1:1189–1201
8. Kerper N, Ashe S, Hebrok M (2022) Pancreatic β-cell development and regeneration. Cold Spring Harb Perspect Biol 14:a040741
9. Zhou Q, Melton DA (2018) Pancreas regeneration. Nature 557:351–358
10. Madácsy T, Pallagi P, Maleth J (2018) Cystic fibrosis of the pancreas: the role of CFTR channel in the regulation of intracellular Ca2+ signaling and mitochondrial function in the exocrine pancreas. Front Physiol 9
11. Logsdon CD, Ji B (2013) The role of protein synthesis and digestive enzymes in acinar cell injury. Nat Rev Gastroenterol Hepatol 10:362–370
12. Berger C, Zdzieblo D (2020) Glucose transporters in pancreatic islets. Pflugers Arch 472:1249–1272
13. Konstantinova I et al (2007) EphA-Ephrin-A-mediated β cell communication regulates insulin secretion from pancreatic islets. Cell 129:359–370
14. Ravier MA et al (2005) Loss of connexin36 channels alters β-cell coupling, islet synchronization of glucose-induced Ca2+ and insulin oscillations, and basal insulin release. Diabetes 54:1798–1807

15. Huising MO (2020) Paracrine regulation of insulin secretion. Diabetologia 63:2057–2063
16. Barreto SG, Carati CJ, Toouli J, Saccone GTP (2010) The islet-acinar axis of the pancreas: more than just insulin. Am J Physiol Gastrointest Liver Physiol 299:G10–G22
17. Gardner JD, Jackson MJ (1977) Regulation of amylase release from dispersed pancreatic acinar cells. J Physiol 270:439–454
18. Williams JA (2010) Regulation of acinar cell function in the pancreas. Curr Opin Gastroenterol 26:478–483
19. Keller J (2005) Human pancreatic exocrine response to nutrients in health and disease. Gut 54:1–28
20. Grapin-Botton A (2005) Ductal cells of the pancreas. Int J Biochem Cell Biol 37:504–510
21. Nauck MA et al (1986) Incretin effects of increasing glucose loads in man calculated from venous insulin and C-peptide responses*. J Clin Endocrinol Metab 63:492–498
22. Li W, Yu G, Liu Y, Sha L (2019) Intrapancreatic ganglia and neural regulation of pancreatic endocrine secretion. Front Neurosci 13
23. Saltiel AR, Kahn CR (2001) Insulin signalling and the regulation of glucose and lipid metabolism. Nature 414:799–806
24. Adeva-Andany MM, Funcasta-Calderón R, Fernández-Fernández C, Castro-Quintela E, Carneiro-Freire N (2019) Metabolic effects of glucagon in humans. J Clin Transl Endocrinol 15:45–53

Diabetes Mellitus

Alena Welters and Eckhard Lammert

Introduction to Diabetes Mellitus

Diabetes mellitus (*diabetes* Greek, pass through/siphon; *mellitus* Latin, honey sweet) is a heterogeneous, multifactorial metabolic disorder characterized by chronic hyperglycemia. This is mostly due to inadequate insulin release on the background of insulin resistance (type 2 diabetes), or due to absolute insulin deficiency following autoimmune β-cell destruction (type 1 diabetes) [1]. Diabetes can be diagnosed based on I) plasma glucose criteria, or II) glycated hemoglobin A1c (HbA1c) criteria, or III) in an individual with classical symptoms of hyperglycemia, that is, polyuria (excessive production of urine) and polydipsia (excessive fluid intake), in the presence of a random plasma glucose level ≥ 11.1 mmol/L. The World Health Organization (WHO) criteria define diabetes as a fasting plasma glucose (FPG) level at or above 7 mmol/L or as a level at or above 11.1 mmol/L 2 h post glucose challenge (2-h plasma glucose level during an oral glucose tolerance test), or an hemoglobin A1c (HbA1c) level $\geq 6.5\%$. The WHO further describes criteria of a prediabetes that comes with a strongly increased risk to develop diabetes. These include an impaired FPG value between 5.6 and 6.9 mmol/L. In addition, an impaired glucose tolerance as indicated by a 2-h plasma glucose between 7.8 and 11 mmol/L or HbA1c between 5.7 and 6.4% point to prediabetes [2].

The International Diabetes Federation (IDF) estimated that in 2021 around 537 million people worldwide had diabetes mellitus, and the incidence continues to increase in both adults and children [3]. The disorder can be divided into two major classes: type 1 diabetes mellitus (T1DM), also known as juvenile diabetes, accounting for around 8.4 million or close to 2% of all cases, and type 2 diabetes mellitus (T2DM), also known as adult-onset diabetes, accounting for more than 480 million or more than 90% of all cases [3, 4]. Additional classes include type 3 diabetes, which is subdivided into, for example, inherited monogenetic forms of diabetes (termed "maturity-onset diabetes of the young," type 3A) and

A. Welters
Heinrich Heine University Düsseldorf, Faculty of Mathematics and Natural Sciences, Institute of Metabolic Physiology, Düsseldorf, Germany

Department of General Pediatrics, Neonatology and Pediatric Cardiology, Medical Faculty and University Hospital Düsseldorf, Heinrich Heine University, Düsseldorf, Germany
e-mail: lammert@hhu.de

E. Lammert (✉)
Heinrich Heine University Düsseldorf, Faculty of Mathematics and Natural Sciences, Institute of Metabolic Physiology, Düsseldorf, Germany

German Diabetes Center, Leibniz Center for Diabetes Research at Heinrich Heine University, Düsseldorf, Germany
e-mail: alena.welters@med.uni-duesseldorf.de

E. Lammert, M. Zeeb (eds.), *Metabolism of Human Diseases*,
https://doi.org/10.1007/978-3-031-96019-2_22

drug- and chemical-induced diabetes (type 3E), and gestational diabetes, defined as the onset of glucose intolerance during pregnancy [5].

However, in recent years, several efforts have been made to refine current diabetes classifications to enable more targeted treatment regimens and to better predict diabetic long-term complications. Approaches range from stratifications using simple clinical characteristics to more complex data-driven cluster analyses on the basis of immunological and metabolic parameters to the identification of risk scores [6, 7]. In all types of diabetes, a combination of genetic predisposition and environmental factors contribute to the onset of diabetes. Moreover, common long-term complications exist, but with different prevalences depending on the (sub)type of diabetes (see, e.g., chapters "Brain: Overview" under the part "Brain," "Heart: Overview" under the part "Heart," and "Blood Vessels: Overview" under the part "Blood Vessels").

Pathophysiology of Diabetes Mellitus and Metabolic Alterations

Type 1 Diabetes

T1DM is regarded as an autoimmune disease, in which the insulin-producing β-cells of the pancreatic islets (see chapter "Pancreas: Overview" under the part "Pancreas") are destroyed. Various immunologic, environmental, and genetic events lead to the activation of immune cells (see chapter "Anatomy and Physiology of the Immune System" under the part "Immune System") that initiate an inflammatory autoimmune process [8], ultimately killing the β-cells.

Different genetic loci associate with an increased or decreased risk to develop T1DM, predominantly within genes encoding for human leukocyte antigens (HLA) that present peptides from professional antigen-presenting cells, such as B cells, to T cells (see chapter "Anatomy and Physiology of the Immune System" under the part "Immune System"). Triggering prenatal and postnatal environmental factors include several maternal or childhood viral infections, possibly by introducing peptides with amino acid sequences similar to islet autoantigens to the developing immune system, thus initiating autoimmunity against β-cells [8]. Dietary factors, such as overall diary intake, timing of cows' milk protein or gluten introduction, or maternal gluten intake, are also implicated in the development of islet autoimmunity and T1DM development. Other factors, including low birth weight, also increase the risk to develop the disease. Notably, the presence of more than one autoantibody against β-cell antigens in childhood is a strong predictor of T1DM later in life, stressing that autoimmune reactions against insulin-producing β-cells occur early in life in order to precipitate in the disease during adolescence or later [9].

Type 2 Diabetes

T2DM is a heterogenous, complex metabolic disease [6, 7]. A crucial step in the development of many (but not all) T2DM subtypes is the decreased biological response to insulin, termed insulin resistance, which affects different organs, including the liver, skeletal muscle, white adipose tissue, and brain (hypothalamus) [9]. Particularly, hepatic insulin resistance and a concomitant increase in endogenous glucose production play a key role in the development of hyperglycemia in patients. In early disease stages, expanding β-cell mass by self-replication of existing β-cells, neogenesis and possibly trans-differentiation of α-cells to β-cells, and increasing insulin secretion compensate for peripheral insulin resistance (hyperinsulinemic phase). However, over time, insulin secretion declines, due to β-cell dysfunction, dedifferentiation, and/or increased β-cell death, while glucagon release from α-cells is enhanced, particularly postprandially, ultimately resulting in hyperglycemia (hypo-insulinemic phase) [9, 10].

Obesity and physical inactivity strongly associate with insulin resistance and the incidence of T2DM (Fig. 1). If the mass of white adipose tissue, particularly visceral and deep subcutaneous depots, increase, adipocytes mainly secrete factors (adipokines; see chapter "Fat Tissue:

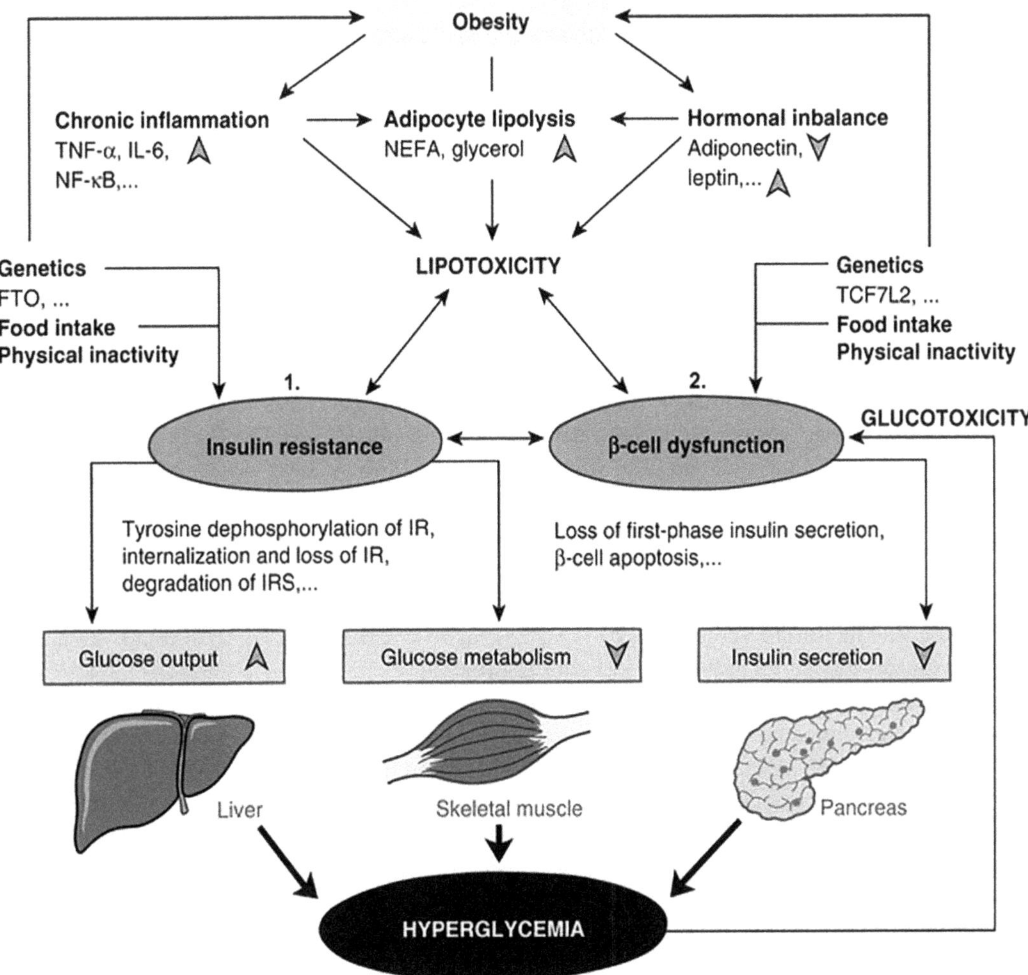

Fig. 1 Pathogenesis of type 2 diabetes mellitus (T2DM). Two main pathophysiological features contribute to the manifestation of T2DM: insulin resistance (*1*) and β-cell dysfunction (*2*). Both develop in the setting of genetic background and environmental factors. Initially, peripheral insulin resistance (*1*) can be overcome by an increased insulin secretion. Over time, this compensatory mechanism can fail due to mild but progressive increase in blood glucose (glucotoxicity) and an inflammatory response of the adipose tissue (lipotoxicity), both leading to β-cell dysfunction (*2*). Thus, peripheral insulin resistance paves the way for the onset of many (but not all) cases of T2DM. *TNF-α* tumor necrosis factor-α, *IL* interleukin, *NF-κB* nuclear factor-κB, *NEFA* nonesterified fatty acids, *FTO* fat mass and obesity-associated protein, *TCF7L2* transcription factor 7-like 2, *IR* insulin receptor, *IRS* IR substrate

Overview" under the part "Fat Tissue") that negatively affect insulin sensitivity, β-cell function, and β-cell survival. In addition, immune cells, such as monocytes and macrophages, infiltrate the adipose tissue, thus contributing to local and systemic inflammation. Altogether, this results in increased levels of proinflammatory cytokines, such as tumor necrosis factor-α (TNF-α), interleukin (IL-)6, IL-1β, monocyte chemoattractant protein-1 (MCP-1), and hormones such as leptin [9, 10]. Leptin regulates food intake and glycemia through the activation of leptin receptors expressed by hypothalamic neurons. Hyperleptinemia in obese patients thus suggests central leptin resistance. Notably, leptin also induces inflammatory responses and activates immune cells to produce proinflammatory cytokines while suppressing the production of anti-inflammatory cytokines [10, 11]. In addition, the release of nonesterified fatty acids (NEFAs) and glycerol from adipocytes is chronically elevated in many obese individuals due to enhanced lipolysis. These fatty acids aggravate insulin resistance, inhibit insulin secretion, and induce β-cell apoptosis, in a process called lipotoxicity (Fig. 1) [11]. Obesity also correlates with a decline in circulating levels of anti-inflammatory adipokines such as adiponectin, secreted frizzled-related protein 5 (SFRP5), visceral adipose tissue-derived serine protease inhibitor (vaspin), and omentin-1 [11]. For instance, adiponectin acts as an insulin sensitizer by suppressing hepatic gluconeogenesis, enhancing glucose uptake in skeletal muscle, and inhibiting lipolysis [12].

As exposure to high glucose and high lipid concentrations are toxic to β-cells, both lipotoxicity and glucotoxicity are suggested to contribute to β-cell dysfunction and death (Fig. 1). Molecular mechanisms underlying gluco- and lipotoxicity include an increased amount of reactive oxygen species (ROS, oxidative stress), endoplasmic reticulum (ER) stress, mitochondrial dysfunction, inflammation, and an impaired autophagic flux in β-cells [13]. Several genes and mutations are also implicated in the development of T2DM. For example, variants of the gene encoding the fat mass and obesity-associated protein (FTO) associate with obesity and metabolic syndrome (see chapter "Metabolic Syndrome"). Moreover, single nucleotide polymorphisms (SNPs) within the gene locus for the transcription factor 7-like 2 (TCF7L2), also known as transcription factor 4, are linked to an approximately 1.5-fold increased risk to develop T2DM. However, nongenetic features such as a large waist circumference appear to have a better predictive value [14].

In general, obesity, insulin resistance, glucotoxicity, and lipotoxicity develop over several years and ultimately result in a decline of functional β-cell mass to 40–60% of normal, thus leading to an overt T2DM in many individuals [10, 15].

Complications

Hyperglycemia is associated with acute life-threatening complications, such as ketoacidosis or hyperosmolar hyperglycemic state, and with serious long-term complications particularly affecting the cardiovascular system (see chapters "Heart Failure," and "Stroke") and kidney (see chapter "Kidney: Overview"). Notably, individuals with diabetes have a substantially increased risk for death from vascular causes as compared to individuals without diabetes, in particular when a metabolic dysfunction-associated steatotic liver disease (MASLD) develops at the same time [9, 16].

Chronic hyperglycemia causes vascular complications in part via an increased formation of advanced glycation end-products (AGEs), production of ROS, or activation of the renin-angiotensin-aldosterone system (RAAS, see chapter "Kidney: Overview") [17, 18]. AGEs are formed by nonenzymatic glycosylation of proteins, lipids, and nucleic acids, which is accelerated in diabetes due to increased blood glucose concentrations. The activation of the receptor for AGE (RAGE) elicits oxidative stress and vascular inflammation, thus leading to endothelial and smooth muscle cell dysfunction [19].

Subsequently, an imbalance in vasoconstriction and vasodilation, failure in the regulation of angiogenesis and blood flow, and changes in platelet function and coagulation contribute to microvascular disease resulting in retinopathy (see chapter "Eye: Overview"), nephropathy, neuropathy, and macrovascular disease, such as atherosclerosis, thrombosis, and thromboembolism (see chapters "Blood Vessels: Overview" and "Stroke") [20].

Diabetes Treatment and Its Influence on Metabolism

A main goal in the treatment of diabetes is to maintain the blood glucose level within a physiological range, for example, by decreasing insulin resistance, by activating endogenous insulin secretion, or by exogenous administration of recombinant insulin. In T1DM, where the functional β-cell mass is already significantly reduced once symptoms occur, patients immediately require insulin replacement therapy. Traditionally, exogenous insulin is administered by subcutaneous injection. Today, insulin therapy is individualized and varies regarding insulin preparation (rapid- and short-acting insulin vs. intermediate- and long-acting insulin), application system (pen vs. insulin pump), and regimen (conventional vs. intensified regimen; see below).

To mimic physiological insulin secretion, an intensified regimen is recommended for T1DM that combines the application of a long-acting insulin, mimicking basal insulin secretion, with pre- or post-meal bolus application of a rapid- or short-acting insulin adjusted for the amount of carbohydrate intake and current blood glucose level. Over the last two decades, advanced insulin delivery systems have been developed and are in use, such as continuous glucose monitoring devices, and hybrid closed-loop systems that automatically adapt insulin delivery in response to the current sensor glucose value [21]. Although remarkable improvements in glycemic control and cardiovascular outcomes have been observed, still, across all age groups, the majority of individuals with T1DM do not reach target HbA1c [9].

By contrast, current treatment guidelines in T2DM often recommend a stepwise approach starting with lifestyle changes, such as physical activity and weight reduction, and follow up with drug treatment (see below and Fig. 2). Treatment also includes diabetes self-management education and support (DSMES). Notably, lifestyle interventions are highly effective in preventing diabetes manifestation in at-risk individuals and in returning T2DM to normal glucose homeostasis or a prediabetes stage [23]. Initially, many

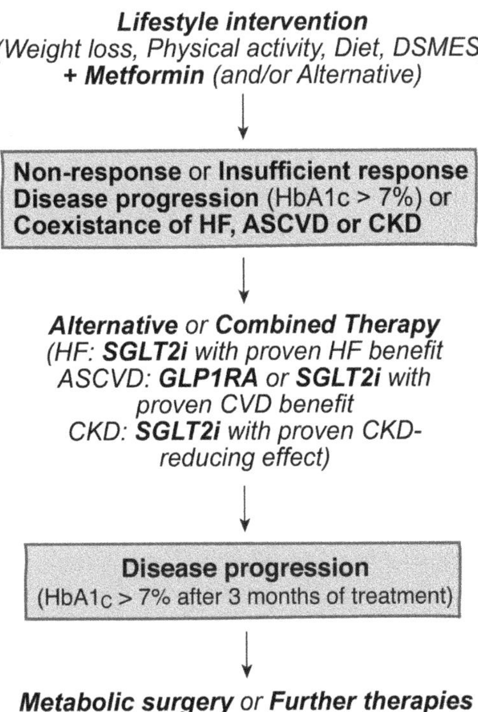

Fig. 2 Long-term management of hyperglycemia in T2DM. Treatment often starts with lifestyle intervention and an oral antidiabetic drug, typically metformin as well as diabetes self-management education and support (DSMES). If glycated hemoglobin (HbA1c) continues to be >7% after 3–6 months of therapy, treatment is intensified by addition of a second oral antidiabetic drug or alternative treatment is initiated (e.g., GLP1RA or SGLT2i). Notably, treatment of T2DM is individualized for each patient, considering further diseases, such as heart failure (HF), ASCVD, or CKD. If the disease further progresses (as indicated by HbA1c measurement after 3–6 months of therapy) and obesity-associated comorbidities progress, metabolic surgery or other therapies (including insulin injections) can be applied. For a comprehensive view, please read the consensus reports from the American Diabetes Association/European Association for the Study of Diabetes (ADA/EASD) [16, 22]

patients can be treated with an oral glucose-lowering drug, typically metformin (see below and Fig. 2). However, over time, due to the progressive nature of the disease or when the drug is neither efficient nor well-tolerated, patients require a combination of antidiabetic drugs or an alternative therapy (Fig. 2). Underlying comorbidities such as atherosclerotic cardiovascular disease (ASCVD) or chronic kidney disease

(CKD) should be considered when choosing a second or third antidiabetic agent [9] or even from the beginning of the treatment. Insulin therapy or metabolic surgery is recommended in individuals with very high blood glucose levels (>16.7 mmol/L), when the HbA1c level exceeds 10% or when obesity and MASLD cannot otherwise be managed, respectively. The benefit of glucose-lowering treatment is judged by the concentration of the HbA1c level representing a patient's average blood glucose level over the past 4 months. A level ≤ 7% is often recommended, even though elderly patients often benefit from having higher levels.

To date, there are several glucose-lowering, oral drugs on the market that differ in their modes of action and safety profiles. These include oral drugs that directly stimulate insulin secretion (sulfonylureas, meglitinides), reduce hepatic glucose output and increase insulin sensitivity (biguanides such as metformin), delay the digestion and absorption of intestinal carbohydrates (α-glucosidase inhibitors), improve insulin action (thiazolidinediones), increase endogenous concentrations of the incretin glucagon-like-peptide-1 (GLP-1, see chapter "Pancreas: Overview") by inhibiting the protease dipeptidyl peptidase-4 (DPP-4), which degrades GLP-1 (DPP-4 inhibitors), or that inhibit renal glucose reabsorption, thus leading to increased glucose excretion and reduction in hyperglycemia (sodium glucose transporter-2 inhibitor or SGLT2i). In addition, homologues or mutated forms of GLP-1 (GLP-1 receptor agonists or GLP1RA) can be injected, increasing insulin secretion in response to food intake and promoting β-cell survival. A first gastric inhibitory polypeptide (GIP) and GLP-1 receptor coagonist has been recently introduced with substantial effects on body weight and blood glucose concentrations [24].

However, many approved antidiabetic drugs can cause serious or, at least, unpleasant adverse effects, such as hypoglycemia (insulin, sulfonylureas, meglitinides), weight gain (insulin, sulfonylureas, meglitinides, thiazolidinediones), gastrointestinal disturbances (α-glucosidase inhibitors, biguanides, GLP1RA), peripheral edema, fractures, cardiac issues (thiazolidinediones), or urogenital infections (SGLT2i).

Perspectives

In recent years, dual GLP-1 and GIP co-agonists, so-called twincretins, have been developed to reduce obesity and improve blood glucose control, and more co-agonists as well as multiple agonists are likely to reach market stage. However, identification of a curative drug to prevent β-cell destruction and trigger β-cell regeneration is the long-term goal of current diabetes research, in particular concerning T1DM. A more detailed understanding of the molecular mechanisms leading to progressive β-cell death and dysfunction in diabetes is required to develop such treatments. In the context of T1DM, some progress has recently been achieved with a non-Fcγ receptor-binding/non-activating monoclonal anti-CD3 antibody in delaying the onset of T1DM in at-risk individuals. A single 14-day course of antibody treatment delayed median time to diagnoses of clinical diabetes by >2 years [25]. Finally, preclinical trials on glucose-sensitive forms of insulin that act at high rather than low blood glucose concentrations, thus attenuating life-threatening hypoglycemic episodes, seem particularly valuable for the future treatment of T1DM and insulin-dependent T2DM [26].

Questions and Answers

Question 1 How is diabetes diagnosed?

Answer 1 Fasting blood glucose concentrations at or above 7 mmol/L or blood glucose levels at or above 11.1 mmol/L 2 h post glucose challenge during an oral glucose tolerance test (2-h plasma glucose), or by a glycated hemoglobin A1c (HbA1c) level ≥6.5%

Question 2 What is DSMES?

Answer 2 Diabetes self-management education and support, representing an important part of diabetes therapy

Question 3 Name a few common comorbidities of diabetes?

Answer 3 Atherosclerotic cardiovascular disease (ASCVD), chronic kidney disease (CKD), heart failure (HF)

References

1. Bluestone JA, Herold K, Eisenbarth G (2010) Genetics, pathogenesis and clinical interventions in type 1 diabetes. Nature 464:1293–1300
2. WHO;IDF (2006) Definition and diagnosis of diabetes mellitus and intermediate hyperglycemia; Report of a WHO/IDF consultation; ISBN: 978-92-4-159493-6
3. Sun H et al (2022) IDF diabetes atlas: global, regional and country-level diabetes prevalence estimates for 2021 and projections for 2045. Diabetes Res Clin Pract 183:109119
4. Gregory GA et al (2022) Global incidence, prevalence, and mortality of type 1 diabetes in 2021 with projection to 2040: a modelling study. Lancet Diabetes Endocrinol 10:741–760
5. American Diabetes Association (2012) Diagnosis and classification of diabetes mellitus. Diabetes Care 35(Suppl 1):S64–S71
6. Ahlqvist E et al (2018) Novel subgroups of adult-onset diabetes and their association with outcomes: a data-driven cluster analysis of six variables. Lancet Diabetes Endocrinol 6:361–369
7. Zaharia OP et al (2019) Risk of diabetes-associated diseases in subgroups of patients with recent-onset diabetes: a 5-year follow-up study. Lancet Diabetes Endocrinol 7:684–694
8. Rewers M, Ludvigsson J (2016) Environmental risk factors for type 1 diabetes. Lancet 387: 2340–2348
9. Abel ED et al (2024) Diabetes mellitus—progress and opportunities in the evolving epidemic. Cell 187:3789–3820
10. Kahn SE, Hull RL, Utzschneider KM (2006) Mechanisms linking obesity to insulin resistance and type 2 diabetes. Nature 444:840–846
11. Coppari R, Bjorbaek C (2012) Leptin revisited: its mechanism of action and potential for treating diabetes. Nat Rev Drug Discov 11:692–708
12. Stumvoll M, Goldstein BJ, van Haeften TW (2005) Type 2 diabetes: principles of pathogenesis and therapy. Lancet 365:1333–1346
13. Robertson RP, Harmon J, Tran PO, Poitout V (2004) Beta-cell glucose toxicity, lipotoxicity, and chronic oxidative stress in type 2 diabetes. Diabetes 53(Suppl 1):S119–S124
14. Siren R, Eriksson JG, Vanhanen H (2012) Waist circumference a good indicator of future risk for type 2 diabetes and cardiovascular disease. BMC Public Health 12:631
15. Weir GC, Cavelti-Weder C, Bonner-Weir S (2011) Stem cell approaches for diabetes: towards beta cell replacement. Genome Med 3:61
16. Davies MJ et al (2022) Management of hyperglycemia in type 2 diabetes, 2022. A consensus report by the American Diabetes Association (ADA) and the European Association of the Study of Diabetes (EASD). Diabetes Care 45:2753–2786
17. Yamagishi S (2011) Role of advanced glycation end products (AGEs) and receptor for AGEs (RAGE) in vascular damage in diabetes. Exp Gerontol 46:217–224
18. Furukawa M, Gohda T, Tanimoto M, Tomino Y (2013) Pathogenesis and novel treatment from the mouse model of type 2 diabetic nephropathy. Sci World J 2013:928197
19. Sho-ichi Y (2010) Role of advanced glycation end products (AGEs) and receptor for AGEs (RAGE) in vascular damage in diabetes. Exp Gerontol 46:217–224
20. Costa PZ, Soares R (2013) Neovascularization in diabetes and its complications. Unraveling the angiogenic paradox. Life Sci 92:1037–1045
21. Leelarathna L et al (2021) Hybrid closed-loop therapy: where are we in 2021? Diabetes Obes Metab 23:655–660
22. Nathan DM et al (2009) Medical management of hyperglycemia in type 2 diabetes: a consensus algorithm for the initiation and adjustment of therapy: a consensus statement of the American Diabetes Association and the European Association for the Study of Diabetes. Diabetes Care 32:193–203
23. Lean MEJ et al (2019) Durability of a primary care-led weight-management intervention for remission of type 2 diabetes: 2-year results of the DiRECT open-label, cluster-randomised trial. Lancet Diabetes Endocrinol 7:344–355
24. Karagiannis T et al (2024) Subcutaneously administered tirzepatide vs semaglutide for adults with type 2 diabetes: a systematic review and network meta-analysis of randomized controlled trials. Diabetologia 67:1206–1222
25. Sims EK et al (2021) Teplizumab improves and stabilizes beta cell function in antibody positive high risk individuals. Sci Transl Med 13:eabc8980
26. Hoeg-Jensen T et al (2024) Glucose-sensitive insulin with attenuation of hypoglycaemia. Nature 634:944–951

Liver: Overview

Dieter Häussinger

Introduction to Anatomy and Physiology

The liver is one of the metabolically most active and versatile organs.

It has a dual blood supply: about 25% comes via the hepatic artery and about 75% is delivered via the portal vein, which drains blood coming from the intestine. Accordingly, it is the first organ to get in contact with intestinally absorbed nutrients, ingested toxins, and products from intestinal microorganisms. Thus, major tasks of the liver are (a) to process and store nutrients contained in the intestinal or splanchnic blood and to guarantee an adequate nutrient supply for other organs during both, the absorptive and post-absorptive state; (b) to detoxify and excrete xeno- and endobiotics into bile; (c) to participate in pathogen defense, immune functions and to trigger acute phase responses in inflammation and (d) to fulfill many other homeostatic functions such as maintenance of acid–base homeostasis, synthesis of most plasma proteins, triglyceride and cholesterol metabolism and transport, hormone processing and secretion. Furthermore, bile acids are synthesized in the liver, which aid in triglyceride digestion in the intestine, and are increasingly recognized as important signaling molecules and coordinators of interorgan metabolism.

Seventy percent of the hepatic cell mass are made up by liver parenchymal cells ("hepatocytes," PC), whereas the remainder comprises different non-parenchymal cell types. These include the fenestrated sinusoidal endothelial cells, Kupffer cells as liver-resident macrophages, vitamin A storing hepatic stellate cells, large granular lymphocytes called Pit cells, cholangiocytes, and progenitor cells (called "oval cells" in rodents). The latter are located at the canal of Hering and can differentiate into hepatocytes or cholangiocytes. Hepatic stellate cells are mesenchymal stem cells and are located in the space of Disse, which has characteristics of a stem cell niche [1, 2]. Following liver injury, liver regeneration is primarily achieved by division of pre-existing hepatocytes, but under conditions of impaired replication ability of hepatocytes, stem cell-based liver regeneration comes into play, which involves oval cells and hepatic stellate cells [2, 3]. Mechanosensing by integrins and mechanotransduction events in sinusoidal endothelial cells and hepatic stellate cells are critically involved in liver regeneration [2, 4].

The various cell types in the liver are embedded in the liver acinus into a structural–functional

D. Häussinger (✉)
Heinrich Heine University Düsseldorf,
Düsseldorf, Germany
e-mail: haeussin@uni-duesseldorf.de

E. Lammert, M. Zeeb (eds.), *Metabolism of Human Diseases*,
https://doi.org/10.1007/978-3-031-96019-2_23

organization with complex intra- and intercellular communication. The acinus represents the functional unit of the liver and extends from the terminal portal venule along the sinusoid to the terminal hepatic venule (Fig. 1). Along the acinus, the portal–venous blood, which mixes with blood from the hepatic artery in the inflow segment of the acinus, passes 20–30 hepatocytes, which are morphologically very similar, but differ in their enzyme and transporter equipment (so-called hepatocyte heterogeneity or metabolic zonation) [5–7]. Metabolite, hormone, and oxygen gradients, but also signals from neighboring cells are thought to be responsible for this metabolic zonation. Periportal hepatocytes, that is, those located at the sinusoidal inflow, are primarily engaged in gluconeogenesis, fatty acid oxidation, glutamine hydrolysis, and urea synthesis, whereas glycolysis, lipogenesis, and biotransformation predominate in perivenous hepatocytes (located at the sinusoidal outflow). Glutamine synthetase is exclusively localized in a small perivenous hepatocyte subpopulation, the so-called perivenous scavenger cells [7], which eliminate ammonia, eicosanoids, and other signaling molecules with high affinity before the sinusoidal blood enters the systemic circulation.

Tissue-Specific Metabolic/Molecular Pathways and Processes

Plasma Protein Synthesis

Except for immunoglobulins, most circulating plasma proteins are synthesized in the liver. These include albumin, which is responsible for transport of some lipophilic substances (in the

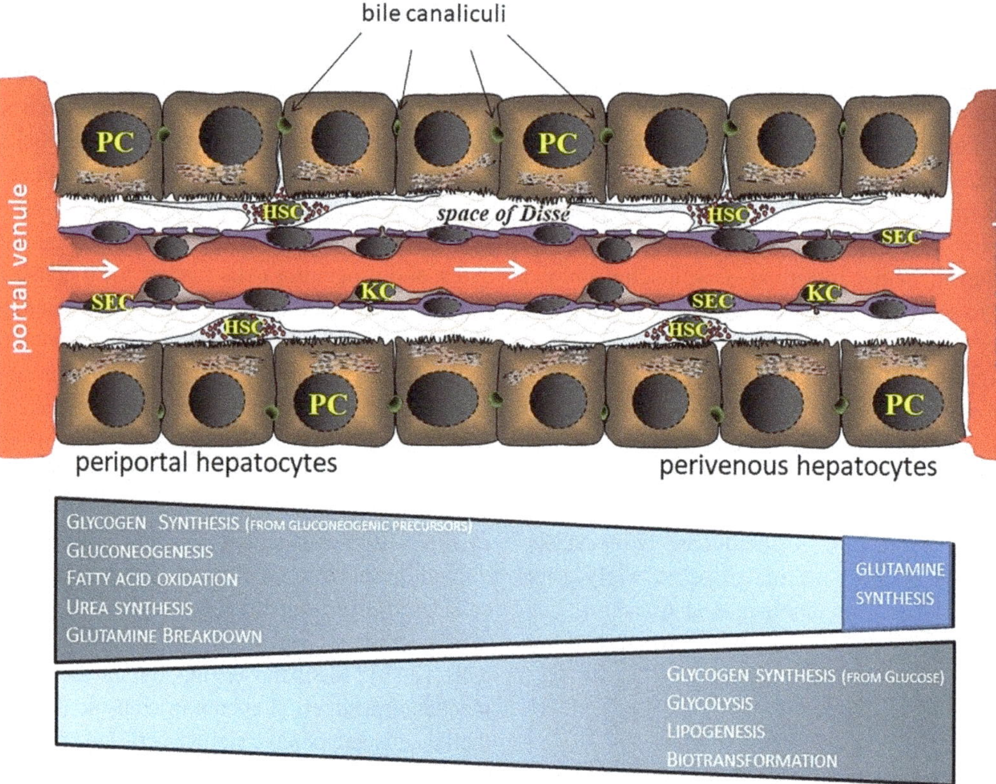

Fig. 1 Acinar organization and metabolic zonation. The liver acinus is the functional unit of the liver and extends from the terminal portal venule to terminal hepatic venule. Schematic presentation of the cells constituting the acinus, that is, parenchymal cells (PC), liver macrophages (Kupffer cells, KC), sinusoidal endothelial cells (SEC) and hepatic stellate cells (HSC), which are located in the space of Disse. PC are polar cells, and adjacent PC form with their apical membrane the bile canaliculus. Along the acinus metabolic pathways show gradients or are heterogeneously distributed ("metabolic zonation"). For details see text

blood stream) and regulation of oncotic pressure, and acute phase proteins, which are synthesized and secreted by PCs in response to cytokines such as tumor necrosis factor-α, interleukin-1 and interleukin-6, which are produced by macrophages including Kupffer cells, endothelial cells, and fibroblasts at sites of injury. The plasma concentrations of acute phase proteins can increase within hours after a local inflammatory reaction up to several hundred fold and apart from opsonization, their role mainly resides in a local restriction of inflammatory processes.

Nutrient Metabolism

The liver is the central organ of glucose homeostasis and acts as a "glucostat." In the absorptive phase, when plasma insulin levels increase, the liver synthesizes and stores glycogen from gluconeogenic precursors in periportal hepatocytes and from absorbed glucose in perivenous hepatocytes, whereas glucose becomes mobilized from glycogen in the postabsorptive state, when insulin levels are low and glucagon levels are high. After exhaustion of glycogen stores, gluconeogenesis mainly from glucogenic amino acids, glycerol, and lactate is stimulated through induction of enzymes of amino acid metabolism, gluconeogenesis, and the urea cycle, whereas glycolytic enzymes become repressed. In line with this, the liver has a high capacity for autophagic proteolysis. The shift from net glucose consumption to net glucose output in the postabsorptive state is accomplished by increasing the flux through the periportal gluconeogenic pathway and a simultaneous decrease in glycolytic flux in perivenous hepatocytes [5]. The liver is also a major organ for synthesis of triacylglycerol, cholesterol, and sphingolipids and secretes very low-density lipoproteins (VLDL). In the postabsorptive state, fatty acid oxidation provides energy for the liver; ketogenesis and ketone body release from the liver provide energy for other organs.

Urea Synthesis and Acid–Base Homeostasis

One liver-specific pathway is urea synthesis [8] from HCO_3^- and NH_4^+, which are generated in almost stoichiometric amounts during complete amino acid oxidation. Urea synthesis can be viewed as an energy-driven neutralization of the strong base HCO_3^- by the weak acid NH_4^+:

$$2\,NH_4^+ + 2\,HCO_3^- \rightarrow UREA + CO_2 + 3\,H_2O$$

Thus, the role of urea synthesis resides not only in the removal of potentially toxic ammonium ions but also in the removal of the base bicarbonate (HCO_3^-), which is neutralized by NH_4^+. Through a pH-regulated partitioning of hepatic ammonia disposal via either bicarbonate-consuming urea synthesis or via glutamine synthesis (from glutamate and ammonia), the liver can adjust the rate of bicarbonate disposal to the needs of systemic acid–base homeostasis. The structural–functional organization of urea and glutamine synthesis in the liver acinus allows for this role: in periportal hepatocytes, ammonia is disposed via urea synthesis according to the needs of bicarbonate homeostasis, whereas downstream perivenous scavenger cells maintain ammonia homeostasis by high-affinity NH_4^+ disposal via glutamine synthesis. In the kidney, NH_4^+ is liberated again from glutamine by renal glutaminase and excreted into urine. Glutamine uptake, glutaminase and carbonic anhydrase V reactions in periportal hepatocytes are major sensitively acid–base regulated steps which adjust flux through the HCO_3^- disposing urea cycle. Selective destruction of perivenous scavenger cells or specific knockdown of glutamine synthetase in these cells triggers systemic hyperammonemia, which may result in brain dysfunction (so-called hepatic encephalopathy [9]).

Bile Formation and Bile Acid Secretion

Another liver-specific pathway is biliary excretion of endo- and xenobiotics and bile formation [10, 11]. PCs are polar cells, in which the basolateral (sinusoidal) membrane faces the blood stream, whereas the apical (canalicular) membrane of two adjacent hepatocytes forms the bile canaliculus which is sealed by tight junctions. Bile formation is an osmotic process, which is driven by the coordinated action of transport systems in sinusoidal and the canalicular membrane of the PC and subsequent water flow. PCs metabolize cholesterol to lipid-soluble, unconjugated bile acids, which are later conjugated to become water-soluble. At the sinusoidal membrane, conjugated bile acids are taken up by the Na^+-taurocholate cotransporting protein (NTCP), whereas sinusoidal uptake of unconjugated bile acids, bilirubin (a catabolite of heme), and other anions is accomplished by the organic anion transporting protein (OATP) family (Fig. 2). Canalicular secretion is achieved by transport ATPases, such as the bile salt export pump (BSEP, ABCB11), and the bilirubin-transporting multidrug resistance-related protein (MRP2, ABCC2), the aminophospholipid transporter FIC1 (ATP8B1), the cholesterol transporter ABCG5/G8, and the phospholipid transporter MDR3 (ABCB4), which act as a floppase and transports phospholipids from the inner to the outer leaflet of the canalicular membrane (Fig. 2).

In addition to bilirubin and bile acids, cholesterol, phospholipids, and other substances are also secreted into the canaliculi via specific transporters (Fig. 2). The BSEP and MRP2 are not only regulated at the level of gene expression but also on short-term time scale by dynamic insertion/retrieval of transporter into/from the canalicular membrane. High hepatocellular bile acid concentration (overload) activates the nuclear transcription factor farnesoid X receptor (FXR), which triggers an upregulation of the bile salt export pump and MRP2 expression and down-regulation of expression of Na^+-taurochalate cotransporting protein and cholesterol-7α-hydroxylase (Cyp7A1), a rate-controlling step of bile acid synthesis. In cholestasis, a condition of impaired bile formation, compensatory bile acid efflux pathways via MRP3 and MRP4 located at the sinusoidal part of the PC membrane become activated (Fig. 2). These responses protect hepa-tocytes against intracellular bile acid accumulation, which is toxic and can lead to hepatocyte apoptosis [10]. In addition to FXR, bile acids can activate Takeda G-protein-coupled receptor 5 (TGR5), a G-protein coupled bile acid receptor in the plasma membrane of cholangiocytes, Kupffer cells, and sinusoidal endothelial cells and other cell types. This triggers cyclic adenosine monophosphate (cAMP) formation, which protects sinusoidal endothelial cells and cholangiocytes against bile acid-induced apoptosis, increases bile flow through stimulation of Cl^- secretion by cholangiocytes, and ameliorates cytokine formation by Kupffer cells. The anion exchanger 2 (AE-2) in the apical cholangiocyte membrane allows for bicarbonate secretion in exchange with chloride and provides a protective bicarbonate-rich milieu in the glycocalyx of cholangiocytes (so-called bicarbonate umbrella [12]).

Detoxification

Endo- and xenobiotics are detoxified by biotransformation of these compounds. Such reactions convert lipophilic compounds into polar, water-soluble metabolites. In a first step (biotransformation phase I), such compounds become hydroxylated or N- or O-dealkylated at cytochrome P450 enzymes or undergo oxidative deamination or hydrolysis. Such reactions introduce or expose reactive groups, which are used in phase II of biotransformation for conjugation reactions, leading to the formation of hydrophilic and water-soluble compounds, which can be excreted into bile or into the urine. Such conjugation reactions include glucuronidation, sulfation, or coupling to glutathione or amino acids such as glycine, taurine, or glutamine. Phase III of biotransformation describes the excretion of such conjugates into bile or blood via canalicular transporters (e.g., MRP2) or sinusoidal OATPs, respectively.

Phase I–III enzyme activities can be induced by endo- and xenobiotics after their binding to the nuclear receptors constitutive androstane receptor or the pregnane X receptor (PXR; syn.:

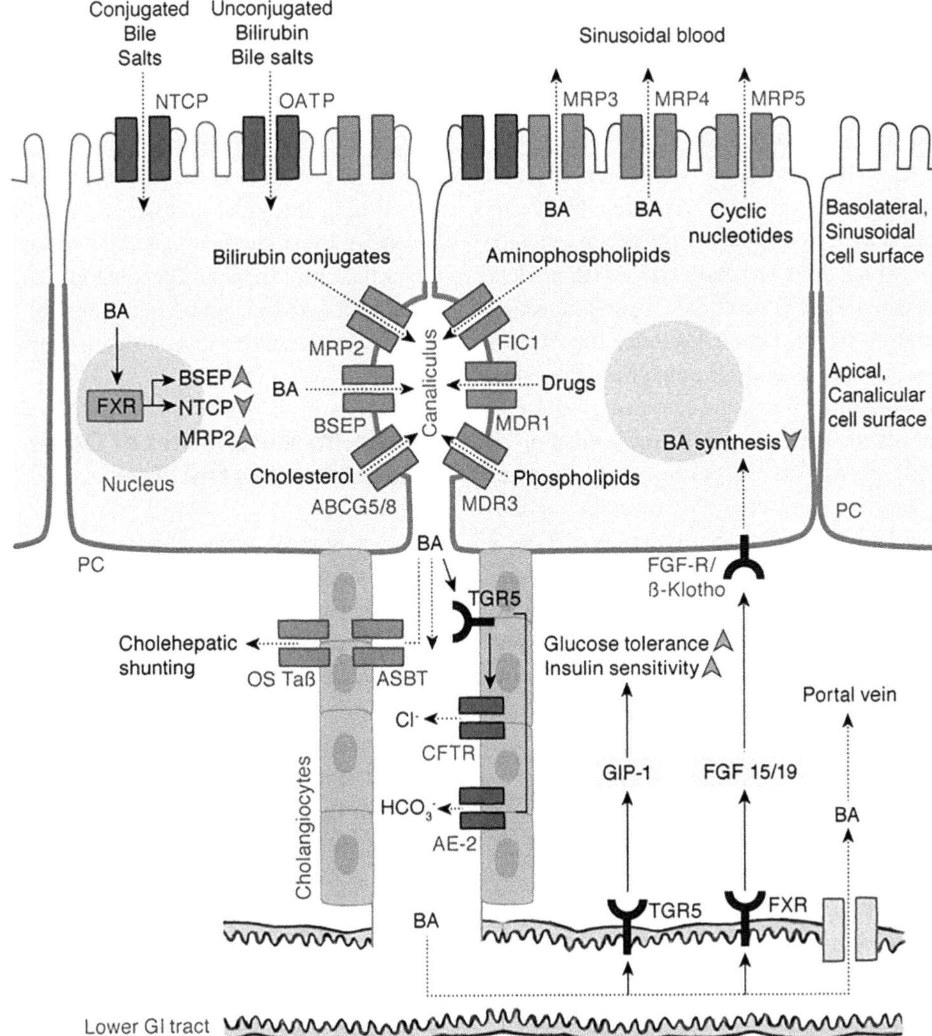

Fig. 2 Hepatobiliary transport and bile acid signaling. Schematic representation of hepatobiliary transport systems, bile formation and bile acid signaling. Biliary excretion is accomplished by transport ATPases in the canalicular membrane of the hepatocyte. In the intestine, bile acids activate not only their receptors in order to signal to the liver via FGF15/19 and to secrete glucagon-like peptide-1 (GLP-1), but are also reabsorbed in the terminal ileum and transported back to the liver ("enterohepatic circulation of bile acids"). In the bile duct, a small fraction of bile acids is reabsorbed by ASBT, a sodium-dependent bile acid uptake system, excreted into the blood via the organic solute and steroid transporter OST αβ and recirculated to the liver ("cholehepatic shunting"). TGR5 in cholangiocytes senses the bile acid concentration in the bile ducts and triggers bicarbonate and chloride excretion into the lumen through activation of the anion exchanger-2 (AE-2) and the cystic fibrosis transmembrane conductance regulator (CFTR), respectively. Activation of the AE-2 provides a protective bicarbonate-rich milieu in the glycocalyx of cholangiocytes (so-called bicarbonate umbrella [12]). Multidrug resistance-associated proteins 3-5 (MRP3-5) can transport bile acids and cyclic nucleotides from the hepatocytes into the blood. Bile acid exit from the hepatocyte via MRP3 and 4 is important for prevention of bile acid-induced liver damage during cholestasis. Downregulation of NTCP and upregulation of BSEP is triggered by FXR following bile acid overload of PC. For further details see text

SXR, steroid, and xenobiotic sensing nuclear receptor). After heterodimerization with the retinoid X receptor, these act as transcription factors for a variety of genes involved in biotransformation, such as cytochrome P450 family 3A (CYP3A), glutathione-S-transferases, or organic anion transporting protein 2. Aromatic hydrocarbons are sensed by the arylhydrocarbon receptor (AHR), which together with the AHR nuclear translocator (ARNT), acts as a ligand-activated transcription factor, which regulates the expression of phase I and II enzyme activities. Biotransformation reactions can also give rise to toxic products. One example is the formation of genotoxic derivatives of benzopyrene. Another example is paracetamol (acetaminophen) toxicity. Paracetamol is partly converted to a highly toxic quinone derivative, which is immediately detoxified by S-conjugation with glutathione. However, after depletion of glutathione stores this highly reactive intermediate forms protein adducts, which can produce acute liver necrosis.

Inside-Out: Metabolites of the Liver Affecting Other Tissues

Bile acids also serve as signaling molecules in other tissues because the bile acid receptors FXR and TGR5 are also found in extrahepatic tissues. TGR5 activation in enteroendocrine intestinal cells triggers release of glucagon-like peptide-1 from the ileal L-cells, thereby increasing glucose tolerance and improving insulin sensitivity [13]. Activation of FXR in the ileum by bile acids triggers formation and release of fibroblast growth factor 15/19, which in the liver activates the fibroblast growth factor receptor/ß-Klotho complex and triggers downregulation of hepatic de novo bile acid synthesis and protects the liver in cholestasis (see also Fig. 2).

Another example for inside-out signaling is the hepatorenal reflex [14], which is activated by amino acids most likely through induction of PC swelling and an increase in the sinusoidal pressure and which triggers a decrease in glomerular filtration rate in the kidney via afferent vagal nerves and sympathetic efferent nerves. Renal water retention triggered by this reflex may counteract splanchnic blood pooling in the absorptive state. Failure of the liver to eliminate NH_4^+, which arises during intestinal and hepatic metabolism, can lead to hyperammonemia and ammonia toxicity in the brain (hepatic encephalopathy, which is characterized by a low-grade cerebral edema and an oxidative/nitrosative stress response [9].

Outside-In: Metabolites of Other Tissues Affecting the Liver

There are several ways how metabolites from other tissues can affect liver function. These include hormones and cytokines released from extrahepatic sites. For example, insulin from pancreatic β-cells binds to insulin receptors on PCs in the post-absorptive state and increases nutrient anabolism, such as glycogen synthesis. In addition, nutrients, hormones, and toxins directly affect liver function through alterations of hepatocellular hydration. Hepatocellular hydration is a dynamic parameter and is controlled by a variety of transport systems in the plasma membrane of PCs, which can create or dissipate osmotic gradients. For example, an increased amino acid load to the liver leads to an osmotic water shift into the PC due to a cumulative amino acid uptake driven by the transmembrane sodium gradient. This increase in hepatocellular volume represents an independent signal regulating liver function. Hepatocyte swelling increases bile flow (choleresis), inhibits protein and glycogen breakdown, stimulates protein and glycogen synthesis, and acts as an antiapoptotic and proliferative signal. Opposite responses are triggered by hepatocyte shrinkage, which for example occurs under oxidative stress. Control of liver cell functions by fluctuations of hepatocyte hydration or hepatocyte volume is mediated by osmosensing and osmosignaling pathways [15]. Integrins were identified as important osmosensors in response to hepatocyte swelling, activating osmosignaling pathways [16], which involves focal adhesion kinase, c-Src tyrosine kinase, the epidermal growth factor receptor, and the mitogen activated protein kinases Erk and p38^{MAPK}. On the other hand, cell shrinkage is sensed by intracellular chloride, and involves endosomal acidification, ceramide formation, activation of protein kinase Cζ and NADPH oxidase with subsequent formation of reactive oxygen species, activation of the Src family kinases Yes and Fyn and c-Jun-N-terminal kinases as signaling events. Of note, certain bile acids can activate osmosensing and osmosignaling pathways in a non-osmotic manner [16]. For example, tauroursodeoxycholate can directly activate volume-sensing ß$_1$-integrins and thereby trigger osmosignalling pathways towards choleresis, whereas proapoptotic effects of glycochenodesoxycholate are triggered by an increase in the intracellular chloride concentration [16].

For more in-depth surveys on liver function, the reader is referred to textbooks of hepatology [17–19].

Final Remarks and Perspectives

The liver plays a major role in intermediary metabolism, maintenance of homeostatic functions, immune responses, and endo- and xenobiotics excretion into bile and is a paradigm for tissue regeneration. These functions are impaired in a variety of diseases, such as hepatitis or cirrhosis. Modern cell and molecular biology techniques will continue to identify the heterogeneity of previously phenotypically described liver disorders and provide new perspectives for personalized molecular medicine.

Acknowledgements Own studies reported herein were supported by Deutsche Forschungsgemeinschaft through Collaborative Research Centers **SFB 154** *Clinical and Experimental Hepatology*, **SFB 575** *Experimental Hepatology*, **SFB 974** *Communication and Systems Relevance of Liver Damage and Regeneration* **and the Clinical Research Group KFO 217** *Hepatobiliary Transport in Health and Disease.*

Question and Answer

Question 1 What are the major liver functions?

Answer 1 Major functions of the liver are (a) nutrient processing and storage and allowing for an adequate nutrient supply for other organs; (b) detoxification and excretion of xeno- and endobiotics into bile; (c) pathogen defense and acute phase responses; (d) acid–base homeostasis, plasma protein and bile acid synthesis, triglyceride and cholesterol metabolism and transport, hormone processing and secretion.

Question 2 What are the main substrates of the canalicular transport ATPases ABCB11 (BSEP), ABCC2 (MRP2), ABCG5/G8, and ABCB4 (MDR3)?

Answer 2 ABCB11 (BSEP): conjugated bile acids; ABCC2 (MRP2): bilirubin glucuronide; ABCG5/G8: cholesterol; ABCB4 (MDR3): phospholipids.

Question 3 Describe the three phases of biotransformation and their regulation.

Answer 3 The three phases of biotransformation are hydroxylation or N- or O-dealkylation, oxidative deamination or hydrolysis of the compound (phase I), conjugation reactions (phase II), and conjugate excretion (phase III). Ligand-induced activation of nuclear receptors such as CAR, PXR, FXR, AHR, regulate expression of proteins involved in phase I-III reactions.

Question 4 Describe the functional consequences of ileal FXR activation by bile acids.

Answer 4 FXR activation by bile acids in the ileum releases the fibroblast growth factor 15/19, which activates in the liver the fibroblast growth factor receptor/ß-Klotho complex which triggers downregulation of bile acid synthesis.

References

1. Kordes C, Häussinger D (2013) Hepatic stem cell niches. J Clin Invest 123:1874–1880
2. Kordes C, Bock HH, Reichert D, May P, Häussinger D (2021) Hepatic stellate cells: current state and open questions. Biol Chem 402:1021–1032
3. Häussinger D (2011) Liver Regeneration, 1st edn. DeGruyter, Berlin/Boston
4. Lorenz L, Axnick J, Buschmann T, Henning C, Urner S, Fang S, Nurmi H, Eichhorst N, Holtmeier R, Bódis K, Hwang JH, Müssig K, Eberhard D, Stypmann J, Kuss O, Roden M, Alitalo K, Häussinger D, Lammert E (2018) Mechanosensing by β_1 integrin induces angiocrine signals for liver growth and survival. Nature 562:128–132
5. Jungermann K, Katz N (1997) Physiol Rev 69:708–764
6. Ben-Moshe S, Shapira Y, Moor AE, Manco R, Veg T, Bahar Halpern K, Itzkovitz S (2019) Spatial sorting enables comprehensive characterization of liver zonation. Nat Metab 1:899–911
7. Paluschinski M, Jin CJ, Qvartskhava N, Görg B, Wammers M, Lang J, Lang K, Poschmann G, Stühler K, Häussinger D (2021) Characterization of the scavenger cell proteome in mouse and rat liver. Biol Chem 402:1073–1085
8. Häussinger D (2007) Ammonia, urea production and pH regulation. In: Benhamou JP, Rizetto M, Rodes J, Blei A, Reichen J (eds) Textbook of hepatology: from basic science to clinical practice. Wiley-Blackwell, Oxford, pp 181–190
9. Häussinger D, Dhiman RK, Felipo V, Görg B, Jalan R, Kircheis G, Merli M, Montagnese S, Romero-Gomez M, Schnitzler A, Taylor-Robinson SD, Vilstrup H (2022) Hepatic encephalopathy. Nat Rev Dis Primers 8:43
10. Häussinger D, Keitel V, Kubitz R (2012) Hepatobiliary transport in health and disease. DeGruyter, Berlin/Boston
11. Gertzen CGW, Gohlke H, Häussinger D, Herebian D, Keitel V, Kubitz R, Mayatepek E, Schmitt L (2021) The many facets of bile acids in the physiology and pathophysiology of the human liver. Biol Chem 402:1047–1062
12. Hohenester S, Wenniger LM, Paulusma CC, van Vliet SJ, Jefferson DM, Elferink RP, Beuers U (2012) A biliary HCO_3^- umbrella constitutes a protective mechanism against bile acid-induced injury in human cholangiocytes. Hepatology 55:173–183
13. Thomas C, Gioiello A, Noriega I, Strehle A, Oury J, Rizzo G, Macchiarulo A, Yamamoto H, Mataki C, Pruzanski M, Pelliciari R, Auwerx J, Schoonjans K (2009) TGR5-mediated bile acid sensing controls glucose homeostasis. Cell Metab 10:167–177
14. Lang F, Tschernko E, Schulze E, Ottl I, Ritter M, Völkl H, Hallbrucker C, Häussinger D (1991) Hepatorenal reflex regulating kidney function. Hepatology 14:590–594
15. Häussinger D, Sies H (eds) (2007) Osmosensing and osmosignaling, Methods enzymol, vol 428. Academic, San Diego/London
16. Bonus M, Häussinger D, Gohlke H (2021) Liver cell hydration and integrin signaling. Biol Chem 402:1033–1045
17. Rodes J, Benhamou JP, Blei A, Reichen J, Rizetto M (eds) (2007) Textbook of hepatology. Blackwell Publ, Oxford
18. Arias I, Alter HJ, Boyer JL, Cohen DE, Shafritz DA, Thorgeirsson SS (2020) Wolkoff AW eds. Biology and Pathobiology. Wiley-Blackwell New York, The Liver
19. Häussinger D, Löffler G. Leber—Zentrales Stoffwechselorgan In: Biochemie und Pathobiochemie (Heinrich PC et al., eds) pp. 1027–1047 Springer Verlag Berlin 2022

Cirrhosis

Matteo Rosselli and Massimo Pinzani

Introduction to Cirrhosis

Chronic liver diseases are characterized by a chronic wound healing reaction following reitered hepatocellular damage that if not withdrawn leads to cirrhosis regardless of the underlying etiology. Cirrhosis is characterized by the accumulation of fibrillar extracellular matrix, nodular regeneration [1], neoangiogenesis, and the establishment of portal hypertension (PH), that is, high blood pressure in the portal vein, its branches, and tributaries. PH represents the most important clinical outcome of cirrhosis since from its severity depend the risk of gastroesophageal variceal bleeding, ascites, hepatic encephalopathy, and other decompensating events. In addition, it has a direct impact on the overall prognosis including its correlation with the development of hepatocellular carcinoma. In advanced stages of cirrhosis, the hemodynamic derangement, due to severe PH and a systemic inflammatory reaction, leads to a hyperdynamic circulation characterized by a decrease in systemic vascular resistances and blood pressure and an increase in heart rate and cardiac output. Inflammation, altered hemodynamics, and tissue perfusion, together with parenchymal extinction,

are also accompanied by a profound metabolic derangement that characterizes cirrhosis in its more advanced stages.

Pathophysiology of Cirrhosis and Metabolic Alterations

Portal Hypertension

PH is the hemodynamic consequence of liver cirrhosis. It is initially caused by two main pathophysiological events. First, collagen deposition and nodular regeneration increase intrahepatic vascular resistance by mechanically compressing vessels. Second, progressive sinusoidal endothelial dysfunction is associated with increased vasoconstrictors leading to the contraction of hepatic myofibroblasts around the sinusoids thus further increasing intrahepatic resistance to portal inflow. In addition, extrahepatic vasodilation that typically occurs in more advanced stages, raises the splanchnic inflow further contributing to the worsening of portal pressure.

PH is defined by a hepatic venous pressure gradient (HVPG) above 5 mm Hg. HPVG is the difference between pressure in the portal vein and the intra-abdominal portion of the inferior vena cava. Under normal conditions, substances absorbed by the intestine follow the enterohepatic circulation, flowing through the portal venous system to be processed by the liver.

M. Rosselli · M. Pinzani (✉)
Institute for Liver and Digestive Health, Division of Medicine, University College London, Royal Free Hospital UP3, London, UK
e-mail: m.pinzani@ucl.ac.uk

© The Author(s), under exclusive license to Springer Nature Switzerland AG 2026
E. Lammert, M. Zeeb (eds.), *Metabolism of Human Diseases*,
https://doi.org/10.1007/978-3-031-96019-2_24

In cirrhosis, once PH increases the HVPG beyond 10 mm Hg, low-resistance vascular sites are used to create alternative circulatory pathways [2] (e.g., gastroesophageal varices, paraumbilical vein, retroperitoneal venous collaterals, splenorenal shunts) allowing a bypass of the "obstructed" liver. As a consequence, there is a reduced hepatic clearance of gut-derived vasodilating agents, such as endogenous gastrointestinal hormones (glucagon, vasoactive intestinal peptide, calcitonin gene-related peptide) and intestinal bacterial products [3].

In cirrhosis, intestinal transit time is prolonged and the intestinal mucosa is often edematous due to low oncotic pressure (as a consequence of hypoalbuminemia) and increased portal pressure. In addition, biliary secretion and gut luminal biliary content are reduced, so that bile acids no longer exert their antimicrobial effects or contribute to the integrity of intestinal mucosa to a sufficient extent (see chapter "Liver: Overview" under the part "Liver") [4]. This can lead to bacterial overgrowth and translocation [5]. The presence of bacterial products in the systemic circulation further activates the immune system, increasing the inflammatory response and leading to a functional immune paralysis that characterizes the typical susceptibility of cirrhotic patients to infections [6]. More specifically, endotoxins (such as lipopolysaccharides) activate inflammatory cells, which release cytokines (such as tumor necrosis factor-α, interleukins 1 and 6) and express specific enzymes such as inducible nitric oxide synthase (iNOS) and heme oxygenase (HO) that produce high levels of nitric oxide and carbon monoxide, respectively [7]. Both of these molecules stimulate soluble guanylate cyclase. The resultant cyclic Guanosine MonoPhosphate (cGMP) then activates protein kinase G and lowers intracellular Ca^{2+} levels in smooth muscle cells (SMCs) thus causing vasodilation. Consequently, the portal pressure increases, whereas the systemic blood pressure decreases (see chapter "Blood Vessels: Overview" under the part "Blood Vessels").

As a compensatory response, the adrenergic system and the renin-angiotensin-aldosterone system (RAAS) are activated (see chapter "Kidney: Overview" under the part "Kidney"). However, despite high levels of catecholamines and other vasoconstrictors such as angiotensin II, the splanchnic and systemic vasodilation persist due to a vascular hyporesponse to the vasoconstrictors. Because of increased sodium and water retention (initially driven by RAAS activity), fluid volume is overall increased but inappropriately distributed and pooled in the splanchnic compartment, in the interstitium, and eventually in the peritoneal space (ascites), thus leading to relative hypovolemia (i.e., decrease in blood plasma volume) [8] (Fig. 1). In response, vasopressin (also called antidiuretic hormone, ADH) is secreted with consequent free water reabsorption, further fluid overload, and dilutional hyponatremia that characterizes cirrhosis in its more advanced stages [9]. Moreover, the low vascular resistances, fluid overload, and high cardiac output characterize the hyperdynamic circulatory syndrome of cirrhosis. Hyperdynamic circulation is sustained by the persistent liver–gut inflammatory interactions, hyperactivation of neurohormonal systems, and reduced renal perfusion. The overall effect is a further increase in portal inflow and portal hypertension, which maintains the vicious cycle [10].

Reduced Parenchymal Metabolic Function

The progressive decline of functional liver parenchyma is accompanied by reduced albumin synthesis. Hypoalbuminemia leads to low colloid-osmotic pressure and extravasation of fluid in the extravascular spaces or interstitium. Transport of endogenous (unconjugated bilirubin, transferrin, apoproteins, lipid-soluble hormones) and exogenous molecules (e.g., antibiotics, diuretics, nonsteroidal anti-inflammatory drugs (NSAIDs)) in the blood is also impaired. Moreover, albumin is the main extracellular source of reduced sulfhydryl groups, and therefore, hypoalbuminemia is accompanied by increased oxidative stress and inflammation [11]. In cirrhosis, the lipid profile is often abnormal due to low synthesis of apoproteins and cholesterol (see chapter "Hyperlipidemia") [12].

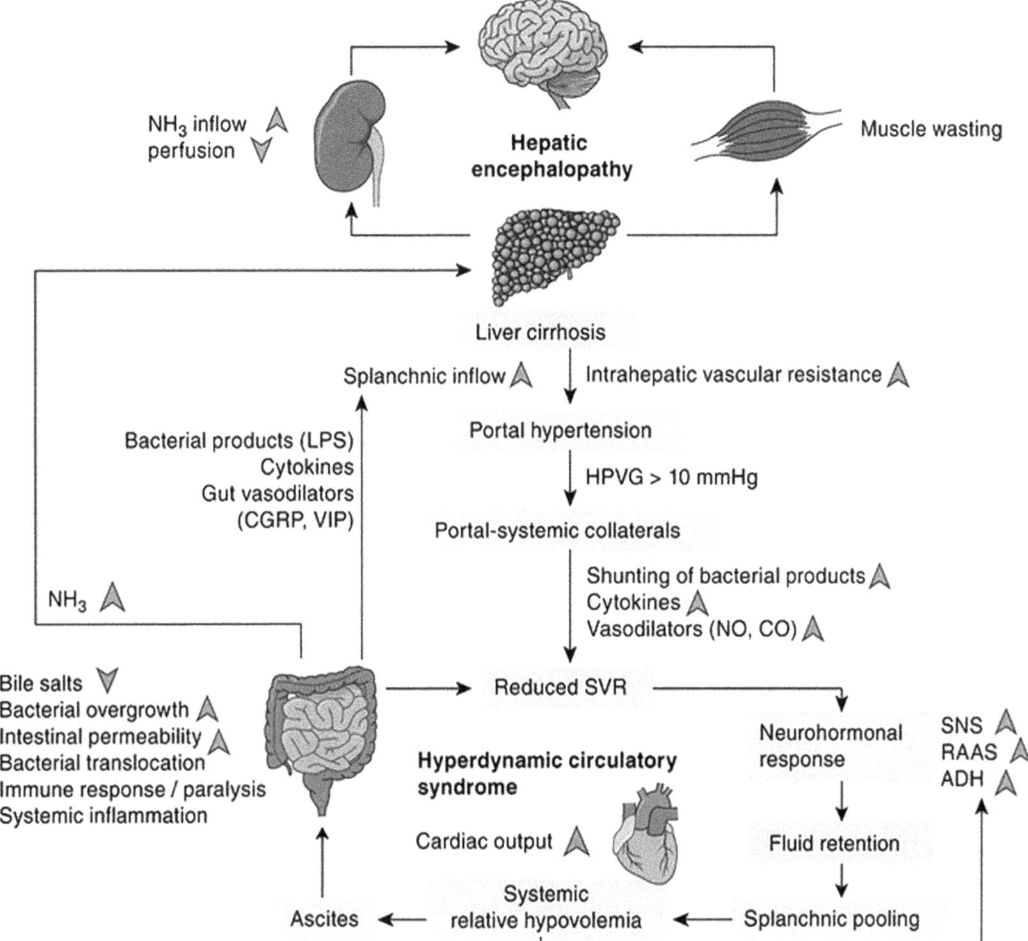

Fig. 1 Pathophysiology of cirrhosis and associated complications. Increased splanchnic inflow and hepatic vascular resistance in cirrhosis lead to portal hypertension. When portal pressure (measured as hepatic venous pressure gradient, HVPG) rises above 10 mm Hg, portal-systemic collaterals reroute inflammatory cytokines, vasodilators, and gut-derived bacterial products to the systemic circulation. Systemic inflammation leads to vasodilation and reduced systemic vascular resistance (SVR) triggering a neurohormonal response. Retained fluid pools in the splanchnic compartment causing relative hypovolemia throughout the body maintaining the vicious cycle. Ascites develops, aggravating bacterial translocation and therefore systemic inflammation. Intestinal bacterial overgrowth and reduced transit time increase gut ammonia (NH₃) production that is not appropriately metabolized by the liver, skeletal muscle, and kidney. Toxic plasma concentrations of NH₃ can then cross the blood–brain barrier causing hepatic encephalopathy. *NO* nitric oxide, *CO* carbon monoxide, *SNS* sympathetic nervous system, *RAAS* renin-aldosterone-angiotensin system, *ADH* antidiuretic hormone (vasopressin), *LPS* lipopolysaccharide, *CGRP* calcitonin gene-related peptide, *VIP* vasoactive intestinal peptide

Hepatocytes are responsible for the first hydroxylation of inactive cholecalciferol to calcidiol (see chapters "Teeth and Bones: Overview" under the part "Teeth and Bones" and "Osteoporosis"). During cirrhosis, this step is impaired and vitamin D synthesis is reduced together with a low production of vitamin D-binding protein [13].

Moreover, in severe forms of cholestatic liver disease (i.e., when bile cannot flow from the liver to the duodenum), for example, as a result of primary biliary cholangitis (i.e., an autoimmune disease accompanied by a progressive destruction of the small bile ducts), vitamin D absorption is also impaired (together with other lipid-soluble vitamins such as A, E, and K). This depletion may severely affect bone metabolism leading to osteopenia (i.e., a condition where bone mineral density is reduced) or even osteoporosis (see chapter "Osteoporosis").

Hepatocyte dysfunction is associated with a reduced insulin clearance. Subsequent hyperinsulinemia may contribute to downregulation of insulin receptors and may lead to glucose intolerance and accelerate the insorgence of type 2 diabetes [14]. In contrast, impaired gluconeogenesis or lack of glycogen stores characterizes overt liver failure and can lead to hypoglycemia.

Under physiological conditions, the nitrogenous products of amino acid catabolism are metabolized by the liver through the urea cycle and glutamine synthesis (see chapter "Liver: Overview" under the part "Liver") [15]. However, in cirrhosis, nitrogen homoeostasis is disrupted, either because ammonia production exceeds urea cycle capacity (mainly by increased gut bacterial ammoniogenesis) or because the liver is unable to metabolize ammonia due to parenchymal insufficiency and portal blood shunting [16]. The skeletal muscle and kidneys are also regulators of ammonia concentration (by ammonia uptake and excretion). However, during cirrhosis, their compensation progressively fails due to muscle wasting and renal impairment, and subsequently, the blood–brain barrier is crossed by an excess of ammonia that accumulates within the astrocytes. The consequent osmotic and inflammatory damage as well as neurotransmission impairment characterizes hepatic encephalopathy (HE), one of the most important expressions of metabolic dysfunction in acute and chronic liver disease.

Treatment of Cirrhosis and Related Complications

Treatment of chronic liver disease is manifold and primarily aimed at counteracting the etiological agent (e.g., a hepatitis virus, steatosis, autoimmunity or excessive alcohol consumption) in order to reduce liver inflammation, preventing the formation of scarring tissue and nodular regeneration that characterize the final and irreversible stages of cirrhosis. However, due to delayed diagnosis, poor response, or low compliance to treatment, progression to cirrhosis is often inevitable. When signs of clinically significant PH occur, treatment is aimed to delay and counteract cirrhosis complications. In cases of advanced cirrhosis, liver transplantation remains the only treatment option.

Portal Hypertension

The management of PH is aimed at preventing further pressure increase and related complications, particularly gastroesophageal bleeding. When PH rises beyond a HVPG of 12 mm Hg, a critical tension within the vascular wall results, increasing the risk of vascular rupture. Nonselective β-blockers (NSBBs, Table 1) represent the first-line treatment to directly reduce portal pressure and to allow unopposed α-adrenergic activity. This produces mesenteric arterial vasoconstriction and therefore reduces portal venous load. Moreover, NSBBs reduce cardiac output and systemic blood pressure contributing to a further decrease in portal flow.

If bleeding occurs despite NSBBs and endoscopic banding of gastroesophageal varices, there may be indication to decompress the portal system by positioning a transjugular intrahepatic portal-systemic shunt (TIPS) [17], that is, an artificial shunt between the (inflow) portal vein and one of the (outflow) hepatic veins. This invasive procedure is used as a treatment option in severe, selected cases of patients with recurrent variceal bleeding, refractory ascites, hydrothorax, and/or hepatorenal syndrome, as in 30–50% of cases, the shunt leads to portal-systemic HE.

Ascites and Spontaneous Bacterial Peritonitis

Ascites is treated by reducing fluid retention via anti-aldosterone and loop diuretics and by a salt restriction diet (<90 mmol/day). In poor responses to medical treatment, there is indication for paracentesis, a puncturing of the peritoneal cavity to drain the intraperitoneal fluid and

Table 1 Principles of standard treatment in cirrhosis

	Clinical presentation	Treatment option	Therapeutic objectives	Negative metabolic effects
Medical	Portal hypertension (PH)	NSBB, diuretics	NSBB: reduce HR, splanchnic inflow, and portal hypertension Diuretics: reduce fluid retention	*NSBB*: suppressed adrenergic response *Loop diuretics*: dehydration, hyponatremia, HA and HE
	Ascites	Loop diuretics, AA diuretics, low sodium diet	Loop diuretics: reduce fluid retention AA diuretics: counteract aldosterone	*Loop diuretics*: dehydration, hyponatremia, HA and HE *AA diuretics*: hyperkalemia and gynaecomastia (*spironolactone*)
	Hepatorenal syndrome (HRS)	Vasoconstrictors (Terlipressin, Octreotide, Midodrine)	*Vasoconstrictors*: counteract vasodilation	*Terlipressin*: hypertension, hyponatremia, ischemia may be triggered in patients with underlying cardiovascular disease, bradycardia
		Albumin	*Albumin*: increases effective volume to improve renal perfusion	
	Spontaneus bacterial peritonitis (SBP)	3rd generation cephalosporins Albumin	Antibiotics: counteract infection Albumin: increases arterial blood volume to prevent renal impairment	
	Hepatic encephalopathy	Osmotic laxatives (lactulose) enemas, non-absorbable antibiotics (*rifaximin/ metronidazole*)	Laxatives: reduce colonic transit time and bacterial ammonia production	Laxatives: diarrhea, dehydration and secondary renal impairment, hypokalemia
		Dietary measures	Antibiotics and enemas: gut decontamination Dietary measures: reduce catabolism and muscle wasting prevention	
Interventional	Gastroesophageal haemorrhage	TIPS	Recurrent bleeding despite medical treament and variceal band ligation	PSE
	Ascites	TIPS	Refractory ascites	PSE

NSBB nonselective β-blockers, *HR* heart rate, *HA* hyperammonemia, *HE* hepatic encephalopathy, *TIPS* transjugular intrahepatic portal-systemic shunt, *PSE* portal-systemic encephalopathy), *AA* anti-aldosterone

relieve abdominal pressure. This fluid removal must be accompanied with albumin infusion to prevent hypotension and renal hypoperfusion, especially in cases of large volume (>5 l) paracentesis [18].

Complications of anti-aldosterone diuretics are hyperkalemia, whereas loop diuretics may lead to hyponatremia, hypokalemia, and hyperammonemia with related HE. A specific side effect of some anti-aldosterones such as spironolactone is gynecomastia, the benign enlargement of breast tissue in males (see chapter "Prostate Cancer").

Bacterial translocation, especially in patients with tense ascites, may lead to infection of the ascitic fluid and spontaneous bacterial peritonitis, a life-threatening complication that calls for prompt treatment with antibiotics, such as oral norfloxacin or intravenous third-generation cephalosporin, and albumin infusion to improve circulatory function and reduce the risk of hepatorenal syndrome.

Hepatorenal Syndrome

Hepatorenal syndrome is defined as the occurrence of renal failure in patients with advanced liver disease in the absence of an identifiable cause [19]. There are two types of HRS. Type 1 HRS is a rapidly progressive acute renal failure that conventionally is diagnosed when creatinine increases more than 100% from baseline to a level $\geq$ 2.5 mg/dL. It usually develops as a consequence of worsening liver function in response to a precipitating event and other organ impairment. HRS 1 can partially be explained by an alteration of renal vascular tone and hypoperfusion as a consequence of the imbalance of vasoactive molecules secondary to PH, or decompensating events that exacerbate such imbalance. Type 2 HRS occurs in patients with refractory ascites with consequent renal hypoperfusion due to renal hypovolemia and increased intra-abdominal pressure. It is of note that patients with HRS 2 may develop HRS 1 [19].

The most effective treatments are splanchnic vasoconstricting drugs, which are based on the fact that kidneys do not receive adequate blood flow due to fluid compartmentalization, relative hypovolemia, and increased renal resistances. Vasopressin analogues, such as terlipressin, together with albumin are the treatment of choice to improve circulatory function. TIPS may be beneficial in selected cases. Liver transplantation remains the best option since it resolves the relative impairment of liver and kidney. Renal replacement therapy may be used as bridging therapy before liver transplantation.

Hepatic Encephalopathy

The objective of HE treatment is to counteract the precipitating causes of hepatic decompensation and target the production and absorption of ammonia [16].

A high-calorie diet maintaining a protein intake of 1.5 g/kg body weight and the use of branched-chain amino acids to counteract muscle wasting are recommended. Laxatives to loosen the stool, such as lactulose, are cathartic (i.e., accelerating defecation), acidify the gut lumen thus inhibiting ammoniogenic bacteria, and trap ammonia in its ionized form ammonium, which cannot be absorbed and is excreted with stool. Nonabsorbable antibiotics, such as neomycin and metronidazole, are used to reduce bacterial derived toxins. However, rifaximin has fewer side effects and a broad-spectrum acting on both gram-positive and gram-negative bacteria [20]. The most common side effects of cathartics are dehydration that may exacerbate HE and renal impairment and electrolyte imbalance, especially hypokalemia (as the intestinal fluid contains high amounts of K^+ due to secretion of K^+ and Cl^- in the stomach).

Table 1 summarizes the principles of standard treatment in cirrhosis, PH, and associated complications, highlighting the possible negative and positive metabolic consequences.

Conclusions and Perspectives

Chronic liver disease and its progression to cirrhosis are secondary to the interplay of many factors such as the etiological agent, chronic inflammation, and the immune system response. The development of PH, hemodynamic derangement, neurohormonal and metabolic dysfunction characterize cirrhosis in its more advanced stages. However, systemic inflammation and the immune system dysregulation are not only responsible of fibrosis progression but also of the liver's susceptibility to its further dysfunction. Along these lines, one of the most important physiopathological roles is played by gut dysbiosis and bacterial translocation that are a source of microbial particles (pathogen-associated molecular patterns) that interact with pattern recognition receptors on cells of the immune system (see chapter "Anatomy and Physiology of the Immune System" under the part "Immune System"). As a result, a cytokine cascade is released with consequent systemic inflammatory reaction. The influence of endogenous inducers of inflammation, such as damage-associated molecular patterns, is instead released as a consequence of liver injury itself, and is responsible for triggering the inflammasome pathway with "sterile inflammation," further tissue injury and organ failure [21]. These mechanisms seem to represent the physiopathological basis of a life-threatening clinical syndrome known as acute on chronic liver failure (ACLF). This is characterized by acute liver failure on the background of cirrhosis triggered by a precipitating event (infection, bleeding, toxins exposure) leading to severe liver dysfunction and one or more organ failures with a high risk of mortality [22]. ACLF should be distinguished from acute decompensated chronic liver disease because it is influenced by a severe systemic inflammation, multiple organ dysfunction, increased risk of infection, and poor prognosis. Molecular interactions, inflammation, and immune dysregulation represent the basis and target for further research and treatment in order to prevent both progression of liver disease and hepatic dysfunction in its more advanced stages.

Questions and Answers

Question 1 Why is portal hypertension related to prognosis in cirrhosis?

Answer 1 Portal hypertension is the consequence of parenchymal and vascular remodeling and expresses the severity of liver disease since these modifications lead to portal-systemic shunting with hepatic encephalopathy, esophageal varices (with a bleeding risk and mortality that is directly proportional to the severity of portal hypertension), ascites, bacterial translocation, and hepatorenal syndrome. All these complications occur in the presence of severe portal hypertension and hence it is the most important prognostic factor in patients with cirrhosis.

Question 2 Is cirrhosis-related metabolic derangement related only to its low synthetic function?

Answer 2 No, metabolic derangement in cirrhosis depends on two main pathophysiological mechanisms that, however, are linked to each other especially in the more advanced stages. One of these mechanisms mainly depends on hepatic synthetic function (Urea's biochemical cycle, vitamin D hydroxylation, protein synthesis, glucose and lipid dysmetabolism), while the other mainly depends on portal hypertension and the hemodynamic modifications that influence, via a complex interplay, the RAAS activity and the adrenergic system.

Question 3 What distinguishes acute-on-chronic liver failure from acutely decompensated chronic liver disease?

Answer 3 Acute decompensation of chronic liver disease is usually a limited complication of cirrhosis that highlights the severity of cirrhosis, but responds to treatment. Acute-on-chronic liver failure seems to be a distinct acute clinical complication of cirrhosis, the severity of which is defined by liver failure, the number of organs involved, and has poorer prognosis.

References

1. Anthony PP, Ishak KG, Nayak NC, Poulsen HE, Scheuer PJ, Sobin LH (1978) The morphology of cirrhosis. Recommendations on definition, nomenclature, and classification by a working group sponsored by the World Health Organization. J Clin Pathol 31:395–414
2. Rodríguez-Vilarrupla A, Fernández M, Bosch J, García-Pagán JC (2007) Current concepts on the pathophysiology of portal hypertension. Ann Hepatol 6:28–36
3. Ramachandran A, Balasubramanian KA (2001) Intestinal dysfunction in liver cirrhosis: its role in spontaneous bacterial peritonitis. J Gastroenterol Hepatol 16:607–612
4. Inagaki T, Moschetta A, Lee YK, Peng L, Zhao G, Downes M, Yu RT, Shelton JM, Richardson JA, Repa JJ, Mangelsdorf DJ, Kliewer SA (2006) Regulation of antibacterial defense in the small intestine by nuclear bile acid receptor. Proc Natl Acad Sci USA 103:3920–3925
5. Wiest R, Garcia-Tsao G (2005) Bacterial translocation (BT) in cirrhosis. Hepatology 41:422–433
6. Bonnel AR, Bunchorntavakul C, Reddy KR (2011) Immune dysfunction and infections in patients with cirrhosis. Clin Gastroenterol Hepatol 9:727–738
7. Martell M, Coll M, Ezkurdia N, Raurell I, Genescà J (2010) Physiopathology of splanchnic vasodilation in portal hypertension. World J Hepatol 2:208–220
8. Pinzani M, Rosselli M, Zuckermann M (2011) Liver cirrhosis. Best Pract Res Clin Gastroenterol 25:281–290
9. Ginés P, Berl T, Bernardi M, Bichet DG, Hamon G, Jiménez W, Liard JF, Martin PY, Schrier RW (1998) Hyponatremia in cirrhosis: from pathogenesis to treatment. Hepatology 28:851–864
10. Moller S, Henriksen JH (2006) Cardiopulmonary complications in chronic liver disease. World J Gastroenterol 12(526–538):526
11. Bernardi M, Maggioli C, Zaccherini G (2012) Human albumin in the management of complications of liver cirrhosis. Crit Care 16:211
12. Janičko M, Veselíny E, Leško D, Jarčuška P (2013) Serum cholesterol is a significant and independent mortality predictor in liver cirrhosis patients. Ann Hepatol 12:413–419
13. Malham M, Jørgensen SP, Ott P, Agnholt J, Vilstrup H, Borre M, Dahlerup JF (2011) Vitamin D deficiency in cirrhosis relates to liver dysfunction rather than aetiology. World J Gastroenterol 17:922–925
14. García-Compean D, Jaquez-Quintana JO, Maldonado-Garza H (2009) Hepatogenous diabetes. Current views of an ancient problem. Ann Hepatol 8:13–20
15. Frederick RT (2011) Current concepts in the pathophysiology and management of hepatic encephalopathy. Gastroenterol Hepatol (NY) 7:222–233
16. Khungar V, Poordad F (2012) Hepatic encephalopathy. Clin Liver Dis 16:301–320
17. Krajina A, Hulek P, Fejfar T, Valek V (2012) Quality improvement guidelines for Transjugular intrahepatic portosystemic shunt (TIPS). Cardiovasc Intervent Radiol 35:1295–1300
18. Moore CM, Van Thiel DH (2013) Cirrhotic ascites review: pathophysiology, diagnosis and management. World J Hepatol 5:251–263
19. European Association for the Study of the Liver (2010) EASL clinical practice guidelines on the management of ascites, spontaneous bacterial peritonitis and hepatorenal syndrome in cirrhosis. J Hepatol 53:397–417
20. Bass NM, Mullen KD, Sanyal A, Poordad F, Neff G, Leevy CB, Sigal S, Sheikh MY, Beavers K, Frederick T, Teperman L, Hillebrand D, Huang S, Merchant K, Shaw A, Bortey E, Forbes WP (2010) Rifaximin treatment in hepatic encephalopathy. N Engl J Med 362:1071–1081
21. Kubes P, Mehal WZ (2012) Sterile inflammation in the liver. Gastroenterology 143:1158–1172
22. Arroyo V, Moreau R, Jalan R (2020) Acute-on-chronic liver failure. N Engl J Med 382:2137–2145

Fat Tissue: Overview

Raja Gopal Reddy Mooli,
Sadeesh K. Ramakrishnan, and Erin E. Kershaw

Introduction

Obesity is a complex multifactorial disease that contributes to several metabolic disorders, including type 2 diabetes, cardiovascular diseases, and metabolic-dysfunction associated steatotic liver disease (MASLD). The prevalence of obesity is rising worldwide and is expected to reach 17% in men and 22% in women by 2030 [1]. Several risk factors, including genetics, epigenetics, and lifestyle, make significant contributions to the development of obesity [2]. Obesity results from positive energy balance, i.e. when energy intake exceeds expenditure. In response to positive energy balance, adipose tissue mass increases and may become dysfunctional. This dysfunctional adipose tissue results in ectopic fat deposition within peripheral tissues that regulate metabolic homeostasis and insulin resistance [3]. It also augments inflammation, fibrosis, hypoxia, and impaired adipokine and lipokine secretion [4]. Notably, non-obese individuals with loss of adipose tissue due to lipodystrophy exhibit similar metabolic derangements.

[4–6]. Lipodystrophy also affects adipose tissue function by impairing the adipogenic program and adipose progenitor cell (APC) function [7]. Thus, adipose tissue is a highly adaptable and essential organ with critical functions that are required to maintain health and prevent disease.

Adipose Tissue: Anatomy and Physiology

Adipose tissue, commonly known as "fat," has evolved into a highly specialized tissue for storing energy in the form of triacylglycerols (TAGs). However, adipose tissue function extends beyond fat storage to various other processes necessary for energy/metabolic homeostasis, immune homeostasis, and reproductive function [8]. Adipose tissue is heterogeneous in cellular composition, location, and function—reflecting its complex role in normal physiology and disease [9]. It is comprised of different cell types, including adipocytes, adipocyte progenitor cells, endothelial cells, immune cells, and many other cell types, to form a truly multicellular organ [10]. The heterogeneity of adipose tissue is reflected by the variety of clinical disorders that result from adipose tissue dysfunction [11, 12]. Indeed, both adipose tissue excess (obesity) and deficiency (lipodystrophy) result in profound physiological impairments that promote metabolic syndrome (i.e., insulin resistance, diabetes, dys-

R. G. R. Mooli · S. K. Ramakrishnan · E. E. Kershaw (✉)
Division of Endocrinology and Metabolism,
Department of Medicine, University of Pittsburgh,
200 Lothrop Street, Biomedical Science Tower
W1055, Pittsburgh, PA, USA
e-mail: RAJA2019@pitt.edu; RAMAKS@pitt.edu;
kershawe@pitt.edu

lipidemia, and hypertension) and cardiovascular disease. Adipose tissue dysfunction also contributes to a myriad of other diseases affecting virtually all organ systems, including liver disease (i.e., steatotic liver, cirrhosis), kidney disease (diabetic and hypertensive nephropathy), pulmonary disease (i.e., sleep apnea), musculoskeletal disease (i.e., arthritis, back pain), reproductive disease (i.e., infertility), and psychological disease (i.e., depression), and even cancer [13]. The distribution of fat is highly variable and affected by sex, genetics, age, drugs, and many other factors [14]. Fat distribution contributes to overall health. For example, expansion of subcutaneous fat is associated with more favorable metabolic health than expansion of visceral fat [15]. These heterogeneous characteristics of fat explain the existence of "metabolically healthy obese" and "metabolically unhealthy normal weight" individuals [16]. Thus, adipose tissue is not simply an inert tissue for storing fat but a highly dynamic tissue required for health and survival. Only by understanding the unique characteristics of adipose tissue, we can begin to exploit its complexities to treat or prevent disease.

Adipose Tissue: Three Shades of Adipocytes

According to recent studies, mammalian adipose tissue includes three main types of adipocytes—white, beige, and brown [17] (Fig. 1). White adipose tissue (WAT) is composed of 'white' adipocytes characterized by large, unilocular lipid droplets and sparse mitochondria (Fig. 1A). The high ratio of fat to mitochondria gives WAT its characteristic white appearance. WAT is located throughout the body in "depots" but is also spread within and around other tissues exhibiting location-specific characteristics [18]. The major WAT depots are located in the subcutaneous (upper superficial abdominal and lower gluteal–femoral depots) and visceral regions (omental, mesenteric, mediastinal, and epicardial). In addition, smaller amounts of adipose tissue are present in the muscle, breast, bone

marrow, face, and orbits [4, 18]. Within WAT, adipocytes occupy approximately 50% of the cellular content with the remaining 50% containing non-adipocyte stromal vascular fraction (SVF) cells composed of adipocyte progenitor cells (APCs; 55%), immune cells (37%), and endothelial cells (8%) [19]. WAT is highly specialized for storing large amounts of fat as TAGs, but it also has several other critical functions, including mechanical protection, thermal insulation, energy homeostasis, and endocrine factor production [17].

Brown adipose tissue (BAT) is comprised of "brown" adipocytes characterized by small, multi-locular lipid droplets surrounded by copious large mitochondria. The high ratio of mitochondria to fat gives BAT its characteristic brown appearance [20] (Fig. 1B). BAT is located in the interscapular region, with small depots in cervical, axillary, perivascular, and perirenal regions. Human infants also have interscapular BAT, which regresses in adulthood [21]. The mitochondria of brown adipocytes express uncoupling protein 1 (UCP1), which dissipates energy in the form of heat by uncoupling the mitochondrial proton gradient from ATP synthesis [22]. Thus, BAT is involved in energy expenditure via non-shivering thermogenesis.

Within WAT, a subset of inducible adipocytes known as "beige" adipocytes transform into adipocytes with more brown-like features. The well-studied inducers of beige adipocytes are cold exposure, exercise, calorie restriction, β-adrenergic receptor treatment, cancer cachexia, and tissue injury, and this phenomenon is referred to as browning or "beiging" of white fat [23–25]. Beige adipocytes are also found in neonates, and their mechanism of induction is not well understood [26]. Beige adipocytes are characteristically and functionally similar to brown adipocytes [27]. However, recent findings show a distinct feature of beige adipocytes regarding their function; that is, they can maintain whole-body energy homeostasis via UCP1-independent and UCP1-dependent mechanisms [28]. These characteristics make beige adipocytes a focus of intense investigation for treating obesity and metabolic disease.

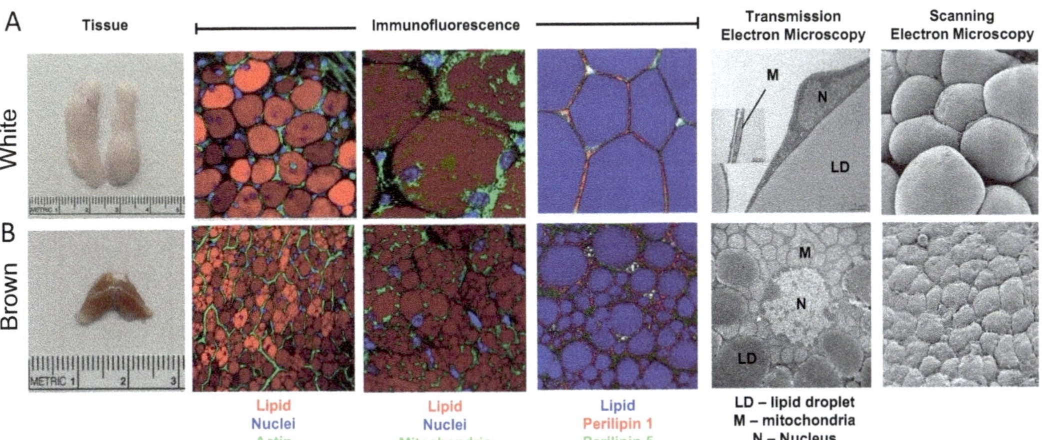

Fig. 1 Morphology in three shades of adipose tissue. Gross and microscopic images distinguishing (**a**) white (WAT) from (**b**) brown (BAT) adipose tissue under physiological conditions. (Images by Erin E. Kershaw, and Donna B. Stolz, University of Pittsburgh's Centre for Biological Imaging)

Metabolic Plasticity of Adipose Tissue

Adipose tissue is exquisitely designed for the regulated storage and release of lipid substrates and possesses all the cellular machinery for both fat synthesis (lipogenesis) and fat breakdown (lipolysis) (Fig. 2). During periods of energy excess (i.e., after a meal), energy substrates (i.e., glucose or fatty acids), adipocytes take up the fatty acids (FAs) via the fatty acid transporters (FATPs and CD36). FAs are converted to acyl-CoAs by the enzyme acyl-CoA synthetase and are sequentially esterified to a glycerol backbone by acyltransferases to form TAGs [29]. Thus, lipogenesis plays a critical role in energy homeostasis by converting excess energy from carbohydrates and lipids into TAGs. The major signal that promotes triglyceride storage is insulin, which increases glucose uptake, promotes lipogenesis, and inhibits lipolysis in WAT [6]. Adipocyte de novo lipogenesis (DNL) also improves insulin sensitivity by generating several anti-inflammatory and insulin-sensitizing lipid mediators/factors [30]. These fatty acids mainly belong to branched fatty acid esters of hydroxy fatty acids (FAHFA) and palmitic acid esters of hydroxy stearic acid esters (PAHSA) [31, 32]. Lipid droplets containing TAGs are highly dynamic organelles associated with various lipid droplet proteins such as those of perilipin (PLIN) family [33]. For example, perilipin 1 (PLIN1) is primarily found in adipocytes, where it is integrally involved in lipolysis, whereas perilipin 5 (PLIN5) is primarily expressed in oxidative tissues such as BAT, where it is integrally involved in lipid oxidation [33].

In the setting of increased energy demand (i.e., fasting, exercise), free fatty acids (FFA) are sequentially released from TAGs by the lipolytic enzymes adipose triglyceride lipase (ATGL), hormone-sensitive lipase (HSL), and monoglyceride lipase (MGL). Sympathetic nerve-derived factors are shown to play a significant role in inducing lipolysis under nutrient-deprivation states. In particular, epinephrine and norepinephrine act via the adrenergic receptors to activate PKA, which phosphorylates PLIN1, triggering the release of CGI-58 and subsequent activation of ATGL, resulting in the hydrolysis of triglycerides [32, 34]. Several endocrine factors and hormones also regulate lipolysis [35]. In WAT, FFAs are primarily released into the systemic circulation for energy. In BAT, on the other hand, FFAs primarily enter the mitochondria for thermogenesis [4]. Any impairment in the regulation of these fundamental processes contributes to metabolic diseases.

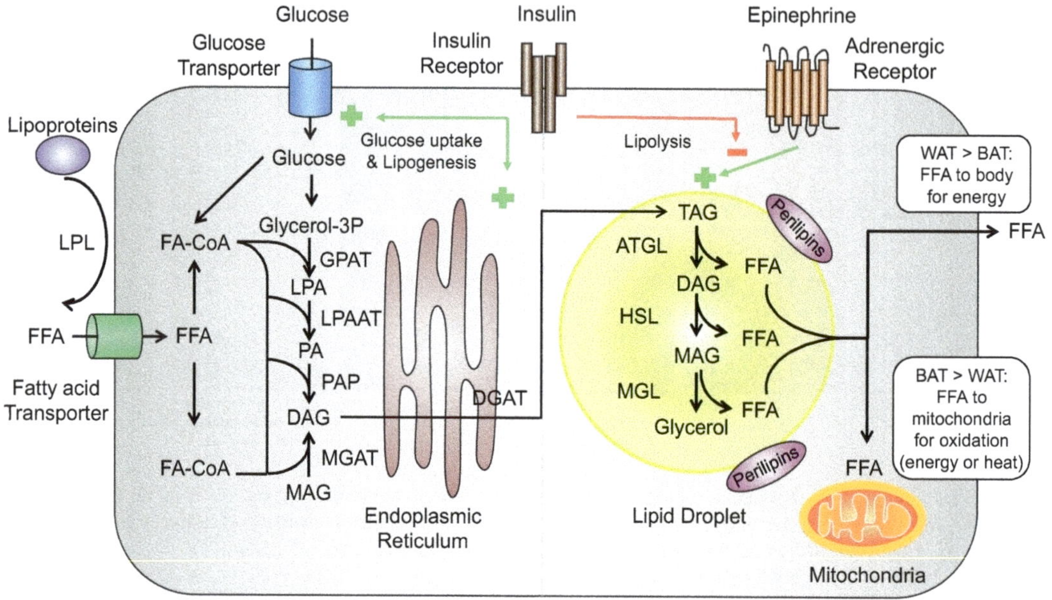

Fig. 2 Metabolic plasticity of white adipose tissue. Anabolic (fat synthesis or lipogenesis) and catabolic (fat breakdown or lipolysis) pathways in adipocytes. Abbreviations: *ATGL* adipose triglyceride lipase, *DAG* diacylglycerol, *DGAT* DAG acyltransferase, *FFA* free fatty acid, *FA-CoA* fatty acyl-CoA, *Glycerol 3-P* glycerol 3-phosphate, *GPAT* glycerol-3P acyltransferase, *HSL* hormone-sensitive lipase, *LPA* lysophosphatidic acid, *LPAAT* LPA acyltransferase, *LPL* lipoprotein lipase, *MAG* monoacylglycerol, *MGL* MAG lipase, *PA* phosphatidic acid, *PAP* PA phosphatase, *TAG* triacylglycerol

Inside-In: Autocrine/Paracrine Effects of Adipose Tissue

Adipose tissue consists not only of adipocytes but also a variety of other cell types, including stromal cells (i.e., fibroblasts, stem cells), vascular cells (i.e., endothelial cells, smooth muscle cells), and immune cells (i.e., macrophages). These cell types interact with each other in an autocrine and paracrine manner by secreting a large number of bioactive substances known as "adipokines" (adipocyte-derived peptides). For example, adipokines, including adiponectin, leptin, and FGF21 play a major role in regulating whole-body homeostasis by regulating adipose tissue energy storage and beiging of white fat [36]. Adipocytes also secrete a variety of cytokines, including tumor necrosis factor α (TNF-α), interleukin-6 (IL-6), angiotensinogen, resistin, and monocyte chemoattractant protein-1 (MCP-

1) [37] (Fig. 3). These adipocyte-derived cytokines are involved in intra- and inter-cellular communication, influencing numerous processes ranging from adipocyte metabolism and development (adipogenesis, maturation, and death) to whole adipose tissue dynamics (i.e., angiogenesis, inflammation). For instance, MCP-1 promotes the recruitment and activation of macrophages that dispose of excess FFAs and dead/dysfunctional adipocytes [38]. Under pathological circumstances, such as severe or prolonged nutritional oversupply, inflammatory factors (e.g. IL-6, TNFα) derived from macrophages and adipocytes induce a vicious cycle of chronic inflammation, adipocyte lipolysis, and adipocyte dysfunction. This loss of adipose tissue homeostasis and the resulting changes in adipokines and FFAs interfere with numerous metabolic processes such as insulin signaling and glucose uptake, thereby causing insulin resis-

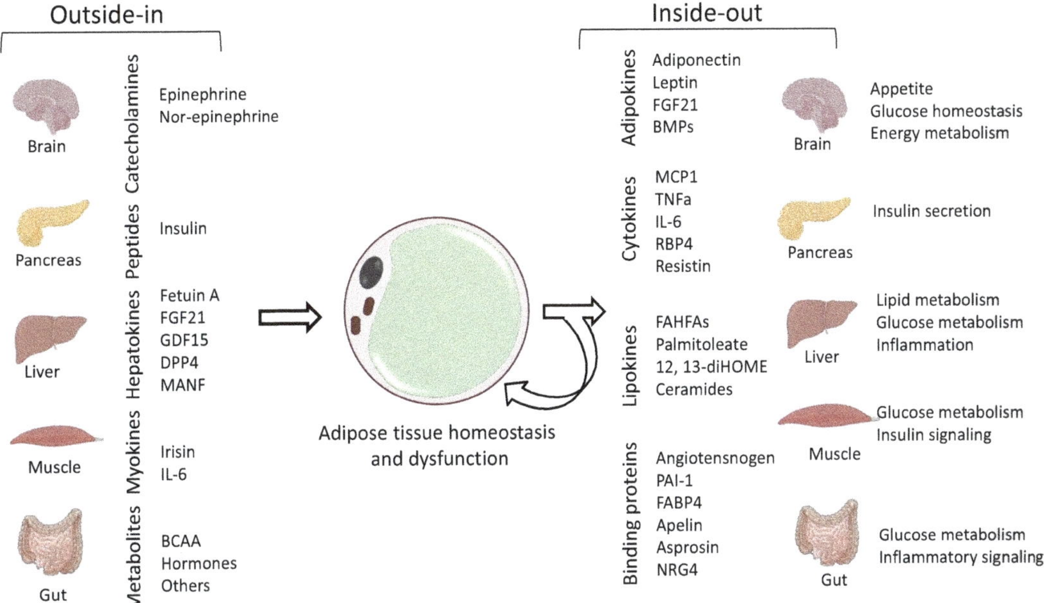

Fig. 3 Interactions between adipose tissue and other tissues. This figure illustrates different classes of adipocyte-derived factors and their varied effects on metabolism. *BCAA* branched-chain amino acids, *NRG4* neuregulin 4, *GDF15* growth differentiation factor 15, *FABP4* fatty acid binding protein 4, *RBP4* retinol binding protein 4, *BMP* bone marrow proteins, *MANF* mesencephalic astrocyte-derived neurotrophic factor. (Figure courtesy of Raja Gopal Reddy Mooli and Sadeesh Ramakrishnan)

tance, glucose intolerance, and other features of the metabolic syndrome [39].

Recent studies show that several lipid mediators known as lipokines, secreted by the WAT and BAT, exert defined physiological effects (Fig. 3). Some documented lipokines include PAHSA, 12,13-dihydroxy- 9Z-octadecenoic acid (12, 13-diHOME), and palmitoleate. PAHSA levels are significantly downregulated in the serum of insulin-resistant mice and humans. On the other hand, PAHSA treatment improves glucose-stimulated insulin secretion and attenuates adipose tissue inflammation [32]. Circulating 12,13-diHOME increases in response to cold in mice and humans, enhancing fatty acid uptake in thermogenic adipocytes [40]. WAT and serum concentrations of the lipokine palmitoleate are significantly decreased in obese humans and mouse models, resulting in decreased adipose tissue de novo lipogenesis (DNL) [41]. Moreover, other classes of lipid species, such as ceramides and phospholipids, are also known to regulate adipose tissue function in an autocrine fashion [42]. For instance, adipose tissue-specific deletion of serine palmitoyl transferase, a key enzyme in the sphingolipid biosynthesis pathway, induces beige adipocyte biogenesis [43]. Thus, adipocyte-derived factors act as signaling molecules regulating adipose tissue homeostasis.

Inside-Out: Endocrine Effects of Adipose Tissue

Adipose tissue also interacts with the rest of the body to orchestrate essential physiological processes (Fig. 3), including energy/metabolic homeostasis, reproductive function, inflammatory responses, and vascular hemodynamics [36] (Fig. 4, left). Adipose tissue communicates with these distant sites through secretory factors (adipokines, lipokines, and cytokines)—making it one of the largest endocrine organs in the body. For example, leptin is a cytokine-like adipokine

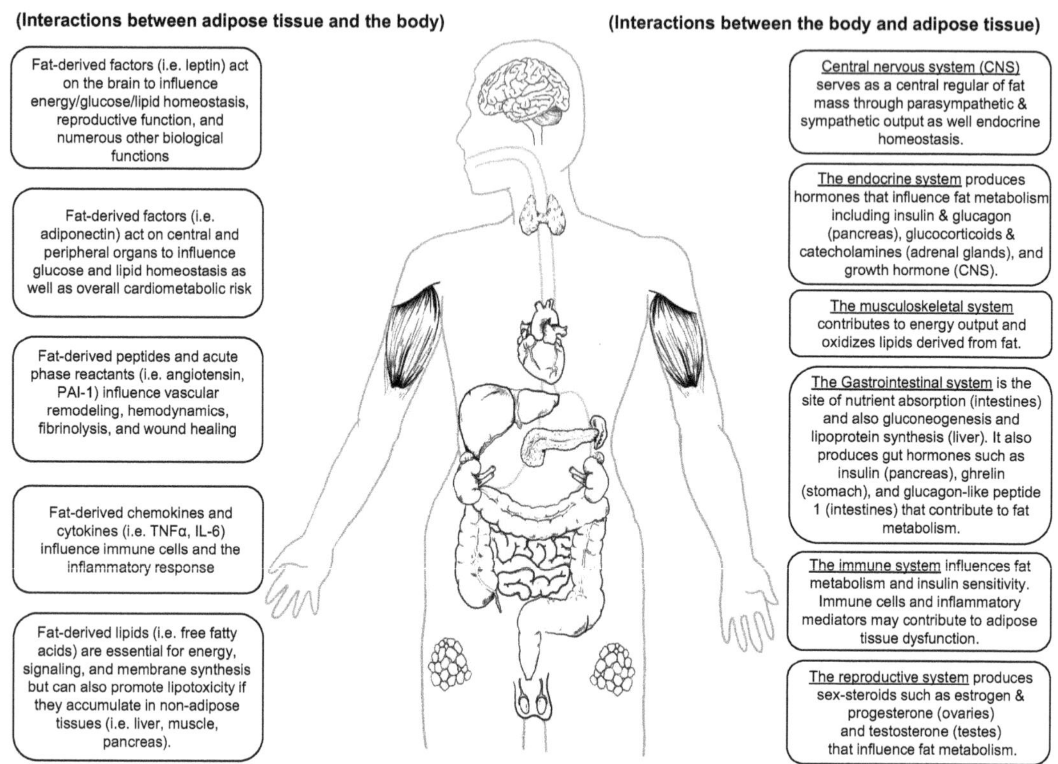

Fig. 4 Functional relationship between adipose tissue and other tissues. The figure illustrates the contribution of adipose tissue in the regulation of non-adipose tissue homeostasis. (Figure courtesy of Erin E. Kershaw and Gianna Paniagua)

that signals the adequacy of adipocyte energy stores to the hypothalamus, where it acts to decrease energy intake, increase energy expenditure, and regulate reproductive function through complex central nervous system circuits that control both neural (i.e., sympathetic and parasympathetic) and endocrine (i.e., gonadal, adrenal, and thyroid axes) output to the whole body [44]. Adiponectin is a multimeric complement-like adipokine that signals a healthy state of adipose tissue and acts at multiple sites via multiple mechanisms to improve cardiometabolic risk [45]. Conversely, the adipose tissue-derived chemokines and cytokines have systemic inflammatory effects that increase cardiometabolic risk by promoting insulin resistance, glucose intolerance, atherosclerosis, and other diseases [39]. In addition to adipokines, adipose tissue-derived lipokines are essential substrates for energy, signaling, and membrane synthesis throughout the body. For example, palmitoleate improves glucose homeostasis by stimulating insulin sensitivity in skeletal muscle [46], suppressing hepatosteatosis by alleviating liver inflammation [47], and reducing atherosclerosis [48]. PAHSA augments insulin and glucagon-like peptide-1 (GLP-1) secretion by stimulating intracellular Ca^{2+} flux through GPR40 [49]. On the other hand, the "spillover" of lipids from adipocytes to other non-adipose tissues, known as "lipotoxicity," causes mitochondrial dysfunction, insulin resistance, and cardiometabolic diseases [50]. For example, certain lipid species, such as ceramides and phospholipids, are increased in obese and insulin-resistant states [51]. Genetic and pharmacological inhibition of their biosynthesis improves insulin sensitivity, liver function, and cardiometabolic functions in mouse models [43, 52]. Moreover, adiponectin (via its receptors adipoR1 and R2) reduces ceramide levels by increasing ceramidase activity [53–55]. Furthermore, neuregulin 4 (NRG4) released from brown adipocytes acts on the liver to decrease lipid accumulation and liver damage [56]. Thus, the endocrine functions of adipose tissue are essential for maintaining whole-body metabolic homeostasis.

Outside-In: Non-Adipose Tissue-Derived Factors Affecting Adipose Tissue Homeostasis

Adipocytes and other cells within adipose tissue express traditional endocrine hormone receptors (i.e., insulin, glucagon, and growth hormone receptors), nuclear hormone receptors (i.e., glucocorticoid, vitamin D, thyroid hormone, androgen, and estrogen receptors), gut hormone receptors (i.e., gastrin and glucagon-like peptide-1 receptors), catecholamine receptors, peptide receptors (i.e., angiotensin II receptors), and FFA receptors (i.e., Toll4 receptors) [8, 36] (Fig. 3). In this way, factors derived from non-adipose tissues, including the liver, muscle, pancreas, intestine, and brain, communicate with adipose tissue to integrate physiological cues. For example, in the setting of increased energy requirements and beiging of white fat, catecholamines (i.e., epinephrine) from the sympathetic nervous system act on adipocyte adrenergic receptors to stimulate lipolysis and promote thermogenesis [57, 58]. Conversely, in the setting of increased energy availability, insulin from the pancreas acts on adipocyte insulin receptors to facilitate glucose uptake and fat synthesis while simultaneously inhibiting fat breakdown and promoting adipogenesis. Interestingly, just as the spectrum of adipocyte-secreted factors varies across adipose tissue depots, so does adipose tissue responsiveness to systemic signals [59]. For example, subcutaneous and breast adipose tissue is responsive to estrogens from the ovaries, whereas visceral adipose tissue is particularly responsive to glucocorticoids from the adrenal glands [60] (Fig. 4, right). Liver and muscle-derived factors, popularly known as hepatokines and myokines,

respectively, play a significant role in regulating adipose tissue homeostasis. For instance, liver-derived factors such as fetuin A, DPP4, FGF-21, and GDF15 regulate adipose tissue inflammation and adipogenesis (Fig. 3). Fetuin A levels are positively associated with adipose tissue inflammation and macrophage accumulation, while FGF21 potently acts as an anti-obesity agent by reducing adipogenesis and promoting thermogenesis [61]. Similarly, myokines such as irisin induce browning of WAT and, thereby, improve thermogenesis and metabolic health [62, 63]. Thus, adipose tissue is a highly adaptive tissue that responds to local and global physiological cues.

The body is exquisitely designed to maintain energy homeostasis and adiposity, and yet the global epidemic of obesity and obesity-associated diseases continues to grow [64, 65]. Numerous intrinsic and extrinsic signals influence fat mass and function, directly by acting on adipocytes or indirectly influencing other peripheral or central organs. Adipose tissue and the central nervous system integrate these signals and communicate with each other via complex neural and hormonal networks to control energy homeostasis and other physiological processes [64]. Generally, genetic and environmental factors have been considered to be the primary determinants of fat mass. Indeed, recent genome-wide association studies suggest that as much as 70% of obesity may be attributed to genetic factors [66]. The main environmental factors contributing to fat mass are diet (energy intake) and exercise (energy expenditure). Lifestyle modification, either by decreasing the former or increasing the latter, promotes weight loss by reducing fat synthesis/storage and/or increasing fat breakdown/oxidation and subsequently improves features of the metabolic syndrome [67]. Various other factors have been implicated in adipose tissue dysfunction, including microorganisms, epigenetics, sleep patterns, pharmacological agents, and endocrine-disrupting chemicals [68, 69]. Despite tremendous progress in understanding adipose tissue biology, many questions remain unanswered.

Conclusions and Future Perspectives

Adipose tissue ("fat") is a highly complex, heterogeneous, and dynamic organ. Both intrinsic and extrinsic factors contribute to the molecular, cellular, and physiological heterogeneity of this complex organ. Thus, gaining mechanistic insights into adipose tissue plasticity and thermogenic capability will help target adipose tissue in our fight against the growing epidemic of obesity, metabolic syndrome, and cardiovascular disease.

Questions and Answers

Question 1 How many types of adipose tissues do mammalians exhibit and what is their role?

Answer 1 Mammalians have at least three types of adipocytes—white, beige, and brown adipocytes. White adipocytes are abundant in white adipose tissue (WAT) and specialized for excess energy storage in the form of lipids such as triglycerides. Beige adipocytes are recruited in WAT under external stimuli such as cold exposure. Brown adipose tissue (BAT) is composed of brown adipocytes. Beige and brown adipocytes improve metabolic health due to their ability to increase whole-body energy expenditure via thermogenesis. Both brown and beige adipocytes share common morphological and biochemical characteristics, including multilocular lipid droplets and dense mitochondria. However, recent studies have identified distinct features between brown and beige adipose tissue, such as their development and function.

Question 2 What are anatomical locations of brown and beige adipocytes?

Answer 2 Classical brown adipose tissue (BAT) is majorly located in the interscapular region, with additional depots located in the cervical, axillary, perivascular, and perirenal regions in mice, infants, and adult humans. Moreover, adult humans possess large amounts of BAT in the paravertebral junctions, along the trachea and vessels, and also in adrenal locations. On the other hand, beige adipose tissue sporadically resides in subcutaneous WAT depots, such as inguinal and anterior subcutaneous WAT in mice. Additionally, studies have isolated beige adipose tissue in paravertebral junctions and perirenal locations in adult humans.

Question 3 What are the various cell types that exist in adipose tissue?

Answer 3 Adipose tissue is composed of various cells, wherein white, beige and brown adipocytes are the major functional cell types in adipose tissue. In addition, adipose tissue is also comprised of adipocyte progenitor cells (APCs), stromal cells, smooth muscle cells, endothelial cells (blood and lymph vessels), and immune cells, including macrophages and T cells. The turnover and proper function of adipose tissue require the coordinated action of different cell types. For example, adipose tissue expansion, beiging, and maintenance of adipocyte number, including de novo adipogenesis, depend on the proper function of APCs. Adipose tissue macrophages play a critical role in removing dead adipocytes by engulfing them. The stromal-vascular fraction has the ability to differentiate into adipocytes. Recent studies have suggested that smooth-muscle-related cells express genes that contribute to beige adipocyte development. Alteration in the plasticity of adipose tissue cell types, while initially adaptive, ultimately contributes to obesity and metabolic diseases.

Question 4 Where is adipose tissue distributed in the body and how does its distribution impact metabolic health?

Answer 4 Adipose tissue is distributed either viscerally or subcutaneously and is known to play a critical role in overall metabolic health. However, adipose tissue distribution is highly variable and is affected by sex, age, development, and genetics, and modified in response to drugs and hormones. For example, increased visceral fat expansion is positively correlated with insulin resistance and risk of cardiometabolic diseases. By contrast, subcutaneous fat expansion is associated with more favorable metabolic health. The concept of fat distribution explains the existence of "metabolically healthy obese" and "metabolically unhealthy normal weight" individuals. The distribution of fat is also affected by adipocyte hypertrophy (increase in fat-cell size) and hyperplasia (increase in fat-cell number). Adipocyte hypertrophy is linked with inflammation, fibrosis, and hypoxia resulting in a metabolic imbalance. On the other hand, hyperplastic growth promotes favorable metabolic health.

Question 5 Why is adipose tissue considered an endocrine organ?

Answer 5 Adipose tissue is considered an essential endocrine organ because it secretes several bioactive factors such as adipokines, cytokines, and lipokines. These adipose tissue-derived factors regulate metabolic health by acting within (autocrine), locally (paracrine), and systemically (endocrine) on the brain, pancreas, liver, and skeletal muscle. Among them, leptin is a well-known adipokine that plays a major role in controlling energy homeostasis by acting on the brain. Moreover, adipokines promote insulin sensitivity (adiponectin), insulin resistance (resistin, RBP4, and lipocalin), and inflammation (TNF-α, IL-6, IL-1β, IL-8, and IL-18). Additionally, lipokines such as palmitoleate, fatty acid esters of hydroxy fatty acids, and ceramides induce beiging in WAT and activation of catabolic enzymes in the liver.

Question 6 What are the hallmarks of adipose-tissue dysfunction?

Answer 6 Adipose-tissue dysfunction is marked by inflammation, hypoxia, reduced angiogenesis, fibrosis, and impaired thermogenesis. For example, adipose tissue macrophages secrete various inflammatory cytokines that promote fibrosis and hypoxia. In addition, excessive adipose tissue expansion and insufficient angiogenesis induce hypoxia, which triggers inflammation. Finally, fibrosis in adipose tissue, particularly APCs, blunts their capacity to expand via hyperplasia and undergo beige adipogenesis. All these processes/mechanisms favor adipose-tissue dysfunction when persists chronically.

References

1. World Obesity Federation (2025) World Obesity Atlas 2025. London: World Obesity Federation.https://data.worldobesity.org/
2. Schwartz MW, Seeley RJ, Zeltser LM et al (2017) Obesity pathogenesis: an Endocrine Society scientific statement. Endocr Rev 38:267–296
3. Smith U, Kahn BB (2016) Adipose tissue regulates insulin sensitivity: role of adipogenesis, de novo lipogenesis and novel lipids. J Intern Med 280:465–475
4. Sakers A, De Siqueira MK, Seale P et al (2022) Adipose-tissue plasticity in health and disease. Cell 185:419–446
5. Chehab FF (2008) Obesity and lipodystrophy--where do the circles intersect? Endocrinology 149:925–934
6. Petersen MC, Shulman GI (2018) Mechanisms of insulin action and insulin resistance. Physiol Rev 98:2133–2223
7. Marcelin G, Ferreira A, Liu Y et al (2017) A PDGFRalpha-mediated switch toward CD9(high) adipocyte progenitors controls obesity-induced adipose tissue fibrosis. Cell Metab 25:673–685
8. Kershaw EE, Flier JS (2004) Adipose tissue as an endocrine organ. J Clin Endocrinol Metab 89:2548–2556
9. Gesta S, Tseng YH, Kahn CR (2007) Developmental origin of fat: tracking obesity to its source. Cell 131:242–256
10. Cinti S (2012) The adipose organ at a glance. Dis Model Mech 5:588–594
11. Garg A (2011) Lipodystrophies: genetic and acquired body fat disorders. J Clin Endocrinol Metab 96:3313–3325
12. Herbst KL (2012) Rare adipose disorders (RADs) masquerading as obesity. Acta Pharmacol Sin 33:155–172
13. Malnick SD, Knobler H (2006) The medical complications of obesity. QJM 99:565–579
14. Chait A, den Hartigh LJ (2020) Adipose tissue distribution, inflammation and its metabolic consequences, including diabetes and cardiovascular disease. Front Cardiovasc Med 7:22
15. Fox CS, Massaro JM, Hoffmann U et al (2007) Abdominal visceral and subcutaneous adipose tissue compartments: association with metabolic risk factors in the Framingham Heart Study. Circulation 116:39–48

16. Smith GI, Mittendorfer B, Klein S (2019) Metabolically healthy obesity: facts and fantasies. J Clin Invest 129:3978–3989
17. Ikeda K, Maretich P, Kajimura S (2018) The common and distinct features of brown and beige adipocytes. Trends Endocrinol Metab 29:191–200
18. Zwick RK, Guerrero-Juarez CF, Horsley V et al (2018) Anatomical, physiological, and functional diversity of adipose tissue. Cell Metab 27:68–83
19. Vijay J, Gauthier MF, Biswell RL et al (2020) Single-cell analysis of human adipose tissue identifies depot and disease specific cell types. Nat Metab 2:97–109
20. Jung SM, Sanchez-Gurmaches J, Guertin DA (2019) Brown adipose tissue development and metabolism. Handb Exp Pharmacol 251:3–36
21. Zhang F, Hao G, Shao M et al (2018) An adipose tissue atlas: an image-guided identification of human-like BAT and beige depots in rodents. Cell Metab 27:252–262.e3
22. Golozoubova V, Hohtola E, Matthias A et al (2001) Only UCP1 can mediate adaptive nonshivering thermogenesis in the cold. FASEB J 15:2048–2050
23. Petruzzelli M, Schweiger M, Schreiber R et al (2014) A switch from white to brown fat increases energy expenditure in cancer-associated cachexia. Cell Metab 20:433–447
24. Wu J, Cohen P, Spiegelman BM (2013) Adaptive thermogenesis in adipocytes: is beige the new brown? Genes Dev 27:234–250
25. Fabbiano S, Suarez-Zamorano N, Rigo D et al (2016) Caloric restriction leads to browning of white adipose tissue through type 2 immune signaling. Cell Metab 24:434–446
26. Yu H, Dilbaz S, Cossmann J et al (2019) Breast milk alkylglycerols sustain beige adipocytes through adipose tissue macrophages. J Clin Invest 129:2485–2499
27. Harms M, Seale P (2013) Brown and beige fat: development, function and therapeutic potential. Nat Med 19:1252–1263
28. Ikeda K, Kang Q, Yoneshiro T et al (2017) UCP1-independent signaling involving SERCA2b-mediated calcium cycling regulates beige fat thermogenesis and systemic glucose homeostasis. Nat Med 23:1454–1465
29. Carpentier AC (2021) 100(th) anniversary of the discovery of insulin perspective: insulin and adipose tissue fatty acid metabolism. Am J Physiol Endocrinol Metab 320:E653–E670
30. Yilmaz M, Claiborn KC, Hotamisligil GS (2016) De novo lipogenesis products and endogenous lipokines. Diabetes 65:1800–1807
31. Zhou P, Santoro A, Peroni OD et al (2019) PAHSAs enhance hepatic and systemic insulin sensitivity through direct and indirect mechanisms. J Clin Invest 129:4138–4150
32. Yore MM, Syed I, Moraes-Vieira PM et al (2014) Discovery of a class of endogenous mammalian lipids with anti-diabetic and anti-inflammatory effects. Cell 159:318–332
33. Bickel PE, Tansey JT, Welte MA (2009) PAT proteins, an ancient family of lipid droplet proteins that regulate cellular lipid stores. Biochim Biophys Acta 1791:419–440
34. Flaherty SE 3rd, Grijalva A, Xu X et al (2019) A lipase-independent pathway of lipid release and immune modulation by adipocytes. Science 363:989–993
35. Langin D (2006) Adipose tissue lipolysis as a metabolic pathway to define pharmacological strategies against obesity and the metabolic syndrome. Pharmacol Res 53:482–491
36. Kahn CR, Wang G, Lee KY (2019) Altered adipose tissue and adipocyte function in the pathogenesis of metabolic syndrome. J Clin Invest 129:3990–4000
37. Karastergiou K, Mohamed-Ali V (2010) The autocrine and paracrine roles of adipokines. Mol Cell Endocrinol 318:69–78
38. Cinti S, Mitchell G, Barbatelli G et al (2005) Adipocyte death defines macrophage localization and function in adipose tissue of obese mice and humans. J Lipid Res 46:2347–2355
39. Gregor MF, Hotamisligil GS (2011) Inflammatory mechanisms in obesity. Annu Rev Immunol 29:415–445
40. Lynes MD, Leiria LO, Lundh M et al (2017) The cold-induced lipokine 12,13-diHOME promotes fatty acid transport into brown adipose tissue. Nat Med 23:631–637
41. Cao H, Gerhold K, Mayers JR et al (2008) Identification of a lipokine, a lipid hormone linking adipose tissue to systemic metabolism. Cell 134:933–944
42. Chaurasia B, Talbot CL, Summers SA (2020) Adipocyte ceramides-the nexus of inflammation and metabolic disease. Front Immunol 11:576347
43. Chaurasia B, Kaddai VA, Lancaster GI et al (2016) Adipocyte ceramides regulate subcutaneous adipose browning, inflammation, and metabolism. Cell Metab 24:820–834
44. Mantzoros CS, Magkos F, Brinkoetter M et al (2011) Leptin in human physiology and pathophysiology. Am J Physiol Endocrinol Metab 301:E567–E584
45. Turer AT, Scherer PE (2012) Adiponectin: mechanistic insights and clinical implications. Diabetologia 55:2319–2326
46. Yang ZH, Miyahara H, Hatanaka A (2011) Chronic administration of palmitoleic acid reduces insulin resistance and hepatic lipid accumulation in KK-ay mice with genetic type 2 diabetes. Lipids Health Dis 10:120
47. Guo X, Li H, Xu H et al (2012) Palmitoleate induces hepatic steatosis but suppresses liver inflammatory response in mice. PLoS One 7:e39286
48. Cimen I, Kocaturk B, Koyuncu S et al (2016) Prevention of atherosclerosis by bioactive palmitoleate through suppression of organelle stress and inflammasome activation. Sci Transl Med 8:358ra126
49. Vijayakumar A, Aryal P, Wen J et al (2017) Absence of carbohydrate response element binding protein in adipocytes causes systemic insulin resistance and impairs glucose transport. Cell Rep 21:1021–1035

50. Shulman GI (2014) Ectopic fat in insulin resistance, dyslipidemia, and cardiometabolic disease. N Engl J Med 371:1131–1141
51. Summers SA (2018) Could ceramides become the new cholesterol? Cell Metab 27:276–280
52. Turner N, Lim XY, Toop HD et al (2018) A selective inhibitor of ceramide synthase 1 reveals a novel role in fat metabolism. Nat Commun 9:3165
53. Hammerschmidt P, Bruning JC (2022) Contribution of specific ceramides to obesity-associated metabolic diseases. Cell Mol Life Sci 79:395
54. Kurek K, Miklosz A, Lukaszuk B et al (2015) Inhibition of ceramide De novo synthesis ameliorates diet induced skeletal muscles insulin resistance. J Diabetes Res 2015:154762
55. Holland WL, Miller RA, Wang ZV et al (2011) Receptor-mediated activation of ceramidase activity initiates the pleiotropic actions of adiponectin. Nat Med 17:55–63
56. Wang GX, Zhao XY, Meng ZX et al (2014) The brown fat-enriched secreted factor Nrg4 preserves metabolic homeostasis through attenuation of hepatic lipogenesis. Nat Med 20:1436–1443
57. Qiu Y, Nguyen KD, Odegaard JI et al (2014) Eosinophils and type 2 cytokine signaling in macrophages orchestrate development of functional beige fat. Cell 157:1292–1308
58. Reitman ML (2017) How does fat transition from white to beige? Cell Metab 26:14–16
59. Wajchenberg BL (2000) Subcutaneous and visceral adipose tissue: their relation to the metabolic syndrome. Endocr Rev 21:697–738
60. Lovejoy JC, Champagne CM, de Jonge L et al (2008) Increased visceral fat and decreased energy expenditure during the menopausal transition. Int J Obes 32:949–958
61. Ameka M, Markan KR, Morgan DA et al (2019) Liver derived FGF21 maintains CORE body temperature during acute cold exposure. Sci Rep 9:630
62. de Oliveira Dos Santos AR, de Oliveira Zanuso B, Miola VFB et al (2021) Adipokines, myokines, and hepatokines: crosstalk and metabolic repercussions. Int J Mol Sci 22:2639
63. Oh KJ, Lee DS, Kim WK et al (2016) Metabolic adaptation in obesity and type II diabetes: myokines, adipokines and hepatokines. Int J Mol Sci 18:E8
64. Flier JS (2004) Obesity wars: molecular progress confronts an expanding epidemic. Cell 116:337–350
65. Malik VS, Willett WC, Hu FB (2013) Global obesity: trends, risk factors and policy implications. Nat Rev Endocrinol 9:13–27
66. O'Rahilly S (2009) Human genetics illuminates the paths to metabolic disease. Nature 462:307–314
67. Magkos F, Yannakoulia M, Chan JL et al (2009) Management of the metabolic syndrome and type 2 diabetes through lifestyle modification. Annu Rev Nutr 29:223–256
68. Grun F, Blumberg B (2009) Endocrine disrupters as obesogens. Mol Cell Endocrinol 304:19–29
69. McAllister EJ, Dhurandhar NV, Keith SW et al (2009) Ten putative contributors to the obesity epidemic. Crit Rev Food Sci Nutr 49:868–913

Metabolic Syndrome

Kana Inoue, Norikazu Maeda,
and Tohru Funahashi

Introduction to Metabolic Syndrome

The metabolic syndrome (MetS) is a cluster of the most dangerous heart attack risk factors, and up to a quarter of the world's adults might have MetS. In general, cardiovascular diseases (CVD) are the most common cause of death in the world. As most patients show no obvious symptoms prior to the first incident, identification of risk factors and early intervention are important. Hypercholesterolemia is a well-established strong risk factor and is a primary target for the prevention of CVD (Fig. 1, see chapter "Hyperlipidemia"). MetS has been identified as the second target. It was previously called syndrome X [1] and insulin resistance syndrome [2], as low insulin sensitivity occurs frequently in this condition (Fig. 1).

Multiple definitions for the diagnosis of MetS exist, yet all require at least three of the following: (1) obesity, especially abdominal obesity (expressed by increased waist circumference); (2) dyslipidemia, expressed by raised serum levels of triglycerides or lowered high-density lipoprotein cholesterol (HDLc); (3) elevated blood pressure; and (4) raised fasting plasma glucose (see also chapters "Diabetes Mellitus") [3]. Moreover, subjects with MetS often show impaired glucose tolerance, insulin resistance, and postprandial abnormalities in lipids.

Abdominal obesity (also termed android obesity [4] or upper body obesity [5]), glucose intolerance, dyslipidemia, and hypertension form "the deadly quartet for CVD" [6]. Studies of abdominal body fat distribution using computed tomography (CT) have revealed that the accumulation of visceral fat in intra-abdominal cavity is more closely related to lipid, glucose, and blood pressure abnormalities rather than absolute body weight or abdominal subcutaneous fat [7]. Thus, the increase in visceral fat area correlates with a cluster of obesity-related risk factors even in mildly obese subjects and is a critical step in MetS development [8]. Finally, from a diagnostic point of view, the waist circumference is an easy-to-measure surrogate marker of visceral fat. MetS usually occurs in association with sedentary lifestyle (overeating and physical inactivity). It has become a global health problem all over the world and is also increasing in Asian countries, where large numbers of people are living. Although the frequency of people with body mass index (BMI) above 30 is still lower in

K. Inoue · N. Maeda
Department of Metabolic Medicine, Graduate School of Medicine, Osaka University, Osaka, Japan
e-mail: kinoue@endmet.med.osaka-u.ac.jp;
norikazu_maeda@endmet.med.osaka-u.ac.jp

T. Funahashi (✉)
Department of Metabolism and Atherosclerosis, Graduate School of Medicine, Osaka University, Osaka, Japan
e-mail: funahashi@endmet.med.osaka-u.ac.jp

231

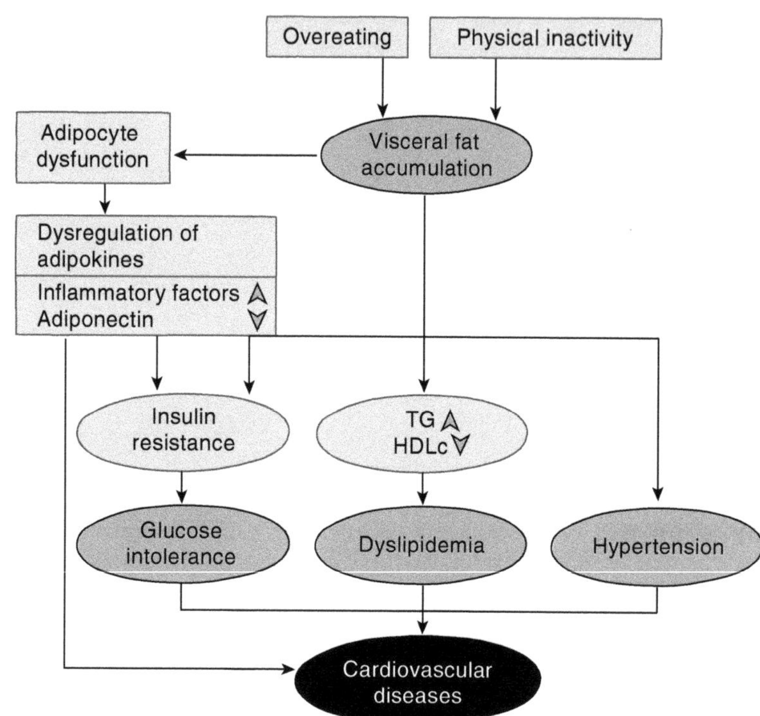

Fig. 1 Changes in metabolic syndrome. Accumulation of visceral fat is associated with a cluster of obesity-related risk factors, including hypertension, dyslipidemia, and glucose intolerance. Overproduction of adipocyte- derived inflammatory factors and reduction in antiatherogenic adiponectin level (summarized as "dysregulation of adipokines") also contribute to the development of cardiovascular disease. *TG,* triglycerides (also called triacylglycerols), *HDLc* high-density lipoprotein cholesterol

Asians compared to Caucasians and Africans, MetS, type 2 diabetes, and CVD have also become a serious health problem in Asia.

Since dyslipidemia, hyperglycemia, and hypertension are common disorders, these risk factors sometimes gather even in a lean individual without visceral fat accumulation. However, MetS is different from a condition randomly clustering multiple risk factors (Fig. 1).

Pathophysiology of the Metabolic Syndrome

Visceral Obesity

Visceral adipose tissue is present in the mesentery and omentum, where innumerable vessels run from the digestive tract to the liver. In response to energy demand, visceral adipose tissue rapidly hydrolyzes triglycerides and delivers the products, free fatty acids (FFA), and glycerol, via the portal vein to the liver, which resynthesizes energy substrates (triglycerides and glucose) for distant tissues (see chapters "Liver:

Overview" under the part "Liver" and "Fat Tissue: Overview" under the part "Fat Tissue") [9]. When the visceral fat accumulates, large amounts of FFA are transferred to liver and systemic circulation, leading to abnormalities in glucose and lipid metabolism, and endothelial dysfunction in MetS.

Another pathogenetic condition in visceral obesity is a dysfunction of adipocytes, especially abnormalities of adipocyte-derived factors, so-called adipokines (see chapter "Fat Tissue: Overview" under the part "Fat Tissue") [10]. Oxidative stress and relative hypoxia due to insufficient blood supply are postulated to cause dysfunction of and damage to hypertrophied adipocytes. The latter secrete various inflammatory adipokines including monocyte chemotactic protein-1 (MCP-1) as an alarm signal, which recruits macrophages into the adipose tissue. Macrophages surround and remove the damaged adipocytes and produce proinflammatory cytokines, such as tumor necrosis factor-α (TNF-α), thus triggering a chronic inflammatory process in the adipose tissue. In addition, hypoxia induces hypoxia-inducible factor 1α, a transcription factor that enhances the expression of plasminogen activator inhibitor 1 in adipocytes. Importantly, in MetS, these proinflammatory and prothrombotic adipokines spread from the adipose tissue to the whole body via the bloodstream, triggering insulin resistance in muscle [11] and thrombus formation in arteries [12]. Adiponectin is a protein specifically produced by adipocytes and abundantly present in plasma. The protein suppresses (1) TNF-α-induced expression of adhesion molecules in vascular endothelial cells, (2) growth factor-induced proliferation of smooth muscle cells, and (3) foam cell transformation of macrophages [12]. In addition to these antiatherogenic activities, adiponectin also has insulin-sensitizing activity and stimulates FFA utilization by activation of AMP-dependent protein kinase (AMPK) and peroxisome proliferator-activated receptor α [13–16]. However, in visceral obesity, plasma levels of adiponectin are decreased. In sum, the dysregulation of adipokines is postulated as a molecular basis of various pathogenetic conditions associated with MetS (Fig. 2) [17].

Dyslipidemia

Triglyceride-rich very low-density lipoprotein (VLDL) is over-synthesized in visceral obesity. Thus, elevation of plasma triglyceride levels is a major feature of MetS.

After hydrolysis of triglycerides, the size of LDLs decreases (see chapter "Hyperlipidemia"). These smaller LDLs are prone to atherogenic oxidative change. Accumulation of VLDL remnants and decrease in HDLc promote atherosclerosis (see below).

Insulin Resistance

Insulin resistance and prediabetic and diabetic conditions are common features of MetS. Impaired metabolism of FFA and dysregulation of adipokines are postulated as mechanisms of insulin resistance in muscle. Muscle cells preferentially metabolize FFA, and glucose uptake is suppressed when the plasma level of FFA is elevated. Additionally, proinflammatory adipokines, such as TNF-α, inhibit insulin signaling in muscle and adiponectin expression. Adiponectin normally activates AMPK and increases FFA utilization and thus acts as insulin-sensitizing hormone. However, in MetS, a decrease in adiponectin leads to insulin resistance. In turn, insulin resistance in muscle causes postprandial hyperinsulinemia. Subsequent rise in fasting blood glucose concentration and hyperinsulinemia occurs due to hepatic insulin resistance. In this condition, suppression of gluconeogenesis in the fasting state is impaired. The influx of large amounts of FFA from accumulated visceral fat to the liver and the decrease in adiponectin may contribute to hepatic insulin resistance.

Hypertension

Hyperinsulinemia promotes urinary reabsorption of sodium chloride in the kidney. Thus, subjects with MetS sometimes develop salt-sensitive hypertension.

Additionally, adipocytes synthesize angiotensinogen (see chapter "Kidney: Overview" under the part "Kidney"), and consequently its levels are high in obesity, further contributing to hypertension.

Elevation of plasma FFA causes vascular endothelial dysfunction, resulting in an impairment of vasodilation activity. Increased sympathetic activity is also reported in subjects with MetS.

Atherosclerosis

Ultimately, MetS increases the risk of atherosclerosis. Hyperglycemia, hypertension, and increased FFA can damage the vascular endothelium (see chapter "Diabetes Mellitus"). Increased levels of VLDL remnants, small-sized LDLs, and low HDLc levels promote the formation of lipid-rich atheromatous plaques. Hypoadiponectinemia, hyperinflammatory-cytokinemia, and high levels

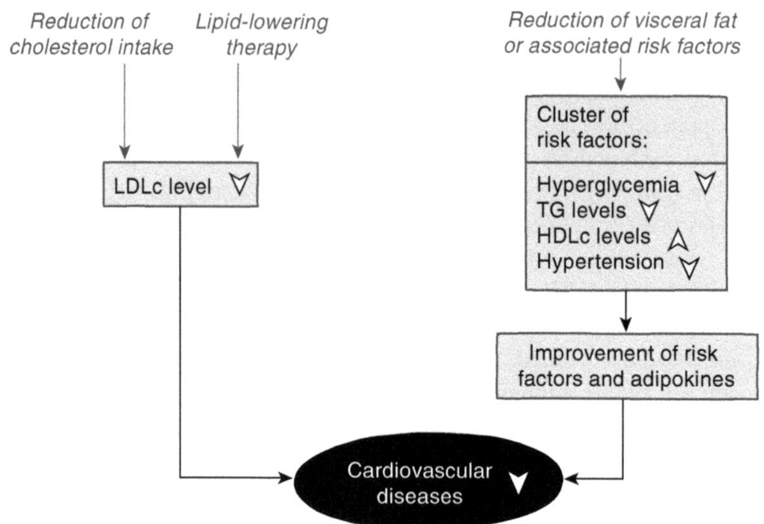

Fig. 2 Effects of treatment of metabolic syndrome on metabolism. Treatment of the multiple risk factors involved in metabolic syndrome apart from increased low-density lipoprotein cholesterol (*LDLc*, left side) aims to reduce visceral fat or directly targets individual risk factors. Reduction in visceral fat improves dysregulation of several adipokines and the whole cluster of risk factors. LDLc levels can be targeted by reducing cholesterol intake and lipid-lowering therapies. *TG* triglycerides (also called triacylglycerols); *HDLc* high-density lipoprotein cholesterol

of thrombotic factors accelerate the atherogenic process and trigger the rupture of plaques with the possibility of a lethal outcome.

Treatment of Metabolic Syndrome and Impact on Metabolism

Since MetS subjects often have type 2 diabetes [18], dyslipidemia, and hypertension, physicians may prescribe medicines according to these conditions (see chapters "Diabetes Mellitus" and "Hyperlipidemia"). As visceral fat accumulation is a key factor in the progression of MetS and underlies multiple risk factors, the primary management strategy is to reduce accumulated visceral fat [17]. Overwhelming evidence indicates that weight reduction alleviates glucose intolerance, hypertriglyceridemia, hypo-HDL-cholesterolemia, and hypertension. Reduction in visceral fat ameliorates the cluster of multiple CVD risk factors simultaneously (Fig. 2). As such, a change in lifestyle is the first-line treatment in MetS. This generally includes cardiovascular exercise, generally increased physical activity, and a restricted calorie intake.

To date, there is no single pharmacological treatment available, specifically tailored for the entire cluster of multiple disturbances involved in MetS.

Perspectives

With increasing comforts in the world, the average lifespan increased. However, reduced physical activity and excessive intake of nutrition lead to increased visceral fat and obesity causing type 2 diabetes, hypertension, dyslipidemia, and CVD. A strategy to combat and prevent MetS is critically required. Therefore, education and promotion of healthy lifestyle remain our most important and effective tool to prevent and treat MetS.

References

1. Reaven GM (1988) Banting lecture 1988. Role of insulin resistance in human disease. Diabetes 37:1595–1607
2. DeFronzo RA, Ferrannini E (1991) Insulin resistance. A multifaceted syndrome responsible for NIDDM, obesity, hypertension, dyslipidemia, and atherosclerotic cardiovascular disease. Diabetes Care 14:173–194
3. Alberti KG, Eckel RH, Grundy SM, Zimmet PZ, Cleeman JI, Donato KA, Fruchart JC, James WP, Loria CM, Smith SC Jr (2009) Harmonizing the metabolic syndrome. Circulation 120:1640–1645
4. Vague J (1947) La différenciation sexuelle: facteur déterminant des formes de l'obésité. Presse Med 30:339–340
5. Kissebah AH, Vydelingum N, Murray R, Evans DJ, Hartz AJ, Kalkhoff RK, Adams PW (1982) Relation of body fat distribution to metabolic complications of obesity. J Clin Endocrinol Metab 54:254–260
6. Kaplan NM (1989) The deadly quartet. Upper-body obesity, glucose intolerance, hypertriglyceridemia, and hypertension. Arch Intern Med 149:1514–1520
7. Matsuzawa Y, Fujioka S, Tokunaga K, Tarui S (1992) Classification of obesity with respect to morbidity. Proc Soc Exp Biol Med 200:197–201
8. Teramoto T, Sasaki J, Ueshima H, Egusa G, Kinoshita M, Shimamoto K, Daida H, Biro S, Hirobe K, Funahashi T, Yokote K, Yokode M (2008) Metabolic syndrome. J Atheroscler Thromb 15:1–5
9. Maeda N, Funahashi T, Shimomura I (2008) Metabolic impact of adipose and hepatic glycerol channels aquaporin 7 and aquaporin 9. Nat Clin Pract Endocrinol Metab 4:627–634
10. Funahashi T, Matsuzawa Y (2007) Metabolic syndrome: clinical concept and molecular basis. Ann Med 39:482–494
11. Hotamisligil GS, Shargill NS, Spiegelman BM (1993) Adipose expression of tumor necrosis factor-alpha: direct role in obesity-linked insulin resistance. Science 259:87–91
12. Shimomura I, Funahashi T, Takahashi M, Maeda K, Kotani K, Nakamura T, Yamashita S, Miura M, Fukuda Y, Takemura K, Tokunaga K, Matsuzawa Y (1996) Enhanced expression of PAI-1 in visceral fat: possible contributor to vascular disease in obesity. Nat Med 2:800–803
13. Matsuzawa Y, Funahashi T, Kihara S, Shimomura I (2004) Adiponectin and metabolic syndrome. Arterioscler Thromb Vasc Biol 24:29–33
14. Kadowaki T, Yamauchi T (2011) Adiponectin receptor signaling: a new layer to the current model. Cell Metab 13:123–124
15. Ouchi N, Walsh K (2012) Cardiovascular and metabolic regulation by the adiponectin/C1q/tumor necrosis factor-related protein family of proteins. Circulation 125:3066–3068
16. Turer AT, Scherer PE (2012) Adiponectin: mechanistic insights and clinical implications. Diabetologia 55:2319–2326
17. Kishida K, Funahashi T, Matsuzawa Y, Shimomura I (2012) Visceral adiposity as a target for the management of the metabolic syndrome. Ann Med 44:233–241
18. Inoue K, Maeda N, Kashine S, Fujishima Y, Kozawa J, Hiuge-Shimizu A, Okita K, Imagawa A, Funahashi T, Shimomura I (2011) Short-term effects of liraglutide on visceral fat adiposity, appetite, and food preference: a pilot study of obese Japanese patients with type 2 diabetes. Cardiovasc Diabetol 10:109

Lung

Lung: Overview

Martin Zeeb, Andreas Schnapp,
and Michael Paul Pieper

Anatomy and Physiology of the Lung

The lung is the human body's respiratory organ, responsible for supply of all tissues and organs with vital oxygen (O_2) and for disposal of carbon dioxide (CO_2), the end-product of internal respiration and of several metabolic pathways.

The lung, located in the upper thorax, consists of two parts, the right lung further comprising three lobes and the left lung comprising two. Left and right lungs are individually surrounded by a pleural cavity, which consists of two pleurae and the cavity in between. The parietal pleura lines the rib cage, whereas the visceral pleura covers the surface of the lungs. The space between the pleurae is filled with pleural fluid. The pleurae are critically important for breathing motions (see below).

The lungs represent the functional anatomical part of the respiratory system consisting of the upper and lower respiratory tract, also called upper and lower airways (Fig. 1a). The upper airways comprise nasal cavity, pharynx (including oropharynx and laryngopharynx), and larynx. Their role is to warm and moisten the inhaled air and to protect from noxious agents, mainly filtering them via the nasal turbinates. The nasal cavity also allows smelling. The pharynx is part of both digestive and respiratory systems (see chapter "Gastrointestinal Tract: Overview") and offers an alternate route of air supply via the mouth. The larynx contains the vocal cords (vocal folds), necessary for human speech. It continues into the lower respiratory tract.

The lower respiratory tract consists of the trachea and the lungs. The trachea bifurcates first into primary bronchi, subsequently into bronchioles, and finally into terminal and respiratory bronchioles, which ultimately give rise to alveolar ductus and sacs. Airways smaller than 2 mm in diameter are referred to as small airways. While small in diameter, their volume makes up >95% of the lung.

The main function of the lung is to allow gas exchange of inhaled air with the blood, which takes place in respiratory bronchioles and subsequent lung regions and is most pronounced in the alveoli.

Histologically, the wall of the conducting airways (trachea, bronchia, and bronchioles) consists of mucosa, submucosa, underlying cartilage, and adventitia. The mucosa contains a mucus layer (see below) and pseudostratified columnar,

M. Zeeb (✉)
Corporate Publishing, Springer Medizin Verlag GmbH, Neu-Isenburg, Germany
e-mail: martin.zeeb@springernature.com

A. Schnapp · M. P. Pieper
Immunology and Respiratory Diseases Research, Boehringer Ingelheim Pharma GmbH & Co KG, Biberach (Riss), Germany
e-mail: andreas.schnapp@boehringer-ingelheim.com; michael_paul.pieper@boehringer-ingelheim.com

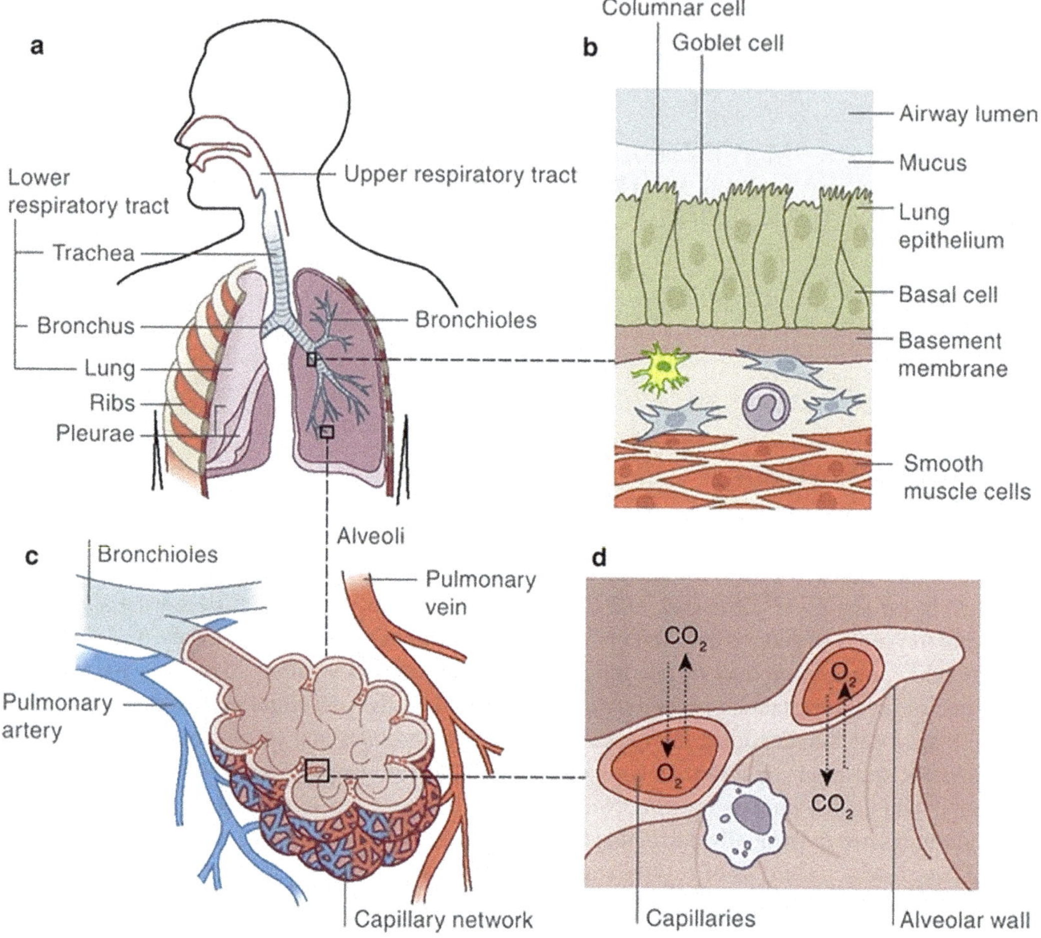

Fig. 1 Anatomy of the respiratory system. (**a**) Macroscopic overview of upper and lower respiratory tract. (**b**) Magnification of a section through a bronchial wall showing the mucosa and the underlying tunica muscularis. In the mucosa, basal, ciliated columnar cells, and mucus-secreting goblet cells are shown. A submucosal gland is not included due to space constraints albeit the glands are present in bronchia. Whereas goblet cells, glands, and elastic fibers decrease in number concomitant with the length of the cilia towards the smaller airways, the muscle layer transiently increases. (**c**) Anatomy of an alveolar sac showing the approximate arrangement of individual alveoli and the surrounding dense capillary network. Note that in the lung vasculature, arteries carry oxygen-deficient blood, whereas veins carry oxygen-rich blood (shown in *blue* and *red*, respectively). (**d**) Detailed view of a section through an alveolar wall showing thin respiratory surface, movement of gases, and an alveolar macrophage

ciliated epithelium with mucus-secreting goblet cells, columnar cells, and basal cells (Fig. 1b). Underneath, it harbors the lamina propria, a layer of connective tissue, containing large amounts of elastin. The submucosa contains submucosal glands, which are connected to the mucosa and also contribute to mucus secretion. In addition, this submucosal layer includes fibroblasts, dendritic cells, and neutrophils, the latter two of which participate in host defense. C-shaped rings of hyaline cartilage provide a semirigid outside support to prevent collapse of the airway during inspiration. Finally, the adventitia, a fibroelastic connective tissue layer, confines the airways towards the outside.

As the bronchi decrease in diameter, a gradual transition in the architecture of the epithelium can be observed. The epithelial layer transforms from pseudostratified, columnar epithelium towards a ciliated, simple columnar, and finally simple cuboidal epithelium (the shape of a typical respiratory epithelium). Concomitantly, a gradual decrease in the number of goblet cells as well as submucosal glands is seen, whereas Club cells emerge. Club cells secret protective substances that are similar to alveolar surfactant (see below), counteract inhaled harmful substances, and may act as stem cells. Furthermore, the amount of elastic tissue decreases and a muscularis mucosae begins to take shape between the lamina propria and the submucosa. The cartilage skeleton is absent from the bronchioles and smaller airways.

Alveoli (Fig. 1c, d) feature an extremely thin wall of specialized respiratory epithelium to facilitate gas exchange. Most of the alveolar epithelium is covered by small, squamous-like type I pneumocytes. It also includes larger cuboidal type II pneumocytes that produce surfactant (see below) and can transdifferentiate into type I pneumocytes. Alveolar macrophages in the alveolar lumen scavenge particulate matter and microorganisms.

In order to perform its function, the lung is extensively vascularized; the alveoli are surrounded by a dense capillary network (Fig. 1c) that can be compared to a columnar hallway. Due to its potential volume and sophisticated regulation, the pulmonary capillaries can act as a large blood reservoir.

The ~300 mio alveoli, each 0.2–0.3 mm in diameter, sum up to a functional area of gas exchange of up to 140 m^2. The alveolar wall, consisting of alveolar epithelium, minimal interstitium, and endothelium (Fig. 1d), totals to less than 1 µm.

Mechanism of Breathing

Breathing can be initiated by two similar actions: upward rib movement (chest breathing) or downward movement of the diaphragm (abdominal breathing). As the outer pleura is physically connected to both, it will follow the movement passively. Consequently, during inspiration, the existing low pressure (below atmospheric pressure) in the intrapleural cavity is further decreased and will cause the inner pleura and eventually the lung tissue to follow (as interpleural fluid cannot be expanded). This effectively increases the lung volume and results in an inward movement of air (inhalation). Consequently, under the pathophysiological condition of a pneumothorax, air leaks into the pleural space after puncture thereby effectively hindering inhalation. In contrast to inhalation, exhalation occurs mainly passively as the lung has the intrinsic tendency to retract due to surface tension in the alveoli and distension of its elastic fibers in the lung tissue. The total lung capacity can amount up to 6 l of air. Yet, even after maximal exhalation, a residual volume of 1.25 l will remain, reflecting the volume of the conducting airways, which do not participate in gas exchange [1].

Any change leading to an obstruction of the airways, like those in asthma and chronic obstructive pulmonary disease (COPD) will have detrimental effects. Ventilation deficiencies can be divided into restrictive and obstructive disorders. The former describes a disturbed expansion of the airways (accompanied by reduced total capacity), whereas the latter describes difficulties in conduction and increased inflow or outflow resistance.

The critical factors of breathing are delivery of oxygenated air (ventilation), passive gas exchange with the blood via diffusion, and propulsion of blood (convection) all of which aim to create a stable and steep gradient of partial pressure of oxygen (pO_2) and pCO_2 over the alveolar membrane (see below) [2].

It should be noted that gas exchange with the blood never reaches complete equilibration. Whereas the inhaled gas mixture contains 21% O_2 and around 0.05% CO_2, the exhaled gas mixture still contains 14–16% O_2 and around 4% CO_2.

The respiratory quotient is the ratio of exhaled CO_2 to consumed O_2 (CO_2/O_2). It is indicative of the primary metabolic fuel (carbohydrates or fatty acids) at the moment of measurement. As fatty acids require more oxygen for complete oxidation of a single carbon atom, preferential metabolism of fatty acids decreases the respiratory quotient [3]. Consequently, measuring the respiratory quotient can help to detect pathological biochemical states.

Lung-Specific Metabolic/Molecular Pathways and Processes

Mucus Production

Mucus is secreted from goblet cells and submucosal glands in the bronchial epithelium as a viscous fluid containing mucins, water, ions, and antimicrobial substances such as immunoglobulin A (IgA, see also chapter "Anatomy and Physiology of the Immune System") and lysozyme. Mucins are heavily glycosylated proteins of high molecular weight.

Mucus is essential to trap inhaled particles and microorganisms and prevent them from reaching and damaging the respiratory epithelium (Fig. 1). Mucus containing bacteria, debris, and inflammatory cells and products is referred to as phlegm. Rhythmic upward propulsion by the cilia on the bronchial epithelium transports mucus and captured particles towards the larynx, where most of it is swallowed or expectorated. Cough is a physiological reflex that aids to eject particles and sputum (coughed-up mucus) or phlegm that would otherwise damage or block the airways. Damage to cilia, for example, by inhalation of cigarette smoke, causes accumulation of mucus and increased ventilatory resistance resulting in obstructive pathology.

Surfactant Production and Function

Alveoli are subject to dramatic surface tension resulting from the attraction between the molecules of the fluid film that covers the alveolar cell surface. This force supports exhalation but would drive all liquids within the airways to retract, the lung to collapse, and the alveoli to merge into larger vesicles (prohibiting functional gas exchange). However, the stability of the alveoli is ensured by surfactant, a surface-active agent decreasing surface tension and lubricating the alveolar epithelium. Surfactant is a mixture of proteins and lipids, mainly consisting of lecithin derivatives. Its main lipid components are dipalmitoylphosphatidylcholine (DPPC) molecules that tend to repel each other, thus counteracting the surface tension of the fluid film on the alveolar cell surface; surfactant also contains $\sim 40\%$ of other phospholipids and $\sim 5\%$ surfactant-associated proteins and is secreted by type II alveolar epithelial cells.

Gas Exchange

Gas exchange across the alveolar wall occurs by passive diffusion; a process that can be described by Fick's law of diffusion. According to the law, a large exchange area, short diffusion distance, and a constant gradient of substances (across the membrane) facilitate diffusion. All these prerequisites are provided by the unique setup of the alveoli (Fig. 1c, d). Nature of the gas also influences diffusion rate, with CO_2 diffusing more freely over the alveolar membrane than O_2.

The substance gradient is expressed by the partial pressure of gases (pO_2, pCO_2) in the alveolar lumen and in the blood. In order to provide a constant gradient that is as steep as possible, constant and coordinated ventilation and perfusion are of utmost importance. The ratio between ventilation and perfusion, called the ventilation-perfusion coefficient (V_A/Q), is relatively constant (0.8–1) among healthy subjects to secure a proper partial pressure gradient and thus gas exchange but can change significantly in some pathologies, for example, when pulmonary shunts are created by perfusion of non-ventilated alveoli (see chapter "Community-Acquired Pneumonia").

Hemoglobin helps to maintain the gradient by binding of O_2 and thus effectively removing it from the pool of free O_2. Interestingly, the binding capacity of hemoglobin for O_2 is dependent on pO_2, with the high partial pressure in the lungs increasing its binding affinity (see chapter "Blood: Overview"). Consequently, the contact time of 0.3–0.7 s between an individual erythrocyte and the alveolar wall is already sufficient for saturation of hemoglobin with O_2.

Outside-In: Metabolites of Other Tissues Affecting the Lung

Regulation of Breathing

Regulation of breathing is mediated primarily via the sympathetic and parasympathetic nervous system. Sympathetic activity causes relaxation of smooth muscle cells and thus bronchodilation supporting inspiration mostly via β2 adrenergic receptors. Parasympathetic activity can constrict the airways, a feature used during exhalation. Overactivation of the parasympathicus often causes pathological constriction. However, constriction is more commonly caused by local inflammation, as occurs during asthma and COPD.

The central nervous control center of breathing is represented by neural oscillators located in the medulla oblongata, called the ventral respiratory group. These neurons trigger breathing autonomously but are adjustable and react to systemic need and metabolism (as indicated by the blood gas status). Detection of the blood gas status occurs mainly by arterial chemosensors in the carotid artery (glomus caroticum) and in the brainstem. Among the three prime indicators triggering breathing (i.e., reduced pO_2, increased pCO_2, and increased H^+/decreased pH), increased pCO_2 is the most important signal. There is an almost linear correlation between pCO_2 and breathing induction. At high concentrations, however, CO_2 acts narcotic and may reduce breathing.

The tight chemosensation of pCO_2 has important metabolic implications. Upon hyperventilation (increased/frequent breathing), pCO_2 can be drastically reduced, reducing or even completely inhibiting further breathing. The latter can lead to hypoxemia (reduced pO_2 in the blood) and might remain undetected until unconsciousness occurs. Additionally, hypoxemia-induced breathing lowers pCO_2, which will in turn decrease breathing activity more efficiently than required based on

pO_2 status. Dramatic alterations of blood gases can occur in divers or at high altitudes [4, 5].

Increased H^+ is compensated by hyperventilation, causing an increased exhalation of CO_2 as pCO_2 and H^+ are interconnected via carbonic acid formation and dissociation (see chapter "Blood: Overview"). By this mechanism, the lung is a major regulator of blood pH along with the kidneys (see below). Again, the expected response to high H^+ is stymied by the "more effective" response to altered pCO_2.

Breathing and Pathologies

Observation of breathing patterns can hint at specific underlying pathologies and disturbances.

For example, in altitude sickness, reduced pO_2 in the atmosphere causes increased heart rate and deep and more frequent breathing (to incorporate sufficient O_2). However, this effectively lowers pCO_2 in the blood causing decreased breathing, which again causes hypoxemia, or even apnea (lack of breathing). The resulting periodic breathing pattern is known as Cheyne-Stokes breathing.

Decreased atmospheric pressure at high altitudes can cause pulmonary edema. If the intrapulmonary pressure is too low compared to the blood pressure, plasma fluid leaks into the alveoli. More frequently though, pulmonary edema is caused by left ventricular heart failure causing blood accumulation in the pulmonary vessels that lead to fluid leakage into the alveoli. In both cases, patients present with superficial breathing, hypoxemia, and orthopnea.

In contrast, hypercapnia (abnormally increased pCO_2 in the blood) and acidosis are characterized by extremely deep breathing called Kussmaul breathing (see below).

Breathing and Treatment

Drug delivery via the lungs, for example, using inhalation devices, is mainly used to treat respiratory disorders topically. Bronchodilatory agents like anticholinergics and β2 adrenoceptor agonists as well as anti-inflammatory agents like inhaled steroids are used in certain obstructive airways diseases. Inhaled delivery of compounds to achieve systemic effects has been utilized to treat diabetes mellitus, multiple sclerosis and others [6]. Inhaled drug delivery harnesses immediate contact to mucosal surfaces and easy translocation to the blood across the alveolar membrane [7]. However, inhaled systemic treatments have not generally replaced the gold standard therapies.

Inside-Out: Metabolites of the Lung Affecting Other Tissues

Function Under Extreme Conditions

Hypercapnia not only increases breathing but also causes arousal and initiates head turning during sleep. It can be caused by lung function deficits (as in several lung diseases), breathing depression, inhalation of increased concentrations of CO_2 (as in rebreathing), and long-term artificial respiration. It is often accompanied by respiratory acidosis due to the formation of carbonic acid (H_2CO_3) and subsequent dissociation to H^+ and bicarbonate (HCO_3^-).

Respiratory acidosis (blood pH <7.35) causes clinical symptoms, such as blue, cyanotic lips; increased heart rate (tachycardia); and pulmonary hypertonia (Fig. 2). Respiratory acidosis is commonly corrected by H^+ excretion in the kidney (see chapter "Kidney: Overview").

Acidosis can also occur due to increased anaerobic glycolysis (causing increased lactic acid concentrations) and ketone body synthesis that may result in ketoacidosis (due to increased acetoacetic acid and 2-hydroxybutyric acid concentrations). This metabolic acidosis occurs during shock or, more commonly, deregulated diabetes (coma diabeticum, see chapter "Diabetes Mellitus"). Other forms of metabolic acidosis originate from decreased H^+ excretion (as occurs

in kidney disease, increased HCO_3^- excretion, or acid intoxication (e.g., by acetylsalicylic acid).

If acidosis exceeds the buffer capacity of the blood, it can be fatal. In general, metabolic deregulation is antagonized by respiratory mechanisms, and vice versa. Thus, severe metabolic acidosis is characterized by Kussmaul breathing, a deep and gasping breathing pattern, which is a sign of life-threatening conditions.

Prolonged reduced oxygen levels (hypoxemia) trigger release of erythropoietin (EPO) via hypoxia-inducible factor (HIF)-1α. This increases production of hemoglobin and thus facilitates oxygen transport (see chapter "Blood: Overview"), for example, during adaptation to high altitudes [8].

Other Functions

The lung also performs functions different from gas exchange. For example, lung endothelium synthesizes angiotensin-converting enzyme (ACE) that converts angiotensin I to angiotensin II (Fig. 2). This conversion is of critical importance to blood pressure homeostasis (see chapter "Kidney: Overview" under the part "Kidney) and a protective role of ACE in the lung has been reported [9]. Finally, the lungs create the necessary air movement across the vocal folds that allows for speech.

Interaction with Other Organs

The function of the lung needs to be coordinated with other organs in particular with the cardiovascular system. For example, increased breathing is only useful if heart rate and blood convection are also increased. Coordination with sensory reactions, such as speech and cough reflex (see above), is also mandatory. These complex coordinating responses involve multiple centers in the central nervous system (CNS) [10].

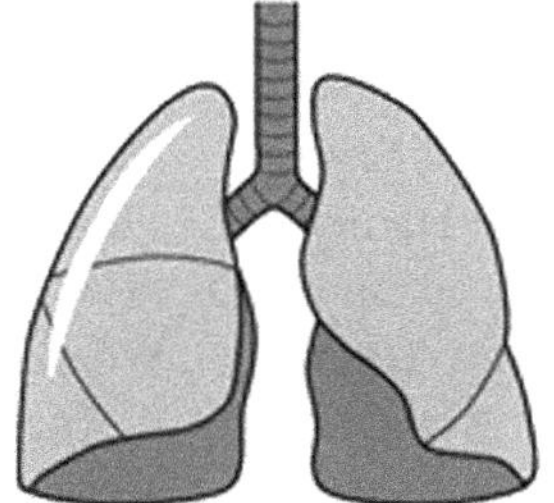

Fig. 2 Specific metabolism of the lung and interaction with other organs and tissues. *pCO₂* partial pressure of CO₂ in the blood, *EPO* erythropoietin, *ACE* angiotensin-converting enzyme, *AT* angiotensin

Inside-In: Metabolites of the Lung Affecting Itself

Lung Perfusion and Usage of Alveoli

As mentioned above, the ventilation-perfusion coefficient is critical for gas exchange. Concomitantly, only ~50% of alveolar capillaries are perfused at rest as only a similar amount of alveoli is ventilated. Increased oxygen demand increases perfusion and activates reserve capillaries. Simultaneously, breathing is intensified to ventilate corresponding alveoli (via central regulatory mechanisms). While standing, the base of the lung is much more perfused than the tips, due to a gradient in hydrostatic blood pressure.

Perfusion is also regulated locally, as less ventilated alveolar regions cause local vaso-constriction (called hypoxic pulmonary vaso-constriction) to prevent perfusion of non-ventilated alveoli and shunting of deoxy-genated blood to the heart. This mechanism is known as the Euler-Liljestrand mechanism and is driven by local hypoxia (Fig. 2). Hypoxia is sensed by oxygen-sensitive potassium channels, which close and thus cause depolarization of smooth muscle cells leading to Ca^{2+} influx and vasoconstriction. However, local release of vasoactive mediators in disease (e.g., during inflammation) may counteract this mechanism (see chapter "Community-Acquired Pneumonia").

Reflexes and Internal Regulation

Several internal reflexes are directed at pulmonary protection. These originate from the lung and are mediated by the nucleus tractus solitarius in the CNS. A major mechanism is the Hering-Breuer inflation reflex, which is induced upon deep inhalation and aims to prevent overstretching of lung tissue. In reflexive manner, it triggers exhalation (Fig. 2). Another protective mechanism is the reflex induction of coughing in response to inhaled particles, fluids, or noxious gases.

Final Remarks

The lungs are critical regulators but also effectors of human metabolism, as they are both starting and end-point of internal respiration. By supplying O_2 and removing CO_2, the lungs are implicated in virtually all metabolic pathways of which energy metabolism surely is the most influential. Due to the unique composition and organization of the conducting and respiratory tracts that permit its efficient function, the lung is also susceptible to both extrinsic and intrinsic pathologies. Microorganisms, particles, and toxic gases can easily enter with the airflow, and pathological disturbance of the delicate balance between ventilation, diffusion, and convection has an immediate and sometimes dramatic effect on respiratory function and associated metabolism.

Questions and Answers

1. What are the lung regions that actively participate in functional gas exchange?

   Only the branches of the terminal bronchioles, the respiratory bronchioles, as well as the alveoli. Gas exchange is most pronounced in the alveoli.

2. What does the structure of alveolar vasculature look like? How is perfusion regulated locally?

   The pronounced vascularization of the alveoli resembles a columnar hallway.

Reduced amounts of oxygen (hypoxia) in less ventilated alveoli cause local vasoconstriction (and thus reduced perfusion). This mechanism is known as the Euler-Liljestrand mechanism and is mediated by oxygen-sensitive potassium channels.

3. What is the diameter of the alveolar wall? What is the significance of that value?

   The alveolar wall is only 0.2–0.3 μm in diameter. The smaller the distance between blood vessels and alveolar lumen, the easier the diffusion of gases. Short distances favor gas exchange.

4. How and why does the respiratory quotient change in relation to the metabolic fuel?

   The respiratory quotient is the ratio of produced carbon dioxide to consumed oxygen. As different metabolic fuels require higher or lower amounts of oxygen for complete oxidation, this ratio changes. Lipids are the least oxygenated molecules among the standard dietary components (resulting in the lowest respiratory quotient), whereas carbohydrates are most oxidized to begin with. Proteins range in between.

5. What is the function of mucus?

   Mucus is a major defense mechanism. It traps inhaled particles (due to its viscous nature) and contains antimicrobial substances (mostly IgA) to fight off invading bacteria, viruses etc.

6. Which forces and components are relevant for the stability of the alveoli?

   On the one hand, surface tension of the liquids lining the inner alveolar wall aim to decrease the surface area aiding in expiration. However, unimpeded, this would cause alveoli to collapse and coalesce. The stability is ensured by surfactant, a lubricating fluid that reduces surface tension thanks to its phospholipid ingredients.

7. What is the main regulator of breathing induction? What are potential disadvantages of this tight regulation?

   An increased level of pCO_2 in the blood is the main inductor of breathing. The CO_2 level is detected in the carotid artery and the brainstem. Usually, there is a balance between pCO_2 and pO_2 in the blood. If one is high, the other tends to be lower. However, if pCO_2 is low despite a reduced level of O_2 in the blood (e.g., due to hyperventilation), the decreased pCO_2 could limit or even prevent breathing causing a detrimental hypoxemia.

References

1. Thews G (2000) Lungenatmung. In: Schmidt RF, Thews G, Lang F (eds) Physiologie des Menschen, 28th edn. Springer, Heidelberg, pp 565–590., German
2. West JB (2011) Respiratory physiology: the essentials, 9th edn. Lippincott Williams & Wilkins, Philadelphia/London
3. Levitzky M (2013) Pulmonary physiology, 8th edn. McGraw-Hill Medical, New York
4. Francis TJR, Mitchell SJ (2003) Pathophysiology of decompression sickness. In: Brubakk AO, Neuman TS (eds) Bennett and Elliott's physiology and medicine of diving, 5th edn. Saunders, Philadelphia, pp 530–556
5. Grocott MP, Martin DS, Levett DZ, McMorrow R, Windsor J, Montgomery HE (2009) Arterial blood gases and oxygen content in climbers on Mount Everest. N Engl J Med 360:140–149
6. Cipolla D (2016) Will pulmonary drug delivery for systemic application ever fulfill its rich promise? Exp Opin Drug Deliv 13(10):1337–1340
7. Patton JS, Byron PR (2007) Inhaling medicines: delivering drugs to the body through the lungs. Nat Rev Drug Discov 6(1):67–74
8. Kenneth NS, Rocha S (2008) Regulation of gene expression by hypoxia. Biochem J 414:19–29
9. Imai Y, Kuba K, Rao S, Huan Y, Guo F, Guan B, Yang P, Sarao R, Wada T, Leong-Poi H, Crackower MA, Fukamizu A, Hui CC, Hein L, Uhlig S, Slutsky AS, Jiang C, Penninger JM (2005) Angiotensin-converting enzyme 2 protects from severe acute lung failure. Nature 436:112–116
10. Feldmann JL (1986) Neurophysiology of breathing in mammals. In: Bloom FE (ed) Handbook of physiology; section I: the nervous system, Volume IV: Intrinsic regulatory systems of the brain. American Physiological Society, Bethesda, pp 463–524

Community-Acquired Pneumonia

Catia Cilloniz and Antoni Torres

Introduction to Community-Acquired Pneumonia

Pneumonia is an acute inflammatory process of the pulmonary parenchyma of infectious origin, characterized by the presence of symptoms of lower respiratory infection (e.g., cough, purulent expectoration, and fever) and it is confirmed by the presence of pulmonary infiltrates in a chest X-ray [1]. This infection can be acquired in the community or in a hospital environment (intra-hospital) and it can be transmitted by aspirated or inhaled microorganisms (the most frequent routes to acquiring the infection). Pneumonia can be caused by different microorganisms, the most frequent being bacteria, especially *Streptococcus pneumoniae* (pneumococcus) and *Haemophilus influenzae*, and respiratory viruses such as influenza virus A/B. Pneumonia is rarely caused by fungi or parasites [2] (Fig. 1).

Community-acquired pneumonia (CAP) is a serious health problem that affects all age groups and it is associated with high rates of morbidity, complications, and short- and long-term mortality [3]. Approximately one in three patients with severe CAP present sepsis. Cardiac complications have also been reported in approximately 30% of hospitalized patients with CAP [1]. These show us that pneumonia not only affects the lungs but is a complex disease that can also affect other organs. The recent coronavirus disease 2019 (COVID-19) pandemic has highlighted the impact of acute respiratory infections on global health. This chapter offers an updated overview of community-acquired pneumonia in adults in terms of epidemiology, microbial etiology, clinical characteristics, management, and prevention.

According to the 2019 Global Burden of Disease (GDB) report [4], lower respiratory tract infections (LRTI), including pneumonia and bronchiolitis, affected 489 million people worldwide. The most vulnerable populations are children under five years old and adults over 70 years old. According to these data, pneumonia was the leading infectious cause of death globally.

C. Cilloniz
Institut d'Investigacions Biomèdiques August Pi i Sunyer (IDIBAPS), University of Barcelona, Barcelona, Spain

Faculty of Health Sciences, Continental University, Huancayo, Peru

University of Barcelona, Barcelona, Spain

Centro de Investigación Biomédica En Red – Enfermedades Respiratorias (CIBERES), Madrid, España

A. Torres (✉)
Institut d'Investigacions Biomèdiques August Pi i Sunyer (IDIBAPS), University of Barcelona, Barcelona, Spain

University of Barcelona, Barcelona, Spain

Centro de Investigación Biomédica En Red – Enfermedades Respiratorias (CIBERES), Madrid, España
e-mail: atorres@clinic.cat

© The Author(s), under exclusive license to Springer Nature Switzerland AG 2026
E. Lammert, M. Zeeb (eds.), *Metabolism of Human Diseases*,
https://doi.org/10.1007/978-3-031-96019-2_28

Streptococcus pneumoniae	**Legionella pneumophiña**	**Pseudomonas aerugionsa**	**MRSA**
Dementia, seizure disorders, congestive heart failure, cerebrovascular disease, chronic obstructive pulmonary disease (COPD), HIV infection, overcrowded living conditions and smoking	Smoking, COPD, compromised immune system, travel to outbreak areas, residence in a health-care facility and proximity to cooling towers or whirlpool spas	Previous *Pseudomonas* infection or colonisation; previous tracheostomy; bronchiectasis; invasive respiratory or vasopressor support; very severe COPD	Previous MRSA infection or colonization, residence in a nursing home or long-term care facility and prior hospitalization within the previous 90 days

Fig. 1 Risk factors for specific pathogens

According to the 2019 GBD, pneumonia caused an estimated 2.5 million deaths across all ages worldwide, with an estimated 672,000 deaths in children under five years old and 1.2 million deaths in adults over 70 years. However, COVID-19 pneumonia deaths could have added up to two million further deaths to the global number of deaths from respiratory infections by 2020.

According to the prospective population-based cohort study examined by Ramirez et al. [5] in the United States, approximately 1.5 million adults are hospitalized with CAP annually, and the incidence of CAP patients requiring admission to an intensive care unit (ICU) was estimated as 145 cases per 100,000 population of adults, meaning that every year 356,000 adults hospitalized with CAP require ICU admission [6]. Moreover, 102,821 deaths occur during hospitalization because of CAP, while half of the patients admitted to the ICU for CAP will die within a year [5]. Significantly, the number of patients who present clinical failure or one-year mortality from unresolved pneumonia may also be as high as 50% [7].

In recent years, we have observed an increase in the proportion of immunocompromised patients with CAP [8, 9], as well as an increase in the proportion of elderly patients with pneumonia who required ICU admission [10]: approximately 18% of hospitalized patients with CAP are immunocompromised [8, 9], and one million elderly patients are hospitalized annually in the United States due to pneumonia [11].

A multicenter study with data from the Community-Acquired Pneumonia Organization (CAPO) database investigated the effects of age and comorbidities on CAP mortality and reported a mortality of 5% in patients under 65 years of age, 8% in those between 65 and 79 years, and 14% in those over 80 years of age [12]; these mortality rates increased to 20, 42, and 43%, respectively, in patients with more than one chronic disease. One-year pneumonia mortality was reported as being 30%, which means that approximately one in every three adults hospitalized with pneumonia will die within 1 year.

Pathophysiology, Risk Factors, Diagnosis, and Biomarkers of Community-Acquired Pneumonia

Risk factors: Several risk factors for CAP have been recognized. Underlying medical conditions associated with an increased risk of CAP include previous episodes of pneumonia; chronic respiratory diseases (chronic obstructive pulmonary disorder (COPD), bronchitis, or asthma); chronic cardiovascular diseases; cerebrovascular disease/stroke and dementia; dysphagia; diabetes; cancer; and chronic liver disease or renal disease [13]. The incidence of CAP increases with age and with the presence of comorbidities. The study by Ramirez et al. reported an incidence in patients ≥65 years old of 20.93 cases per 1,000 individuals per year which mean that in the United States each year one million of elderly people are hospitalized because of pneumonia

[5]. In Spain, the incidence was reported as 14 cases per 1000 individuals per year [14]. A range of lifestyle factors have also been associated with an increased risk of CAP, that is, current smoking habit history of alcohol abuse/alcoholism, being underweight, living in a household of over ten people, and regular contact with children [13] (Fig. 2).

Clinical presentation and diagnosis: The clinical presentation of CAP varies from mild pneumonia characterized by fever and cough, to severe pneumonia with sepsis, respiratory failure, and shock. The interaction between a patient's immune system, associated risk factors, and the virulence of the pathogen play an important role in the clinical presentation of pneumonia. Cough breathlessness, pleuritic pain, and sputum production are the most common symptoms of pneumonia. In elderly patients, the most common symptoms are confusion, falls, fatigue, lethargy, anorexia, and exacerbation or decompensation of a chronic underlying disease [15]. Elderly patients' response to an infection can be reduced by the effect of immunosenescence, while the absence of symptoms such as fever can increase the risk of misdiagnosis in this population [15]. The diagnosis of pneumonia is based on the presence of acute ($\leq$7 days) symptoms of LRTI, such as cough, purulent sputum production, fever, and dyspnea, as well as the presence of new infiltrates on chest X-rays (CXR). Some patients with other lung diseases such as pulmonary embolism and lung cancer can present fever and pulmonary infiltrates that mimic pneumonia, so it is important to review previous CXR, if available, to rule out the disease. A radiographic confirmation is essential for the diagnosis of pneumonia. The standard CXR for CAP consists of posterior–anterior (PA) and lateral images which provide information about the site, extent, and associated features of pneumonia (i.e., the lobes involved, pleural effusion, and cavitation). The presence of pleural fluid or multilobar pneumonia has been clearly demonstrated to be an indicator of severity [16].

Microbial etiology of CAP: Understanding the microbial etiology of pneumonia is crucial to providing more accurately targeted, wide-ranging empiric antibiotic therapy, preventing the emergence of antimicrobial resistance through selection pressure, and reducing complications. The guidelines for microbiological diagnosis in CAP vary according to the severity of the pneumonia [16] (Fig. 3). Generally speaking, microbiological diagnosis should, at the very least, be based on a sputum culture, a urinary antigen test for *S. pneumoniae* and *L. pneumophila*, a blood culture, and a nasopharyngeal swab for respiratory viruses (influenza virus A/B, para-influenza viruses, rhinovirus, adenoviruses, respiratory syncytial virus, human metapneumovirus, and coronaviruses) [17]. In cases of severe CAP, bronchoalveolar lavage (BAS) or tracheal aspirates (TAS) are recommended, while the culture of pleural fluid is also recommended in patients with pleural effusion. The current American Thoracic Society/Infectious Diseases Society of America (ATS/IDSA) recommendation for the

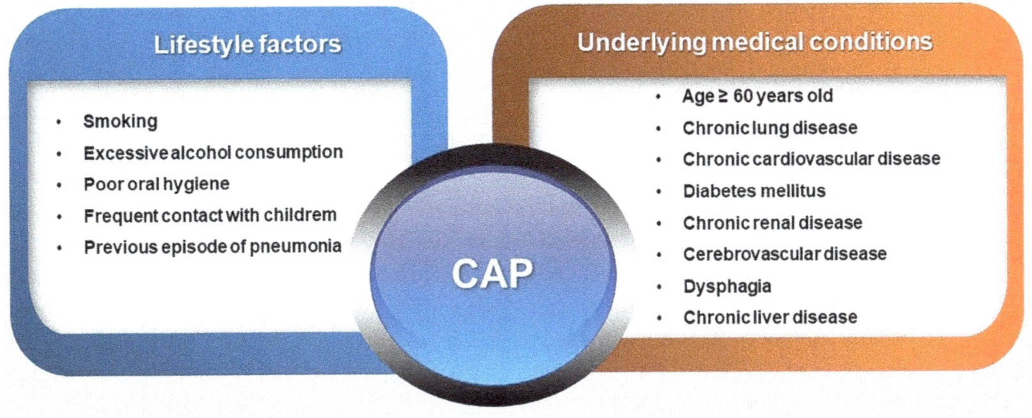

Fig. 2 Risk factors for community-acquired pneumonia

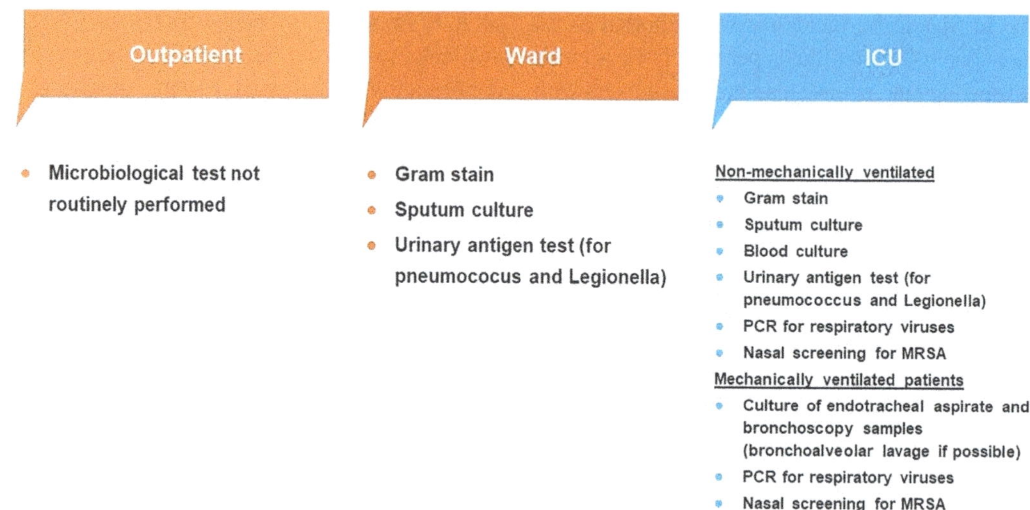

Fig. 3 Microbiological diagnosis of community-acquired pneumonia

use of molecular test in severe CAP is to testing for influenza viruses during seasons with community spread of influenza. Additionally, a recent update of the guidelines recommends in patients with severe CAP or immunocompromised patients the use of molecular tests for respiratory viruses apart from influenza [18].

The cause of CAP can be identified in only 30–50% of patients, despite the availability of various diagnostic methods. Overall, *S. pneumoniae* and respiratory viruses are the most commonly identified pathogens in pneumonia [1, 2]. However, studies from the United States have reported a higher proportion of respiratory viruses as the main cause of CAP, whereas in Europe the main cause is considered to be pneumococcus. The higher rate of pneumococcal vaccination and the reduction in smoking observed in the United States may explain this difference [19]. The proportion of patients with polymicrobial etiology (bacteria+ bacteria/bacteria + respiratory viruses) varies between 6% and 15% of CAP patients with a microbial diagnosis [20–22]. In severe cases of CAP admitted to the ICU, respiratory viruses are often present in combination with bacterial pathogens. In an observational study by Karhu et al. [22] that analyzed the data of 49 mechanically ventilated patients with CAP, respiratory viruses were detected in 49% of the cases (pure or mixed); rhinovirus and adenovirus

were the most frequently detected of these viruses. Similarly, a retrospective study in Spain found that 26% of the patients with severe CAP admitted to an ICU presented with viral sepsis and that the ICU mortality in these cases was 8%. More recently, it has been reported that respiratory viruses were the causal agent in 11% of mechanically ventilated patients with CAP and acute respiratory distress syndrome (ARDS) [23, 48].

In a small proportion of patients with microbial diagnosis (6%), CAP is caused by resistant pathogens such as the so-called PES pathogens (*Pseudomonas aeruginosa*, extended-spectrum β-lactamase (ESBL)-positive Enterobacterales, and methicillin-resistant *Staphylococcus aureus* [MRSA]) [24, 25]. The risk factors for specific pathogens are reported in Fig. 2.

It is important to mention the impact of microbial etiology on the outcomes of CAP patients. One single-center retrospective study that included data from 123 patients with severe CAP admitted to the ICU investigated the impact of etiology (bacterial, viral, or unidentified etiology) on one-year outcomes in this population [26]. They reported that patients with severe bacterial CAP were more seriously ill (Pneumonia Severity Index (PSI) 131) than patients with severe viral CAP (PSI 116) and those with severe CAP with an unidentified etiology (PSI 118).

Furthermore, the one-year mortality was higher in patients with severe bacterial CAP (58%) compared with those with severe viral CAP (27%) and severe CAP with an unidentified etiology (28%). A multivariate analysis showed that one-year mortality in the severe bacterial group was higher compared to the severe viral group, and it also tended to be higher compared to the severe unidentified etiology group, although no difference was found between the viral and the unidentified etiology group. The risk factors related to various pathogens are shown in Fig. 1.

Biomarkers in CAP: Biomarkers are useful for the diagnosis and treatment of CAP, as well as helping to identify and stratify patients with severe infection, detect complications (such as cardiac complications), and guide the discontinuation of antibiotic therapy [27]. Biomarkers are, however, adjunctive tools in the management of CAP and they should be used in conjunction with clinical parameters and data that provide an assessment of severity. The main limitation of biomarkers is that their value can be diminished (particularly in immunocompromised patients) by other factors such as immunomodulatory therapy, the characteristics of the pathogen, and the point at which a biomarker is measured. The most commonly used biomarkers are C-reactive protein (CRP) and procalcitonin (PCT). It is important to bear in mind that their kinetics in response to an infection are different. CRP levels increase after the first 3 days of infection (peak time 36–50 hours), whereas in the case of PCT the levels rise rapidly (peak time 12–24 hours). CRP levels increase in response to inflammation and some treatment agents, such as corticosteroids and antibiotics, may modify their levels, whereas PCT is more susceptible to bacterial infection. The values of both CRP and PCT are lower in cases of viral infection as their production is reduced under the influence of cytokines [1].

The current CAP guidelines do not recommend the use of PCT to begin antimicrobial therapy, which is itself only recommended in patients with clinically suspected and radiographically confirmed CAP [16].

Treatment of Community-Acquired Pneumonia

Empiric antibiotic therapy: The first dose of antibiotics should be given as soon as possible after the diagnosis of CAP, preferably within 4–8 h after a patient's hospital admission; an early start to antibiotic treatment is associated with better outcomes. Generally speaking, antibiotic treatment for CAP is empiric because it is difficult to identify the cause of pneumonia, and data such as individual risk factors, chronic comorbidities, the severity of pneumonia, local epidemiology, etiology, and resistance features must be taken into account [1, 16]. The guidelines for starting empiric treatments of patients hospitalized for CAP are shown in Table 1. Current ATS/IDSA guidelines recommend the use of amoxicillin, doxycycline, or a macrolide (only in areas with <25% of *S. pneumoniae* resistant to macrolides) for healthy outpatients without comorbidities or risk factors for drug-resistant pathogens. For outpatients with comorbidities, they recommended monotherapy with a respiratory fluoroquinolone, or a combination therapy of amoxicillin–clavulanic acid or a cephalosporin and a macrolide or doxycycline [16].

A combination therapy with a β-lactam plus macrolide or monotherapy with a respiratory fluoroquinolone is recommended for hospitalized patients with non-severe CAP, and with no risk factors for MRSA or *P. aeruginosa* [16]. For patients with severe CAP, they recommended a combination of β-lactam plus macrolide or β-lactam plus a respiratory fluoroquinolone for those patients without risk factors for MRSA or *P. aeruginosa*. In patients with risk factors for MRSA and *P. aeruginosa*, they recommend the addition of anti-MRSA or anti-Pseudomonas coverage [16].

Another important point to consider is the switch from intravenous to oral antibiotic therapy in CAP patients who achieve clinical stability. Significant reductions in the length of hospital stays and reductions in adverse drug reactions have been reported in patients with severe CAP who switched their antibiotic therapy early on [30–32].

Table 1 Guidelines for initial empirical treatments of hospitalized patients with CAP

Characteristics	Non- severe CAP	Severe CAP
No special considerations	Standard regimen: a β-lactam and a macrolide; or a respiratory fluroquinolone alone	Standard regimen: a β-lactam plus a macrolide; or a β-lactam plus a respiratory fluoroquinolone
Previous respiratory isolation of MRSA or *Pseudomonas aeruginosa*	Standard regimen based on severity, with additional MRSA/Pseudomonas coverage as indicated; cultures or PCR testing of a nasal sample are recommended to guide de-escalation or continuation of therapy	
Recent hospitalization, parenteral antibiotic and locally validated risk factor for MRSA infection	Standard regimen based on severity; obtain cultures, but withhold MRSA coverage unless culture results are positive. If nasal screening is available, withhold empirical MRSA therapy unless rapid test is positive; obtain respiratory cultures	Standard regimen with MRSA coverage; perform nasal screening and obtain cultures to guide de-escalation or continuation of therapy
Recent hospitalization, parenteral antibiotic and locally validated risk factor for *Pseudomonas aeruginosa* infection	Standard regimen based on severity; include *Pseudomonas aeruginosa* coverage only if culture results are positive	Standard regimen with *Pseudomonas aeruginosa* coverage; obtain cultures to guide de-escalation or continuation of therapy

Abbreviations: *MRSA* methicillin resistant staphylococcus aureus, *PCR* polymerase chain reaction

The use of an influenza antiviral such as oseltamivir is recommended in patients with suspected influenza pneumonia, regardless of the duration of the disease. Several studies have shown a greater benefit of antivirals when they are started within the first 48 h of the onset of infection. It is important to note that antibiotics should be started empirically even when influenza has been diagnosed, to account for any possible bacterial superinfection [16].

Adjunctive therapy: The use of systemic corticosteroids with broad-spectrum anti-inflammatory activity as adjunctive therapy for severe CAP is important, as several studies have demonstrated their effectiveness as adjunctive therapies in reducing treatment failure and ARDS in such cases. Furthermore, some studies have found an association between this adjunctive therapy and lower mortality rates, although the administration of corticosteroids in patients with severe influenza infection is associated with higher mortality [31]. One meta-analysis that included almost 6,000 patients with influenza pneumonia reported that the mortality risk ratio was 1.75 for patients who received corticosteroids [32]. Similarly, a systematic review and meta-analysis that included data from 6,427 patients with severe pneumonia and ARDS, the group of patients who received corticosteroids presented higher mortality and higher incidence of nosocomial infections [33]. The current recommendations of the ATS/IDSA guidelines for the management of CAP advise against the use of corticosteroids in adults with severe influenza pneumonia [34].

In the case of patients with COVID-19, the results from the RECOVERY trial [35] showed that the use dexamethasone (6 mg/day for 10 days) in COVID-19 patients requiring oxygen therapy reduced 28-day mortality. A meta-analysis carried out by the World Health Organization (WHO) on clinical trials of patients critically ill with COVID-19 reported that the use of corticosteroids compared with routine care placebo was associated with a reduction in 28-day mortality [36]. Recently, a multicenter study reported that corticosteroids were administered to COVID-19 patients admitted to an ICU on the basis of age, severity, baseline inflammation, and invasive mechanical ventilation, but its authors reported that their early administration after the onset of symptoms may prove harmful. COVID-19 guidelines recommend the use of corticosteroids only for patients with hypoxemic respiratory failure requiring the administration of oxygen [37]. The National Institutes of Health (NIH) recommends the use of systemic corticosteroids in COVID-19 patients requiring supplementary oxygen [38].

Complications in CAP

Pulmonary and extra-pulmonary complications are common in CAP and contribute to increased mortality in both the short and long term.

Pulmonary complications: Pleural effusion, multilobar involvement, and empyema are the most frequent pulmonary complications in CAP, especially in cases of pneumococcal pneumonia. One Spanish study [39] that included data from 625 patients hospitalized for pneumococcal CAP, the percentages of patients with pleural effusion, multilobar involvement, and empyema were 52, 64, and 8%, respectively. The authors reported that 20% of the CAP patients experienced more than one complication, and that ICU admission and shock were more frequent, and the hospital stay was longer, in patients with pulmonary complications. However, the 30-day mortality rate was similar in patients with and without pulmonary complications. Another study reported that bilateral multilobar pneumonia was an independent risk factor for mortality and was associated with poor outcomes, compared to patients with unilateral multilobar involvement. Of 4,644 patients with CAP, 1,069 (23%) had multilobar pneumonia; of these, 13, 45, and 77% had bilateral, unilateral, and localized infiltrates, respectively. The clinical course was severe in the bilateral group, with more cases requiring ICU admission and mechanical ventilation compared to the other two groups [40].

Cardiovascular complications: Approximately 2% of CAP patients receiving outpatient care develop cardiac complications, and in the case of hospitalized CAP patients these complications were reported in 20–30% of patients [41]. It has also been reported that patients with pneumococcal CAP have a significant risk of a concurrent acute cardiac event that significantly increases mortality, such as myocardial infarction, severe arrhythmia, or new or worsened congestive heart failure [42]. A study by Aldas et al. [43] found that factors such as age, smoking, chronic heart disease, pneumonia severity, and *S. pneumoniae* infection are the main risk factors for early (< 30 days) and late cardiac events (> 30 days). In this study, the presence of cardiac complications

was an independent risk factor for early and late mortality. Interestingly, another study by Chen et al. [44] observed cardiac complications in 24% of patients with influenza-related pneumonia. The presence of hypertension, cerebrovascular disease, coronary artery disease, preexisting heart failure, systolic blood pressure <90 mm Hg, respiratory rates $\geq$30 breaths/min, a lymphocyte count <0.8×10^9/L, PaO_2/FiO_2 < 300 mm Hg, and systemic corticosteroid use were independently associated with the incidence of cardiac complications, while the early neuraminidase inhibitor treatment and angiotensin-converting enzyme inhibitors/angiotensin II receptor blocker treatment were associated with a lower risk of cardiac complications.

A recent study by Africano et al. [45] reported a major adverse cardiovascular event (MACE) in 28% of pneumococcal CAP patients, and this was associated with a significant increase in the incidence of mechanical ventilation and in-hospital mortality. More importantly, the authors found that the pneumococcal serotypes 3 and 9n were associated with the presence of MACE. In recent decades, several studies have examined the possible mechanisms of cardiac complications in patients with CAP, and so far these mechanisms have been related to pneumococcal microlesions in the heart, the presence of pneumolysin, and the platelet activation process that occurs in pneumonia [46].

Perspectives

Despite advances in the management of pneumonia, this disease continues to have a major impact on global health. Pneumonia remains the leading infectious disease cause of mortality among all ages worldwide. The main challenges for clinicians are related to the early identification of patients with severe pneumonia and the impact of pneumonia in patients with multiple chronic underlying diseases. The early identification of severe pneumonia and the adequate antimicrobial treatment are crucial for the best outcomes especially in critically ill patients. Following the recommendation of the local and international guidelines for the management of CAP will ensure better outcomes especially in patients admitted to the ICU with severe pneumonia.

Questions and Answers

Question 1 Is there any change in the epidemiology of community-acquired pneumonia?

Answer 1 In the last decade, we observed a change in the epidemiology of CAP and an increase in elderly patients with severe pneumonia that required admission to the ICU. Also, we observed an increase in patients with some immunosuppressed condition. This change in the epidemiology of CAP is important because it will affect the management of these group patients.

Question 2 Who is most at risk of having pneumonia?

Answer 2 Pneumonia can occur at any age, but its incidence increases significantly with advanced age. Comorbidities that are associated with an increased risk of CAP are chronic lung diseases, diabetes mellitus, cardiovascular disease, and chronic liver disease. Patients with immunocompromised conditions such as human immunodeficiency virus (HIV) and cancer are other risk population for pneumonia. Some lifestyle factors such as cigarette smoking, alcohol abuse, and poor oral hygiene are associated with increased risk of pneumonia and severe forms of CAP.

Question 3 Is the clinical presentation of pneumonia the same in all people?

Answer 3 In younger and immunocompetent patients, the clinical presentation of pneumonia is related to systemic and respiratory symptoms. The most frequent symptoms include, cough, purulent sputum, chills, fever, and pleuritic pain. However, in elderly patients symptoms such as fever may be absent, and symptoms such as confusion, weakness, falling, and decompensation of chronic disease are frequent present.

Question 4 Has there been any change in the microbial etiology of community-acquired pneumonia?

Answer 4 The identification of the microbial etiology of CAP is achieved in approximately 50% of the cases. *Streptococcus pneumoniae* remains the most frequent pathogen in CAP in all settings. Interestingly, studies from North America reported a decrease in *S. pneumoniae* as the main cause of CAP; this change was related to a reduce in the smoking rate and the systemic pneumococcal vaccination. However, these changes were not observed in Europe.

The implementation of new molecular tests has led to an increase in the number of respiratory viruses identified in patients with CAP.

References

1. Torres A, Cilloniz C, Niederman MS, Menéndez R, Chalmers JD, Wunderink RG, van der Poll T (2021) Pneumonia. Nat Rev Dis Primers 7:1–28. Nature Publishing Group
2. Cillóniz C, Ewig S, Polverino E, Marcos MA, Esquinas C, Gabarrús A, Mensa J, Torres A (2011) Microbial aetiology of community-acquired pneumonia and its relation to severity. Thorax 66:340–346
3. Cillóniz C, Polverino E, Ewig S, Aliberti S, Gabarrús A, Menéndez R, Mensa J, Blasi F, Torres A (2013) Impact of age and comorbidity on cause and outcome in community-acquired pneumonia. Chest 144:999–1007
4. GBD 2019 Diseases and Injuries Collaborators (2020) Global burden of 369 diseases and injuries in 204 countries and territories, 1990–2019: a systematic analysis for the Global Burden of Disease Study 2019. Lancet 396:1204–1222
5. Ramirez JA, Wiemken TL, Peyrani P, Arnold FW, Kelley R, Mattingly WA, Nakamatsu R, Pena S, Guinn BE, Furmanek SP, Persaud AK, Raghuram A, Fernandez F, Beavin L, Bosson R, Fernandez-Botran R, Cavallazzi R, Bordon J, Valdivieso C, Schulte J, Carrico RM, University of Louisville Pneumonia Study Group (2017) Adults hospitalized with pneu-

monia in the United States: incidence, epidemiology, and mortality. Clin Infect Dis 65:1806–1812

6. Cavallazzi R, Furmanek S, Arnold FW, Beavin LA, Wunderink RG, Niederman MS, Ramirez JA (2020) The burden of community-acquired pneumonia requiring admission to ICU in the United States. Chest 158:1008–1016

7. Peyrani P, Arnold FW, Bordon J, Furmanek S, Luna CM, Cavallazzi R, Ramirez J (2020) Incidence and mortality of adults hospitalized with community-acquired pneumonia according to clinical course. Chest 157:34–41

8. Jain S, Self WH, Wunderink RG, Fakhran S, Balk R, Bramley AM, Reed C, Grijalva CG, Anderson EJ, Courtney DM, Chappell JD, Qi C, Hart EM, Carroll F, Trabue C, Donnelly HK, Williams DJ, Zhu Y, Arnold SR, Ampofo K, Waterer GW, Levine M, Lindstrom S, Winchell JM, Katz JM, Erdman D, Schneider E, Hicks LA, McCullers JA, Pavia AT et al (2015) Community-acquired pneumonia requiring hospitalization among U.S. N Engl J Med 373:415–427

9. Di Pasquale MF, Sotgiu G, Gramegna A, Radovanovic D, Terraneo S, Reyes LF, Rupp J, González del Castillo J, Blasi F, Aliberti S, Restrepo MI (2019) Prevalence and etiology of community-acquired pneumonia in immunocompromised patients. Clin Infect Dis 68:1482–1493

10. Laporte L, Hermetet C, Jouan Y, Gaborit C, Rouve E, Shea KM, Si-Tahar M, Dequin P-F, Grammatico-Guillon L, Guillon A (2018) Ten-year trends in intensive care admissions for respiratory infections in the elderly. Ann Intensive Care 8:84

11. Arnold FW, Reyes Vega AM, Salunkhe V, Furmanek S, Furman C, Morton L, Faul A, Yankeelov P, Ramirez JA (2020) Older adults hospitalized for pneumonia in the United States: incidence, epidemiology, and outcomes. J Am Geriatr Soc 68:1007–1014

12. Luna CM, Palma I, Niederman MS, Membriani E, Giovini V, Wiemken TL, Peyrani P, Ramirez J (2016) The impact of age and comorbidities on the mortality of patients of different age groups admitted with community-acquired pneumonia. Ann Am Thorac Soc 13:1519–1526

13. Torres A, Peetermans WE, Viegi G, Blasi F (2013) Risk factors for community-acquired pneumonia in adults in Europe: a literature review. Thorax 68:1057–1065

14. Vila-Corcoles A, Ochoa-Gondar O, Rodriguez-Blanco T, Raga-Luria X, Gomez-Bertomeu F, EPIVAC Study Group (2009) Epidemiology of community-acquired pneumonia in older adults: a population-based study. Respir Med 103:309–316

15. Cillóniz C, Dominedò C, Pericàs JM, Rodriguez-Hurtado D, Torres A (2020) Community-acquired pneumonia in critically ill very old patients: a growing problem. Eur Respir Rev 29:190126

16. Metlay JP, Waterer GW, Long AC, Anzueto A, Brozek J, Crothers K, Cooley LA, Dean NC, Fine MJ, Flanders SA, Griffin MR, Metersky ML, Musher DM, Restrepo MI, Whitney CG (2019) Diagnosis and treatment of adults with community-acquired pneumonia. An official clinical practice guideline of the American Thoracic Society and Infectious Diseases Society of America. Am J Respir Crit Care Med 200:e45–e67

17. Torres A, Lee N, Cilloniz C, Vila J, Van der Eerden M (2016) Laboratory diagnosis of pneumonia in the molecular age. Eur Respir J 48:1764–1778

18. Evans SE, Jennerich AL, Azar MM, Cao B, Crothers K, Dickson RP, Herold S, Jain S, Madhavan A, Metersky ML, Myers LC, Oren E, Restrepo MI, Semret M, Sheshadri A, Wunderink RG, Dela Cruz CS (2021) Nucleic acid-based testing for noninfluenza viral pathogens in adults with suspected community-acquired pneumonia. An official American Thoracic Society clinical practice guideline. Am J Respir Crit Care Med 203:1070–1087

19. Gadsby NJ, Musher DM (2022) The microbial etiology of community-acquired pneumonia in adults: from classical bacteriology to host transcriptional signatures. Clin Microbiol Rev:e0001522

20. Cillóniz C, Ewig S, Ferrer M, Polverino E, Gabarrús A, Puig de la Bellacasa J, Mensa J, Torres A (2011) Community-acquired polymicrobial pneumonia in the intensive care unit: aetiology and prognosis. Crit Care 15:R209

21. Cillóniz C, Civljak R, Nicolini A, Torres A (2016) Polymicrobial community-acquired pneumonia: an emerging entity. Respirology 21:65–75

22. Cillóniz C, Calabretta D, Palomeque A et al (2025) Risk Factors and Outcomes Associated With Polymicrobial Infection in Community-Acquired Pneumonia. Arch Bronconeumol 6:S0300-2896(25)00007-9

23. Karhu J, Ala-Kokko TI, Vuorinen T, Ohtonen P, Syrjälä H (2014) Lower respiratory tract virus findings in mechanically ventilated patients with severe community-acquired pneumonia. Clin Infect Dis 59:62–70

24. Cilloniz C, Ferrer M, Liapikou A, Garcia-Vidal C, Gabarrus A, Ceccato A, Puig de La Bellacasa J, Blasi F, Torres A (2018) Acute respiratory distress syndrome in mechanically ventilated patients with community-acquired pneumonia. Eur Respir J 51:1702215

25. Cilloniz C, Videla AJ, Pulido L, Uy-King MJ (2025) Viral community-acquired pneumonia: what's new since COVID-19 emerged? Expert Rev Respir Med 19(4):347–362

26. Prina E, Ranzani OT, Polverino E, Cillóniz C, Ferrer M, Fernandez L, Puig de la Bellacasa J, Menéndez R, Mensa J, Torres A (2015) Risk factors associated with potentially antibiotic-resistant pathogens in community-acquired pneumonia. Ann Am Thorac Soc 12:153–160

27. Cillóniz C, Dominedò C, Nicolini A, Torres A (2019) PES pathogens in severe community-acquired pneumonia. Microorganisms 7:49

28. Sangla F, Legouis D, Marti P-E, Sgardello SD, Brebion A, Saint-Sardos P, Adda M, Lautrette A, Pereira B, Souweine B (2020) One year after ICU admission for severe community-acquired pneumonia of bacterial, viral or unidentified etiology. What are the outcomes? PLoS One 15:e0243762

29. García-Río F, Alcázar-Navarrete B, Castillo-Villegas D, Cilloniz C, García-Ortega A, Leiro-Fernández V, Lojo-Rodriguez I, Padilla-Galo A, Quezada-Loaiza CA, Rodriguez-Portal JA, Sánchez-de-la-Torre M, Sibila O, Martínez-García MA (2022) Biological biomarkers in respiratory diseases. Arch Bronconeumol 58:323–333

30. Oosterheert JJ, Bonten MJM, Schneider MME, Buskens E, Lammers J-WJ, Hustinx WMN, Kramer MHH, Prins JM, Slee PHTJ, Kaasjager K, Hoepelman AIM (2006) Effectiveness of early switch from intravenous to oral antibiotics in severe community acquired pneumonia: multicentre randomised trial. BMJ 333:1193

31. Kohno S, Yanagihara K, Yamamoto Y, Tokimatsu I, Hiramatsu K, Higa F, Tateyama M, Fujita J, Kadota J-I (2013) Early switch therapy from intravenous sulbactam/ampicillin to oral garenoxacin in patients with community-acquired pneumonia: a multicenter, randomized study in Japan. J Infect Chemother 19:1035–1041

32. Castro-Guardiola A, Viejo-Rodríguez AL, Soler-Simon S, Armengou-Arxé A, Bisbe-Company V, Peñarroja-Matutano G, Bisbe-Company J, García-Bragado F (2001) Efficacy and safety of oral and early-switch therapy for community-acquired pneumonia: a randomized controlled trial. Am J Med 111:367–374

33. Cillóniz C, Torres A, Niederman MS (2021) Management of pneumonia in critically ill patients. BMJ 375:e065871

34. Ni Y-N, Chen G, Sun J, Liang B-M, Liang Z-A (2019) The effect of corticosteroids on mortality of patients with influenza pneumonia: a systematic review and meta-analysis. Crit Care 23:99

35. Zhou Y, Fu X, Liu X, Huang C, Tian G, Ding C, Wu J, Lan L, Yang S (2020) Use of corticosteroids in influenza-associated acute respiratory distress syndrome and severe pneumonia: a systemic review and meta-analysis. Sci Rep 10:3044

36. Metlay JP, Waterer GW, Long AC, Anzueto A, Brozek J, Crothers K, Cooley LA, Dean NC, Fine MJ, Flanders SA, Griffin MR, Metersky ML, Musher DM, Restrepo MI, Whitney CG (2019) Diagnosis and treatment of adults with community-acquired pneumonia. An official clinical practice guideline of the American Thoracic Society and Infectious Diseases Society of America. Am J Respir Crit Care Med 200:e45–e67. American Thoracic Society – AJRCCM

37. RECOVERY Collaborative Group, Horby P, Lim WS, Emberson JR, Mafham M, Bell JL, Linsell L, Staplin N, Brightling C, Ustianowski A, Elmahi E, Prudon B, Green C, Felton T, Chadwick D, Rege K, Fegan C, Chappell LC, Faust SN, Jaki T, Jeffery K, Montgomery A, Rowan K, Juszczak E, Baillie JK, Haynes R, Landray MJ (2021) Dexamethasone in hospitalized patients with covid-19. N Engl J Med 384:693–704

38. WHO Rapid Evidence Appraisal for COVID-19 Therapies (REACT) Working Group, Sterne JAC, Murthy S, Diaz JV, Slutsky AS, Villar J, Angus DC, Annane D, Azevedo LCP, Berwanger O, Cavalcanti AB, Dequin P-F, Du B, Emberson J, Fisher D, Giraudeau B, Gordon AC, Granholm A, Green C, Haynes R, Heming N, JPT H, Horby P, Jüni P, Landray MJ, Le Gouge A, Leclerc M, Lim WS, Machado FR, McArthur C et al (2020) Association between administration of systemic corticosteroids and mortality among critically ill patients with COVID-19: a meta-analysis. JAMA 324:1330–1341

39. Chalmers JD, Crichton ML, Goeminne PC, Cao B, Humbert M, Shteinberg M, Antoniou KM, Ulrik CS, Parks H, Wang C, Vandendriessche T, Qu J, Stolz D, Brightling C, Welte T, Aliberti S, Simonds AK, Tonia T, Roche N (2021) Management of hospitalised adults with coronavirus disease 2019 (COVID-19): a European Respiratory Society living guideline. Eur Respir J 57:2100048

40. Information on COVID-19 Treatment, Prevention and Research [Internet]. COVID-19 Treatment Guidelines [cited 2020 Dec 28]. Available from: https://www. covid19treatmentguidelines.nih.gov/

41. Cillóniz C, Ewig S, Polverino E, Muñoz-Almagro C, Marco F, Gabarrús A, Menéndez R, Mensa J, Torres A (2012) Pulmonary complications of pneumococcal community-acquired pneumonia: incidence, predictors, and outcomes. Clin Microbiol Infect 18:1134–1142

42. Liapikou A, Cillóniz C, Gabarrús A, Amaro R, De la Bellacasa JP, Mensa J, Sánchez M, Niederman M, Torres A (2016) Multilobar bilateral and unilateral chest radiograph involvement: implications for prognosis in hospitalised community-acquired pneumonia. Eur Respir J 48:257–261

43. Corrales-Medina VF, Musher DM, Wells GA, Chirinos JA, Chen L, Fine MJ (2012) Cardiac complications in patients with community-acquired pneumonia: incidence, timing, risk factors, and association with short-term mortality. Circulation 125:773–781

44. Musher DM, Rueda AM, Kaka AS, Mapara SM (2007) The association between pneumococcal pneumonia and acute cardiac events. Clin Infect Dis 45:158–165

45. Aldás I, Menéndez R, Méndez R, España PP, Almirall J, Boderías L, Rajas O, Zalacaín R, Vendrell M, Mir I, Torres A, Grupo NEUMONAC (2019) Early and late cardiovascular events in patients hospitalized for community-acquired pneumonia. Arch Bronconeumol 56:551

46. Chen L, Han X, Li Y, Zhang C, Xing X (2021) Comparison of clinical characteristics and outcomes between respiratory syncytial virus and influenza-related pneumonia in China from 2013 to 2019. Eur J Clin Microbiol Infect Dis 40:1633

47. Africano HF, Serrano-Mayorga CC, Ramirez-Valbuena PC, Bustos IG, Bastidas A, Vargas HA, Gómez S, Rodriguez A, Orihuela CJ, Reyes LF (2020) Major adverse cardiovascular events during invasive pneumococcal disease are serotype dependent. Clin Infect Dis 72:e711–e719

48. Feldman C (2020) Cardiac complications in community-acquired pneumonia and COVID-19. Afr J Thorac Crit Care Med 26(2):10.7196. https://doi.org/10.7196/AJTCCM.2020.v26i2.077

Heart

Heart: Overview

Axel Gödecke and André Heinen

Anatomy and Physiology of the Heart

The circulatory system, consisting of the heart and arterial, venous, and capillary blood vessels, delivers oxygen and nutrients to all organs and transports metabolites to sites of further metabolization and excretion. To execute this central role, the heart continuously pumps the blood through the vasculature (see chapter "Blood Vessels: Overview" under part "Blood Vessels"). Although the heart is a single organ, in terms of function, it represents two pumps working in series. Whereas the left heart generates high pressure (basal level 120 mm Hg) to supply all organs (except the lung) with oxygenated blood, the right heart enables blood flow through the pulmonary vasculature by generating a low-pressure gradient (basal level 20 mm Hg). Left and right hearts show a similar gross anatomy: both consist of an atrium and a ventricle, which are separated by a septum. Due to the differences in workload, the muscle mass of the left ventricle exceeds that of the right one. To achieve a directed blood flow, inlet valves (i.e., mitral and tricuspid valves) separate the atria from the ventricles, and outlet valves (aortic and pulmonary valves) separate the ventricles from the aorta and pulmonary artery (Fig. 1).

The cardiac cycle is subdivided into two major parts: During systole, the walls of the blood-filled ventricles rapidly contract, resulting in a steep rise of left ventricular pressure. Since the aortic and mitral valves of the left ventricle are closed at the beginning of systole, the first phase represents an isovolumic contraction. However, when left ventricular pressure exceeds that in the aortic root, the aortic valve opens and the blood is pumped into the aorta. At the beginning of the ejection phase blood pressure still rises but begins to decline when the majority of the blood has left the ventricle. As soon as the left ventricular pressure drops below the aortic pressure, the aortic valve will be closed. This event demarcates the end of the systole and the beginning of diastole. The diastolic phase begins with a rapid isovolumic relaxation leading to a drop of left ventricular pressure below the atrial pressure. In consequence, the mitral valve opens for filling of the left ventricle. The early phase of filling is a passive event; the later phase, however, is mediated by atrial contraction.

A. Gödecke (✉) · A. Heinen
Heinrich-Heine-Universität Düsseldorf, Medical Faculty, Institut für Herz- und Kreislaufphysiologie, Düsseldorf, Germany
e-mail: axel.goedecke@uni-duesseldorf.de; andre.heinen@uni-duesseldorf.de

E. Lammert, M. Zeeb (eds.), *Metabolism of Human Diseases*,
https://doi.org/10.1007/978-3-031-96019-2_29

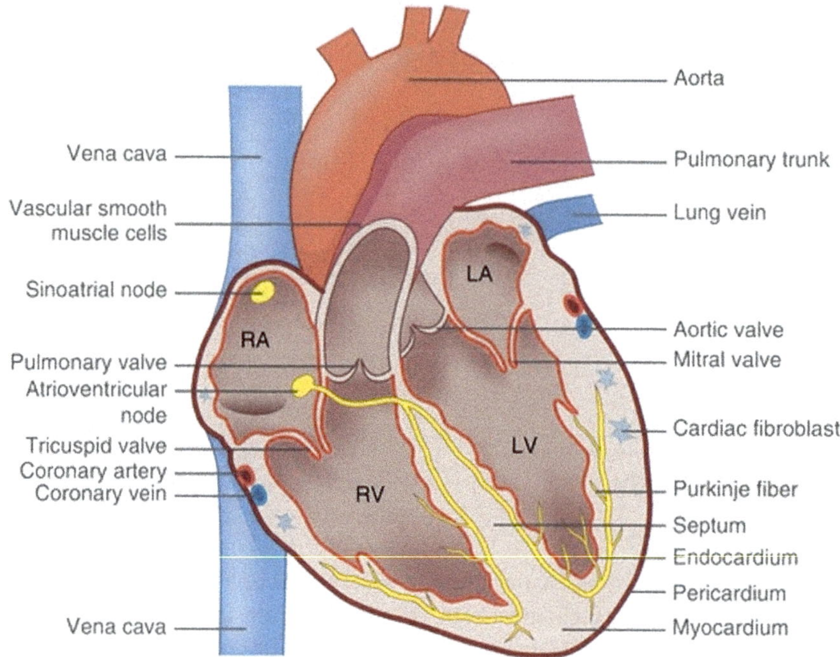

Fig. 1 Anatomic features of the human heart

Cellular Composition

On a cellular scale, cardiac myocytes (CMs), interstitial fibroblasts, and capillary endothelial cells represent the major cell fractions of the heart (Fig. 1). Moreover, the coronary arteries and veins contribute smooth muscle and endothelial cells to the cardiac cell pool.

CMs represent the major cell type in terms of mass and dominate the structure of the myocardium. CMs show a characteristic striated pattern similar to skeletal muscle, which is due to the parallel organization of the contractile actin and myosin filaments. At their ends, CMs are connected via intercalated discs. These structures are rich in gap junctions, which allow the spreading of the electric excitation over the myocardium due to a direct electric coupling of CMs. A highly coordinated sequence of temporal and spatial excitation and contraction of CMs leads to a continued cycle of ventricular contraction and relaxation of the whole heart.

A detailed knowledge of the exact cellular composition of the heart would be of interest from a scientific as well as from a therapeutic point of view. However, although this topic is investigated and discussed for a long time, the relative contribution of each cell type to the complete cardiac cell pool is still not resolved completely. Recent publications indicate that CMs might account for about 30–50% of all cells in ventricular myocardium, and fibroblasts for approximately 10–15% [1, 2]. Cardiac fibroblasts synthesize the extracellular matrix, which provides an extracellular protein scaffold for the attachment of CMs, fibroblasts, and endothelial cells. The amount and composition of the cardiac extracellular matrix are important determinants of myocardial stiffness. Therefore, activation of cardiac fibroblasts and enhanced matrix deposition which occur during aging but also in the context of heart failure (see chapter "Heart Failure") may lead to a reduced ventricular compliance, which impedes diastolic filling.

For the numbers of endothelial cells (ECs), however, highly divergent data with a variation between about 8 and 40% of all cardiac cells were reported [1, 2]. This variety seems to be caused, at least in part, by methodological differences in analytical approaches, and future investigations will continue to address this topic. ECs form not only the inner cell layer of vessels but also of cardiac chambers, termed endocardium. Finally, the heart is surrounded by the pericardium, protecting this vital organ.

The heart is supplied by its own circulatory system, which originates at the aortic root in the form of two main coronary arteries. Since the heart critically depends on aerobic metabolism, there is an extraordinary high capillary density within the myocardium [3].

Cardiac Muscle Contraction

The heartbeat is the consequence of a well-defined temporal and spatial sequence of excitation and contraction of the cardiac muscle. The origin of the electric excitation is the sinoatrial node (SA node), which consists ofspecializedC-Msable to depolarize spontaneously and autonomously. The SA node represents the primary pacemaker of the heart, and upon its depolarization, a wave of depolarization spreads over the atria. The atrioventricular plane which is assembled by the heart valves insulates the ventricles from the atria. The only conducting connection is the atrioventricular node (AV node), which is located in the right atrial septum. The atrial excitation passes the AV node and is then conducted to all regions of the ventricular walls by the specialized CMs forming the conduction system including the bundle of His, branches of Tawara, and the Purkinje fibers. The final step of ventricular excitation involves the conduction between CMs via gap junctions.

Due to an unstable resting potential, SA nodal cells depolarize spontaneously, and the SA node is able to initiate cardiac excitation autonomously. However, the slope of this diastolic depolarization is increased by noradrenaline released by sympathetic nerve fibers, which elevates heart rate (positive chronotropy). Acetylcholine, the parasympathetic transmitter, reduces heart rate by flattening the slope of diastolic depolarization. By similar mechanisms, sympathetic innervation of the AV node increases the AV conduction time (positive dromotropy), whereas acetylcholine decreases it (negative dromotropy).

Heart-Specific Metabolic/Molecular Pathways and Processes

With more than 3 billion contractions, the heart pumps more than 200 million l of blood through the vasculature during a normal human lifespan. In order to continuously perform its function, the heart critically depends on oxidative phosphorylation. Therefore, delivery of oxygen via coronary perfusion must tightly match cardiac oxygen demands even under high workload. As cardiac muscle contains only around 5 µmol ATP/g wet weight, all ATP would be consumed within 10 s if oxidative ATP generation was stopped.

Almost 95% of cardiac ATP generation occurs through oxidative phosphorylation, which represents the most efficient way to generate ATP. This high level of oxidative phosphorylation is reflected by a high mitochondrial content (30–35 Vol %) in the CMs. Glycolysis produces most of the residual ATP (<5%).

Most ATP (60–70%) is spent in CMs to generate mechanical work driving the circulation. Another large part (30–40%) is required to run ion pumps (mostly sarco−/endoplasmic reticulum Ca^{2+}-ATPase), which in a concerted manner generate the Ca^{2+} transients [4]. The periodic release and reuptake of Ca^{2+} are required for electromechanical coupling, linking cardiac action potentials with contraction. When the membrane potential of a CM reaches the threshold (-70 mV), opening of voltage-gated sodium channels depolarizes the membrane potential to $+30$ mV followed by the opening of L-type Ca^{2+} channels which mediate an influx of Ca^{2+} from the interstitial space. This Ca^{2+} current serves dual functions. First, it leads to a long-lasting (200–400 ms) depolarization at a voltage around 0 mV before K^+ conductivity reverses the membrane potential back to the diastolic values (-90 mV). Second, the Ca^{2+} influx gives rise to further Ca^{2+} release from intracellular stores by opening the ryanodine channel (Ryr2) in the membrane of the sarcoplasmic reticulum (SR). Finally, the intracellular Ca^{2+} concentration increases to 10^{-5} mol/L (from initially 10^{-7} mol/L). Subsequent Ca^{2+} binding to troponin C displaces tropomyosin from actin, enabling myosin to bind and to perform contraction. This is terminated by Ca^{2+} reuptake into the

SR and via export by the sarcolemmal Na^+/Ca^{2+} exchanger.

Sympathetic nerve fibers also innervate the myocardium. Norepinephrine increases Ca^{2+} influx, release from the SR, and accelerates reuptake into the SR leading to faster and increased Ca^{2+} transients, which are the basis for enhanced contractile force (positive inotropy).

As only a minor part of the ATP is required for basal biochemical reactions, an experimentally arrested heart consumes only 10% of the oxygen required to run a working heart. The high demand for ATP of a beating heart even under resting conditions results in a high degree of oxygen demand. The heart consumes 10% of the whole-body oxygen, although it contributes only 0.5% to whole-body mass. The high rate of oxygen consumption leads to low PO_2 values in cardiac tissues. The resultant steep O_2 gradient toward the capillaries favors oxygen release from hemoglobin and results in a high oxygen extraction (60–70%) from the perfused blood already under basal conditions. Thus, the elevated oxygen demand under conditions of high workload can be covered only to a small part by an increase in oxygen extraction (+20%). Instead, an increase in coronary flow (up to fivefold) largely ensures cardiac oxygen supply to sustain the oxidative generation of ATP.

Cardiac Substrate Metabolism

Cardiac substrate metabolism occurs in two major variants: The fetal heart depends to a large extent on glucose, whereas the adult heart prefers to consume fatty acids rather than glucose. However, the heart is also able to metabolize lactate and ketone bodies. Therefore, the adult heart has often been named a "metabolic omnivore." All substrates may finally drain into the citric acid cycle that is fueled by acetyl-CoA derived from glycolysis (glucose), β-oxidation (fatty acids), lactate oxidation, and ketone bodies. Interestingly, in heart failure (see chapter "Heart Failure"), the adult heart may reduce fatty acid oxidation in favor of glucose (termed metabolic remodeling) [5].

Fatty acids account for 50–70% of total cardiac energy supply in adults. Fatty acid uptake into CMs is mediated to a large extent by fatty acid translocase (FAT/CD36). This protein is in part localized in storage vesicles in CMs, which may fuse with the sarcolemma to enhance uptake of fatty acids (Fig. 2). Important stimuli of membrane translocation are an increase in cardiac workload (contraction-mediated translocation) and insulin. Fatty acids are further oxidized in the mitochondria during β-oxidation yielding high amounts of reduced nicotineamid adenine dinucleotide ($NADH + H^+$), reduced flavin adenine dinucleotide ($FADH_2$), and CO_2. Transport of acyl-CoA into the mitochondrion involves the carnitine shuttle system (Fig. 2).

Glucose uptake by CMs may be stimulated under anabolic conditions by insulin via Akt (also called protein kinase B) and under catabolic conditions by the adenosine monophosphate (AMP)-dependent protein kinase (AMPK). Both stimulate translocation of the glucose transporter 4 (GLUT4) to the sarcolemma, enhancing glucose uptake (Fig. 2). In part, glucose is used to synthesize glycogen, which serves as an energy store. Upon energy depletion and activation of AMPK, glucose is rapidly mobilized by breakdown of glycogen [6]. Independent of its origin, most glucose enters the glycolytic pathway, leading to the formation of pyruvate.

Lactate also contributes to cardiac ATP generation. It is taken up by the monocarboxylate transporter and converted to pyruvate by the lactate dehydrogenase reaction yielding $NADH + H^+$. The latter enters the respiratory chain, whereas pyruvate is oxidized to acetyl-CoA. Even under conditions of elevated workload, the heart rather consumes than generates lactate. Thus, during exercise, the heart may use lactate produced by anaerobic glycolysis in skeletal muscle.

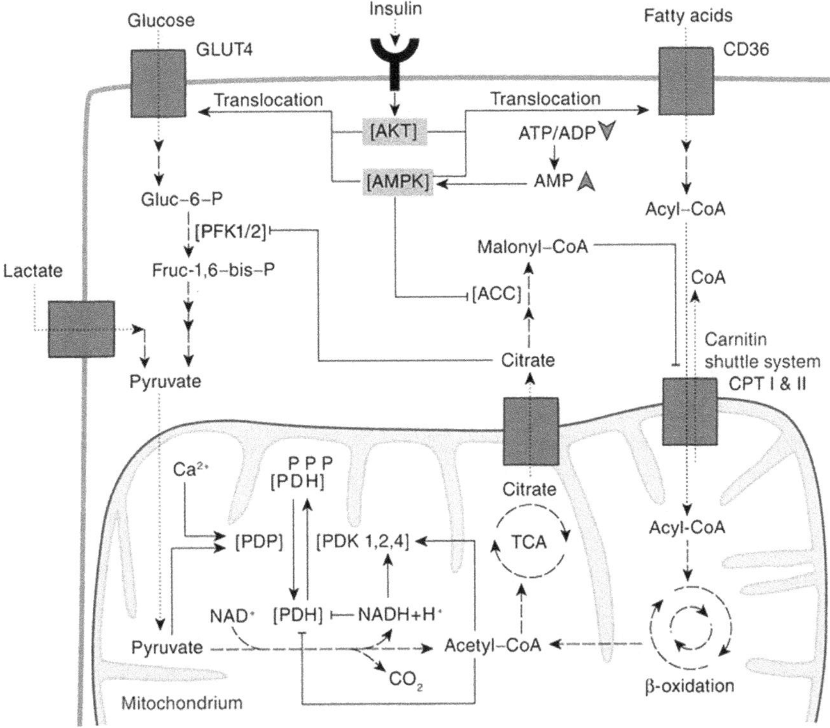

Fig. 2 Regulation of glucose and fatty acid metabolism in the heart. The basic principle of the glucose fatty acid cycle and its modulation by insulin and low energy is shown. A low ATP/ADP ratio leads to formation of AMP, which activates AMP-dependent protein kinase (*AMPK*). Insulin activates protein kinase B (*AKT*) via the insulin receptor. AKT and AMPK induce translocation of the glucose transporter 4 (*GLUT4*) and the long-chain fatty acid translocase (*CD36*) to the plasma membrane. Glucose is metabolized in glycolysis (*left*) to pyruvate, which is subsequently oxidized to acetyl-CoA by pyruvate dehydrogenase (*PDH*) in the mitochondrion. PDH underlies product inhibition by acetyl-CoA and NADH + H⁺ from the β-oxidation. Additionally, they stimulate specific PDH kinases (*PDKs*), which inhibit PDH by phosphorylation. PDH phosphatase (*PDP*) dephosphorylates and thus activates PDH. PDP is activated by pyruvate and Ca²⁺, linking mean elevated cytoplasmic Ca²⁺ under β-adrenergic stimulation to elevated glucose oxidation. Lactate is also converted to pyruvate. Fatty acids are converted to acyl-CoA and transported to the mitochondrion via the carnitine shuttle system consisting of carnitine palmitoyl-transferase (*CPT*) I and II, in the outer and inner mitochondrial membrane, respectively. Within the mitochondrion, acyl-CoAs are degraded to acetyl-CoA during β-oxidation. Acetyl-CoA is metabolized in the tricarboxylic acid cycle (*TCA*, also called citric acid cycle). Citrate from the TCA can be exported to the cytosol, where it inhibits early steps of glycolysis and can be converted to malonyl-CoA. The latter inhibits the carnitine shuttle system and thus fatty acid oxidation. AMPK was shown to inhibit this conversion by blocking acetyl-CoA carboxylase (*ACC*). *Fruc-1,6-bis-P* fructose-1,6-bisphosphate, *PFK1/2* phosphofructokinase 1/2

Inside-In: Metabolites of the Heart Affecting Itself

A hallmark of cardiac metabolism is the ability to switch between carbohydrates and fatty acids as preferred carbon sources (metabolic flexibility). Under conditions when oxygen is short, the heart uses more "oxygen-rich" glucose, which results in a higher P/O ratio (ATP per oxygen) than the use of the "oxygen-poor" fatty acids.

An important pathway is the Randle cycle [7, 8], also termed the glucose–fatty acid cycle, which is substantially involved in regulating the relative contribution of fatty acid vs. carbohydrate utilization by the heart. Pyruvate, acetyl-CoA, malonyl-CoA, and citrate are important metabolic intermediates, which modulate substrate utilization (Fig. 2).

Fatty acid oxidation increases the mitochondrial acetyl-CoA/CoA and NADH/NAD$^+$ ratios. These changes inhibit pyruvate dehydrogenase (PDH), which determines the oxidation of the glycolytic pyruvate to acetyl-CoA. The activity of PDH is controlled by PDH kinases and PDH phosphatases (Fig. 2). In addition, citrate, leaving mitochondria via citrate carriers, inhibits phosphofructokinases 1 and 2 and therefore reduces glycolysis. However, cytoplasmic citrate can also be converted to acetyl-CoA (and oxaloacetate), which is further converted to malonyl-CoA. Malonyl-CoA inhibits the carnitine shuttle system and thereby long-chain fatty acid uptake by mitochondria.

In the heart, the transport capacity of GLUT4 transporters is almost saturated at physiological plasma glucose levels limiting glucose inflow and thus glycolytic flux. Stimulation of GLUT4 translocation (see above) elevates glycolytic flux and increases pyruvate, releasing the blockade of PDH by PDH phosphatases (Fig. 2). In consequence, glucose metabolism is elevated despite the presence of fatty acids. However, the precise interplay of metabolic factors directing cardiac metabolism toward one or the other direction is still not fully understood [9].

The extent of metabolic remodeling is substantially modulated by comorbidities such as diabetes (see chapter "Diabetes Mellitus"), hypertension, and hypercholesterolemia. Thus, identification of the master switches directing cardiac metabolism toward a protective program is of high medical importance.

Further important metabolites released by the heart are nitric oxide (NO) and adenosine. Both of them are only short-lived and exert local functions. These include modulation of coronary vascular tone and cardiac contractility (see below). Moreover, NO and adenosine are able to keep platelets and leukocytes in a quiescent state, acting as antithrombotic and anti-inflammatory factors.

In the heart, NO, which is derived from L-ARGININE, is synthesized in the coronary endothelium by the endothelial NO synthase (type III NOS). In CMs, the endothelial as well as the neuronal NO synthase (type I NOS) are constitutively expressed. Under inflammatory conditions, also the inducible NO synthase (type II NOS) may be upregulated. NO modulates cardiac contractile function, and depending on the NO concentrations, both positive and negative inotropic actions have been described. The underlying mechanism appears to involve modulation of Ca^{2+} homeostasis as well as the availability of cyclic AMP (cAMP) due to the modulation of phosphodiesterases. NO-mediated vasodilation involves a direct activation of the soluble guanylyl cyclase, elevation of cyclic guanosine monophosphate (cGMP), and activation of the cGMP-dependent protein kinase (PKG). PKG in turn activates the myosin light chain (MLC) phosphatase by phosphorylation, which leads to a dephosphorylation of the regulatory myosin light chain in smooth muscle cells and the concomitant reduction of vascular tone.

Besides NO derived from NOS, recent studies suggest that under hypoxic conditions, deoxygenated hemoglobin and myoglobin reduce circulating nitrite to NO. This interesting concept provides a link between cardiac ischemia and the vasodilatory and cardioprotective functions of NO [10].

Under hypoxic or ischemic conditions, ATP breakdown may lead to the formation of adenosine. Adenosine released from the myocardium or formed extracellularly from ATP is able to elevate cardiac perfusion by the activation of smooth muscle cell A_2A receptors and subsequent enhancement of cAMP. cAMP in turn stimulates PKA, which phosphorylates and thereby inhibits the regulatory myosin light chain kinase and stimulates MLC phosphatase, leading to relaxation. Moreover, activation of K_{ATP} channels via PKA-mediated phosphorylation may lead to hyperpolarization of smooth muscle cells and therefore vasodilation. The cardiac release of adenosine is minimal as long as oxygen supply is sufficient to match the demands. Therefore, in contrast to NO, which is important for the setting of the basal vascular tone, adenosine-mediated vasodilation is more important under hypoxic conditions.

Inside-Out: Metabolites of the Heart Affecting Other Tissues

In the context of inside-out signaling, peptide factors released by the heart act on other organs. The natriuretic peptides (NPs), atrial and brain NP, which are expressed predominantly in the atria and the ventricles, respectively, are released from the heart in response to mechanical signals such as stretch and elevated mechanical load, for example, during volume overload or exercise. They activate membrane-bound guanylyl cyclase receptors, which elevate cGMP levels in target cells and regulate fluid homeostasis and reduce blood pressure via increased Na^+ excretion (see chapter "Kidney: Overview" under part "Kidney"). However, recent results indicate that natriuretic peptides also have an important metabolic function. NP receptors are present on adipocytes and their activation stimulates lipolysis. Interestingly, the potency of NP to enhance lipolysis is similar to that of β-adrenergic stimulation, which represents the classical way. Whereas the latter is mediated by the cAMP–PKA pathway, NPs exert their effects via cGMP and PKG. The physiological role of this pathway can be seen in an elevated supply of heart and skeletal muscle with fatty acids during exercise, when mechanic forces stimulate the release of NP from the heart [11].

It is well known that heart failure (see chapter "Heart Failure") may be associated with the loss of skeletal muscle. This interconnection of cardiac dysfunction and skeletal muscle wasting has been termed cardiac cachexia. Chronic heart failure leads to upregulation of myostatin expression in cardiac tissue. Myostatin, a cytokine of the transforming growth factor-β (TGF-β) family, inhibits muscle development and might be involved in cardiac cachexia. Exercise reduces myostatin expression, whereas physical inactivity may enhance it. The mechanisms by which myostatin leads to atrophy of skeletal muscle involves a reduced activation of protein kinase B (AKT), which might diminish protein synthesis via low mTOR activity or enhance protein degradation via induction of atrogin-1.

Outside-In: Metabolites of Other Tissues Affecting the Heart

Although fatty acids represent an important fuel for covering cardiac energy supply, elevated levels of circulating free fatty acids (FFA) may negatively impact cardiac structure and function [4]. Cytosolic lipolysis occurs by an enzymatic cascade, initiated by adipose triglyceride lipase (ATGL), followed by hormone-sensitive lipase (HSL) and monoacyl glycerol lipase (MGL). Numerous studies of recent years on the function of ATGL have demonstrated an enhanced lipolysis in adipose tissue during heart failure and acute myocardial infarction which may be the result of enhanced sympathetic tone and natriuretic peptide release. Inhibition of ATGL in adipocytes by pharmacologic blockade or adipocyte specific deletion, resulted in improved pump function, as well as attenuated inflammation and apoptosis in the context of cardiac hypertrophy and myocardial infarction, respectively [12]. In terms of mechanism, the elevated uptake of fatty acids is frequently associated with insulin resistance and leads to the development of cardiac lipotoxicity characterized by contractile dysfunction, CM apoptosis, and fibrosis. An enhanced FFA uptake may increase the cardiac stores of triacylglycerol or triglyceride (TG). Enhanced deposition of TG in the myocardium results in cardiac steatosis, that is, the ectopic deposition of surplus fatty acids in the form of TG in CMs. The enhanced availability and metabolism of TG also elevate ceramide levels in the heart. Ceramides have been shown to induce apoptosis of CMs by enhancing mitochondrial permeability. Moreover, ceramides inhibit Akt isoforms, which are involved in cellular metabolism, growth, and survival. On the other hand, elevated TGs also enhance diacylglycerol (DAG) known to activate conventional and novel PKC isoforms [13], which may be protective or pathologic. Among the latter, PKCβ1 was shown to phosphorylate the β subunit of the insulin receptor as well as insulin receptor substrate 1 (IRS1) on multiple serine/threonine residues thus contributing to insulin resistance. In addition to these TG- and FFA-related effects on the heart, there is increasing interest in further cross-talk mechanisms between adipose tissue and the heart. For example, a special interest is related to signaling molecules released from brown adipose tissue (BAT) (see chapter "Fat Tissue: Overview"), so-called "batokines" [14]. There is experimental evidence that this heterogeneous group of molecules has an impact on systemic metabolism as well as on cardiac function and health. One example for a member of the batokine group is the linoleic acid derivate 12,13-diHOME that is released from BAT in response to cold exposure and exercise, and improves cardiac pump function, calcium cycling, and fatty acid uptake [15]. However, the (patho-) physiological relevance and therapeutical potential of these fat-heart cross-talk molecules is currently unknown.

Also, enhanced glucose levels can have detrimental effects on cardiac function. For example, hyperglycemia enhances diacylglycerol (DAG) formation and activation of PKC. Interestingly, in the presence of high fatty acid levels, glucose oxidation is blocked. Glucose is redirected to other pathways including the pentose phosphate pathway, which leads to synthesis of NADPH required for the formation of glutathione, an important antioxidant.

Final Remarks

The healthy heart is a metabolic omnivore, which is able to utilize different carbon sources including glucose, lactate, ketone bodies, and fatty acids for oxidative generation of ATP. The hypertrophic and failing heart, respectively, switch their metabolism from the usually preferred fatty acids to higher glucose consumption. Although still not unambiguously clarified, many experimental studies indicate that this metabolic remodeling is an important mechanism contributing to adaptation during the compensated state in heart failure progression and encourage to develop metabolic therapies to treat heart failure [16].

Questions and Answers

Question 1 Name the primary mechanism of cardiomyocytes to generate ATP.

Answer 1 Oxidative phosphorylation

Question 2 Name the two most relevant ATP consuming processes of the myocardium.

Answer 2 Mechanical work (contraction) and ion homeostasis (ion pumps)

Question 3 Name the most relevant substrate of the healthy heart.

Answer 3 Fatty acids

Question 4 How is the disease-related shift in myocardial substrate preference named?

Answer 4 Metabolic remodeling

References

1. Litvinukova M et al (2020) Cells of the adult human heart. Nature 588(7838):466–472
2. Pinto AR et al (2016) Revisiting cardiac cellular composition. Circ Res 118(3):400–409
3. Rakusan K et al (1992) Morphometry of human coronary capillaries during normal growth and the effect of age in left ventricular pressure-overload hypertrophy. Circulation 86(1):38–46
4. Lopaschuk GD et al (2010) Myocardial fatty acid metabolism in health and disease. Physiol Rev 90(1):207–258
5. Lopaschuk GD et al (2021) Cardiac energy metabolism in heart failure. Circ Res 128(10):1487–1513

6. Longnus SL et al (2003) 5-Aminoimidazole-4-carboxamide 1-beta -D-ribofuranoside (AICAR) stimulates myocardial glycogenolysis by allosteric mechanisms. Am J Physiol Regul Integr Comp Physiol 284(4):R936–R944

7. Hue L, Taegtmeyer H (2009) The Randle cycle revisited: a new head for an old hat. Am J Physiol Endocrinol Metab 297(3):E578–E591

8. Randle PJ et al (1963) The glucose fatty-acid cycle. Its role in insulin sensitivity and the metabolic disturbances of diabetes mellitus. Lancet 1(7285):785–789

9. Choi CS et al (2007) Continuous fat oxidation in acetyl-CoA carboxylase 2 knockout mice increases total energy expenditure, reduces fat mass, and improves insulin sensitivity. Proc Natl Acad Sci USA 104(42):16480–16485

10. Deussen, A., et al., *Metabolic coronary flow regulation--current concepts.* Basic Res Cardiol, 2006. 101(6): p. 453–464

11. Moro C et al (2006) Atrial natriuretic peptide stimulates lipid mobilization during repeated bouts of endurance exercise. Am J Physiol Endocrinol Metab 290(5):E864–E869

12. Kintscher U et al (2020) The role of adipose triglyceride lipase and cytosolic lipolysis in cardiac function and heart failure. Cell Rep Med 1(1):100001

13. Geraldes P, King GL (2010) Activation of protein kinase C isoforms and its impact on diabetic complications. Circ Res 106(8):1319–1331

14. Pinckard KM, Stanford KI (2022) The heartwarming effect of Brown adipose tissue. Mol Pharmacol 102(1):460–471

15. Pinckard KM et al (2021) A novel endocrine role for the BAT-released Lipokine 12,13-diHOME to mediate cardiac function. Circulation 143(2):145–159

16. Honka H et al (2021) Therapeutic manipulation of myocardial metabolism: JACC state-of-the-art review. J Am Coll Cardiol 77(16):2022–2039

Heart Failure

Rojda Ipek, Mareike Cramer, and Malte Kelm

Introduction

Heart failure (HF) is a multisystemic disease defined by structural and/or functional cardiac abnormalities. These abnormalities lead to an imbalance of oxygen supply and demand, resulting in elevated intracardiac pressures and/or inadequate cardiac output [1]. HF can arise from various inherited or acquired cardiac and extra cardiac conditions, with coronary artery disease (CAD) and hypertension being the predominant causes in developed nations [1]. Globally, HF affects 64.3 million people (approximately 1–2% of adults), with prevalence exceeding 10% among individuals over 70 years of age [1, 2]. HF poses a significant socioeconomic burden, with annual healthcare costs of up to € 25,000 per patient in the Western world, and is associated with a five-year mortality rate of 50–75% [2].

HF can be classified into three primary phenotypes based on the measurement of left ventricular ejection fraction (LVEF): HF with reduced ejection fraction (HFrEF, LVEF <40%), HF with mildly reduced ejection fraction (HFmrEF, LVEF 40–49%), and HF with preserved ejection fraction (HFpEF, LVEF $\geq$50%) [1]. HFrEF and HFmrEF are designated for impaired systolic function with a higher frequency of an ischemic etiology [1]. In contrast, HFpEF, accounting for approximately half of the HF cases, is characterized by diastolic dysfunction, increased ventricular stiffness, and structural remodeling such as hypertrophy and left atrial enlargement. HFpEF is further associated with older age, female sex, hypertension, obesity, and diabetes (see chapter "Diabetes Mellitus") [1, 3].

The diagnosis of HF relies on clinical assessments, imaging modalities, and laboratory testing. HF can be assessed indirectly through findings on physical examination, such as elevated jugular venous pressure, pulmonary crackles, and peripheral edema, which suggest congestion indicating cardiac dysfunction. Echocardiography is central to the diagnosis, as it provides detailed evaluation of LVEF, wall motion abnormalities, and diastolic function, revealing structural and functional abnormalities [1]. Additionally, blood tests measuring natriuretic peptides, such as B-type natriuretic peptide (BNP) or N-terminal pro-BNP (NT-proBNP), are used to differentiate HF from other causes of dyspnea. BNP is released in response to myocardial stretch and volume overload, with elevated levels supporting the diagnosis of HF [1, 4].

The pathophysiology of HF involves significant functional, systemic, and metabolic dearangements [4–6]. The neurohormonal systems,

R. Ipek · M. Cramer · M. Kelm (✉)
Universitätsklinikum Düsseldorf, Klinik für Kardiologie, Pneumologie und Angiologie, Düsseldorf, Germany
e-mail: rojda.ipek@med.uni-duesseldorf.de;
mareike.cramer@med.uni-duesseldorf.de;
malte.kelm@med.uni-duesseldorf.de

E. Lammert, M. Zeeb (eds.), *Metabolism of Human Diseases*,
https://doi.org/10.1007/978-3-031-96019-2_30

including the sympathetic nervous system (SNS) and renin-angiotensin-aldosterone system (RAAS), play a central role in HF pathophysiology. Whilst SNS and RAAS activation initially compensate for reduced cardiac output, it becomes maladaptive in the long-term, leading to increased oxidative stress and myocyte apoptosis [4, 5]. One of the key metabolic changes in the myocardium is the shift from its reliance on fatty acids to an increased dependence on glucose and ketone bodies for energy production [6]. As a result, the failing heart is characterized by a substrate shift toward glycolysis and ketone oxidation to address the high energetic demand of the constantly contracting heart [7]. This metabolic reprogramming is influenced by the downregulation of peroxisome proliferator-activated receptors (PPARs), which are pivotal in regulating fatty acid oxidation [6]. Concurrently, mitochondrial dysfunction leads to elevated reactive oxygen species (ROS) production, exacerbating oxidative stress and further impairing cardiac function [6, 7]. These pathophysiological changes contribute to the progression of HF and present potential targets for therapeutic intervention [6, 7].

Emerging therapeutic strategies have revolutionized HF management. Sodium-glucose cotransporter-2 (SGLT2) inhibitors, initially developed for diabetes, have demonstrated significant benefits in reducing hospitalization and mortality across all HF phenotypes [1, 8]. Similarly, glucagon-like peptide-1 (GLP-1) receptor agonists show promise, particularly in patients with coexisting overweight or obesity, offering both metabolic and cardiovascular benefits [9]. The combination of these novel therapies, alongside established treatments such as beta-blockers, angiotensin-converting enzyme (ACE) inhibitors, and mineralocorticoid receptor antagonists (MRAs), underscores the importance of a multimodal approach in addressing the diverse pathophysiological pathways in HF [1, 8].

Alongside new pharmacological treatments, ever-advancing imaging techniques, such as cardiovascular magnetic resonance (CMR), may begin to play a pivotal role in improving diagnostics and therapies for HF. CMR provides unparalleled precision in assessing myocardial structure, fibrosis, and function, offering new opportunities for distinguishing between HF phenotypes and guiding personalized treatment strategies [1, 10, 11].

Pathophysiology

HF represents a multifaceted syndrome arising from disturbances in cardiac structure and function, systemic compensatory mechanisms, and metabolic inefficiencies. These processes collectively compromise the heart's ability to maintain adequate cardiac output to meet metabolic demands in both HFrEF and HFpEF. However, the pathophysiological and mechanistic processes underlying these conditions reveal some distinctions that influence the clinical outcomes of the two phenotypes, which include the following:

Functional Impairment

HFrEF and HFpEF are defined by distinct pathophysiological mechanisms, each leading to a different pattern of functional deterioration. In HFrEF, progressive myocardial injury leads to chamber dilation and eccentric remodeling, driven by myocyte loss, reduced contractile function, and increased wall stress, which result in impaired systolic function [4]. Conversely, HFpEF is characterized by diastolic dysfunction resulting from increased myocardial stiffness and fibrosis. This stiffness arises from enhanced extracellular matrix deposition and microvascular inflammation, which restrict ventricular relaxation and filling during diastole. Additionally, endothelial dysfunction and systemic inflammation contribute to interstitial fibrosis, compounding diastolic impairment in HFpEF [3, 12].

Systemic Compensatory Mechanisms

The neurohormonal systems, including the SNS and the RAAS are central to the pathophysiology of HF. In response to the decreased cardiac output, SNS activation is triggered, leading to increased heart rate through β-adrenergic receptor stimulation and myocardial contractility as compensatory mechanisms to maintain circulatory function [4, 5]. However, chronic SNS activation can be maladaptive, leading to increased oxidative stress, and myocyte apoptosis [5].

Decrease in cardiac output further causes impaired renal perfusion, which activates RAAS, promoting vasoconstriction, sodium retention, and increased afterload. Albeit expedient in the acute setting, in the long-term, these responses worsen cardiac hemodynamics with increase in both pre- and afterload (due to increased volume and peripheral resistance, respectively) [4, 5]. Several counter-regulatory biological systems, including natriuretic peptides such as ANP (atrial natriuretic peptide) and BNP, are upregulated to mitigate the harmful effects of SNS and RAAS signaling [4]. These peptides are released in response to increased myocardial and atrial stretch and act to unload the heart by promoting renal sodium and water excretion, vasodilation, and inhibition of renin and aldosterone release from the kidneys [4].

While initially protective, sustained neurohormonal activation contributes to the progression of HF by promoting detrimental cardiac remodeling and functional decline.

Metabolic Inefficiencies

Metabolic remodeling is believed to precede, initiate, and sustain structural and functional changes underlying HF. This process involves significant derangement in all three fundamental steps of energy generation: substrate uptake, metabolic pathways, and mitochondrial adenosine triphosphate (ATP) production [6].

(i) Glucose and fatty acid metabolism

In HFrEF, significant metabolic reprogramming occurs, characterized by a shift away from fatty acid oxidation as the primary energy source toward glucose metabolism [6, 7], in response to limitations in oxygen supply. This shift is driven by the downregulation of peroxisome proliferator-activated receptor alpha (PPARα), a transcription factor pivotal for regulating fatty acid oxidation enzymes in the myocardium [6]. This increased reliance on glucose metabolism leads to enhanced glycolysis, a less efficient process for ATP production compared to fatty acid oxidation, which exacerbates energy deficits in the myocardium [6, 7]. In HFpEF, changes in glucose and fatty acid metabolism are more variable as compared to HFrEF due to the heterogeneity of phenotypes within the HFpEF spectrum [9].

(ii) Ketone utilization

In HF, ketone bodies emerge as an alternative energy source when glucose and fatty acid metabolism are impaired. Ketone bodies (β-hydroxybutyrate and acetoacetate) are produced by the liver during periods of fasting. Upon entering cardiomyocytes and the mitochondrial matrix, ketone bodies provide a more efficient substrate for ATP production compared to glucose, helping to alleviate the inefficiency of other metabolic pathways [6, 9].

(iii) Mitochondrial dysfunction

Mitochondrial dysfunction is a critical contributor to the pathophysiology of HF, as it impairs the heart's ability to generate sufficient energy. A key consequence of mitochondrial dysfunction in HF is the increased production of reactive oxygen species (ROS), which results from impaired oxidative phosphorylation. This elevation in ROS not only impairs mitochondrial function but also contributes to cellular damage, amplifying oxidative stress and promoting further myocardial injury [6, 7]. Another significant aspect of mitochondrial dysfunction in HF is the alteration of calcium handling, particularly the downregulation of sarcoplasmic reticulum Ca^{2+}-ATPase (SERCA2a). Reduced SERCA2a activity disrupts calcium homeostasis, impairing myocardial relaxation and contributing to diastolic dysfunction in HF [6]. Additionally, the phosphocreatine/ATP (PCr/ATP) ratio, an important indicator of cellular energetic balance, is both reduced in HFrEF and HFpEF, reflecting impaired mitochondrial ATP production and further energy depletion in the myocardium [6, 9, 13].

Treatment

Due to the complex pathophysiological mechanisms of HF, therapeutic management requires a multifaceted approach, including pharmacological therapies targeting neurohormonal and metabolic pathways, and device-based interventions for advanced HF.

Neurohormonal Pathway: Beta-Blockers and RAAS Blockade

Mainstay of current HF treatment is the inhibition of the neurohormonal systems. Beta-blockers effectively counteract SNS activation in HF by inhibiting beta-adrenergic receptors. This reduces the adverse effects of sustained adrenergic stimulation on the myocardium, including maladaptive remodeling and increased myocardial oxygen demand [1]. These agents exhibit dose-dependent selectivity for beta-1 adrenergic receptors, particularly in the myocardium. Their administration has been shown to improve LVEF and diastolic filling time in HF patients [1, 14]. Furthermore, chronic sympathetic activation contributes to increased catecholamine levels and heightened vasoconstriction, which beta-blockers mitigate, thereby improving hemodynamic stability [14]. Beta-blockers also reduce the risk of arrhythmias by lowering heart rate (negative chrono- and dromotropy), contributing to their overall beneficial effects in HF management [1, 14].

Rather, RAAS inhibition forms a cornerstone of HF therapy by mitigating maladaptive neurohormonal activation [15]. Key pharmacological strategies include angiotensin-converting enzyme (ACE) inhibitors, angiotensin receptor blockers (ARBs), mineralocorticoid receptor antagonists (MRAs), and angiotensin receptor-neprilysin inhibitors (ARNIs), all of which target the pathophysiological consequences of RAAS activation.

ACE inhibitors work by inhibiting the conversion of angiotensin I to angiotensin II, leading to reduced vasoconstriction, aldosterone release, and sodium retention, thus reducing myocardial remodeling [15]. ARBs block the angiotensin II type 1 receptor, preventing the vasoconstrictor effects of angiotensin II. Without inhibiting ACE itself, ARBs offer an alternative to ACE inhibitors in patients who experience intolerable side effects like cough [1, 15]. MRAs block the effects of aldosterone, reducing sodium retention, myocardial fibrosis, and vascular damage, which are significant contributors to the progression of HF [15]. Lastly, ARNIs, combining ARB and a neprilysin inhibitor, offer additional benefits by increasing natriuretic peptide levels, which promote vasodilation and reduce fluid overload [1, 15]. Both beta blockers and RAAS inhibition were shown to clearly reduce hospitalization and improve long-term morbidity and mortality in HFrEF. These therapies form the cornerstone of modern guideline-directed medical therapy for HF [1, 4, 15]. However, the use of RAAS inhibitors carries a risk of adverse effects such as hypotension and hyperkalemia (through their inhibition of aldosterone effects), which is a particular problem in patients with pre-existing renal dysfunction, necessitating regular monitoring [1, 4, 15].

Cardiorenal Pathway: Loop Diuretics

Loop diuretics are widely used to treat fluid retention in HF patients. These drugs primarily alleviate symptoms such as dyspnea and edema by promoting natriuresis and diuresis [16]. Loop diuretics are recommended for the use in patients with HFrEF who exhibit signs and/or symptoms of congestion, as they effectively alleviate HF symptoms, reduce hospitalizations, and improve exercise capacity [1]. Furthermore, diuretics are also utilized in the management of HFpEF to reduce congestion and improve patient symptoms [8]. They act by inhibiting the sodium-potassium-chloride cotransporter in the thick ascending limb of the loop of Henle, thereby increasing sodium and water excretion [16]. However, loop diuretics are considered symptomatic therapies rather than disease-modifying treatments, as they do not influence the progression of HF [16]. Their use can lead to electrolyte imbalance, dehydration, and worsening renal function if not closely monitored, necessitating careful monitoring and individualized dosing strategies [16].

Metabolic Pathway: SGLT2 Inhibitors

Sodium-glucose cotransporter-2 (SGLT2) inhibitors, such as empagliflozin and dapagliflozin, have significantly impacted the management of HF in recent years. These drugs inhibit SGLT2, reducing glucose reabsorption and promoting natriuresis. In both HFrEF and HFpEF, SGLT2 inhibitors have been shown to reduce hospitalization for HF and improve survival, independent of diabetes status. Thus, their use is now recommended for the treatment of both HFrEF and HFpEF [8, 17, 18]. However, recent research challenges the potential for these agents to improve myocardial metabolism. No significant improvements in myocardial PCr/ATP ratio or β-hydroxybutyrate concentrations was observed after treatment with empagliflozin in both HFrEF and HFpEF patients [19]. Despite these findings, SGLT2 inhibitors remain integral to the therapeutic regime, as they offer proven benefits in reducing HF hospitalizations and mortality [8].

Chronotropic and Inotropic Pathway: Ivabradine and Digitalis Glycosides

Ivabradine is a treatment option for patients with HFrEF who remain symptomatic and have a resting heart rate of ≥ 70 beats per minute despite optimal medical therapy (OMT) with beta-blockers, ACE inhibitors/ARBs, or ARNI [1]. Ivabradine selectively inhibits the I_f current, known as pacemaker current, in the sinoatrial node (see chapter "Heart: Overview"), resulting in a reduced heart rate. This leads to improved diastolic filling, which alleviates symptoms, decreases hospitalizations, and lowers the risk of cardiovascular death. [1, 20, 21]

Digitalis glycosides, such as digoxin, are used in HFrEF patients who remain symptomatic despite OMT, particularly those with atrial fibrillation and rapid ventricular response or persistent symptoms [1]. By inhibiting the Na^+/K^+-ATPase pump, digoxin increases cytosolic Ca^{2+} via the Na^+/Ca^{2+} exchanger, thereby enhancing myocardial contractility (positive inotropy) [22]. Although digoxin improves symptoms, it does not modify disease progression or reduce mortality and is associated with a narrow therapeutic index. To avoid toxicity, close monitoring of digoxin levels, renal function, and potassium levels is essential, particularly in older patients or those with impaired renal function [1].

Electrical Device Therapy: ICD and CRT

In advanced HF, device therapies like implantable cardioverter defibrillators (ICDs) and cardiac resynchronization therapy (CRT) become relevant because pharmacological treatments alone may not adequately manage symptoms or prevent sudden cardiac death. Sudden cardiac death can occur in advanced HF due to arrhythmias caused by electrical disturbances in the heart, which are common in patients with severely reduced ejection fraction [1].

For ICDs, primary prevention is recommended for patients with symptomatic HF, an LVEF ≤35%, and ischemic etiology, who have been on OMT for at least 3 months. ICDs are also indicated for secondary prevention in patients who have survived a life-threatening arrhythmia causing hemodynamic instability [1].

CRT improves ventricular synchronization and myocardial contractility through biventricular pacing, which helps alleviate symptoms such as dyspnea and fatigue. It is recommended for patients with symptomatic HF and an LVEF ≤35%, who remain symptomatic despite OMT. CRT-Pacemaker (CRT-P) implantation, providing biventricular pacing, is indicated for patients with left bundle branch block (LBBB) and a QRS duration of 130–149 ms. In contrast, CRT-Defibrillator (CRT-D) implantation, combining biventricular pacing with defibrillator function, is recommended for patients with non-LBBB or a QRS duration of ≥150 ms, who are at high risk for sudden cardiac death [1].

Perspectives

Many HF patients continue to have progressive disease despite optimal guideline-directed therapy. Moreover, as HF progresses, many patients become refractory or intolerant to conventional medical therapy [4, 23]. Thus, there is a need to develop advanced treatment options that do not have overlapping side effects or intolerances (hypotension, renal dysfunction, hyperkalemia) with current therapies. Recent therapeutic approaches include:

Finerenone, a non-steroidal MRA, has shown promise in HF with mildly reduced or preserved ejection fraction (HFmrEF/HFpEF) by reducing adverse cardiovascular outcomes, including hospitalization due to HF. Its dual action on inflammation and fibrosis, without the typical side effects of hyperkalemia, makes it a potential alternative to traditional therapies for patients with these conditions [24].

GLP-1 receptor agonists, such as semaglutide, have gained attention for their potential benefits in HF due to their effects on metabolic regulation, particularly in enhancing mitochondrial efficiency and improving metabolic flexibility. These agents help improve energy utilization in the myocardium. They reduce HF symptoms and cardiovascular events, particularly in patients with HFpEF or coexisting obesity, independent of their effects on glycemic control [9, 25, 26].

Recently, intralipid infusion has been proposed as a strategy to support metabolic flexibility in HFrEF, helping cardiomyocytes utilize fatty acids more efficiently. Intralipid infusion has been associated with improved LVEF, enhanced diastolic function, and a more favorable PCr/ATP ratio, indicating better mitochondrial energetics and overall myocardial performance [27].

Ketone supplementation presents a promising therapeutic approach by providing an alternative energy substrate to the heart, particularly in patients with HF who have impaired glucose oxidation. This shift in metabolism may improve myocardial efficiency and reduce symptoms, offering a potential benefit when conventional therapies are insufficient to address the metabolic dysregulation associated with HF [9].

In addition to novel pharmacological therapies, advancements in imaging techniques are playing a pivotal role in enhancing the precision of HF diagnosis and guiding personalized treatment strategies. Cardiovascular magnetic resonance (CMR) imaging has evolved into a multiparametric modality in the characterization of HF, providing detailed insights into myocardial structure, function, hemodynamics, and metabolism. It may help in differentiating HFrEF and HFpEF phenotypes by assessing myocardial fibrosis, functional abnormalities, and evaluating myocardial energetics as PCr/ATP ratios [10, 11]. This comprehensive phenotyping enables the identification of pathophysiological mechanisms underlying HF in individual patients, creating opportunities to develop personalized therapies. As a result, CMR is increasingly becoming relevant in HF management guidelines [1, 10, 11], with the potential to significantly influence clinical decision-making and advance personalized treatment approaches for HF patients.

Overall, the prevalence of HF is projected to further increase in the coming decades, driven by demographic changes and increasing comorbidities [1, 2]. This highlights the critical need for continued advancements in HF therapies and precision medicine to address the growing burden of global HF.

Questions and Answers

Question 1 Please name the three types of heart failure!

Answer 1 HF with reduced ejection fraction (HFrEF, LVEF <40%), HF with mildly reduced ejection fraction (HFmrEF, LVEF 40–49%), and HF with preserved ejection fraction (HFpEF, LVEF ≥50%).

Question 2 What is the major pathophysiologic difference between HFrEF and HFpEF?

Answer 2 In HFrEF, progressive myocardial injury leads to chamber dilation and eccentric remodeling, driven by myocyte loss, reduced contractile function, and increased wall stress, which result in impaired systolic function [4]. In contrast, HFpEF is characterized by diastolic dysfunction resulting from increased myocardial stiffness and fibrosis; the latter arising from enhanced extracellular matrix deposition and microvascular inflammation, which restrict ventricular relaxation and filling during diastole.

Question 3 What kind of energy source does the healthy heart use and how does it change in HFrEF?

Answer 3 The healthy heart prefers fatty acids as key nutrients. In contrast, the HFrEF heart switches to glucose in response to reduced oxygen supply and to ketone bodies.

Question 4 What is the benefit of SGLT2 inhibitors for HFrEF and HFpEF?

Answer 4 In both HFrEF and HFpEF, SGLT2 inhibitors have been shown to reduce hospitalization for HF and improve survival, independent of diabetes status.

References

1. McDonagh TA et al (2021) 2021 ESC guidelines for the diagnosis and treatment of acute and chronic heart failure. Eur Heart J 42:3599–3726
2. Savarese G et al (2023) Global burden of heart failure: a comprehensive and updated review of epidemiology. Cardiovasc Res 118:3272–3287
3. Redfield MM, Borlaug BA (2023) Heart failure with preserved ejection fraction: a review. JAMA 329:827–838
4. Mann DL, Felker GM (2021) Mechanisms and models in heart failure: a translational approach. Circ Res 128:1435–1450
5. Floras JS, Ponikowski P (2015) The sympathetic/parasympathetic imbalance in heart failure with reduced ejection fraction. Eur Heart J 36:1974–1982b
6. Ng SM, Neubauer S, Rider OJ (2023) Myocardial metabolism in heart failure. Curr Heart Fail Rep 20:63–75
7. Honka H et al (2021) Therapeutic manipulation of myocardial metabolism: JACC state-of-the-art review. J Am Coll Cardiol 77:2022–2039
8. Force M et al (2024) 2023 Focused update of the 2021 ESC guidelines for the diagnosis and treatment of acute and chronic heart failure: developed by the task force for the diagnosis and treatment of acute and chronic heart failure of the European Society of Cardiology (ESC) With the special contribution of the Heart Failure Association (HFA) of the ESC. Eur J Heart Fail 26:5–17
9. Ng YH, Koay YC, Marques FZ, Kaye DM, O'Sullivan JF (2024) Leveraging metabolism for better outcomes in heart failure. Cardiovasc Res 120:1835–1850
10. Pan J, Ng SM, Neubauer S, Rider OJ (2023) Phenotyping heart failure by cardiac magnetic resonance imaging of cardiac macro- and microscopic structure: state of the art review. Eur Heart J Cardiovasc Imaging 24:1302–1317
11. Ipek R, Holland J, Cramer M, Rider O (2024) CMR to characterize myocardial structure and function in heart failure with preserved left ventricular ejection fraction. Eur Heart J Cardiovasc Imaging 25:1491–1504
12. Borlaug BA, Sharma K, Shah SJ, Ho JE (2023) Heart failure with preserved ejection fraction: JACC scientific statement. J Am Coll Cardiol 81:1810–1834

13. Henry JA, Couch LS, Rider OJ (2024) Myocardial metabolism in heart failure with preserved ejection fraction. J Clin Med 13

14. Silke B (2006) Beta-blockade in CHF: pathophysiological considerations. European Heart J Suppl 8:C13–C18

15. Leong DP, McMurray JJV, Joseph PG, Yusuf S (2019) From ACE inhibitors/ARBs to ARNIs in coronary artery disease and heart failure (part 2/5). J Am Coll Cardiol 74:683–698

16. Felker GM, Ellison DH, Mullens W, Cox ZL, Testani JM (2020) Diuretic therapy for patients with heart failure: JACC state-of-the-art review. J Am Coll Cardiol 75:1178–1195

17. Packer M et al (2020) Cardiovascular and renal outcomes with Empagliflozin in heart failure. N Engl J Med 383:1413–1424

18. Solomon SD et al (2022) Dapagliflozin in heart failure with mildly reduced or preserved ejection fraction. N Engl J Med 387:1089–1098

19. Hundertmark MJ et al (2023) Assessment of cardiac energy metabolism, function, and physiology in patients with heart failure taking Empagliflozin: the randomized, controlled EMPA-VISION Trial. Circulation 147:1654–1669

20. Ide T et al (2019) Ivabradine for the treatment of cardiovascular diseases. Circ J 83:252–260

21. Swedberg K et al (2010) Ivabradine and outcomes in chronic heart failure (SHIFT): a randomised placebo-controlled study. Lancet 376:875–885

22. Patocka J, Nepovimova E, Wu W, Kuca K (2020) Digoxin: pharmacology and toxicology-a review. Environ Toxicol Pharmacol 79:103400

23. Sabbah HN (2017) Silent disease progression in clinically stable heart failure. Eur J Heart Fail 19:469–478

24. Solomon SD et al (2024) Finerenone in heart failure with mildly reduced or preserved ejection fraction. N Engl J Med 391:1475–1485

25. Kosiborod MN et al (2023) Semaglutide in patients with heart failure with preserved ejection fraction and obesity. N Engl J Med 389:1069–1084

26. Kosiborod MN et al (2024) Effects of Semaglutide on symptoms, function, and quality of life in patients with heart failure with preserved ejection fraction and obesity: a prespecified analysis of the STEP-HFpEF trial. Circulation 149:204–216

27. Watson WD et al (2023) Retained metabolic flexibility of the failing human heart. Circulation 148:109–123

Blood Vessels

Blood Vessels: Overview

Victor W. M. van Hinsbergh, Rick Meijer,
and Etto C. Eringa

Anatomy and Physiology of Blood Vessels

The blood vessels conduct blood from the heart to the tissues and back, thus achieving continuous supply of oxygen and nutrients, removal of waste products, and—when needed—delivery of leukocytes to the organs. In one overall circulation cycle, the heart is passed two times in order to pump the blood through the other tissues and lungs (see chapter "Heart: Overview" under part "Heart"). Starting in the left ventricle of the heart, oxygenated blood flows into the systemic circulation through the aorta, and then into the large conduit arteries, which subsequently divide into smaller conduit arteries, resistance arteries, and the microcirculation. There, arterioles branch out into capillaries, the smallest blood vessels and site of solute and gas exchange. Capillaries merge into venules, and those merge into veins, conducting the blood toward the right heart. From the right heart, blood flows into the pulmonary artery to enter the pulmonary circulation, where it is reoxygenated. The blood then returns to the left heart.

All large blood vessels are composed of three layers: the tunica intima, tunica media, and tunica adventitia. The intima of healthy arteries is mainly a continuous layer of endothelial cells (ECs), which prevents intravascular clotting of the blood, regulates fluid and solute transport from the blood to the tissues and back, and controls leukocyte recruitment and vascular tone [1]. The media consists of multiple layers of smooth muscle cells (SMCs) embedded in collagen, alternated by layers of elastin. The adventitia is a loose fibrous tissue that contains fibroblasts as well as small vessels (*vasa vasorum*) that nourish the outer cells of large arteries and can harbor leukocytes, mast cells, and mesenchymal stem cells. At its outside, it continues diffusely into a layer of perivascular adipose tissue (PVAT) that provides vasoregulatory adipokines to the arteries and arterioles (Fig. 1). In atherosclerotic arteries, the intima is thickened by accumulation of lipoproteins and lipid-laden cells underneath the endothelium.

The aorta and large arteries are characterized by a thick layer of SMCs in their media, which maintains a rather constant blood pressure. The elasticity (Windkessel function) of arteries helps not only to dampen the blood pulse generated by each heart beat but also—by recoil—to continue propelling the blood toward the tissues during diastole.

V. W. M. van Hinsbergh (✉) · E. C. Eringa
Institute for Cardiovascular Research, VU University
Medical Center, Amsterdam, BT, The Netherland
e-mail: v.vanhinsbergh@vumc.nl; e.eringa@vumc.nl

R. Meijer
Department of Internal Medicine, Institute for
Cardiovascular Research, VU University Medical
Center, Amsterdam, MB, The Netherlands
e-mail: r.meijer@vumc.nl

E. Lammert, M. Zeeb (eds.), *Metabolism of Human Diseases*,
https://doi.org/10.1007/978-3-031-96019-2_31

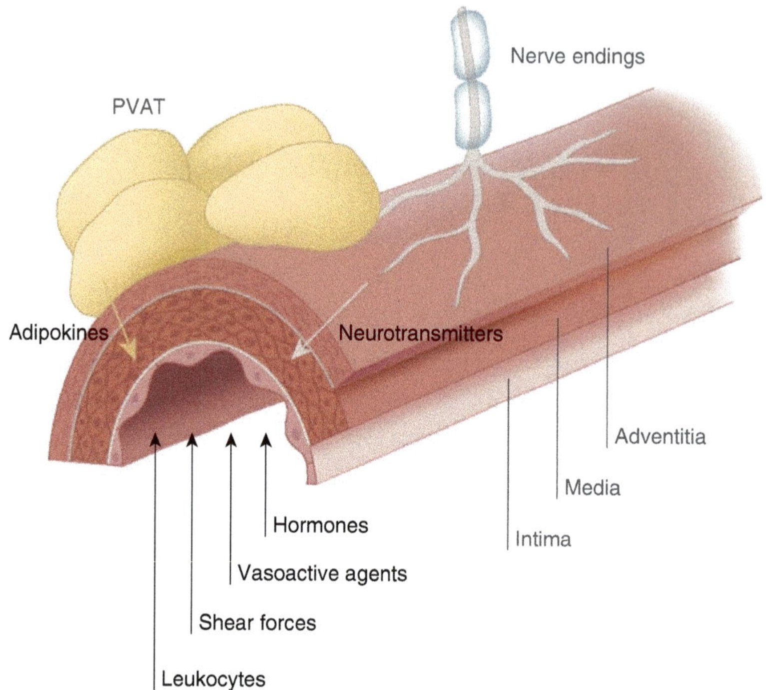

Fig. 1 External mediators acting on blood vessels. The arterial wall consists of three main layers: the intima, which in healthy vessels mainly consists of the endothelium; the media that harbors concentric smooth muscle cells, alternated by collagen and elastin layers; and the adventitia. Many exogenous factors act on the wall of arteries. From the luminal side, hormones and vasoactive agents bind to cellular receptors and regulate the interplay between endothelium and smooth muscle cells; shear forces influence the endothelial behavior; and leukocytes, platelets, and their products also interact with the vessel wall. From the outside, sympathetic nerve endings release neurotransmitters and the perivascular adipose tissue (*PVAT*) releases adipokines that contribute to vasoregulation

The small arteries and arterioles are the major site of resistance to blood flow. The collective diameter of all resistance vessels together is a main determinant of blood pressure, together with cardiac output. In addition to regulation of blood pressure, these "resistance vessels" determine and regulate the perfusion of the connected capillary bed, depending on the local demand, for example, preferential perfusion through the skeletal muscle during exercise or to the splanchnic bed after meals. Neuronal factors, hormones, and tissue-derived paracrine factors modulate the perfusion of a tissue in order to meet its metabolic demand.

Delivery of oxygen and nutrients as well as removal of waste products occur in the capillary bed [2], as the surface area of the capillary endothelium available for diffusion is large and transport distance from blood to the tissue is low. In most tissues, the capillary ECs are in contact with pericytes and form a continuous endothelium. However, in specific tissues, such as liver and adrenal glands, they have large pores (so-called fenestrae) to allow rapid penetration of cholesterol-containing lipoproteins required for bile and steroid production, respectively. In contrast, in the brain, a tight endothelial barrier known as the blood–brain barrier is formed by interplay between endothelium, pericytes, and astrocyte foot ends [3]. However, upon a thrombotic stroke, the site distal to the occluded vessel becomes hypoxic and leaky (see chapter "Stroke").

The walls of postcapillary venules, which collect the blood from the capillaries, consist of endothelium only and are the first to respond to vasoactive agents and noxious stimuli by temporarily allowing protein leakage to the interstitium and facilitating the first recruitment of phagocytes after injury or infection.

Walls of veins are considerably thinner than those of arteries, especially their media. Limb veins contain valves, facilitating conduction of the blood back toward the heart despite a low blood pressure. No valves are encountered in the smallest veins, the great collecting veins, and the veins of viscera and the brain. Veins are easily distended and contain the larger portion of blood in the circulatory system. In chronically overdistended veins in the legs, so-called varicose veins, the valves are no longer competent to sustain blood movement toward the heart, and, subsequently, a higher pressure on the distal valves arises, creating a vicious cycle, eventually resulting in stasis and ankle edema.

Inside-In: Paracrine Signals Acting Within the Vessel

SMC contraction occurs after stimulation of Ca^{2+} influx and subsequent activation of the enzyme myosin light chain (MLC) kinase [4]. The phosphorylated MLC initiates movement of myosin along F-actin fibers leading to cell contraction. MLC phosphatase activity undoes MLC phosphorylation and prevents contraction [4].

Many vasoactive agents, such as norepinephrine and substance P released from neurons (see below) and bradykinin (formed from a blood plasma protein), enhance cytoplasmic Ca^{2+} levels in SMCs leading to contraction, when applied to SMCs in the absence of endothelium. However, when a healthy endothelium is present, these vasoactive agents often also activate the endothelium prompting the generation of nitric oxide (NO) by endothelial NO synthase (eNOS), prostacyclin or prostaglandin E2 via cyclooxygenase, and occasionally endothelium-derived hyperpolarization factor (Fig. 2a). These factors, of which NO is the most potent one, cause "endothelium-dependent" relaxation of SMCs. NO activates guanylate cyclase in SMCs and the resulting cyclic guanosine monophosphate (cGMP) activates protein kinase G (PKG), which, among

others, limits the influx of Ca^{2+} ions into the cytoplasm and subsequent contraction of SMCs [5]. Prostacyclin and prostaglandin E2 cause elevation of cellular cyclic adenosine monophosphate (cAMP) level, which by protein kinase A (PKA)-mediated phosphorylation and inhibition of myosin light chain kinase (MLCK) also leads to reduced actin-myosin interaction. In contrast, endothelin-1 is secreted by activated endothelium to stimulate vascular contraction (Fig. 2b).

The effects of NO and prostacyclin extend beyond SMC contraction. These mediators also reduce the activation and aggregation of blood platelets (Fig. 2). Furthermore, NO reduces inflammatory activation (see below) in the healthy arterial endothelium itself by interference with nuclear factor κB signaling that is required for the transcription of inflammation-specific genes [6].

SMCs also respond to pressure and radial strain, directly. Prolonged changes in blood pressure can induce remodeling of SMCs and adaptation of the vessel diameter. Another physical factor that affects vascular functioning is vascular stiffening. Calcification due to deposition of calcium phosphate in arteries causes media stiffening in conduit arteries of elderly people, by which the vessel wall becomes less compliant. However, arterial stiffening can also be caused by formation of advanced glycation end products (see chapter "Diabetes Mellitus") that cross-link proteins within the vessel wall. Stiffening results in a reduced dampening of the pulse wave and thus a higher pulse pressure. This can promote vascular damage in brain and microvasculature of diabetic patients (see chapter "Diabetes Mellitus").

Inside-out: Vascular Factors Affecting Other Tissues

In addition to local vasoregulation, ECs in specific organs can also affect distant tissues by the conversion and catabolism of vasoactive agents [7]. In particular, the lung vasculature plays an important role through the production of angiotensin-converting enzyme (ACE), which converts angiotensin I into angiotensin II, thus

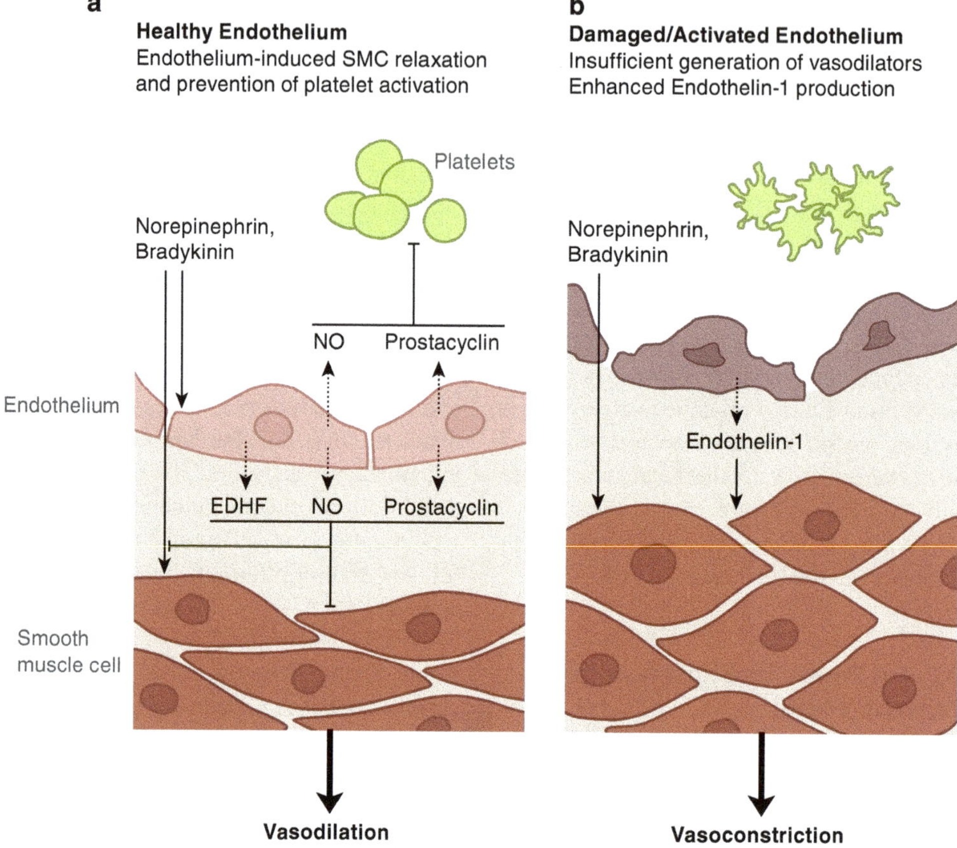

Fig. 2 Paracrine interaction between endothelium and smooth muscle cells. (**a**) After stimulation with vasoactive agents or neural factors (e.g., norepinephrine and bradykinin), the healthy endothelium releases nitric oxide (*NO*), prostacyclin, and endothelium-derived hyperpolarization factor (*EDHF*), which counteract smooth muscle contraction by various intracellular mechanisms and therewith influencing systemic blood pressure and volume contribute to vasodilation. Prostacyclin and NO also counteract platelet activation. (**b**) In disease, the production of these vasodilating agents can decrease, while production of the protein endothelin-1 can increase. Additionally, direct effects of vasoactive agents on smooth muscle cells are no longer suppressed, inducing vasoconstriction

influencing systemic blood pressure and volume via the renin-angiotensin-aldosterone-system (see chapter "Kidney: Overview" under part "Kidney"). The lung vasculature also inactivates bradykinin, an important vasodilator, and takes up and degrades serotonin, a vasoconstrictor terminating their effects.

The endothelium of healthy vessels produces several proteins that prevent thrombus formation (see chapter "Blood: Overview" under part "Blood") [8]. It interrupts the coagulation cascade by providing antithrombin III that neutralizes thrombin and by thrombomodulin-facilitated activation of the anticoagulant protein C. It binds a metalloproteinase called ADAMTS13 that proteolytically cleaves von Willebrand factor multimers limiting platelet adhesion and activation. Furthermore, the endothelium reduces platelet activation and aggregation (Fig. 2a). Finally, it releases tissue-type plasminogen activator, a fibrinolysis-catalyzing enzyme. These antithrombotic activities ensure undisturbed circulation. If the balance between pro- and antithrombotic/coagulant factors is disturbed, thrombus formation or bleeding will occur. While thrombus formation is required for limiting blood loss after wounding, it can also lead to adverse events, such as deep vein thrombosis or stroke (see chapter "Stroke").

Inflammatory cells in atherosclerotic vessels produce cytokines, such as tumor necrosis factor-α (TNF-α), that alter the properties of ECs. This results not only in a reduction in antithrombotic properties and activity of eNOS but also in the expression of leukocyte adhesion molecules, such as vascular cell adhesion molecule-1 (VCAM1) and E-selectin, as well as chemo- and cytokines, such as monocyte chemotactic protein-1 and inter-leukin-8 that stimulate influx of various leukocyte types [9]. Moreover, atherosclerotic arteries are associated with a subtle elevation of circulating C-reactive protein (CRP), reflecting a weak activation of the acute phase response in the liver (see chapter "Liver: Overview" under part "Liver").

Outside-in: Factors from Other Tissues Affecting Conduit and Resistance Vessels

Neural factors, such as norepinephrine and substance P, and hormones, such as epinephrine, insulin, and estrogen, are important regulators of vascular tone (Fig. 1). However, the effect of a specific factor can vary among the vascular beds of different organs because of different receptor isoforms, receptor sensitivities, or tissue-specific receptor distributions. In addition to primary effects on the vasculature, inflammatory cytokines and hypoxia can change gene expression in ECs [9]. Furthermore, shear stress by laminar flowing blood on arterial ECs causes a gene induction pattern within these cells that reduces inflammatory activation, while disturbance of the laminar flow pattern in arteries alters endothelial functioning and contributes to atherosclerotic lesion generation [10].

Hormones have to pass the endothelium to reach tissue cells and usually also act directly on blood vessels. In muscle or heart tissue, insulin has to pass the endothelium, while activating insulin receptors on the endothelium simultaneously. Insulin receptor activation in the proximal resistance vessels of tissues that store nutrients usually causes vasodilation by activation of eNOS via a pathway that involves insulin recep-

tor susbstrate-1 (IRS1), phosphatidylinositol-3-kinase (PI3K), and Akt (also called protein kinase B). In obesity, this effect is diminished, while endothelin-1 favors vascular contraction. Other hormones also act on blood vessels. For example, estrogens interact with an endothelial membrane-bound estrogen receptor variant, which induces NO production and vasodilation (see chapter "Reproductive System: Overview" under part "Reproductive System") [11].

The vessel wall is approached not only from its luminal side by endocrine mediators but also from its outside by products of the surrounding PVAT acting in a paracrine fashion (Fig. 1). In obesity and type 2 diabetes (see chapters "Metabolic Syndrome" and "Diabetes Mellitus," respectively), PVAT expands and becomes inflamed. While lean PVAT produces adipokines like adiponectin that facilitate vasodilation both in conduit arteries and insulin-stimulated resistance arteries, the expanded fatty PVAT loses this ability and shifts toward the production of nonesterified fatty acids (NEFA), the pro-inflammatory cytokine TNF-α, leptin, and other factors that favor the contractile pathway induced by insulin [12]. These events affect, for example, the white adipose tissue and illustrate the mutual interrelationship between metabolism, inflammation, and vessel functioning.

Final Remarks

Blood vessels distribute fuel for metabolism, control tissue perfusion by vasoregulation, and deliver tissue products and leukocytes to the sites where they are needed. Being critically important for the organism, blood vessels have a high capacity to adapt to various stresses and local needs. This can occur by short-term responses, such as acute adaptation of the vessel tone or release of factors that contribute to hemostasis, or by the induction of new genes, as in inflammation or hypoxia. Normally, this response is temporary. However, this adaptability can be overstretched by chronic activation or injury, leading to chronic or acute adverse responses.

References

1. Aird WC (2007) Phenotypic heterogeneity of the endothelium: I. Structure, function, and mechanisms. Circ Res 100:158–173
2. Mehta D, Malik AB (2006) Signaling mechanisms regulating endothelial permeability. Physiol Rev 86:279–367
3. Armulik A, Genové G, Mäe M, Nisancioglu MH, Wallgard E, Niaudet C, He L, Norlin J, Lindblom P, Strittmatter K, Johansson BR, Betsholtz C (2010) Pericytes regulate the blood-brain barrier. Nature 468:557–561
4. Somlyo AP, Somlyo AV (2003) Ca2+ sensitivity of smooth muscle and nonmuscle myosin II: modulated by G proteins, kinases, and myosin phosphatase. Physiol Rev 83:1325–1358
5. Bian K, Doursout MF, Murad F (2008) Vascular system: role of nitric oxide in cardiovascular diseases. J Clin Hypertens (Greenwich) 10:304–310
6. De Caterina R, Libby P, Peng HB, Thannickal VJ, Rajavashisth TB, Gimbrone MA Jr, Shin WS, Liao JK (1995) Nitric oxide decreases cytokine-induced endothelial activation. Nitric oxide selectively reduces endothelial expression of adhesion molecules and proinflammatory cytokines. J Clin Invest 96:60–68
7. Fishman AP (1982) Endothelium: a distributed organ of diverse capabilities. Ann N Y Acad Sci 401:1–8
8. van Hinsbergh VWM (2012) Endothelium—role in regulation of coagulation and inflammation. Semin Immunopathol 34:93–106
9. Pober JS, Sessa WC (2007) Evolving functions of endothelial cells in inflammation. Nat Rev Immunol 7:803–815
10. Davies PF (2009) Hemodynamic shear stress and the endothelium in cardiovascular pathophysiology. Nat Clin Pract Cardiovasc Med 6:16–26
11. Mendelsohn ME, Karas RH (2010) Rapid progress for non-nuclear estrogen receptor signaling. J Clin Invest 120:2277–2279
12. Aghamohammadzadeh R, Withers S, Lynch F, Greenstein A, Malik R, Heagerty A (2012) Perivascular adipose tissue from human systemic and coronary vessels: the emergence of a new pharmacotherapeutic target. Br J Pharmacol 165:670–682

Stroke

Arunmozhiarasi Armugam,
Dwi Setyowati Karolina, and Kandiah Jeyaseelan

Introduction to Stroke

Stroke is an acute cerebrovascular disease that requires immediate medical attention. It is characterized by rapid loss in brain function due to declining blood supply to the brain. Obstruction to cerebral blood flow caused by the presence of a clot in the blood vessel results in an ischemic stroke, whereas a rupture of the brain blood vessel leads to a hemorrhagic stroke. Though adults above 40 years of age are more prone to stroke, children with sickle cell disease have also been found to be contributing up to 8% of the stroke patients globally [1].

Ischemic stroke, accounting for >80% of all cases, occurs when the afferent vasculature to the brain is occluded by local arteriosclerosis and/or by a localized or a moving thrombus [2]. Hemorrhagic stroke (~15% of the cases) occurs when the blood vessel ruptures and bleeds into the brain [3]. Stroke is a medical emergency because the brain tissue ceases to function within minutes of oxygen deprivation and infarction occurs as the cells die, causing an irreversible injury (necrosis) at the core of the infarct. This core region is surrounded by a penumbra that consists of recoverable tissue damage (apoptosis). Activation of inflammatory cells causes the release of cytokines, matrix metalloproteinases (MMPs), and reactive oxygen species (ROS), eventually leading to blood–brain barrier (BBB) breakdown and cerebral edema [4].

In hemorrhagic stroke, edema is induced by the buildup of pressure in the cranium due to internal bleeding. Upon red blood cell lysis, hemoglobin is degraded into iron, carbon monoxide, and biliverdin. Carbon monoxide and iron can cause tissue damage via the formation of free radicals. Altogether, hemoglobin and its degradation products may exacerbate brain edema [5]. Risk factors of stroke include hypertension, hypercholesterolemia (see chapter "Hyperlipidemia"), and hyperglycemia (see chapter "Diabetes Mellitus"), as these can exert stress on the walls of the blood vessels causing them to thicken, forming plaques on the arterial walls and narrowing them further. These plaques could break away and occlude the cerebral blood vessels resulting in stroke. Chronic hypertension may weaken the vasculature leading to rupture and hemorrhagic stroke.

A. Armugam (✉)
Institute of Bioengineering and BioImaging,
Singapore, Singapore
e-mail: armugama@ibb.a-star.edu.sg

D. S. Karolina · K. Jeyaseelan (✉)
Department of Biochemistry and Centre for
Translational Medicine, Yong Loo Lin School of
Medicine, National University of Singapore,
Singapore, Singapore

Pathological Changes in Metabolism Following Stroke Onset

Reduced cerebral blood flow exerts a direct effect on the cerebral metabolism. During the early phase of acute ischemic stroke, an inflammatory response is initiated alongside leukocyte activation. Leukocyte influx (neutrophils in particular) induces release of destructive enzymes and pro-inflammatory cytokines with neurotoxic effects resulting in hyperthermia [6]. Hyperthermia in the brain augments the tissue's metabolic rate (hypermetabolism), causing rapid exhaustion of the limited energy and oxygen supplies, increased production of free radicals and toxic substances such as glutamate, and Ca^{2+} overload, thereby advancing the conversion of ischemic but viable tissue to infarction (Fig. 1) [7]. Oxygen deprivation induces anaerobic cerebral glycolysis, and the increased lactate production serves as a (clinical) marker of energy failure and the degree of damage [8]. Chronic lactate production leads to lactic acidosis and reduced extracellular pH, which are detrimental to the brain. Nevertheless, it was suggested that glucose rather than lactate in combination with acidosis is the main culprit that aggravates brain damage during stroke [9]. Glucose toxicity could be mediated via formation of advanced glycation end products (see chapter "Diabetes Mellitus"), which may cause protein dysfunction or enhance oxidative stress [9]. In fact, cerebral hyperglycolytic lactate production may indicate increased alternative energy substrate usage and is now emerging as a good rather than bad marker indicating favorable long-term recovery [10].

Specific neuronal circuitry perceives the metabolic status of peripheral tissues via hormones (e.g., leptin, insulin, and ghrelin) as well as through direct macronutrient sensing. The center of these brain networks is localized within the mediobasal hypothalamus interconnected with brainstem areas and the mesolimbic reward circuitry [11]. Disruption of brain activity impairs these networks thereby affecting regulatory control over glucose metabolism. A high proportion of stroke patients develop hyperglycemia even in the absence of pre-existing diabetes [12]. During stroke, excitotoxicity due to excess glutamate, inflammation, as well as release of free radicals induces stress to the cells. In response to this, the hypothalamic–pituitary–adrenal system (see chapter "Brain: Overview" under part "Brain") triggers the release of neurohormones such as cortisol and catecholamines [13]. This in turn promotes glycogenolysis and gluconeogenesis in the liver (see chapter "Liver: Overview") while inhibiting insulin sensitivity and glucose uptake in the skeletal muscle [14]. They also stimulate lipolysis, resulting in accumulation of nonesterified fatty acids (NEFA) [14]. Hyperglycemia fuels anaerobic metabolism, lactic acidosis, and free radical production, which then exert direct membrane lipid peroxidation and cell lysis in the metabolically challenged tissues [15]. Such positive feedback leads to a further derangement in pH homeostasis, which can trigger the production of free radicals and affects the Ca^{2+} balance in neurons, possibly resulting in apoptosis and neuronal cell death [16]. These free radicals can cause functional and morphological changes to the endothelium, thus increasing permeability of the BBB, edema formation, as well as risk of hemorrhagic transformation.

Treatment of Strokes

Medical intervention depends on the type of stroke. Ischemic stroke requires restoration of blood flow to the affected region, while hemorrhagic stroke requires controlling the bleeding from the ruptured blood vessels. Thrombolytic drugs such as alteplase and reteplase, which are synthetic enzymes with similar function to the endogenously produced tissue plasminogen activator (tPA), are the first line of treatment against ischemic stroke [17]. tPA is a serine protease found on vascular endothelial cells and is involved in the lysis of blood clots (see chapter "Blood: Overview" under part "Blood"). Although thrombolysis can improve blood flow, these clot-dissolving drugs are only effective within 4.5 h of stroke onset, after which they pose increased risk of hemorrhage. Hence, patients with hemorrhagic stroke or severely elevated blood

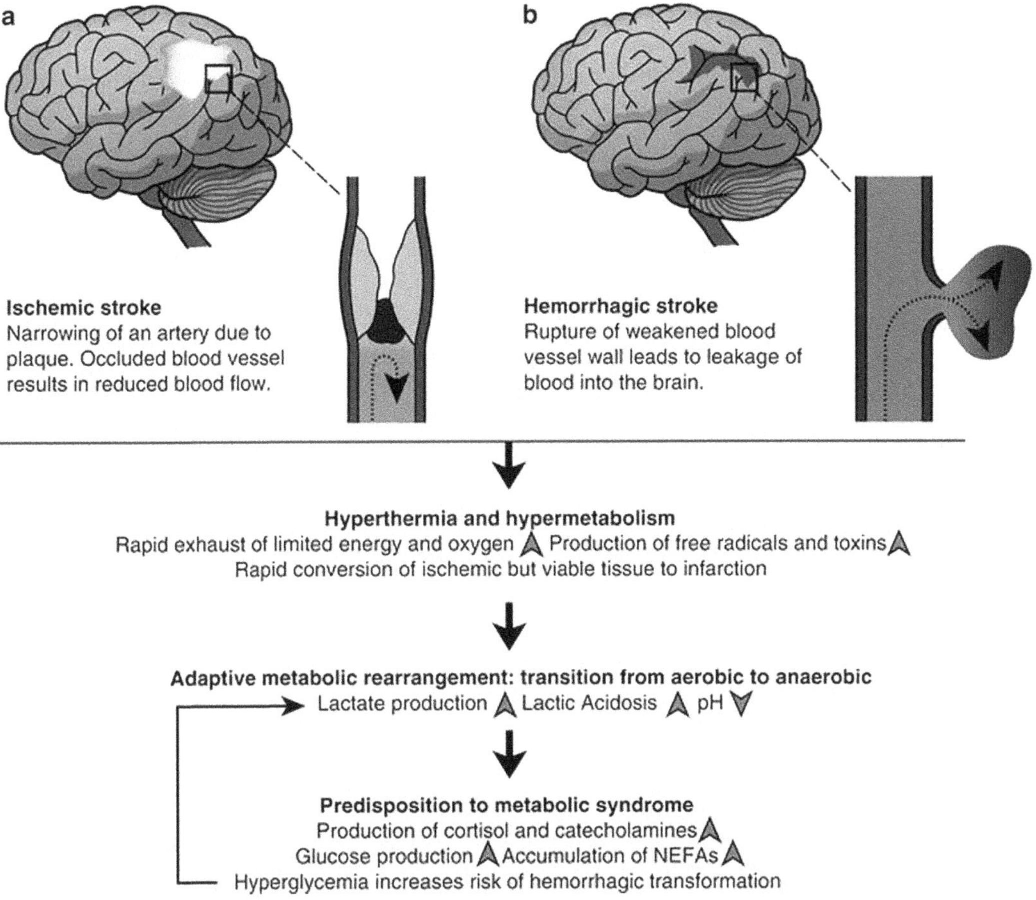

Fig. 1 (**a, b**) Metabolic effects of stroke. *NEFA* nonesterified fatty acids

pressure are not eligible for such treatment. A more invasive procedure, mechanical thrombectomy (MERCI) may be performed to remove or break up the clot [18]. An alternative mechanical approach is the Penumbra System, an embolectomy device specifically designed to remove the clot by aspiration and extraction [18]. Treatment for hemorrhagic stroke aims to stop the bleeding by surgical clipping (closing the base of a leakage or aneurysm by constriction) or endovascular coiling (insertion of a metallic coil into the site of leakage followed by blood clotting) and to drain the blood (hematoma) that has accumulated in the brain [19]. Apart from these, stroke treatment also involves medications that control inflammation, brain swelling, blood pressure, hyperglycemia, and hypercholesterolemia (Table 1).

Influence of Treatment on Metabolism

Early phase damage in the brain can be reversed when the disrupted blood flow is restored in time by medical interventions. Yet, reperfusion at the ischemic region can activate astrocytes, microglia, and endothelial cells to produce cytokines and chemokines, which are potent mediators of inflammatory responses. This results in activation of matrix metalloproteinases (MMPs), which disrupt the BBB by digesting extracellular matrix proteins within the basal lamina, such as type IV collagen, laminin, and fibronectin. This consequently causes BBB disruption, edema formation, and hemorrhagic transformation [20].

Table 1 Effects of stroke treatments

Immediate treatment for ischemic stroke	Immediate treatment for hemorrhagic stroke
Thrombolytic drugs	Surgical approaches
Alteplase, reteplase, tenecteplase	Surgical clipping
Anistreplase, streptokinase, urokinase	Endovascular coiling Evacuation of hematoma
Surgical approaches MERCI retriever Penumbra system	
Advantages: Restoration of blood flow	*Advantages:* Stops bleeding
Disadvantages: Reperfusion induces inflammatory response and increases risk of hemorrhagic transformation	*Disadvantages:* Risk of death during procedures
Prescribed medications to prevent recurring stroke/ complications	
Anticoagulant or antiplatelet Prevents blood clotting Reduces platelet aggregation Inhibits thrombus formation	
Antihypertensive drugs Lower blood volume Reduce systemic vascular resistance Lower cardiac output	
Cholesterol-lowering drugs Lower cholesterol synthesis in liver also has antihypertensive effects	
Antihyperglycemic drugs Insulin therapy maintains strict glycemic control and improves blood circulation to the ischemic areas Improved prognosis	

Secondary Treatment Options

The majority of stroke patients have a combination of medical disorders such as hypertension, dyslipidemia (see chapter "Hyperlipidemia"), and diabetes (see chapter "Diabetes Mellitus"). Hence, treatment for stroke also includes the management of these conditions. Antihypertensive drugs such as thiazide diuretics and Ca^{2+} channel blockers are prescribed to control blood pressure by reducing blood volume, systemic vascular resistance, and cardiac output. Cholesterol-lowering drugs such as statins are used to control cholesterol synthesis in the liver (see chapter "Hyperlipidemia").

Insulin therapy is usually recommended for treating hyperglycemia in stroke patients. Besides lowering blood glucose levels, insulin also exerts antioxidant and anti-inflammatory effects by suppressing several proinflammatory transcription factors, such as nuclear factor-κB (NF-κB), early growth response protein 1 (Egr-1), and activator protein 1 (AP-1), as well as generation of reactive oxygen species (ROS), thereby minimizing stroke complications [21]. Moreover, insulin therapy has also been shown to significantly reduce systolic blood pressure and improve stroke outcome [21].

Perspectives

Pathogenesis of stroke involves multiple factors, which coordinately impair the metabolic status of the body. Recombinant tPAs, the only therapeutic arsenal available today, have several limitations such as short therapeutic window and the risk of hemorrhagic transformation. Even recombinant tPA therapy in combination with other drugs to reduce stroke complications is not always effective, and novel therapeutic agents are necessary.

Recently, microRNAs (miRNAs) have gained attention as biomarkers and therapeutic agents against several diseases including stroke [22]. MicroRNAs are endogenous gene regulators that are recognized as the guardians of the genome. They regulate the entire biological processes including those that are impaired in pathological conditions. Restoration of dysregulated miRNAs to their natural levels has been shown to alleviate the symptoms of the disease. For example, miR-320, a key player in edema formation, is upregulated in stroke conditions [23]. Hence, downregulation of miR-320 expression has been associated with favorable outcomes in stroke patients [24]. Recently, Mens et al. [25] have also proposed that circulatory miRNAs could potentially serve as diagnostic, prognostic and therapeutic molecules in stroke pathogenesis. Several miRNA based clinical trials are currently ongoing for cancer, infectious disease, heart failure, and vascular diseases [26]. Nevertheless, more and extensive studies are needed before miRNAs could reach clinical trials in stroke therapy. Yet, the development of miRNAs as multitarget drugs holds great potential.

Questions and Answers

Question 1 What are the major types of stroke and how are they been currently treated?

Answer 1 Ischemic and hemorrhagic stroke constitute the two major types of stroke. Thrombolytic drug therapy and mechanical thrombectomy (MERCI) as well as an embolectomy device to remove the clot by aspiration and extraction have been used for the treatment of ischemic stroke. Treatment of hemorrhagic stroke involves stopping the bleeding by surgical clipping or endovascular coiling and draining the blood (hematoma) that has accumulated in the brain. Apart from these, stroke treatment also involves medications that control inflammation, brain swelling, blood pressure, hyperglycemia, and hypercholesterolemia.

Question 2 What are microRNAs and describe their roles in human diseases?

Answer 2 MicroRNAs (miRNAs) are endogenous short (20–22 nucleotides) non-coding RNA molecules synthesized from their own genes to mediate cellular gene expression both at the transcription and translation level. Thus, they are able to modulate fundamental cellular processes such as differentiation, proliferation, death, metabolism, and pathophysiology of many diseases. miRNAs bind to their target mRNAs via partial or perfect complementarity resulting in degradation and/or translational repression of the transcripts. This regulatory control enforced by miRNAs makes them intriguing candidates especially in human diseases. Their involvement in diseases such as cancer, heart failure, and diabetes has so far been quite well established.

Question 3 Do microRNAs have potential of being developed as biomarkers and therapeutic agents against stroke?

Answer 3 The specific regulation at both the transcription and the translation level (inhibition or mRNA degradation) opens an avenue to use these small RNA molecules as potential targets for the development of novel drugs as well as for the diagnosis of several human diseases. Important information about the role of a miRNA in disease can be deduced by mimicking or inhibiting its activity and examining its impact on the phenotype/behavior of the cell or organism. Modulating the activity of a miRNA is expected to lead to improvement in disease symptoms and this implies that the target miRNA plays an important role in the disease. It is also now possible to develop miRNA-based therapeutic products that can either increase or decrease the levels of proteins in pathophysiological conditions such as cancer, cardiovascular diseases, viral diseases, metabolic disorders, and programmed cell death. Specific miRNA expression has also been shown in both brain tissue and blood following ischemic stroke. Besides, circulating miRNA expression varies significantly in stroke patients as well as for the different stroke subtypes. Thus, circulating miRNAs manifest the potential to be developed as ischemic stroke biomarkers or therapeutic agents. Though several groups have reported on altered expression of miRNAs during ischemic stroke, the specificity to acute stroke pathology or exclusion of confounding risk factors have not been established. These are critical factors that need to be addressed in order to identify stroke specific miRNAs with clinical potential.

References

1. Stroke, Cerebrovascular accident WHO – http://www.emro.who.int/health-topics/stroke/cerebrovascular-accident/index.html
2. Hademenos GJ, Massoud TF (1997) Biophysical mechanisms of stroke. Stroke 28:2067–2077
3. Qureshi AI, Mendelow AD, Hanley DF (2009) Intracerebral haemorrhage. Lancet 373:1632–1644
4. Danton GH, Dietrich WD (2003) Inflammatory mechanisms after ischemia and stroke. J Neuropathol Exp Neurol 62:127–136
5. Hua Y, Keep RF, Hoff JT, Xi G (2007) Brain injury after intracerebral hemorrhage: the role of thrombin and iron. Stroke 38:759–762

6. Zaremba J (2004) Hyperthermia in ischemic stroke. Med Sci Monit 10:RA148–RA153

7. Karaszewski B, Wardlaw JM, Marshall I, Cvoro V, Wartolowska K, Haga K, Armitage PA, Bastin ME, Dennis MS (2009) Early brain temperature elevation and anaerobic metabolism in human acute ischaemic stroke. Brain 132:955–964

8. Valenza F, Aletti G, Fossali T, Chevallard G, Sacconi F, Irace M, Gattinoni L (2005) Lactate as a marker of energy failure in critically ill patients: hypothesis. Crit Care 9:588–593

9. Kikuchi S, Shinpo K, Takeuchi M, Yamagishi S, Makita Z, Sasaki N, Tashiro K (2003) Glycation: a sweet tempter for neuronal death. Brain Res Brain Res Rev 4:306–323

10. Oddo M, Levine JM, Frangos S, Maloney-Wilensky E, Carrera E, Daniel RT, Levivier M, Magistretti PJ, LeRoux PD (2012) Brain lactate metabolism in humans with subarachnoid hemorrhage. Stroke 43:1418–1421

11. Farooqui AA, Farooqui T, Panza F, Frisardi V (2012) Metabolic syndrome as a risk factor for neurological disorders. Cell Mol Life Sci 69:741–762

12. Melamed E (1976) Reactive hyperglycaemia in patients with acute stroke. J Neurol Sci 29:267–275

13. Anne M, Juha K, Timo M, Mikko T, Olli V, Kyösti S, Heikki H, Vilho M (2007) Neurohormonal activation in ischemic stroke: effects of acute phase disturbances on long-term mortality. Curr Neurovasc Res 4:170–175

14. Zafari AM, Wang SS, Song Q (2006) Genetics of metabolic syndrome. Clin Rev:51–61. www.turner-white.com/pdf/hp_oct06_genetic.pdf

15. Lindsberg PJ, Roine RO (2004) Hyperglycemia in acute stroke. Stroke 35:363–364

16. Orlowski P, Chappell M, Park CS, Grau V, Payne S (2011) Modelling of pH dynamics in brain cells after stroke. Interface Focus 6:408–416

17. The National Institute of Neurological Disorders and Stroke rt-PA Stroke Study Group (1995) Tissue plasminogen activator for acute ischemic stroke. N Engl J Med 333:1581–1587

18. Tenser MS, Amar AP, Mack WJ (2011) Mechanical thrombectomy for acute ischemic stroke using the MERCI retriever and penumbra aspiration systems. World Neurosurg 76:S16–S23

19. Lapchak PA, Araujo DM (2007) Advances in hemorrhagic stroke therapy: conventional and novel approaches. Expert Opin Emerg Drugs 12:389–406

20. Visse R, Nagase H (2003) Matrix metalloproteinases and tissue inhibitors of metalloproteinases: structure, function, and biochemistry. Circ Res 92:827–839

21. Garg R, Chaudhuri A, Munschauer F, Dandona P (2006) Hyperglycemia, insulin, and acute ischemic stroke: a mechanistic justification for a trial of insulin infusion therapy. Stroke 37:267–273

22. Tan JR, Koo YX, Kaur P, Liu F, Armugam A, Wong PT, Jeyaseelan K (2011) microRNAs in stroke pathogenesis. Curr Mol Med 11:76–92

23. Sepramaniam S, Armugam A, Lim KY, Karolina DS, Swaminathan P, Tan JR, Jeyaseelan K (2010) MicroRNA 320a functions as a novel endogenous modulator of aquaporins 1 and 4 as well as a potential therapeutic target in cerebral ischemia. J Biol Chem 285:29223–29230

24. Tan KS, Armugam A, Sepramaniam S, Lim KY, Setyowati KD, Wang CW, Jeyaseelan K (2009) Expression profile of MicroRNAs in young stroke patients. PLoS One 4:e7689

25. Mens MMJ, Heshmatollah A, Fani L, Ikram MA, Ikram MK, Ghanbari M (2021) Circulatory miRNAs as potential biomarkers for stroke risk: the Rotterdam study. Stroke 52(3):945–953

26. Chakraborty C, Sharma AR, Sharma G, Lee SS (2020) Therapeutic advances of miRNAs: a pre-clinical and clinical update. J Adv Res 28:127–138

Blood

Blood: Overview

Norris Igbineweka, Deena Iskander,
and Barbara J. Bain

Introduction to Blood

Blood is made up of cells suspended in plasma. The plasma is composed of electrolytes, proteins, glucose, and lipids dissolved or suspended in water. The cellular compartment comprises red blood cells (erythrocytes, RBCs), platelets (thrombocytes), and white blood cells (leukocytes), which are all produced in the bone marrow (Fig. 1).

Blood is the major transport and delivery system in the body. It transports the end-product of internal respiration, carbon dioxide (CO_2), via the vena cava and the right side of the heart (see chapter "Heart: Overview" under part "Heart") to the lungs for expiration (see chapter "Lung: Overview" under part "Lung"). It delivers oxygenated blood from the lungs to the tissues, via the left side of the heart and the arteries (see chapter "Blood Vessels: Overview" under part "Blood Vessels"). Blood also transports nutrients from the gut to the tissues (see chapter "Gastrointestinal Tract: Overview" under part "Gastrointestinal Tract"). The waste products of tissue metabolism are transported via the blood to the site of detoxification, generally the liver (see chapter "Liver: Overview" under part "Liver"), or excretion, generally the kidney (see chapter "Kidney: Overview" under part "Kidney"). These processes are essential for the metabolic processes that sustain life.

Due to the function of blood as the major transport system of the body, changes in blood constituents influence all organs and tissues. In turn, many diseases can be diagnosed from biomarkers in the blood.

In this chapter, we shall describe the constituents and metabolic activities of blood and highlight diseases resulting from aberrations in these pathways.

Blood-Specific Metabolic/Molecular Pathways and Processes

The constituents of blood are required for its own homeostasis and that of other bodily systems and have many functions. The salient blood-specific metabolic processes are outlined below.

Many diseases are caused by or involve deregulated plasma proteins that are involved in a plethora of metabolic pathways. Some of these diseases are mentioned below.

N. Igbineweka · D. Iskander (✉) · B. J. Bain
Centre for Haematology, Department of Immunology and Inflammation, Imperial College London, London, UK
e-mail: n.igbineweka@imperial.ac.uk; d.iskander@imperial.ac.uk; b.bain@imperial.ac.uk

E. Lammert, M. Zeeb (eds.), *Metabolism of Human Diseases*,
https://doi.org/10.1007/978-3-031-96019-2_33

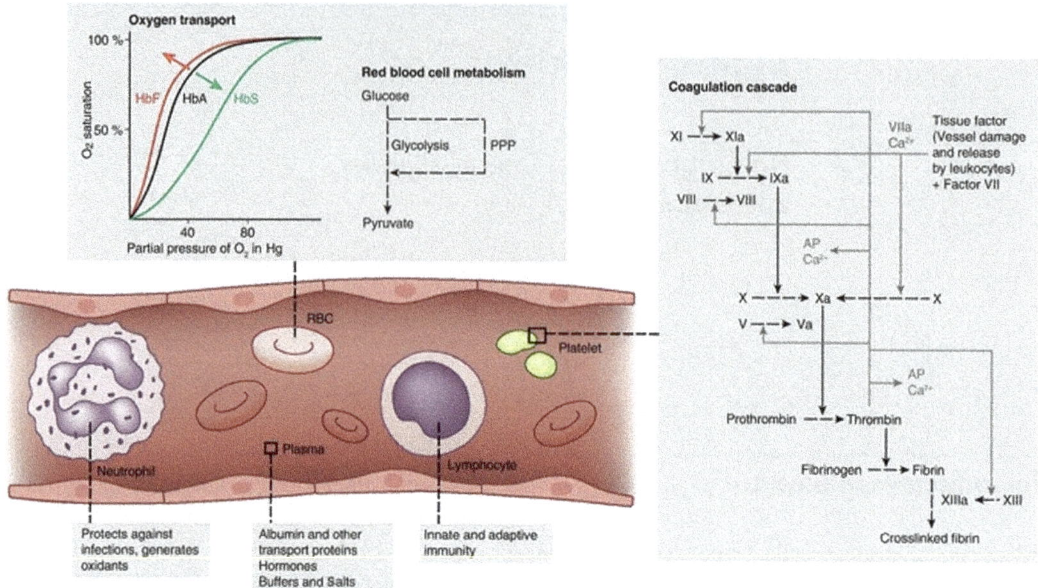

Fig. 1 Important components of the blood and their physiological functions. Erythrocytes (red blood cells, *RBCs*) are the major cellular component of the blood. RBCs transport oxygen to the tissues via hemoglobin (*Hb*). Oxygen binding is mainly regulated by partial pressure of O_2 in the surrounding tissues. The major metabolic pathways in RBCs are glycolysis and the pentose phosphate pathway (*PPP*) as well as adenosine nucleotide metabolism and the Rapoport-Luebering shunt. Platelets are small anucleate cell fragments in the blood that play a major role in blood clotting. Blood clotting is regulated by a complex, sequential, cascade-like activation of proteases resulting in cross-linked fibrin to seal an injury. Lymphocytes, components of the immune system, are also transported by the blood. Similarly, neutrophils aid in host defense. Additionally, the blood plasma transports various proteins, hormones, ions, and buffers. *AP* activated platelets, *HbF* fetal hemoglobin, *HbA* adult hemoglobin, *HbS* sickle-cell hemoglobin

Oxygen Transport

Oxygen (O_2) transport depends upon RBCs and hemoglobin, which are specially adapted for gas transport. The affinity of hemoglobin for oxygen at different partial pressures is represented by a sigmoid-shaped oxygen dissociation curve (Fig. 1). Oxygen binding depends on the structure of hemoglobin in terms of its subunits and on external factors such as acidity, temperature, and partial pressure of oxygen. Hemoglobin is a tetrameric protein consisting of two alpha chains combined with, in adult Hb, 2 beta chains (HbA) or two delta chains (HbA2) and in fetal Hb, 2 gamma chains (HbF) [1]. In sickle cell anemia (see chapter "Sickle Cell Disease"), a variant called HbS replaces HbA. This has reduced oxy-gen affinity compared with HbA, partially explaining the lower hemoglobin concentration observed in patients [1]. A recent development in sickle cell disease therapeutics was a small molecule allosteric effector, voxelotor, which maintained hemoglobin in the oxygenated state. This reversed its reduced oxygen affinity and thereby reduced polymerization of HbS and red blood cell sickling [2].

Oxygen is transported by binding to the Fe^{2+} present in the heme component of hemoglobin. If the iron is oxidized to Fe^{3+}, oxygen does not bind. Binding of oxygen to one heme molecule increases the oxygen affinity of other hemes by a mechanism called cooperativity. CO_2 and protons (H^+) decrease oxygen affinity (known as the "Bohr effect" [3]) as does 2,3-diphosphoglycerate (2,3-DPG), the intermediate metabolite in the

Luebering-Rapoport glycolytic pathway and therefore a side product of glycolysis. Together the result is that O_2 is more easily released in metabolically active tissues.

Carbon Dioxide (CO_2) Transport

This occurs by three mechanisms: CO_2 dissolved in plasma (5–10%), CO_2 bound to proteins such as hemoglobin as a carbamate (5–10%), and, principally, CO_2 combined with water to form carbonic acid (H_2CO_3, 70–80%) as part of a bicarbonate buffer system. The latter reaction is catalyzed by carbonic anhydrase in the RBC. In an equilibrium reaction, carbonic acid then dissociates to form bicarbonate ions (HCO_3^-) and H^+. This not only allows transport of CO_2 but also favors the release of O_2 from hemoglobin at sites of high aerobic demand [4].

Buffering and Homeostasis

Buffers in the blood maintain its pH within a narrow range of 7.35–7.45. Hemoglobin is the most important buffer for carbonic acid because of its high concentration and large number of histidine residues that bind hydrogen ions (H^+). Deoxyhemoglobin binds hydrogen ions even more easily than oxyhemoglobin. This synchronizes buffering action with the generation of additional H^+, increasing the amount of CO_2 that can be carried back to the lungs. General acidosis is buffered by other buffering systems such as bicarbonate, which binds metabolic acids, and phosphate, which serves a minor role in buffering due to the low concentration of phosphate ions in the blood [4]. Plasma proteins such as albumin may also contribute to buffering, but their main role is the regulation of blood volume by maintaining oncotic pressure. Disease states leading to hypoalbuminemia, for example, liver disease (see chapter "Liver: Overview") and nephrotic syndrome, therefore result in edema [3].

Iron Transport

This occurs by iron binding to the plasma protein transferrin, although <0.5% total body iron is present in the plasma. On average, two thirds of total body iron is incorporated into hemoglobin, and the remainder is mainly stored in the liver, as ferritin, or in body macrophages, as hemosiderin. Iron deficiency anemia is preceded by a depletion of body iron stores and a fall in serum ferritin concentration [5].

Coagulation Pathway and Anticoagulant Properties

Hemostasis depends upon platelets and a series of coagulation factors that cooperate in a series of complex interactions. The coagulation factors are inactive zymogen proteins (proteolytic precursor enzymes), mainly serine proteases. On activation by proteolytic cleavage, they can activate one or more other components, ultimately leading to clot formation and hemostasis (Fig. 1). Physiological coagulation occurs following injury and precedes wound healing, whereas pathological clotting occurs in intravascular sites, for example, in coronary artery occlusion and thrombotic stroke (see chapter "Stroke"). Hemophilia—a failure of hemostasis—is caused by mutations in one of the genes encoding coagulation factors; for example, hemophilia A is due to a mutation in the gene encoding factor VIII.

The plasma contains naturally occurring anticoagulants, such as protein C, protein S, and antithrombin, which regulate the activity of the coagulation cascade. Deficiency of these factors can lead to a thrombotic tendency. The plasma also contains proteins that lead to fibrinolysis, for example, tissue plasminogen activator, and proteins that regulate fibrinolysis, for example, plasminogen activator inhibitor and α2-antiplasmin. The former re-establish vascular patency after thrombosis, while the latter prevent excess breakdown of thrombi at the site of injury [6].

Leukocytes

Leukocytes, namely, neutrophils, eosinophils, basophils, lymphocytes, and monocytes, are an important component of blood as they form part of the body's defense mechanism against infection (see chapter "Anatomy and Physiology of the Immune System" under part "Immune system"). Overactivation of white blood cells can also exert a pathological role in autoimmunity and allergy.

Hormonal Transport for Regulation of Whole Body Metabolism

As the body's primary transport system, the blood carries endocrine signals from the secretory glands to the target. These chemical messengers, or hormones, are either dissolved in the plasma (in the case of hydrophilic hormones such as epinephrine) or bound to carriers such as albumin (in the case of lipophilic hormones such as glucocorticoids). As hormones are major regulators of metabolism and homeostasis, disturbances in the blood can significantly impair this communication system [3].

Lipoprotein Metabolism

Lipoproteins are complex, micellar-like aggregates used to transport lipophilic substances within the bloodstream, delivering various lipids to target tissues and recycling them to the liver (see chapter "Hyperlipidemia").

Inside-In: Metabolites of the Blood Affecting the Blood Itself

The biconcave shape of the RBC optimizes its flexibility and oxygen exchange across the cell surface. Mature RBCs lack nuclei, mitochondria, and ribosomal machinery, maximizing the storage capacity for hemoglobin. At the same time, the lack of these organelles limits the RBC's metabolic capacity. Two of the major four metabolic pathways, anaerobic glycolysis and to a lesser extent the pentose phosphate pathway, contribute to most of the RBC glucose metabolism. While anaerobic glycolysis provides the energy required for the RBC membrane ion pumps, the pentose phosphate pathway provides the reduction potential to protect against oxidant damage. Both contribute to maintaining Fe^{2+} in its reduced state. The latter two metabolic pathways, adenosine nucleotide metabolism and Rapoport-Luebering shunts, play smaller roles. In adenosine nucleotide metabolism, adenosine diphosphate (ADP) and adenosine-5′-triphosphate (ATP) release into the circulation induce platelet aggregation. The Rapoport-Luebering shunt is regarded as a glycolysis waste pathway for energy not needed through its maintenance of 2,3- DPG.

The Anaerobic Glycolytic (Embden-Meyerhof) Pathway

This pathway metabolizes glucose to generate energy and has three important products: (i) ATP mainly powers ion transport across the RBC membrane to maintain normal homeostasis despite exposure to osmotic stress, for example, in the renal circulation. In particular, the red cell membrane sodium pump maintains the osmotic potential within red cells, preventing their lysis. (ii) Nicotinamide adenine dinucleotide (NADH) is a cofactor in the methemoglobin reductase reaction. This regulates the conversion of methemoglobin (containing Fe^{3+}) back to hemoglobin (containing Fe^{2+}), as only the latter can bind oxygen (see above). (iii) 2,3-DPG is an allosteric effector of hemoglobin, binding to deoxyhemoglobin, thus favoring oxygen release.

Inherited deficiency of glycolytic enzymes (most commonly pyruvate kinase) can lead to chronic hemolytic anemia by decreasing ATP levels and impairing the aforementioned ion pumps leading to cell swelling and lysis. Recently, pyruvate kinase activators such as mitapivat, which "re-energize the red blood cell," have been shown to improve anemia in pyruvate kinase deficiency, hemoglobinopathies such as sickle cell disease and thalassemias [7].

The Pentose Phosphate Shunt (or Pentose Phosphate Pathway)

As oxygen concentration in the RBC is very high, heme iron is in danger of oxidation, deactivating its function and producing reactive oxygen species such as superoxide O_2^-. The red cell is also subject to exogenous oxidant stress, for example, from drugs and dietary constituents. The pentose phosphate pathway protects the RBC against oxidant damage in order both to maintain the iron in hemoglobin in a reduced, ferrous state (Fe^{2+}) and also to limit oxidant attack on RBC membrane proteins and lipids. The pentose phosphate shunt generates NADPH (nicotinamide adenine dinucleotide phosphate hydrogen), which, in turn, maintains the supply of reduced glutathione in the cells that is used to mop up free radicals that cause oxidative damage.

Inherited deficiency of enzymes of the pentose phosphate shunt renders RBC susceptible to oxidant stress. The most common of these worldwide is X-linked glucose-6-phosphate dehydrogenase (G6PD) deficiency, which is manifest as acute hemolysis following oxidant stress [5].

Inside-Out: Metabolites of the Blood Affecting Other Tissues

At the end of their natural lifespan of about 120 days, red blood cells are mainly removed by splenic macrophages. In pathological states, intravascular hemolysis can occur. This releases hemoglobin that binds to plasma haptoglobin. This complex is metabolized in the liver, which degrades hemoglobin to iron and protoporphyrin (Fig. 2). The latter is converted to bilirubin and then bilirubin glucuronide, which is excreted in the bile. Iron is recycled. Within the intestine, bacteria convert bilirubin to urobilinogen, which either is reabsorbed and excreted in the urine or passes further down the intestinal tract forming stercobilinogen [3]. Both products contribute to the color of the excretions.

If the rate of hemolysis exceeds the plasma supply of haptoglobin, free hemoglobin is excreted in the urine with some being reabsorbed and the iron incorporated into ferritin and then hemosiderin in the renal tubules. Initial hemoglobinuria is thus followed by hemosiderinuria, as renal tubule cells (containing the hemosiderin) are lost in the urine. Acute intravascular hemolysis can cause renal failure. Cell-free plasma hemoglobin also acts as a nitric oxide scavenger, which can lead to adverse effects on blood vessels (e.g., vasoconstriction, see chapter "Blood Vessels: Overview" under part "Blood Vessels"). The accumulation of bilirubin in the circulation also leads to a yellow discoloration of the skin, that is, jaundice. Chronic hemolysis leads to the formation of pigment gallstones [3].

Outside-in: Metabolites of Other Tissues Affecting the Blood

Impaired function of many organs, including the lungs and kidneys, involved in the regulation of blood oxygen levels, pH, and/or temperature can affect blood metabolism by shifting the oxygen dissociation curve (Fig. 1).

Normal hematopoiesis requires a supply of nutrients, including iron, vitamin B_{12}, and folic acid. Their availability is dependent on normal intestinal absorption. Vitamin B_{12} absorption additionally requires that the stomach secretes intrinsic factor, which combines with B_{12}, permitting its absorption by the small intestine (see chapter "Gastrointestinal Tract: Overview" under part "Gastrointestinal Tract" and Fig. 2). Thus, celiac disease and autoimmune gastric atrophy can lead indirectly to anemia. Erythropoiesis requires normal levels of growth hormone and adrenal and thyroid hormones, so that hypopituitarism, adrenal insufficiency, and hypothyroidism all lead to anemia. Erythropoiesis likewise requires normal renal function, including production of erythropoietin (see chapter "Kidney: Overview" under part "Kidney"). In renal failure, there is reduced synthesis of erythropoietin and thus anemia. In addition, nitrogenous waste products, including urea, accumulate in the blood, suppressing erythropoiesis and leading to platelet dysfunction and hemorrhage. Acid/base disturbance can also impact upon the oxygen dissociation curve.

In liver disease, abnormal metabolism can lead to specific forms of hemolytic anemia.

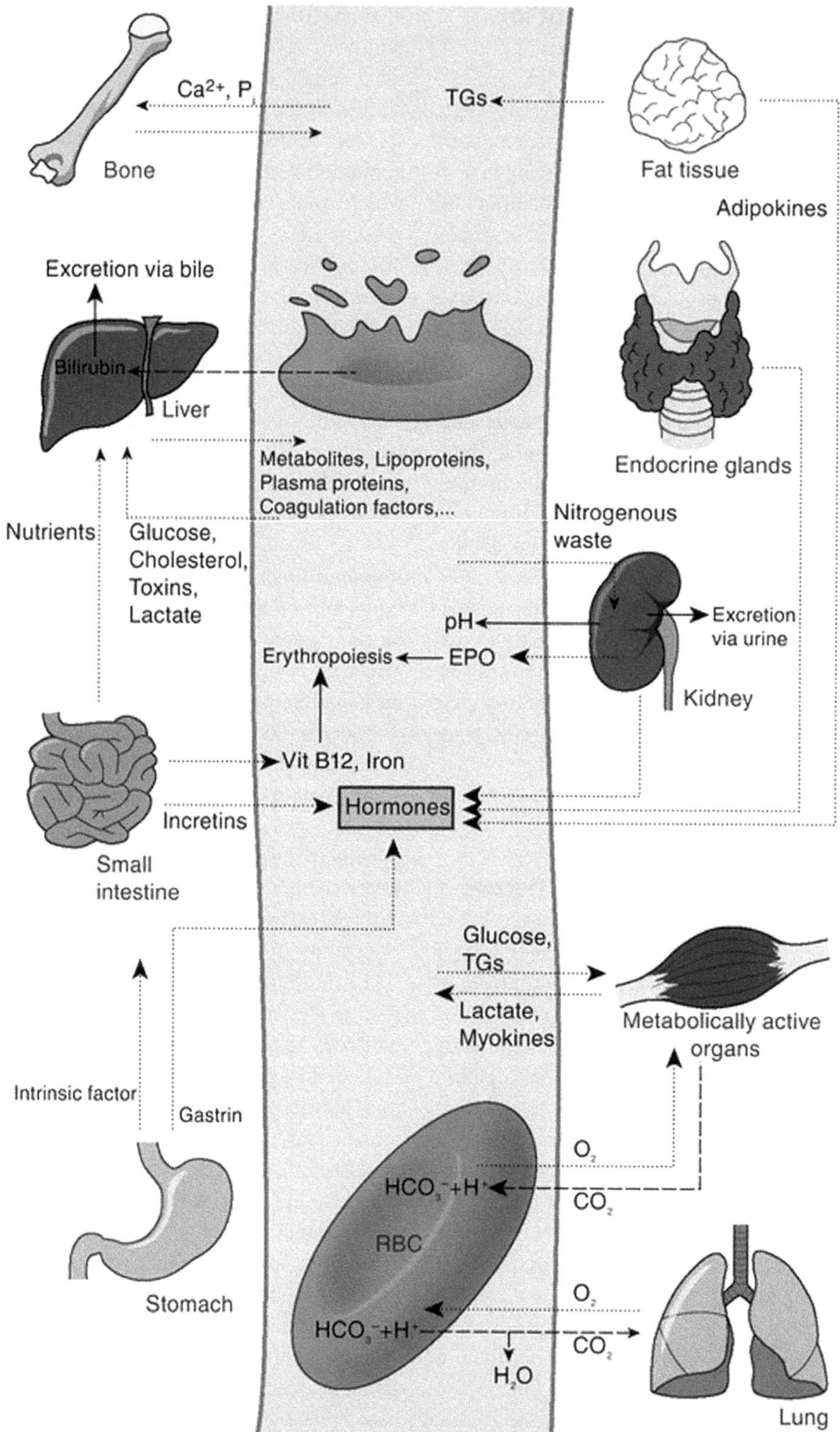

Fig. 2 Interactions of blood with other organs. As the major transport system in the body, the blood interacts with virtually all organs and tissues. Nutrients absorbed from the small intestine are transported to the liver and further to the whole body (e.g., to skeletal muscle). Recycling and redistribution of nutrients and their

Zieve's syndrome is characterized by acute hemolysis, irregularly contracted red cells, and hyperlipidemia associated with alcoholic fatty liver. Spur cell hemolytic anemia shows marked acanthocytosis (RBCs with a spiked, thorny cell membrane) occurring in liver failure of any etiology. Acute oxidant-induced Heinz body hemolytic anemia can occur in advanced Wilson's disease as a result of the oxidant effect of copper released from damaged or dying liver cells. Liver disease also leads to impaired synthesis of coagulation factors and thus hemorrhage [8].

Acute infection can cause anemia, neutrophilia, platelet consumption, and activation of coagulation. Infection generates oxidant substances and can lead to hemolysis in G6PD-deficient subjects (see above). Chronic infections and inflammatory states lead to generation of cytokines, causing anemia of chronic disease, in which iron is not mobilized from macrophages.

Perspectives

The blood is the major transport system in the body, delivering oxygen and nutrients to the tissues for metabolic processes and removing the waste products of metabolic reactions. Impaired function of organs and tissues can adversely impair the homeostatic mechanisms of the blood. In turn, measurement of the levels of blood constituents can assist in the diagnosis and monitoring of many diseases and their therapies.

Questions and Answers

Question 1 Where is blood produced?

Answer 1 Blood is produced in the bone marrow.

Question 2 What is the major function of blood?

Answer 2 Blood serves as a transport and delivery system for carbon dioxide, oxygen, nutrients, and waste products.

Question 3 What does the oxygen-dissociation curve represent?

Answer 3 The affinity of hemoglobin for oxygen at different partial pressures.

Question 4 Iron deficiency results from depletion of which body stores of iron?

Answer 4 Iron deficiency anemia is preceded by a depletion of body iron stores from the liver and macrophages and a decrease in serum ferritin.

Question 5 What important cofactor does the Pentose Phosphate Shunt generate and why is this important?

Answer 5 NADPH (nicotinamide adenine dinucleotide phosphate hydrogen), which protects RBC from oxidative stress.

Fig. 2 (continued) products (e.g., from skeletal muscle and adipocytes) also occur via the blood. The liver contributes a multitude of metabolic products to the bloodstream to be transported to the target tissues. This includes glucose from gluconeogenesis or glycogen stores and triglycerides (also called triacylglycerols, *TGs*) and cholesterol being transported in lipoproteins, amino acids, and ketone bodies during prolonged fasting. Ions transported in the blood are critical for bone stability (mainly calcium) and erythropoiesis (mainly iron). Hormones, such as erythropoietin (*EPO*) from the kidney, are secreted into the blood by a large variety of endocrine organs. "Cleaning" of the blood occurs in the kidney, where unwanted contents such as nitrogenous waste can be excreted and pH changes can be regulated. Red blood cells (*RBCs*) transport O_2 and aid in CO_2 transport from and to the lungs, respectively. Hemoglobin degradation products released from "dying" erythrocytes are generally removed via the liver. P_i inorganic phosphate (PO_4^{3-}), *Vit B12* vitamin B12

References

1. Bain BJ and Leach M (2025) Blood cells a practical guide, 8th edn. Wiley-Blackwell, Oxford
2. Howard J, Hemmaway CJ, Telfer P et al (2019) A phase 1/2 ascending dose study and open-label extension study of voxelotor in patients with sickle cell disease. Blood 133(17):1865–1875
3. Haslett C, Chilvers ER, Boon NA, Colledge NR (2002) Davidson's principles and practice of medicine, 19th edn. Churchill Livingstone, Edinburgh
4. Brandis, K. Anaesthesia education website. Available from URL: http://www.anaesthesiaMCQ.com
5. Bain BJ, Gupta R (2003) A-Z of haematology, 1st edn. Blackwell Publishing Ltd., Oxford
6. Laffan MA, Manning R (2017) Investigation of haemostasis. In: Bain BJ, Bates I, Laffan MA (eds) Dacie and Lewis practical haematology, 12th edn. Churchill Livingstone, Edinburgh
7. Al-Samkari H, Galactéros F, Glenthøj A et al (2022) Mitapivat versus placebo for pyruvate kinase deficiency. N Engl J Med 386:1432–1442
8. Mamun-al-Mahtab SR (2009) Liver: a complete book on hepato-pancreato-biliary diseases, 1st edn. Elsevier, India, Gurgaon

Sickle Cell Disease

Martin H Steinberg

Introduction

The genetic basis of sickle cell disease (SCD) is a mutation in the β-hemoglobin gene (*HBB*:c.20A > T). This mutation codes for the sickle β-globin chain, $β^{E7V}$, that together with normal α globin results in the hemoglobin tetramer, sickle hemoglobin (HbS; $α_2β^{E7V}_2$), in which the glutamate (E) at position 7 of the β-globin chain is replaced by valine (V). Carriers of the HbS gene who have sickle cell trait account for nearly half the population in some parts of Africa, the Middle East, and India, where the HbS mutation originated and endemic malaria provided a selective advantage for carriers. In North America, where the HbS gene was introduced by slave trading, sickle cell trait is common. For example, among African Americans, about 8% carry the sickle cell trait. Sickle cell trait, with some exceptions, is benign because each red blood cell (RBC) has <40% HbS. Carriers are normal hematologically and have a normal lifespan but have a small increased risk of thromboembolic disease and chronic renal disease. They should be coun-

seled especially for the risks of bearing or fathering children with SCD and for the rare adverse consequences of engaging in vigorous exertion. SCD is usually a severe disease because each RBC has 50–95% HbS. Worldwide, about 300,000 infants with sickle cell disease are born each year, mainly in Africa and India. An estimated 100,000 Americans, mainly of African descent, have SCD. SCD is a phenotype caused by several common and many rare genotypes. The most common genotypes are homozygosity for the HbS mutation or sickle cell anemia, compound heterozygosity for HbS and HbC or HbSC disease, and compound heterozygosity for HbS and one of many different β-thalassemia mutations [1, 2]. HbS homozygotes have the most severe symptoms, but phenotypic heterogeneity within all genotypes of disease is the rule. There are two major modulators of phenotypic heterogeneity of SCD. First is the presence of coincident α-thalassemia that is usually due to α-globin gene deletions. α-Thalassemia decreases RBC density and HbS concentration, which inhibits the highly concentration-dependent polymerization of HbS and reduces hemolysis. Second is the level of fetal hemoglobin (HbF, $α_2γ_2$). HbF and its mixed hybrid tetramer, $α_2γβ^{E7V}$, are excluded from the polymer phase reducing the cellular polymer content and cell injury [3, 4]. Many other genetic modulators of disease are likely to exist but are less well characterized and have smaller effects on the phenotype.

M. H. Steinberg (✉)
Department of Medicine, Division of Hematology and Medical Oncology, Boston University Chobanian & Avedisian School of Medicine, Center of Excellence in Sickle Cell Disease, Boston, MA, USA
e-mail: mhsteinb@bu.edu

E. Lammert, M. Zeeb (eds.), *Metabolism of Human Diseases*,
https://doi.org/10.1007/978-3-031-96019-2_34

Pathophysiology of Sickle Cell Disease and Metabolic Alterations

HbS and its polymer damage the RBC membrane (Fig. 1) culminating in erythrocyte (RBC) injury and deformation, termed sickling. Cellular dehydration and rigidity are among the key injuries to the sickle RBC. RBC damage causes the clinical and laboratory features or the phenotype of SCD. The phenotype includes occlusion of small and sometimes large blood vessels, hemolytic anemia, inflammation, and reperfusion injury, all leading to widespread acute and chronic organ damage and a reduced lifespan. Abnormal adhe-sive interactions among RBCs, leukocytes, and endothelial cells mediated by a variety of adhesion molecules and their ligands cause an inflammatory response secondary to endothelial activation, damage, and reperfusion injury [5, 6]. Activation of three ion channels, the Gardos channel, which is a Ca^{2+}-activated K^+ channel, the K^+:Cl^- cotransport channel, and PIEZO1, a mechanosensitive ion channel, by combinations of deoxygenation, acidification, cell swelling, Ca^{2+} influx, and cell sickling leads to dense, dehydrated sickle RBCs in which HbS concentration is increased enhancing its polymerization tendency. Perturbed endothelial cells display

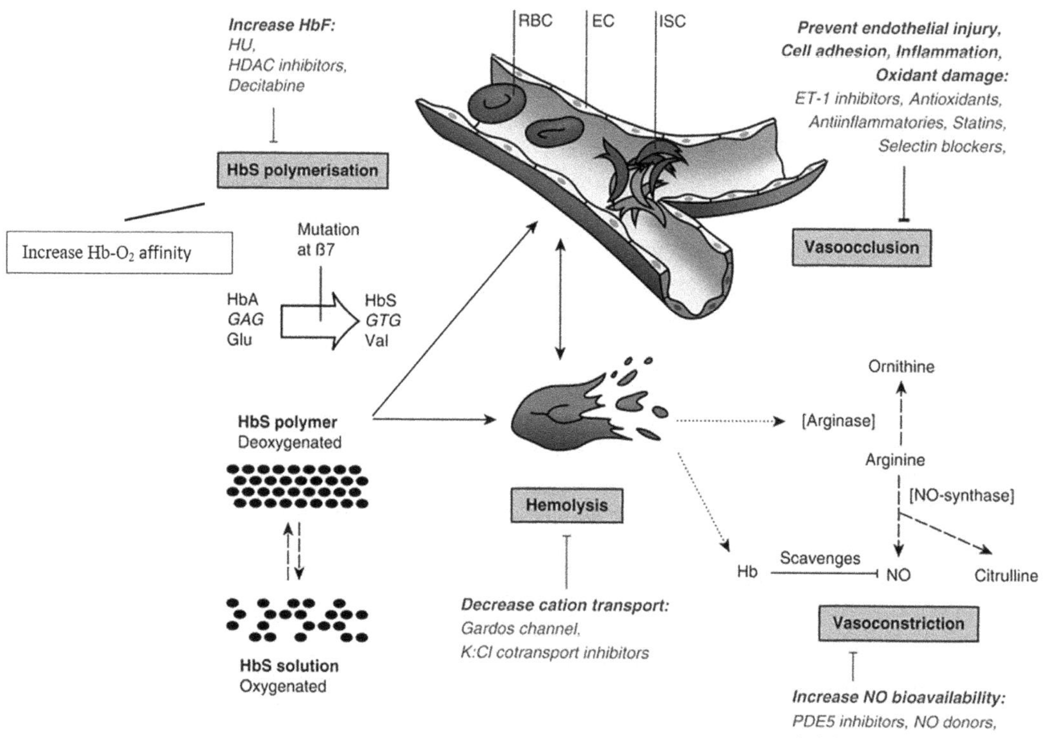

Fig. 1 Pathophysiology of SCD and pathophysiological-based pharmacologic treatment approaches. DeoxyHbS (sickle cell hemoglobin) polymerizes, deforms and injures the sickle RBC, leading to vasoocclusion and hemolytic anemia. Intravascular hemolysis of sickle cells releases heme into the blood that scavenges nitric oxide (*NO*), and the liberated arginase destroys arginine, the substrate of the NO synthases. Each facet of the pathophysiology could be targeted to interfere with its adverse effects. Agents that induce HbF expression, such as hydroxyurea and gene therapy (*HU, upper left*), can reduce the polymerization tendency of HbS. HbS polymerization can also be decreased by agents that "lock" hemoglobin in its oxy conformation or *PKLR* agonists that decrease RBC 2,3 BPG. Cell density (and thus hemolysis) could be reduced by inhibiting cation transport channels (*bottom center*), adhesive interactions with the endothelium could be interrupted by blocking receptor-ligand docking such as the P-selectin inhibitor crizanlizumab (*upper right*), and finally, availability and effects of NO could be increased (*bottom right*). *HbA* adult hemoglobin, *Glu* glutamate, *Val* valine, *EC* endothelial cell, *ISC* irreversible sickled cell, *Hb* hemoglobin, *ET-1* endothelin-1, *PDE* phosphodiesterase

adhesion molecules like integrins, selectins, and vascular cell adhesion molecule-1 (VCAM-1); produce cytokines and inflammatory mediators; and develop a procoagulant phenotype, thereby attracting adhesive RBCs, leukocytes, and platelets. Together, intrinsic features in the sickle RBC, like polymer content, and extrinsic factors in their environment, like endothelial injury, may result in decreased blood flow and vasoocclusion. Vasoocclusion and hemolysis provoke the majority of clinical complications of SCD [7]. Sickle vasoocclusive events that are most common and often dominate the life of patients include the acute painful episode, acute chest syndrome, and osteonecrosis. Acute painful episodes ("sickle crises") are excruciatingly painful, last hours to days, and occur in most patients although their frequency and severity can vary greatly amongst and within patients. Exactly why these attacks occur is not known, but they are presumed to originate from cellular adhesive interaction and vasoocclusion with subsequent ischemia and inflammation. The acute chest syndrome, a vasoocclusive episode of the lung, is often precipitated by necrotic bone marrow emboli and infection among other causes, is characterized by fever, chest pain, wheezing, cough, hypoxia, lung infiltrate, and a mortality rate of about 5%. Osteonecrosis of the hip and shoulder joints affects about half of all patients and can progress to loss of joint functionality. Osteonecrosis and bone marrow emboli are likely caused by microvascular occlusion. Hemolysis, or premature destruction of RBCs, is a result of membrane damage to the sickle RBC due to HbS polymer, oxidant radical generation, and the exposure of epitopes that facilitate the interaction of sickle cells with endothelium and macrophages. Intravascular hemolysis, which makes up 33% of all hemolysis, is accompanied by the liberation of free hemoglobin that scavenges nitric oxide (NO) promoting vasoconstriction and inflammation. Certain complications of disease are strongly correlated with the intensity of intravascular hemolysis, NO depletion, and proliferative vasculopathy [8,9]. Complications associated with intravascular hemolysis include pulmonary hypertension, which affects up to 10% of adults and is a leading cause of SCD mortality; skin ulceration, usually around the ankles; priapism, a prolonged undesirable painful erection (e.g., due to venous outflow occlusion in the penis because of reduced NO bioavailability), seen in 40% of men; stroke; renal failure [1]. Stroke, a major complication of SCD, is seen mainly in HbS homozygotes and is caused by stenosis and vessel occlusion in children; hemorrhagic stroke is more common in adults.

Introduction to Treatment, Influence on Metabolism, and Consequences for Patients

Two drugs with disease-modifying potential are approved for treatment in many countries: hydroxyurea (HU, hydroxycarbamide) and crizanlizumab. HU increases HbF concentration. Its cytotoxicity causes erythroid regeneration and the selection of erythroid progenitors that synthesize high levels of HbF; it may also lead to augmented γ-globin gene expression via a NO-mediated increase in cyclic guanosine monophosphate (cGMP) [10]. HU can reduce painful episodes and acute chest syndrome, decrease hemolysis, increase hemoglobin levels, and decrease mortality [11–19]. It should be given at maximal tolerated doses to nearly all patients with the sickle cell anemia beginning in the first year of life [20]. Although adverse effects in adults are minor, the long-term consequences of decades of use in children and the sustainability of its excellent effects when started early in life remain to be assessed.

Inducing about 10 pg of HbF in most sickle RBCs should "cure" SCD [21].While this is not possible using HU, it can be accomplished by gene therapy. Gene therapy trials modify autologous CD34+ progenitor cells by (1) adding a *HBB* engineered to prevent HbS polymerization (HbA^{T87Q}); (2) downregulating the activity of HbF gene repressors like *BCL11A* using CRISPR/Cas9 to disrupt its erythroid enhancer or a short hairpin mRNA to inhibit *BCL11A* transcription. Three trials have shown that >40% HbF or a "HbF-like" *HBB* can be induced in most RBCs nearly normalizing hematologic parameters and abrogating symptoms of SCD [22–24]. Two agents are now widely approved.

Adhesive interactions among endothelial cells, leukocytes, platelets, and RBCs are responsible for vasoocclusive complications of SCD. P-selectin is involved in these interactions; blocking selectins prevents sickle cell-endothelial adhesion [25]. Monthly intravenous infusions of crizanlizumab, a P-selectin blocking monoclonal antibody, reduced acute painful episodes by ~45%, a reduction similar to that of hydroxyurea [26]. Perhaps long-term use of crizanlizumab can prevent complications arising from vasoocclusive disease.

New drugs have changed the treatment landscape for SCD. Their combinatorial use with HU will not result in cure but should alter the course of disease. Presently there is no consensus on how these drugs should be integrated into practice and what combinations are most efficacious. Until the technologically formidable but potentially curative cell-based therapies are proven safe, effective, and long-lasting, widely available, and inexpensive oral drug therapy will be paramount.

Symptomatic treatments that do not alter the pathophysiology of disease include fluid replacement and opioid analgesics for pain [1, 2]. Blood transfusions are used for certain acute complications like severe anemia, acute chest syndrome and stroke and are used preoperatively to prevent severe postoperative complications like acute chest syndrome. An important prophylactic use of transfusion is to prevent stroke in children who on screening tests that estimate cerebral blood flow using laser-Doppler velocimetry are deemed at high risk for a stroke. Transfusion can cause iron overload in heavily transfused patients while alloimmunization (an immune reaction against the "foreign" blood) makes transfusion in about 20% of patients difficult [28]. Allogeneic hematopoietic stem-cell transplantation from HLA-identical siblings "cures" >90% of recipients but <20% of patients will have a donor, hence the attractiveness of gene therapy using autologous stem cells [29, 30].

Perspectives

Drugs that target (1) RBC dehydration by preventing cation and water efflux, (2) NO dysregulation by increasing its production or reducing its destruction, and (3) inflammation might provide useful adjunct therapies, should they prove to be efficacious. Earlier and more widespread use of HU, which took decades to be widely accepted, and better adherence to this treatment, along with improvement in supportive care, might further improve the outlook. Where medical resources are adequate, nearly all children survive to age 20 years, but during the transition from pediatric to adult care mortality rates increase. The impact of newer therapeutic approaches on chronic organ damage and mortality cannot yet be assessed.

Quesitons and Answers

Question 1 A woman with sickle cell trait is planning to start a family with a man who also has sickle cell trait. You should counsel the prospective parents that:

A. The risk of a fetus with sickle cell disease is 100% for each pregnancy.
B. The risk of a fetus with sickle cell disease is 50% for each pregnancy.
C. The risk of a fetus with sickle cell disease is 25% for each pregnancy.
D. The risk of a fetus with sickle cell disease is 0% for each pregnancy.

Answer 1 C

Question 2 Homozygosity for the mutation coding for sickle hemoglobin (HbS):

A. causes decreased synthesis of the abnormal hemoglobin.

B. decreases the ability of the abnormal hemoglobin to deliver oxygen.
C. causes irreversible polymerization of the abnormal hemoglobin.
D. causes red cell membrane damage.

Answer 2 D

Question 3 The presence of high levels of fetal hemoglobin (HbF) in sickle cell disease:

A. prevents HbS from polymerizing
B. decreases the affinity of hemoglobin for oxygen.
C. directly inhibits the adherence of sickle cells to endothelium.
D. is a sign that a patient is younger than apparent.

Answer 3 A

References

1. Gladwin MT, Kato GJ, Novelli EM (eds) (2021) Sickle cell disease. McGraw Hill, NY
2. Steinberg MH, Forget BG, Higgs DR, Weatherall DJ (2009) Disorders of hemoglobin: genetics, pathophysiology, clinical management, 2nd edn. Cambridge University Press, Cambridge
3. Steinberg MH, Sebastiani P (2012) Genetic modifiers of sickle cell disease. Am J Hematol 87:824–826
4. Steinberg MH (2020) Fetal hemoglobin in sickle cell anemia. Blood 136(21):2392–2400
5. Joiner CH, Gallagher PG (2009) The erythrocyte membrane. In: Steinberg MH, Forget BG, Higgs DR, Nagel RL (eds) Disorders of hemoglobin: genetics, pathophysiology, and clinical management. Cambridge University Press, Cambridge
6. Conran N, Costa FF (2009) Hemoglobin disorders and endothelial cell interactions. Clin Biochem 42:1824–1838
7. Piel FB, Steinberg MH, Rees DC (2017) Sickle cell disease. N Engl J Med 376(16):1561–1573
8. Kato GJ, Gladwin MT, Steinberg MH (2007) Deconstructing sickle cell disease: reappraisal of the role of hemolysis in the development of clinical subphenotypes. Blood Rev 21:37–47
9. Kato GJ, Steinberg MH, Gladwin MT (2017) Intravascular hemolysis and the pathophysiology of sickle cell disease J Clin Invest. Mar 1;127(3):750–760
10. Almeida CB, Scheiermann C, Jang JE, Prophete C, Costa FF, Conran N, Frenette PS (2012) Hydroxyurea and a cGMP-amplifying agent have immediate benefits on acute vaso-occlusive events in sickle cell disease mice. Blood 120:2879–2888
11. Dover GJ, Humphries RK, Moore JG, Ley TJ, Young NS, Charache S, Nienhuis AW (1986) Hydroxyurea induction of hemoglobin F production in sickle cell disease: relationship between cytotoxicity and F cell production. Blood 67:735–738
12. Steinberg MH, Lu ZH, Barton FB, Terrin ML, Charache S, Dover GJ (1997) Fetal hemoglobin in sickle cell anemia: determinants of response to hydroxyurea. Multicenter Study Hydroxyurea Blood 89:1078–1088
13. Bridges KR, Barabino GD, Brugnara C, Cho MR, Christoph GW, Dover G, Ewenstein BM, Golan DE, Guttmann CR, Hofrichter J, Mulkern RV, Zhang B, Eaton WA (1996) A multiparameter analysis of sickle erythrocytes in patients undergoing hydroxyurea therapy. Blood 88:4701–4710
14. Cokic VP, Andric SA, Stojilkovic SS, Noguchi CT, Schechter AN (2008) Hydroxyurea nitrosylates and activates soluble guanylyl cyclase in human erythroid cells. Blood 111:1117–1123
15. Charache S, Terrin ML, Moore RD, Dover GJ, Barton FB, Eckert SV, McMahon RP, Bonds DR (1995) Effect of hydroxyurea on the frequency of painful crises in sickle cell anemia. N Engl J Med 332:1317–1322
16. Steinberg MH, Barton F, Castro O et al (2003) Effect of hydroxyurea on mortality and morbidity in adult sickle cell anemia: risks and benefits up to 9 years of treatment. JAMA 289:1645–1651
17. Steinberg MH, McCarthy WF, Castro O, Pegelow CH, Ballas SK, Kutlar A, Orringer E, Bellevue R, Olivieri N, Eckman J, Varma M, Ramirez G, Adler B, Smith W, Carlos T, Ataga K, DeCastro L, Bigelow C, Saunthararajah Y, Telfer M, Vichinsky E, Claster S, Shurin S, Bridges K, Waclawiw M, Bonds D, Terrin M (2010) The safety and effectiveness of hydroxyurea in sickle cell anemia: a 17.5 year follow-up. Am J Hematol 85:403–408
18. Wang WC, Ware RE, Miller ST, Iyer RV, Casella JF, Minniti CP, Rana S, Thornburg CD, Rogers ZR, Kalpatthi RV, Barredo JC, Brown RC, Sarnaik SA, Howard TH, Wynn LW, Kutlar A, Armstrong FD, Files BA, Goldsmith JC, Waclawiw MA, Huang X, Thompson BW (2011) Hydroxycarbamide in very young children with sickle-cell anaemia: a multicentre, randomised, controlled trial (BABY HUG). Lancet 377:1663–1672
19. Voskaridou E, Christoulas D, Bilalis A, Plata E, Varvagiannis K, Stamatopoulos G, Sinopoulou K, Balassopoulou A, Loukopoulos D, Terpos E (2009) The effect of prolonged administration of hydroxyurea on morbidity and mortality in adult patients with

sickle-cell syndromes: results of a 17-year, single center trial (LaSHS). Blood 115:2354–2363

20. McGann PT, Niss O, Dong M, Marahatta A, Howard TA, Mizuno T, Lane A, Kalfa TA, Malik P, Quinn CT, Ware RE, Vinks AA (2019) Robust clinical and laboratory response to hydroxyurea using pharmacokinetically guided dosing for young children with sickle cell anemia. Am J Hematol 94(8):871–879

21. Steinberg MH, Chui DH, Dover GJ, Sebastiani P, Alsultan A (2014) Fetal hemoglobin in sickle cell anemia: a glass half full? Blood 123(4):481–485

22. Kanter J, Walters MC, Krishnamurti L, Mapara MY, Kwiatkowski JL, Rifkin-Zenenberg S, Aygun B, Kasow KA, Pierciey FJ Jr, Bonner M, Miller A, Zhang X, Lynch J, Kim D, Ribeil JA, Asmal M, Goyal S, Thompson AA, Tisdale JF (2022) Biologic and clinical efficacy of LentiGlobin for sickle cell disease. N Engl J Med 386(7):617–628

23. Esrick EB, Lehmann LE, Biffi A, Achebe M, Brendel C, Ciuculescu MF, Daley H, MacKinnon B, Morris E, Federico A, Abriss D, Boardman K, Khelladi R, Shaw K, Negre H, Negre O, Nikiforow S, Ritz J, Pai SY, London WB, Dansereau C, Heeney MM, Armant M, Manis JP, Williams DA (2021) Post-transcriptional genetic silencing of bcl11a to treat sickle cell disease. N Engl J Med 384(3):205–215

24. Frangoul H, Altshuler D, Cappellini MD, Chen YS, Domm J, Eustace BK, Foell J, de la Fuente J, Grupp S, Handgretinger R, Ho TW, Kattamis A, Kernytsky A, Lekstrom-Himes J, Li AM, Locatelli F, Mapara MY, de Montalembert M, Rondelli D, Sharma A, Sheth S, Soni S, Steinberg MH, Wall D, Yen A, Corbacioglu S (2021) CRISPR-Cas9 gene editing for sickle cell disease and beta-thalassemia. N Engl J Med 384(3):252–260

25. Embury SH, Matsui NM, Ramanujam S, Mayadas TN, Noguchi CT, Diwan BA, Mohandas N (2004) Cheung AT the contribution of endothelial cell P-selectin to the microvascular flow of mouse sickle erythrocytes in vivo. Blood 104:3378–3385

26. Ataga KI, Kutlar A, Kanter J, Liles D, Cancado R, Friedrisch J, Guthrie TH, Knight-Madden J, Alvarez OA, Gordeuk VR, Gualandro S, Colella MP, Smith WR, Rollins SA, Stocker JW, Rother RP (2016) Crizanlizumab for the prevention of pain crises in sickle cell disease. N Engl J Med 376:429–439

27. Yazdanbakhsh K, Ware RE, Noizat-Pirenne F (2012) Red blood cell alloimmunization in sickle cell disease: pathophysiology, risk factors, and transfusion management. Blood 120:528–537

28. Hirtz D, Kirkham FJ (2019) Sickle cell disease and stroke. Pediatr Neurol 95:34–41

29. Furstenau DK, Tisdale JF (2021) Allogenic hematopoietic stem cell transplantation in sickle cell disease. Transfus Apher Sci 60(1):103057

30. de la Fuente J, Gluckman E, Makani J, Telfer P, Faulkner L, Corbacioglu S; Paediatric Diseases Working Party of the European Society for Blood and Marrow Transplantation (2020) The role of haematopoietic stem cell transplantation for sickle cell disease in the era of targeted disease-modifying therapies and gene editing. Lancet Haematol. 7(12):e902-e911

Hyperlipidemia

Paul Durrington, Bilal Bashir, and Handrean Soran

Introduction to Hyperlipidemia

The lipids in the body are mainly represented by cholesterol, triglycerides (TGs, also called triacylglycerols), and phospholipids. Lipids are transported in the blood as lipoproteins, which are mixed micellar-like particles with a central droplet containing their most hydrophobic components, cholesteryl esters and TGs, and an outer layer comprising amphiphilic phospholipid molecules interspersed with free (non-esterified) cholesterol, giving the particle a hydrophilic surface [1–3]. The protein components of lipoproteins are mainly enzymes, such as lecithin-cholesterol acyltransferase (LCAT), and apolipoproteins. The latter play important roles in lipid and lipoprotein secretion by the liver and gut, lipoprotein structure and transport in the lymph and blood, and uptake by various tissues, as they serve as receptor ligands. Also, they have structural and regulatory functions, modifying the activity of enzymes relevant to lipoprotein metabolism. They are oriented similarly to the lipids, with hydrophobic regions toward the core and polar regions to the outside. The lipoproteins are classified into four major classes (chylomicrons, very low-density lipoproteins [VLDLs], low-density lipoproteins [LDLs], and high-density lipoproteins [HDLs]), which differ in size, density, composition, and function (Fig. 1).

In this chapter, we will give an overview of lipoprotein physiology and metabolism followed by discussion of its major disturbances, the hyperlipidemias.

P. Durrington (✉)
Division of Cardiovascular Science, University of Manchester Core Technology Facility,
Manchester, UK
e-mail: pdurrington@manchester.ac.uk

B. Bashir
Division of Cardiovascular Science, University of Manchester Core Technology Facility,
Manchester, UK

Department of Diabetes, Endocrinology and Metabolism, Manchester University NHS Foundation Trust, Manchester, UK
e-mail: bilal.bashir@nhs.net

H. Soran
Department of Diabetes, Endocrinology and Metabolism, Manchester University NHS Foundation Trust, Manchester, UK
e-mail: Handrean.soran@mft.nhs.uk

Physiological Lipoprotein Metabolism

Chylomicrons

The largest lipoproteins are the chylomicrons produced by the enterocytes of the small intestine. They permit dietary fat to enter the body. The fatty acid and monoglyceride products of digestion absorbed through the intestinal villi are resynthesized into TG and packaged into chylo-

© The Author(s), under exclusive license to Springer Nature Switzerland AG 2026
E. Lammert, M. Zeeb (eds.), *Metabolism of Human Diseases*,
https://doi.org/10.1007/978-3-031-96019-2_35

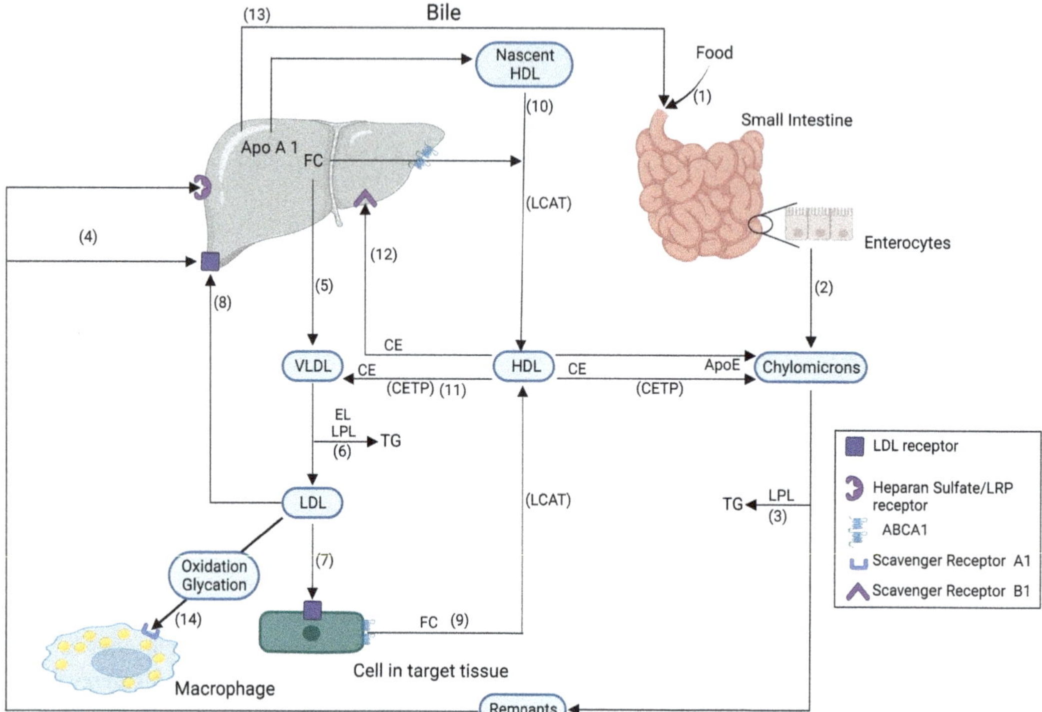

Fig. 1 An outline of lipoprotein metabolism. (1) Fatty acid, monoglyceride, and cholesterol products of food digestion are absorbed. (2) Triglyceride (TG), cholesterol, phospholipid, and apolipoprotein B48 (apoB48) are released from enterocytes as chylomicrons into the gut lymphatics. (3) Chylomicrons enter the blood circulation via the thoracic duct (i.e., the largest lymphatic vessel). Then, in the capillary endothelium particularly of muscle and adipose tissue, lipoprotein lipase (LPL) releases TGs from chylomicrons by hydrolysis to free fatty acids (FA) and glycerol. Continuation of this process yields remnants rich in cholesterol. (4) These remnants are taken up by the liver via heparin sulfate/LDL receptor-like protein (heparan sulfate/LRP) receptors or low-density lipoprotein (LDL) receptors (both binding apoE). (5) The liver synthesizes TG-rich very-low-density lipoproteins (VLDL) to carry TG to target tissues. VLDL contains apoB100. (6) Removal of TG from VLDL by both LPL and the more widely distributed endothelial lipase (EL) together with uptake of cholesteryl esters from high-density lipoprotein (HDL) via cholesteryl ester transfer protein (CETP) creates LDL rich in cholesterol. (7) LDL can be taken up by cells. This requires binding of apoB100 to LDL receptors expressed in tissues requiring cholesterol. (8) Generally, in adults most LDL receptor-mediated uptake is by the liver. (9) Excess cholesterol accumulating in extrahepatic tissues is transferred to high-density lipoprotein (HDL) through ATP binding cassette transporter A1 (ABCA1). (10) HDL contains lecithin-cholesterol acyltransferase (LCAT), which esterifies free cholesterol (FC) to form the hydrophobic core of HDL. HDL originates from the liver and small intestine as small apoA1-rich particles (nascent HDL), which enlarge in the circulation as they acquire cholesterol. (11) HDL transfers cholesterol to VLDL, LDL, and chylomicrons as cholesteryl esters (CE) with the assistance of CETP. (12) CE can be cleared from the circulation from HDL by hepatic scavenger receptors or from LDL or chylomicron remnants via hepatic LDL receptors or heparan sulfate/LRP receptors. (13) Cholesterol re-entering the liver can be secreted in bile directly or after its conversion to bile acids. Bile is secreted into the small intestine where it also aids in the uptake of lipids including cholesterol. (14) LDL cholesterol can enter macrophage-type cells derived from blood monocytes and uncommitted smooth muscle cells in the arterial wall to form foam cells, the basis of atheroma. Modifications of LDL, such as oxidation and glycation, permit its rapid uptake through macrophage scavenger receptors

microns, a process that is highly efficient (if it is only minimally compromised, steatorrhea results). Cholesterol entering the intestine from the diet and as bile is also packaged into chylomicrons. However, its absorption is incomplete despite the presence in the luminal surface of intestinal cells of the cholesterol transporter proteins, Niemann-Pick C1-like 1 (NPC1L1) and ATP-binding cassette (ABC) transporter G5 and G8. The gut thus constitutes a means of excretion of cholesterol as well as its entry into the body.

Chylomicrons enter the blood circulation via the lymph. Once in the blood circulation, particularly as they pass through adipose tissue, skeletal muscle and cardiac muscle, TGs are hydrolyzed by lipoprotein lipase (LPL) located on the capillary endothelium. They are converted to fatty acids and glycerol, which are then used as respiratory substrates or reconstituted (into TGs) as an energy store.

LPL is highly regulated. It has a binding site for sulfated glycosaminoglycans. This allows it to be anchored to the capillary endothelium from where it protrudes into the current of circulating blood. It thus comes into contact with TG-rich lipoproteins, chylomicrons, and very-low-density lipoprotein (VLDL). It preferentially acts on chylomicrons and larger VLDL. LPL is activated by apoCII and by insulin. It is downregulated by insulin resistance, by angiopoietin-like protein 3 (ANGPTL3), and by apoCIII. Chylomicrons are converted by LPL to TG-depleted remnants. These remnants carry cholesterol absorbed from the gut and transferred to them from smaller lipoproteins by cholesteryl ester transfer protein (CETP). They are removed from the circulation via the liver (Fig. 1). This receptor-mediated clearance occurs through the heparan sulfate/LDL receptor-related protein (heparan sulfate/LRP) pathway, which binds remnants through apolipoprotein E (apoE) that has been transferred to them from high-density lipoprotein (HDL) during their circulation. In health, the whole process occurs within a few hours of ingesting food. Once in the liver, cholesterol can leave to enter the blood circulation as a component of VLDL or in bile as cholesterol and bile salts to which it has been converted. The fraction of cholesterol and bile salts lost from the gut by incomplete absorption is not replaced by absorption of dietary cholesterol, and

is balanced by cholesterol synthesis in the liver. The physiological rate-limiting enzyme for this is 3-hydroxy-3-methylglutaryl-CoA reductase (HMG-CoA reductase). Hepatic HMG-CoA reductase is the target of statin therapy.

Very-Low-Density Lipoprotein

VLDL is produced by the liver. It contains the TG and cholesterol secreted by the liver. It is rich in TG and comprises most of the TG in serum. It is often advised to measure serum TG in the fasting state to represent VLDL concentration and avoid the inclusion of TG in chylomicrons, but in health chylomicron clearance from the circulation is so rapid that this is unnecessary at least in screening for hypertriglyceridemia (HTG). Key to VLDL assembly in the liver is microsomal triglyceride transfer protein (MTP). The major role of VLDL is to transport endogenously produced hepatic TG to the target tissues and to be the precursor of low-density lipoprotein (LDL). Its TG, like that of chylomicrons, is removed during circulation by LPL. However, as VLDL particle size is diminished by LPL, two other lipases become critical for hydrolysis of its TG and thus for its conversion to LDL. These are endothelial lipase and hepatic lipase. Endothelial lipase has a wider tissue distribution than LPL, including organs such as the liver, lungs, kidney, thyroid, ovary, testis, and placenta. It acts on VLDL once its size is reduced by LPL to a particle intermediate between VLDL and LDL [4]. It can hydrolyze both TG and phospholipids, increasing LDL production from its VLDL precursor. The inhibition of ANGPTL3 has recently been discovered to lower circulating LDL concentrations by a mechanism likely to involve endothelial lipase. A third lipase, hepatic lipase, removes TG and phospholipid from the LDL produced by endothelial lipase to produce even smaller LDL particles, which are believed to participate most readily in atherogenesis.

Low-Density Lipoprotein

Removal of TGs and/or uptake of cholesterol via CETP converts VLDL to the cholesterol-rich smaller lipoprotein, LDL. LDL crosses the capillary endothelium and supplies cholesterol to the tissues by endocytosis after binding to cell-surface LDL receptors [5], which are expressed when cholesterol is required for membrane growth, for repair, or for steroid hormone synthesis. Thus, LDL delivers cholesterol to the tissues.

The liver, which is the major organ of LDL catabolism, also expresses LDL receptors in abundance. Cholesterol entering liver by this route is secreted into the gut as biliary cholesterol, converted to bile salts or re-secreted into the blood circulation as a component of VLDL. The LDL receptor binds to apoB100 present on LDL and apoE present on chylomicron remnants. Its upregulation in the liver lowers serum LDL cholesterol (LDLc). On the other hand, increased LDL receptor catabolism is promoted by proprotein convertase subtilisin/kexin type 9 (PCSK9) raising serum LDL cholesterol [6]. The mechanism for this downregulation of the LDL receptor is that PCSK9 in the circulation binds to LDL and when the LDL/PCSK9 complex thus formed docks with the LDL receptor, instead of the LDL receptor being released after internalization into the hepatocyte to resume a position on the cell surface (recycling), it is retained to undergo lysosomal degradation.

LDL can also enter macrophages located in the arterial wall contributing to atherosclerosis. These macrophages are derived from blood monocytes crossing the vascular endothelium. In the arterial subintima, they take up LDL and their cytoplasm becomes loaded with cholesterol droplets giving the appearance of foam. Accumulation of these foam cells leads to fatty streak formation [3, 7]. In addition, progenitor smooth muscle cells from the tunica media can enter the developing lesion and commit to differentiation to either fibroblast or macrophage/foam cell phenotypes. Subsequently, an atheroma develops with a fibrous cap overlying a cholesterol lake formed by necrosis and apoptosis of foam cells. Foam cells can also produce collagenase, which contributes to the likelihood of atheromatous lesions rupturing in regions where foam cells are active. Thrombosis on the ruptured surface of atheroma is a major cause of acute cardiovascular events, such as acute coronary syndrome and stroke.

While LDL uptake by macrophage/monocytes can occur through the LDL receptor, more rapid uptake is required for foam cell formation.

Macrophage scavenger receptors (e.g., scavenger receptor A1; SRA1) [8] mediate LDL uptake of sufficient velocity, but, before acting as a ligand for these, LDL must undergo atherogenic modification such as oxidation, glycation, or glycoxidation. Smaller denser LDL that has circulated longer than less dense subfractions is more susceptible to these modifications. So also, as the result of its retention in the arterial wall, may be a subfraction of LDL, termed lipoprotein (a) or Lp(a), in which apolipoprotein (a), a member of the plasminogen family of proteins, is covalently linked to its apoB [9].

High-Density Lipoprotein and Reverse Cholesterol Transport

High-density lipoprotein (HDL) is the smallest lipoprotein and, although contributing less to total serum cholesterol than LDL, is more abundant, particularly in tissue fluid. It originates as small disk-like particles containing apolipoprotein A1 (apoA1) secreted by the liver and gut. These nascent HDL particles progressively enlarge and become globular as they receive free cholesterol from the liver and extra-hepatic tissues [10]. Free cholesterol arriving in the outer layer of HDL is esterified by LCAT. The more hydrophobic cholesteryl ester thus produced then enters the particle core, permitting further uptake of free cholesterol at the surface. Without this enlargement, HDL cannot escape filtration and loss through the kidneys. Thus, the acquisition of free cholesterol from the liver by nascent HDL is essential. This is accomplished via the hepatic ATP binding cassette A1 transporter A1 (ABCA1) (also known as "cholesterol efflux regulating protein"). In an alphalipoproteinemia (also known as Tangier disease) loss-of-function mutations of ABCA1 lead to a virtual absence of HDL, splenomegaly, and occasionally orange-yellow cholesterol deposits in the tonsils and rectal mucosa. LCAT deficiency is also a cause of low HDL, which is accompanied by corneal opacity, and proteinuria.

High-density lipoprotein (HDL) is inversely related to atherosclerotic cardiovascular disease (ASCVD) risk in epidemiological studies. It was thought that this was principally because it constituted the route for returning excess cholesterol from tissues to the liver (reverse cholesterol transport). Certainly, HDL may be critical for the passage of excess cholesterol out of cells. Excess cholesteryl ester within cells once converted back to free cholesterol can cross the outer cell membrane aided by ABCA1 to be incorporated into HDL when it can be cleared from the circulation by the liver that it enters through the scavenger receptor B1 (SR-B1). However, cholesterol can also be transferred from HDL to chylomicron remnants and LDL via CETP and then be cleared by the liver through heparan sulfate/LRP and LDL receptors. The contribution of this mechanism to reverse cholesterol transport is shown by familial CETP deficiency and by pharmacological inhibition of CETP in both of which HDL cholesterol (HDLc) levels are high and the HDL particles enlarged as cholesterol can no longer be transferred from HDL to other lipoproteins. Raised HDL from this cause does not appear to reduce ASCVD risk appreciably. Recently, to explain the association between decreased ASCVD risk and HDL, attention has therefore been most directed toward the early uptake of cellular cholesterol by HDL (cholesterol efflux capacity) and by the recently discovered protection HDL affords against atherogenic modifications of LDL [11].

Apolipoproteins

The apoB proteins are essential for the assembly and release of chylomicrons and VLDL by gut and liver, respectively. More specifically, apoB48 is produced by the small intestine and is present in chylomicrons, where it is important for efficient chylomicron formation and secretion. ApoB100 produced by the liver is present in VLDL and LDL. It is not only essential for VLDL formation, but also mediates binding to the LDL receptor on hepatocytes leading to LDL clearance. There is one apoB molecule in each LDL particle and measurement of serum apoB concentration is often a better reflection of LDL particle concentration than LDLc [12], but the latter remains the parameter most commonly used clinically. ApoE present in chylomicron remnants also binds to the LDL receptor, allowing them, despite their lack of apoB100, also to be cleared by LDL receptors as well as by a multiligand receptor (heparan sulfate/LRP; see Fig. 1). ApoA1 is the main apolipoprotein in HDL and is secreted by hepatocytes and intestinal enterocytes. ApoA1 is important for the structural integrity and function of HDL, including the receptor-mediated release of cholesteryl ester to hepatocytes via SR-B1 during circulation of HDL through the liver.

Various Hyperlipidemias, Molecular Origin, and Disturbed Metabolism

Several disorders are grouped under the term "hyperlipidemia" (Table 1) [1–3]. What can be defined as increased LDLc needs careful thought [13]. Concentrations above 2 mmol/L increase the incidence of ASCVD and lowering LDLc therapeutically from say 3 to 2 mmol/L can reduce ASCVD risk by one-fifth [14] (more if it is higher before treatment, for example lowering it from 4 to 2 mmol/L can reduce ASCVD risk by almost two-fifths and from 5 to 2 mmol/L by half [14]). In many western societies in which ASCVD incidence is high, LDLc has risen by middle-age to levels that are typically around 4 mmol/L. Thus, even the mean values are unhealthy [13] (common hypercholesterolemia; see Table 1). When LDLc exceeds 4.9 mmol/L, it demands treatment [15, 16]. When to treat levels lower than this, particularly with medication, is usually decided on the basis of an assessment of ASCVD risk [14–16]. This is made on the basis firstly of

Table 1 The hyperlipidemias

Lipoprotein disorder	Frequency	Cause	Inheritance	Clinical features
Common hypercholesterolemia (LDLc >2 mmol/L)	In >50% of adults, significance depends on cardiovascular disease (CVD) risk	Obesity, high fat diet	Polygenic	Generally none; sometimes xanthelasmata, corneal arcus, ASCVD[a]
Combined hyperlipidemia/ metabolic syndrome (TGs >1.7; LDLc >2 mmol/L)	1 in 50, significance depends on CVD risk	Obesity, high fat diet, insulin resistance	Polygenic	Generally none; xanthelasmata, corneal arcus, ASCVD[a]
Heterozygous familial hypercholesterolemia (LDLc >2 mmol/L; typically >5 mmol/L)	1 in 250–500	Decreased LDL catabolism	Monogenic autosomal dominant (LDL receptor, ApoB100, or PCSK9 mutations)	None in younger people; corneal arcus, xanthelasmata, tendon and sub-periosteal xanthomata; early onset ASCVD[b]
Remnant removal disease (rare cause of combined hyperlipidemia)	1 in 5,000	Decreased catabolism of chylomicron remnants	Monogenic (generally recessive *ApoE* variant)	Striate palmar and tubero-eruptive xanthomata; early onset ASCVD[b]
Severe hypertriglyceridemia (TGs >10 mmol/L)	1 in 1,000	Decreased chylomicron and VLDL catabolism	Commonly polygenic; rarely homozygous mutation of LPL	Milky serum, xanthomata, hepatosplenomegaly, eruptive xanthoma, acute pancreatitis, ASCVD[b]

[a]Risk of ASCVD frequently multifactorial
[b]Risk generally increased

whether ASCVD is already clinically evident, secondly of whether a high-risk clinical syndrome, such as familial hypercholesterolemia, is present, and thirdly whether other risk factors, such as age, gender, smoking history, serum cholesterol, HDL cholesterol, blood pressure, and diabetes (although diabetes is often regarded in its own right as an indication for cholesterol-lowering therapy; see chapter "Diabetes Mellitus"), combined together in a particular individual represent an ASCVD risk sufficient for the inconvenience and the possibility of drug side effects to be outweighed by benefit in terms of reduced ASCVD risk. Algorithms for this purpose are readily available providing estimates of ASCVD risk calibrated as events per 100 people (%) in the next 10 years [14–16]. When ASCVD risk exceeds 7.5–10% in the next 10 years, treatment is considered.

Hypertriglyceridemia (HTG; increase in serum TGs to ≥1.7 mmol/L) occurs independently or combined with raised LDLc.

Polygenic Hyperlipidemias

The great majority of hyperlipidemia patients harbor polymorphisms predisposing them to raised LDLc, which only manifests when combined with nutritional excess [1–3, 13, 17, 18]. Both, obesity and high fat/cholesterol intake increase the likelihood of raised LDLc.

When HTG occurs in some members of a family in combination with hypercholesterolemia, it is termed familial combined hyperlipidemia. HTG is a major component of metabolic syndrome and type 2 diabetes mellitus (T2DM; see chapters "Diabetes Mellitus" and "Metabolic Syndrome"). People with raised TGs typically have low levels of HDL cholesterol (HDLc), raised blood pressure, and an increased likelihood of developing T2DM. Frequently, they also have increased levels of a small dense LDL. Hyperlipidemia in T2DM was formerly thought to be secondary, but now T2DM should be viewed as part of a dyslipidemic syndrome [3, 19].

Familial Hypercholesterolemia

The most common monogenic cause of raised serum cholesterol is heterozygous familial hypercholesterolemia (HeFH; Table 1). It is dominantly inherited, and in affected people, cholesterol levels are doubled compared to healthy relatives, right from birth, and thus HeFH can be diagnosed in childhood [1–3, 5, 13, 20]. Untreated, it results in xanthomata (deposits of cholesterol and fibrous tissue) in tendons. HeFH also increases cardiovascular disease (CVD) risk. Many patients die from ASCVD before the age of 60 years [1, 3, 5, 13, 20].

HeFH results from defective hepatic LDL catabolism, most often due to a mutation of the LDL receptor, more rarely an apoB100 or PCSK9 gene mutation [3, 5, 13, 20]. Gain-of-function mutations of PCSK9, which accelerate the degradation of hepatic LDL receptors, cause an unusually severe HeFH phenotype. In contrast, PCSK9 variants with impaired function result in longevity due to lower LDLc and reduced ASCVD risk.

Expression of HeFH does not require obesity. Affected individuals often appear lean and physically fit.

In homozygous familial hypercholesterolemia (HoFH), both alleles of the LDL receptor gene have a mutation, and LDL cholesterol is greatly increased. However, this disease is extremely rare (1 in 10^6 individuals) unless there is consanguinity.

Remnant Removal Disease

Hypercholesterolemia associated with marked HTG occurs in remnant removal disease (also called familial dysbetalipoproteinemia) (Table 1), in which chylomicron remnants accumulate in the circulation producing subcutaneous xanthomas located over the tuberosities and in palmar skin creases. It greatly increases the risk of ASCVD [1–3, 21]. Remnant removal disease is a monogenic autosomal recessive disorder, resulting from genetic variants of apoE with diminished receptor binding. ApoE normally plays a major role in the hepatic uptake and catabolism of remnant particles. Expression of remnant removal disease is more likely with obesity and diabetes. It is rare in women before the menopause.

Severe Hypertriglyceridemia (HTG)

Severe HTG occurs when the capacity of LPL to clear TGs from chylomicrons and VLDL in the circulation is exceeded because of their increased production or due to mutations of one LPL gene or of its regulators such as LMF1, ApoAV, GPIHBP1, apoCII, or ANGPTL3 [1–4, 17, 22–24]. Most commonly, it appears as a combination of such genetic variants with factors raising TGs (high-fat diet, obesity, type 2 diabetes, high alcohol consumption) and/or further compromising LPL function (insulin deficiency or resistance, hypothyroidism, β-adrenoceptor blockade). More rarely, it is an autosomal recessive condition with mutations in both LPL genes.

Serum and plasma appear milky. Severe HTG is associated with increased risk of acute pancreatitis [24], hepatosplenomegaly, and eruptive xanthomata.

Secondary Hyperlipidemia

Hyperlipidemia secondary to other diseases is common. T1DM and T2DM are associated with HTG (see chapter "Diabetes Mellitus"). In T1DM, insulin treatment tends to restore TG levels to normal. HTG is also caused by high alcohol consumption, by chronic renal insufficiency, and by parenchymal liver disease. Hyperuricemia and gout (see chapter "Gout") are strikingly associated with most forms of HTG. The reason is not entirely clear. Hypothyroidism can cause both hypercholesterolemia, because of decreased receptor-mediated LDL catabolism, and HTG, because of decreased TG clearance, probably mediated via decreased LPL activity. Nephrotic syndrome raises LDLc and can decrease HDL when the leak of protein is large enough to permit its urinary loss. Obstructive liver disease is also associated with high cholesterol, but this is because of an abnormal, pathological lipoprotein (called LpX), which interferes with the uptake of chylomicron remnants. LpX occurs when there is reflux of biliary phospholipids into the circulation as a result of obstruction.

Treatment of Hyperlipidemias

Treatment generally aims at a reduction of LDLc (below 2 mmol/L, or even further in high-risk patients) to avoid ASCVD events [2, 3, 13–17]. Lowering TG serum levels below 10 mmol/L is important to avoid acute pancreatitis, and increasingly the importance of achieving levels lower than this is recognized in ASCVD prevention.

Dietary Treatment

All obese hyperlipidemic patients benefit from weight loss. Dietary saturated fat and cholesterol should also be avoided; in severe hypertriglyceridemia (HTG), all fat should be restricted to avoid chylomicron formation.

Drug Treatment

Initiation of lipid-lowering medication in primary prevention is mainly based on LDLc levels and future ASCVD risk [14–17]. In patients with established ASCVD, diabetes mellitus, monogenic hyperlipidemia, and severe HTG, drug treatment is indicated without multifactorial risk evaluation. Lipid modification therapies are generally contraindicated during pregnancy and breastfeeding.

Lipid-Lowering Drugs and Their Mechanism of Action (Table 2)

Statins

Statins inhibit 3-hydroxy-3-methylglutaryl-CoA reductase mostly in the liver. This is the rate-limiting enzyme for cholesterol biosynthesis. The resulting decrease in intrahepatic cholesterol leads to enhanced hepatic LDL receptor

Table 2 Lipid-lowering drugs

Drugs decreasing circulating LDL
3-Hydroxy-3-methylglutary-CoA reductase (HMG-CoA reductase) inhibition (statins)
Inhibit cholesterol biosynthesis: statins (atorvastatin, fluvastatin, lovastatin, pitavastatin, pravastatin, rosuvastatin, simvastatin)
Niemann-pick C1-like 1 (NPC1L1) inhibition
Binds to intestinal transport protein inhibiting cholesterol absorption: ezetimibe
Blocking bile salt reabsorption from intestine
Bile acid sequestrating agents deplete bile salts, which stimulates hepatic synthesis from their cholesterol precursor, which in turn upregulates LDL receptor-mediated uptake of LDLc (cholestyramine, colestipol, colesevelam)
ATP citrate lyase (ACL) inhibitor
Inhibits cholesterol biosynthesis: bempedoic acid
Proprotein convertase subtilisin/kexin type 9 (PCSK9) inhibition
Decreases circulating PCSK9 thus decreasing degradation of LDL receptors: monoclonal antibodies, alirocumab, and evolocumab, and small interfering RNA (siRNA), inclisiran
Microsomal triglyceride transfer protein (MTP) inhibition
Decreases VLDL and chylomicron assembly: lomitapide[a]
Apolipoprotein B synthesis inhibition
Antisense oligonucleotide (ASO) blocking hepatic *APOB* translation: mipomersen[a]
Drugs predominantly lowering circulating triglycerides
Peroxisome proliferator-activated receptor alpha (PPARα) agonists (fibric acid derivatives)
Increase TG clearance by decreasing apoCIII, thus stimulating lipoprotein lipase, and decrease hepatic triglyceride synthesis: bezafibrate, ciprofibrate, fenofibrate, gemfibrozil, pemafibrate[b]
Omega 3 fatty acids
Decrease TG synthesis: icosapent ethyl, omega 3 acid ethyl esters. Crude fish oil poorly tolerated in TG-lowering doses.
Angiopoietin-like protein 3 (ANGPTL3) inhibition
Stimulates lipoprotein lipase: monoclonal antibody (evinacumab)
Apolipoprotein CIII (apoCIII) synthesis inhibition
ASO blocking *APOC3* translation, thus stimulating lipoprotein lipase: volanesorsen

[a]Restricted to homozygous FH (risk of steatohepatitis)
[b]Limited use in statin intolerance and severe hypertriglyceridemia

expression and thus a decrease in circulating LDLc [14, 25, 26]. They decrease ASCVD risk by about one-fifth for every 1 mmol/L decrease in LDL cholesterol, irrespective of pre-treatment LDL cholesterol levels. They are the most evidence-based and cost-effective means of preventing ASCVD [25]. Myositis (muscle inflammation) can occasionally be a side effect, as can muscle aching and elevations of creatine kinase. Co-treatment with agents such as macrolide antibiotics, which compete for the same degrading enzymes as statins, should be avoided. However, it is important to realize that severe myositis (rhabdomyolysis) is extremely rare and that muscle aches and pains unrelated to statin treatment are common, but if wrongly attributed to a statin may affect patient compliance. Reports of myositis in statin-treated patients, which do not compare its likelihood with that in controls not taking statins, greatly exaggerate its true incidence [26, 27]. There has also been much discussion about statins as a cause of increased blood glucose. The effect is small and conversion to diabetes occurs mainly in people who already have metabolic syndrome (see chapter "Metabolic Syndrome") associated with impaired glucose tolerance. There is no evidence that any statin-induced rise in glucose is associated with diabetic complications. People with diabetes receiving statins have a substantially lower incidence of ASCVD than those not receiving such treatment [26].

Ezetimibe

Ezetimibe is generally well tolerated but is a less effective LDL-lowering agent than most statins [2, 3, 15, 17]. It acts by inhibiting intestinal cholesterol absorption, which includes both the dietary cholesterol and that entering the intestine in the bile, which would otherwise largely be reabsorbed. It does so by blocking Niemann-Pick C1-like 1, a mediator of cholesterol absorption in the gut. It has its greatest clinical utility as an adjunct to statin therapy in patients whose LDLc remains high.

Bempedoic Acid

Bempedoic acid inhibits ATP citrate lyase in the hepatic cholesterol biosynthetic pathway upstream of HMG-CoA reductase. Like ezetimibe it is less efficacious in lowering LDLc than statin treatment but can be used as an adjunct to statin and/or ezetimibe [2, 3, 15, 17]. It can raise serum uric acid and may be best avoided in people prone to gout.

PCSK9 Inhibitors

PCSK9 inhibitors are at least as efficacious as statin treatment. They can be used in combination with other cholesterol-lowering drugs or as monotherapy in particularly resistant hypercholesterolemia [2, 3, 6, 15, 17, 28–30]. At present they are available as monoclonal antibodies and small interfering RNA (siRNA) that impede *PCSK9* translation. They are not first-line drugs because they are currently expensive. They require subcutaneous administration (once every 2–4 weeks in the case of monoclonal antibodies and every 6 months for siRNA).

Bile Acid Sequestrating Agents

Bile acid sequestrating agents impede the reabsorption of bile acids from the terminal ileum thereby increasing the hepatic demand for cholesterol for their synthesis to replenish them. To do this LDL receptor-mediated LDL catabolism is upregulated and circulating LDL thus lowered. Bile acid sequestrating agents are poorly tolerated and nowadays with the exception of colesevelam rarely prescribed [2, 3, 15, 17].

Drugs Used in Homozygous FH

Because of the extraordinary precocity of coronary artery disease and aortic stenosis in HoFH, often manifest in childhood, and its resistance to treatment with the previously mentioned medications, newer drugs untested in large trials or that risk side effects making them unsuitable for general use, may be used in specialist centers where frequent monitoring is available. Inhibition of ANGPTL3 with the monoclonal antibody, evinacumab, lowers both TG-rich lipoproteins and LDL [31]. Currently it is proving valuable both in HoFH and severe hypertriglyceridemia (HTG). Inhibition of MTP with lomitapide impairs chylomicron and VLDL assembly and decreases both circulating TG and LDL. Its use is limited to people with HoFH and only when strictly supervised, because it can cause hepatic steatosis due to intrahepatic accumulation of TG, which cannot be assembled into VLDL and then secreted. Inhibition of apoB synthesis by blocking hepatic *APOB* translation with the antisense oligonucleotide, mipomersen, is effective in decreasing both VLDL and LDL. However, it is even more likely to be hepatotoxic, again because lipids that would normally be transported out of the liver in VLDL can accumulate leading to steatosis and disturbed liver function.

Drugs Predominantly Lowering Circulating Triglycerides

Peroxisome Proliferator-Activated Receptor Alpha (PPARα) Agonists (Fibric Acid Derivatives)

Before the advent of statins, PPARα agonists (also known as fibric acid derivatives, fibrates; current examples are bezafibrate, ciprofibrate, fenofibrate, gemfibrozil, pemafibrate) were used extensively [2, 3, 15, 17]. However, many issues surrounding their effects remain unresolved, perhaps because PPARα itself has multiple metabolic effects. Nowadays, it is considered preferable to develop pharmaceutical agents that have specific targets. Fibric acid derivatives increase TG clearance from the circulation by decreasing apoCIII, thus stimulating lipoprotein lipase, and they decrease hepatic TG synthesis by diverting fatty acids into oxidative pathways.

Their effect in decreasing circulating LDLc is smaller than that of statins and, although meta-analysis of clinical trials confirms their capacity to decrease the incidence of ASCVD, they have not shown a decrease in overall mortality. In combination with statin treatment, the risk of myositis is increased, particularly with gemfibrozil.

Omega 3 Fatty Acids

Present in phytoplankton and marine animals living in cold regions, when consumed by humans in the diet or as pharmacological agents they decrease VLDL but have little or no effect on LDLc. Crude fish oil is poorly tolerated in TG-lowering doses, but purer preparations such as icosapent ethyl and omega 3 acid ethyl esters are more acceptable [32]. Omega 3 fatty acids also possess anticoagulant properties due to reduced platelet aggregation and decreases in various clotting factors. There is evidence that they can decrease recurrent ASCVD, but this effect may be related at least in part to protection against ventricular dysrhythmias, rather than their lipid-lowering or antithrombotic activity.

ANGPTL3 Inhibition

A monoclonal antibody against ANGPTL3, evinacumab, is proving valuable in the treatment of HoFH. Its action on LPL and endothelial lipase [4] increases both chylomicron and VLDL clearance and, importantly in HoFH, lowers LDL production from VLDL [31].

Inhibition of Apolipoprotein CIII (apoCIII) Synthesis

An antisense oligonucleotide, volanesorsen, by blocking *APOCIII* translation, provides another means of stimulating LPL. This mechanism has the potential to treat both severe hypertriglyceridemia and homozygous familial hypercholesterolemia [33]. Volanesorsen can cause thrombocytopenia. Tagging with asialoglycoprotein to reduce extra-hepatic uptake is being investigated to decrease this complication. In the meantime, careful monitoring is important.

Lp(a) Lowering

Lp(a) is not decreased by statin treatment and could be a source of residual ASCVD risk in some people receiving such therapy. Drugs to lower Lp(a) using antisense oligonucleotide/small interfering RNA directed against apo(a) are in development.

Other Approaches to Lowering Cholesterol and TG

LDL Apheresis

Several techniques are available to remove lipoproteins from the blood by apheresis. Resembling hemodialysis, this involves circulating the patient's blood through an extracorporeal machine. Treatment of HoFH generally involves regular visits to a specialist center for LDL apheresis [34].

Bariatric Surgery

Bariatric surgery, such as laparoscopic jejuno-ileal bypass, not only dramatically reduces body weight, but also improves glucose tolerance and insulin resistance and restores lipoprotein metabolism toward normal [35].

Perspectives

Our knowledge of lipoprotein metabolism and the array of drugs to regulate it are fast growing largely because of the discovery of key genes through DNA sequencing in rare clinically identified monogenic disorders. This is providing new means of treatment for more common disorders of lipid and lipoprotein metabolism, which is increasingly important in clinical practice as the enormous global burden of ASCVD continues to expand.

Questions and Answers

Question 1 LDL cholesterol in the blood circulation:

(a) Is partly derived from recycling.
(b) Is initially secreted from the liver as chylomicrons.
(c) Cannot be removed from the blood circulation.
(d) Is deposited in atheromatous lesions.

Answer 1 LDL cholesterol in the blood circulation:

(a) Is partly derived from recycling.
   True: After uptake by the liver, it can be re-secreted as a component of VLDL.
(b) Is initially secreted from the liver as chylomicrons.
   False: The precursor of LDL in which cholesterol is secreted from the liver is VLDL. Chylomicrons are secreted from the gut and are not LDL precursors.
(c) Cannot be removed from the blood circulation.
   False: LDL cholesterol can be removed from the blood circulation, for example by hepatic LDL receptors.
(d) Is deposited in atheromatous lesions.
   True: The cholesterol in atheromatous lesions is derived from LDL taken up by monocyte-macrophages and smooth muscle precursor cells to create "foam cells" in the arterial wall.

Question 2 In addition to advice about diet, cholesterol-lowering medication should be considered:

(a) In type 2 diabetes
(b) In familial hypercholesterolemia
(c) In survivors of stroke or myocardial infarction

(d) When estimated ASCVD risk is 15% over the next 10 years

Answer 2 In addition to advice about diet, cholesterol-lowering medication should be considered:

(a) In type 2 diabetes
   True: Cholesterol-lowering medication is indicated in nearly all people with type 2 diabetes.
(b) In familial hypercholesterolemia
   True: Cholesterol-lowering medication should be considered even in children with familial hypercholesterolemia.
(c) In survivors of stroke or myocardial infarction
   True: Cholesterol-lowering medication is indicated in people with established ASCVD (secondary prevention).
(d) When estimated ASCVD risk is 15% over the next 10 years
   True: However, most guidelines recommend that cholesterol-lowering medication is considered at lower degrees of risk than 15%, certainly by the time ASCVD risk has reached 10% over the next 10 years.

Question 3 Statins

(a) For each 1 mmol/L decrease in LDLc they achieve, they reduce ASCVD incidence by approximately one-fifth.
(b) Should never be prescribed in combination with ezetimibe.
(c) Do not decrease ASCVD risk in type 2 diabetes.

(d) Are first-line cholesterol lowering treatment.

Answer 3 Statins

(a) Reduce ASCVD incidence by approximately one-fifth for each 1 mmol/L decrease in LDLc they achieve.
   True: Meta-analyses of randomized controlled clinical trials show this regardless of the presence of other ASCVD risk factors [25].
(b) Should never be prescribed in combination with ezetimibe.
   False: When statin treatment has not achieved a satisfactory LDLc target, its combination with ezetimibe is safe and effective in achieving additional LDLc-lowering and decreasing ASCVD risk.
(c) Do not decrease ASCVD risk in type 2 diabetes.
   False: Statins are at least as effective in lowering ASCVD risk in diabetes as in non-diabetic people.
(d) Are first-line cholesterol-lowering treatment.
   True: They are clinically effective, cost-effective, and generally safe.

References

1. Durrington P (2003) Dyslipidaemia Lancet 362(9385):717–731. https://doi.org/10.1016/S0140-6736(03)14234-1
2. Feingold KR. Introduction to lipids and lipoproteins 2021. https://www.ncbi.nlm.nih.gov/books/NBK305896/
3. Sniderman A, Durrington P (2021) Hyperlipidemia, 6th edn. Karger, Oxford, p 166
4. Adam RC, Mintah IJ, Alexa-Braun CA, Shihanian LM, Lee JS, Banerjee P, Hamon SC, Kim HI, Cohen JC, Hobbs HH, Van Hout C, Gromada J, Murphy AJ, Yancopoulos GD, Sleeman MW, Gusarova V (2020) Angiopoietin-like protein 3 governs LDL-cholesterol levels through endothelial lipase-dependent VLDL clearance. J Lipid Res 61(9):1271–1286. https://doi.org/10.1194/jlr.RA120000888. Epub 2020 Jul 9. PMID: 32646941; PMCID: PMC7469887
5. Goldstein JL, Brown MS (2009) The LDL receptor. Arterioscler Thromb Vasc Biol 29(4):431–438.

https://doi.org/10.1161/ATVBAHA.108.179564. PMID: 19299327; PMCID: PMC2740366

6. Lambert G, Sjouke B, Choque B, Kastelein JJ, Hovingh GK (2012) The PCSK9 decade. J Lipid Res 53(12):2515–2524. https://doi.org/10.1194/jlr.R026658. Epub 2012 Jul 17. PMID: 22811413; PMCID: PMC3494258

7. Libby P, Buring JE, Badimon L, Hansson GK, Deanfield J, Bittencourt MS, Tokgözoğlu L, Lewis EF (2019) Atherosclerosis. Nat Rev Dis Primers 5(1):56. https://doi.org/10.1038/s41572-019-0106-z. PMID: 31420554

8. Mineo C (2020) Lipoprotein receptor signalling in atherosclerosis. Cardiovasc Res 116(7):1254–1274. https://doi.org/10.1093/cvr/cvz338. PMID: 31834409; PMCID: PMC7243280

9. Durrington PN, Bashir B, Bhatnagar D, Soran H (2022) Lipoprotein (a) in familial hypercholesterolaemia. Curr Opin Lipidol 33(4):257–263. https://doi.org/10.1097/MOL.0000000000000839

10. Allard-Ratick MP, Kindya BR, Khambhati J, Engels MC, Sandesara PB, Rosenson RS, Sperling LS (2021) HDL: fact, fiction, or function? HDL cholesterol and cardiovascular risk. Eur J Prev Cardiol 28(2):166–173. https://doi.org/10.1177/2047487319848214. Epub 2019 May 13

11. Soran H, Schofield JD, Durrington PN (2015) Antioxidant properties of HDL Front Pharmacol 6:222. https://doi.org/10.3389/fphar.2015.00222. PMID: 26528181; PMCID: PMC4607861

12. Glavinovic T, Thanassoulis G, de Graaf J, Couture P, Hegele RA, Sniderman AD (2022) Physiological bases for the superiority of Apolipoprotein B over low-density lipoprotein cholesterol and non-high-density lipoprotein cholesterol as a marker of cardiovascular risk. J Am Heart Assoc 11(20):e025858. https://doi.org/10.1161/JAHA.122.025858. Epub 2022 Oct 10

13. Goldstein JL, Brown MS (2015) A century of cholesterol and coronaries: from plaques to genes to statins. Cell 161(1):161–172. https://doi.org/10.1016/j.cell.2015.01.036. PMID: 25815993; PMCID: PMC4525717

14. Soran H, Adam S, Iqbal Z, Durrington P (2022) Mathematical modelling of the most effective goal of cholesterol-lowering treatment in primary prevention. BMJ Open 12(5):e050266. https://doi.org/10.1136/bmjopen-2021-050266. PMID: 35613766; PMCID: PMC9131112

15. Visseren FLJ, Mach F, Smulders YM, Carballo D, Koskinas KC, Bäck M, Benetos A, Biffi A, Boavida JM, Capodanno D, Cosyns B, Crawford C, Davos CH, Desormais I, Di Angelantonio E, Franco OH, Halvorsen S, Hobbs FDR, Hollander M, Jankowska EA, Michal M, Sacco S, Sattar N, Tokgozoglu L, Tonstad S, Tsioufis KP, van Dis I, van Gelder IC, Wanner C, Williams B, National Cardiac Societies ESC, ESC Scientific Document Group (2021) ESC guidelines on cardiovascular disease prevention in clinical practice. Eur Heart J 42(34):3227–3337. https://doi.org/10.1093/eurheartj/ehab484. Erratum in: Eur heart J. 2022 Sep 09; PMID: 34458905

16. Grundy SM, Stone NJ, Bailey AL, Beam C, Birtcher KK, Blumenthal RS, Braun LT, de Ferranti S, Faiella-Tommasino J, Forman DE, Goldberg R, Heidenreich PA, Hlatky MA, Jones DW, Lloyd-Jones D, Lopez-Pajares N, Ndumele CE, Orringer CE, Peralta CA, Saseen JJ, Smith SC Jr, Sperling L, Virani SS, Yeboah J (2019) 2018 AHA/ACC/AACVPR/AAPA/ABC/ACPM/ADA/AGS/APhA/ASPC/NLA/PCNA Guideline on the Management of Blood Cholesterol: A Report of the American College of Cardiology/American Heart Association Task Force on Clinical Practice Guidelines. Circulation 139(25):e1082–e1143. https://doi.org/10.1161/CIR.0000000000000625. Epub 2018 Nov 10. Erratum in: Circulation. 2019 Jun 18;139(25):e1182–e1186. PMID: 30586774; PMCID: PMC7403606

17. Berberich AJ, Hegele RA (2022) A modern approach to dyslipidemia. Endocr Rev 43(4):611–653. https://doi.org/10.1210/endrev/bnab037. PMID: 34676866; PMCID: PMC9277652

18. Gill PK, Hegele RA (2022) Familial combined hyperlipidemia is a polygenic trait. Curr Opin Lipidol 33(2):126–132. https://doi.org/10.1097/MOL.0000000000000796

19. Soran H, Schofield JD, Adam S, Durrington PN (2016 Aug) Diabetic dyslipidaemia. Curr Opin Lipidol 27(4):313–322. https://doi.org/10.1097/MOL.0000000000000318

20. Loh WJ, Watts GF (2022) The inherited Hypercholesterolemias. Endocrinol Metab Clin N Am 51(3):511–537. https://doi.org/10.1016/j.ecl.2022.02.006. Epub 2022 Jul 4

21. Mahley RW, Huang Y, Rall SC Jr (1999) Pathogenesis of type III hyperlipoproteinemia (dysbetalipoproteinemia). Questions, quandaries, and paradoxes. J Lipid Res 40(11):1933–1949

22. Chait A (2022) Hypertriglyceridemia. Endocrinol Metab Clin N Am 51(3):539–555. https://doi.org/10.1016/j.ecl.2022.02.010. Epub 2022 Jul 4

23. Borén J, Taskinen MR, Björnson E, Packard CJ (2022) Metabolism of triglyceride-rich lipoproteins in health and dyslipidaemia. Nat Rev Cardiol 19(9):577–592. https://doi.org/10.1038/s41569-022-00676-y. Epub 2022 Mar 22

24. Sanchez RJ, Ge W, Wei W, Ponda MP, Rosenson RS (2021) The association of triglyceride levels with the incidence of initial and recurrent acute pancreatitis. Lipids Health Dis 20(1):72. https://doi.org/10.1186/s12944-021-01488-8. PMID: 34275452; PMCID: PMC8286611

25. Collins R, Reith C, Emberson J, Armitage J, Baigent C, Blackwell L, Blumenthal R, Danesh J, Smith GD, DeMets D, Evans S, Law M, MacMahon S, Martin S, Neal B, Poulter N, Preiss D, Ridker P, Roberts I, Rodgers A, Sandercock P, Schulz K, Sever P, Simes J, Smeeth L, Wald N, Yusuf S,

Peto R (2016) Interpretation of the evidence for the efficacy and safety of statin therapy. Lancet 388(10059):2532–2561. https://doi.org/10.1016/S0140-6736(16)31357-5. Epub 2016 Sep 8. Erratum in: Lancet. 2017 Feb 11;389(10069):602

26. Soran H, France M, Adam S, Iqbal Z, Ho JH, Durrington PN (2020) Quantitative evaluation of statin effectiveness versus intolerance and strategies for management of intolerance. Atherosclerosis 306:33–40. https://doi.org/10.1016/j.atherosclerosis.2020.06.023. Epub 2020 Jul 6

27. Cholesterol Treatment Trialists' Collaboration (2022) Effect of statin therapy on muscle symptoms: an individual participant data meta-analysis of large-scale, randomised, double-blind trials. Lancet 400(10355):832–845. https://doi.org/10.1016/S0140-6736(22)01545-8. Epub 2022 Aug 29. Erratum in: Lancet. 2022 Oct 8;400(10359):1194. PMID: 36049498; PMCID: PMC7613583

28. Coppinger C, Movahed MR, Azemawah V, Peyton L, Gregory J, Hashemzadeh M (2022) A comprehensive review of PCSK9 inhibitors. J Cardiovasc Pharmacol Ther 27:10742484221100107. https://doi.org/10.1177/10742484221100107

29. Banerjee Y, Pantea Stoian A, Cicero AFG, Fogacci F, Nikolic D, Sachinidis A, Rizvi AA, Janez A, Rizzo M (2022) Inclisiran: a small interfering RNA strategy targeting PCSK9 to treat hypercholesterolemia. Expert Opin Drug Saf 21(1):9–20. https://doi.org/10.1080/14740338.2022.1988568. Epub 2021 Oct 14

30. Katzmann JL, Packard CJ, Chapman MJ, Katzmann I, Laufs U (2020) Targeting RNA with antisense oligonucleotides and small interfering RNA: JACC state-of-the-art review. J Am Coll Cardiol 76(5):563–579. https://doi.org/10.1016/j.jacc.2020.05.070

31. D'Erasmo L, Bini S, Arca M (2021) Rare treatments for rare dyslipidemias: new perspectives in the treatment of homozygous familial hypercholesterolemia (HoFH) and familial Chylomicronemia syndrome (FCS). Curr Atheroscler Rep 23(11):65. https://doi.org/10.1007/s11883-021-00967-8. PMID: 34468855; PMCID: PMC8410715

32. Chapman MJ, Zamorano JL, Parhofer KG (2022) Reducing residual cardiovascular risk in Europe: therapeutic implications of European medicines agency approval of icosapent ethyl/eicosapentaenoic acid. Pharmacol Ther 237:108172. https://doi.org/10.1016/j.pharmthera.2022.108172. Epub 2022 Mar 15

33. Calcaterra I, Lupoli R, Di Minno A, Di Minno MND (2022) Volanesorsen to treat severe hypertriglyceridaemia: a pooled analysis of randomized controlled trials. Eur J Clin Investig 52(11):e13841. https://doi.org/10.1111/eci.13841. Epub 2022 Jul 28

34. Thompson GR (2022) The scientific basis and future of lipoprotein apheresis. Ther Apher Dial 26(1):32–36. https://doi.org/10.1111/1744-9987.13716. Epub 2021 Aug 6

35. Iqbal Z, Bashir B, Adam S, Ho JH, Dhage S, Azmi S, Ferdousi M, Yusuf Z, Donn R, Malik RA, Syed A, Ammori BJ, Heald A, Durrington PN, Soran H (2022) Glycated apolipoprotein B decreases after bariatric surgery in people with and without diabetes: a potential contribution to reduction in cardiovascular risk. Atherosclerosis 346:10–17. https://doi.org/10.1016/j.atherosclerosis.2022.01.005. Epub 2022 Jan 19

Immune System

Anatomy and Physiology of the Immune System

Christian Münz

Introduction

The immune system is a complex network of metabolic pathways and cells, which are designed to distinguish harmful insults from harmless changes or fluctuations in metabolism, and to mount an appropriate response to these harmful insults without compromising the affected tissue. Therefore, it walks a fine line to combat pathogens and cellular transformation on the one hand and tolerate commensals on mucosal surfaces (such as normal gut bacteria), and food components, on the other hand.

Moreover, it has evolved to specifically meet the challenges of its respective host species concerning pathogens encountered in the ecological niche that the host occupies, and during the time to reproduction that has to be protected to guarantee propagation of the species. Therefore, the differences even between closely related mammalian species are considerable, placing the immune system in third position of the most divergent organs between mouse and man [1–3].

These challenges are met by the immune system with stringent education of its components to ignore the physiological state (which happens in so-called primary lymphoid organs such as the thymus or bone marrow), and by utilization/integration of afferent information (which happens in immunological decision centers, the secondary and tertiary lymphoid tissues such as spleen or lymph nodes). Concomitantly, efferent responses (cellular and humoral) to target harmful insults are mounted. All soluble factors that are directed at these insults are called humoral responses. These include invariant molecules like antimicrobial peptides or the alternative pathway of complement activation, which establishes pores in targeted cells, and molecules of the adaptive immune system, mainly antibodies, which are selected from a large repertoire that is generated by somatic DNA recombination and then further shaped to recognize the targeted antigen with higher affinity and additional effector mechanisms by somatic hypermutation and class switch recombination, respectively. The cellular responses include activation of cytotoxic effector cells and phagocytes. Again, these can be innate, like natural killer cells and pathogen recognition by scavenger receptors on phagocytes, including macrophages, or they can be adaptive with cytotoxic T-cells and T-cell activation of phagocytes. Therefore, the immune system has a large armamentarium to restore the healthy steady state.

C. Münz (✉)
Institute of Experimental Immunology, University of Zürich, Zürich, Switzerland
e-mail: christian.muenz@uzh.ch; muenzc@immunology.uzh.ch

E. Lammert, M. Zeeb (eds.), *Metabolism of Human Diseases*, https://doi.org/10.1007/978-3-031-96019-2_36

Tissue-Specific Pathways and Metabolic Processes of the Immune System and Cells

Primary Lymphoid Tissues

The main primary lymphoid tissues are the thymus and the bone marrow, where T-cells and B-cells (also known as T- and B-lymphocytes) are educated, respectively. T-cells can detect changes (such as the presence of foreign proteins) inside cells, whereas B-cells secrete effector molecules, mainly antibodies to target extracellular pathogens. Selectivity of T- and B-cells is achieved by somatic recombination of their respective antigen receptor genes, followed by a stringent selection process to ensure they carry functional receptors, which do not recognize self-structures, such as endogenous proteins or sugar moieties on the surface of host cells [4, 5].

B-cells develop from hematopoietic precursors in the bone marrow and are deleted by apoptosis if they fail to generate a functional antibody on their surface (see below) and also if this antibody recognizes self-structures in the bone marrow.

T-cells originate from precursors that also develop in the bone marrow, but then migrate to the thymus. There, only T-cells continue to develop, whose T-cell receptors recognize major histocompatibility complex (MHC) molecules, which in humans are also called human leukocyte antigen (HLA) molecules. These scaffolding proteins display products of the protein and lipid catabolism of thymic epithelial cells to the T-cells.

After this positive selection, T-cells are eliminated by negative selection, if they strongly react to MHC molecules that present self-structures, thus generating a central tolerance. In general, both cluster of differentiation 4-positive (CD4$^+$) helper T-cells and CD8$^+$ cytotoxic T-cells develop that recognize foreign structures, in particular peptides of extracellular and intracellular sources, on MHC class II and class I molecules, respectively.

Secondary Lymphoid Tissues

Once the mature and educated T- and B-cells emerge from primary lymphoid organs, they home to secondary lymphoid organs like spleen, lymph nodes, tonsils, and gut mucosa-associated lymphoid tissues via the blood stream. In the secondary lymphoid organs, they extravasate from the blood in specialized endothelia, called high endothelial venules, in response to gradients of chemokines, attractants for migration, such as CC motif chemokine ligands CCL19 and CCL21 (Fig. 1).

For activation of T-cells, processed foreign structures (mostly peptides) are presented on MHC or HLA molecules by dendritic cells that have picked up these antigens at various sites of the body in order to carry them to secondary lymphoid tissues via afferent lymphatic vessels (Fig. 1) [6].

This antigen transport occurs from all organs and therefore a dense network of secondary lymphoid tissues weaves through the body to keep the antigen transport times short. Once a T-cell detects a specific antigen, it proliferates and differentiates into an effector cell.

T-cells differentiate into effector cells that secrete different sets of cytokines (e.g., Th1, Th2, Th17, or regulatory T-cells) to communicate with other immune and somatic cells, or into memory cells, which will continue to migrate through secondary lymphoid tissues and promptly respond by proliferation and defense mechanisms, in case the antigen comes back. Some effector cells also settle into the originally affected organ and establish long-lived tissue residency, in case the antigen returns to the very same site.

B-cells enter germinal centers upon activation by cognate antigen recognition [7]. At these sites they affinity mature their antigen receptor by somatic hypermutation in order to produce antibodies that bind foreign structures with higher avidity. At the same time, they isotype switch their antibody molecules, for example, from IgM to IgG, for these humoral effectors to acquire additional effector functions, like binding to acti-

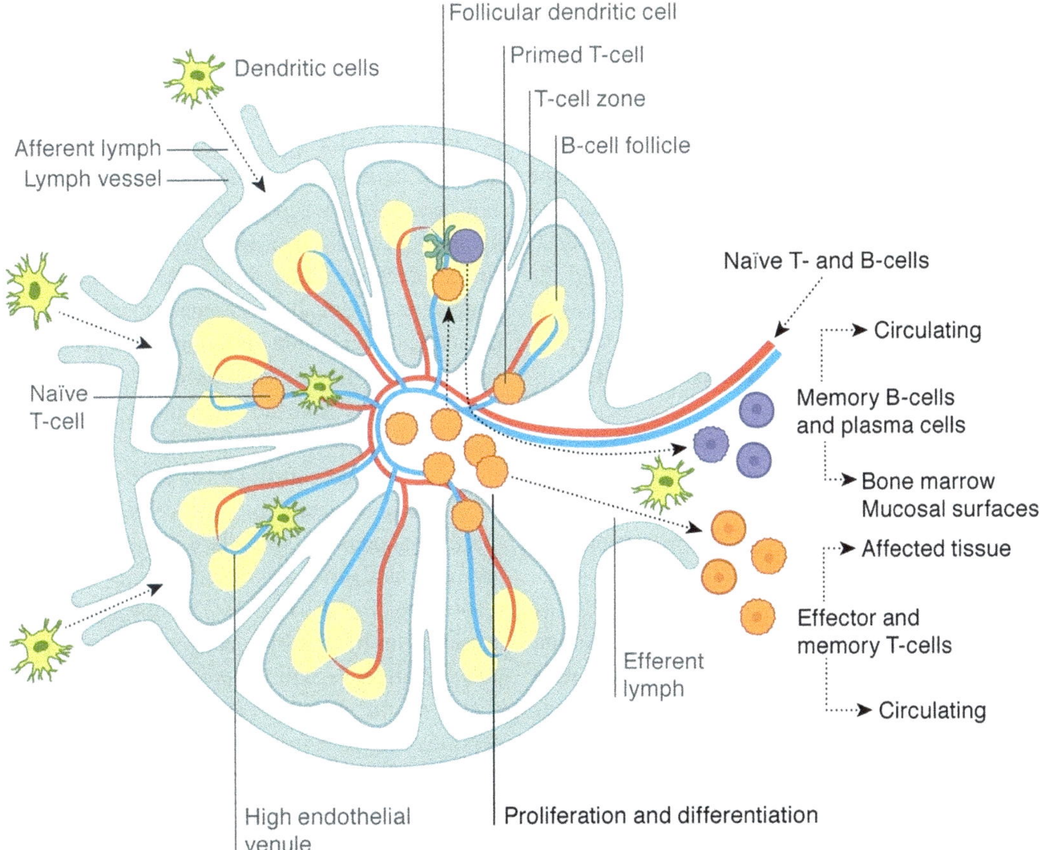

Fig. 1 Lymph nodes are paradigmatic secondary lymphoid organs. Information on the health of peripheral tissues is continuously reported to these in the form of the degree and quality of dendritic cell activation and of processed antigen presented on the dendritic cells. Naïve and memory T-cells circulate through these organs via extravasation at high endothelial venules and exit through efferent lymph vessels. These T-cells get activated via the presentation of their cognate antigens by activated dendritic cells. T-cells get primed (see text), proliferate, and differentiate into effector or memory T-cells. Effector T-cells then home back to the diseased tissue to fight the infection, or assist B-cells in the germinal center reaction. There, antigen-stimulated B-cells affinity-mature their B-cell receptor, which will serve as blueprint for the antibody that will be later secreted by this B-cell. The somatically mutated B-cell receptor has to still recognize antigen, which is bound on the surface of follicular dendritic cells and receives T-cell help. Only if these two check-points are passed, will the activated B-cell go on to develop into a memory B-cell or an antibody producing plasma cell. These can also leave the lymph node via efferent lymph, and plasma cells often home to the bone marrow or mucosal surfaces

vating receptors (FcRs) on phagocytes as well as on cytotoxic cells and for complement fixation. Only if the altered antibody still recognizes antigen (signal 1) and receives T-cell help of follicular helper T-cells that in most cases have to recognize the same antigen (signal 2), the B-cell survives this germinal center reaction and can go on to develop into memory or antibody secreting plasma cells.

T- and B-cells emigrate from secondary lymphoid tissues via the efferent lymphatics back into the blood stream and to the sites of the harmful insult guided by chemokine gradients, like CXC motif chemokine ligands CXCL9 and CXCL10 (Fig. 1). This process of immune response initiation is called priming.

Tertiary Lymphoid Tissues

In order to keep the distances for immune cell migration short and therefore the response time to a minimum, tertiary lymphoid tissues develop at sites of chronic immune cell infiltrates and inflammation [8, 9]. These are similar in structure and function to secondary lymphoid tissues.

Outside-In: Communication of Stromal Cells with Immune Cells

The afferent arm of immune responses is mainly represented by dendritic cells, which continuously report the immunological health of organs to secondary lymphoid tissues by transporting tissue constituents and by reporting the conditions, under which they have acquired these as their surface molecule phenotype and cytokine secretion pattern. In all organs they can detect pathogens directly via receptors for pathogen-associated molecular patterns (PAMPs), such as bacterial cell wall components, viral unmethylated DNA, and viral RNA, which activate them. Alternatively, they can also detect tissue destruction via the release of danger-associated molecular patterns (DAMPs) [10, 11], like urate crystals, high-mobility group B1 protein, and ATP release (Fig. 2). Some of these are recognized by inflammasomes, of which the NLRP3-containing protein complex is the best studied [12]. Their activation allows interleukin 1 (IL-1) production, which is the main mediator of inflammation causing heat, redness, pain, swelling, and loss of tissue function. Both, PAMPs and DAMPs, thus activate dendritic cell migration and immune response priming in secondary lymphoid organs. In addition (to PAMPs and DAMPs), stromal cells can communicate with dendritic cells via chemokines (i.e., special cytokines that attract immune cells) and cytokines (i.e., secreted, small protein-based signals of the immune system). Therefore, the input by the organ environment is crucial for the afferent communication of the immune system with secondary lymphoid organs.

Inside-In: Communication Between Immune Cells

Chemokines, such as CXCL9 and CXCL10, are produced to build gradients in tissues to attract immune cells such as effector T-cells. Immune cells (such as macrophages and T-cells) communicate with each other and stromal cells through surface receptors that accumulate at membrane contact areas, socalled immunological synapses. Cytokines (including interferon [IFN]-γ) are secreted into these synapses or to neighboring cells to further refine the communication between immune cells.

Three signals constitute the core of the communication between immune cells, which primarily happens in secondary lymphoid tissues between antigen carrying dendritic cells and responding T-cells (Fig. 2). The first signal is the presentation by dendritic cells of catabolic products of antigens (mainly peptides) on MHC molecules to the T-cell receptor of T-cells. These peptides originate from the two main proteolytic machineries of the cell, that is, lysosomes and proteasomes [13, 14]. Proteasomal products are presented on MHC class I molecules to cytotoxic CD8+ T-cells, whereas lysosomal products are presented on MHC class II molecules to helper CD4+ T-cells, which assist in maintenance and differentiation of both primed CD8+ T- and B-cells. This first signal induces proliferation of T-cells, if co-stimulatory signals (see below) are present. If co-stimulatory signals are absent (after little or no activation of dendritic cells), these antigen-specific T-cells are eliminated after a few cell divisions, a process contributing to peripheral tolerance [6].

Thus, activated dendritic cells save the proliferating T-cells from dying by releasing cytokines and other co-stimulatory molecules, like IL-12 and IL-15, toward them, shaping their profile [6]. These cytokines imprint information about the conditions, under which dendritic cells have been activated, onto the responding T-cell, for example, IL-12 favors the development of Th1 polarized T-cell responses and is mainly secreted by dendritic cells after virus encounter. This so-

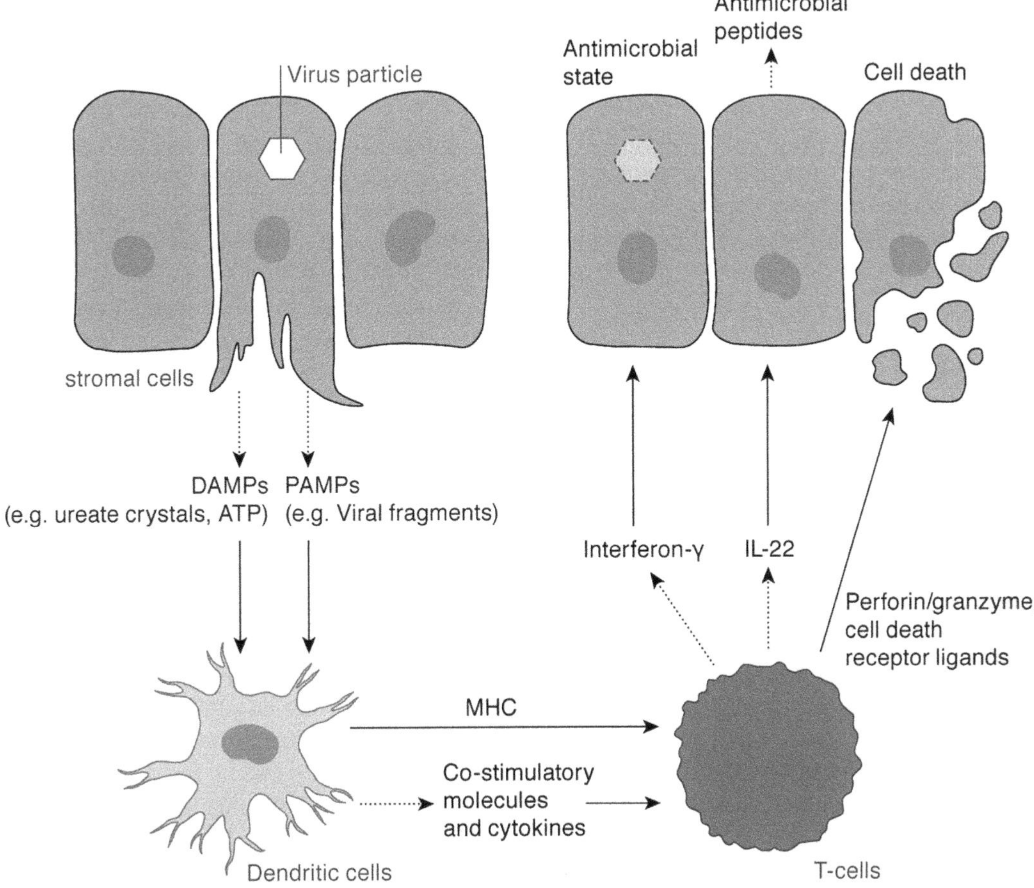

Fig. 2 The immune system receives cues from all organs and changes its metabolism via inflammatory infiltrates. Stromal cells can activate dendritic cells through the release of danger- (DAMPs) and, if infected, pathogen-associated molecular patterns (PAMPs), which are for example contained in necrotic cellular debris. Once activated through these PAMPs and DAMPs, they can prime T-cells, which home to inflamed tissues to change their intracellular milieu and their secretome via interferon and interleukin secretion, for example, IL-22 and IFN-γ, and influence the survival of dangerous or infected cells via cell-contact-dependent cytotoxicity, for example, via perforin/granzyme or apoptosis inducing ligands

called polarization allows primarily CD4$^+$ T-cells to return to the site of DC activation by acquiring a certain profile of chemokine receptors that will direct them to certain tissues, like CCR9 for homing to the gut [15, 16], while CD8$^+$ T-cells acquire a less variant chemokine repertoire that will direct them to sites of inflammation. There, a tailored response is mounted, including the expression of cytokines, acting on the infected or transformed tissue, for example, IFN-γ to inhibit viral replication, and the expression of cytotoxic molecules to directly destroy the diseased cells.

Expression of these molecules requires distinct metabolic pathways with anaerobic glycolysis being necessary for immune cell expansion, for example, T-cell proliferation [17], and macroautophagy [18, 19] as well as oxidative phosphorylation being required for T-cell memory maintenance [20]. This suggests that proliferating immune cells require rapidly larger ATP amounts that they generate via anaerobic glycolysis, while memory immune cells are more energy efficient and maintain themselves with oxidative phosphorylation. Interestingly, glycolytic enzymes, such as glyceraldehyde-3-phosphate dehydrogenase (GAPDH), moonlight as mRNA binding proteins and upon engagement in glycolysis release the bound mRNAs, resulting in rapid cytokine production, for example, IFN-γ [21]. Furthermore, cytotoxicity as another main effector function of T-cells depends on adenosine monophosphate (AMP)-activated protein kinase (AMPK) [22].

Inside-Out: Communication of Immune Cells with Stromal Cells

The effector molecules from polarized T-cells are then able to change tissue homeostasis in the inflamed and/or infected organs, from which the dendritic cells carried the antigens to the secondary lymphoid organs. A wide variety of responses can be elicited, which in target organs range from inducing metabolic changes in target cells, which lead to reduced pathogen replication, to release of antimicrobial peptides or to induction of apoptosis. Metabolic changes aim to make cells less hospitable for infectious agents. For example, interferons can induce an antiviral state [23]. Secretion of antimicrobial peptides by epithelial cells at mucosal surfaces is stimulated by interleukin 22, which is either secreted by innate lymphoid cells or Th17 polarized CD4+ T-cells [24, 25]. Cell death can be triggered by immune cells via perforin-mediated granzyme delivery, during which at least one of the cell death initiating proteases of the granzyme family enter target cells through pores that are formed by perforin, or by activation of cell death receptors on infected or transformed cells [26] (Fig. 2). Thus, immune cells can dramatically change the physiology of the inflamed organ.

Miscommunication in the Immune System as the Basis of Disease

The borderline between hypo-responsiveness of the immune system resulting in susceptibility to disease [27] and hyper-responsiveness leading to immunopathology and autoimmunity [28–30] is not easily defended by the immune system, and

drawn by the genetic make-up of the individual. Any significant insult that releases DAMPs can change the organ environment and metabolism so that it is no longer recognized as self by the immune system, as it was educated toward a different steady-state. Sometimes the resulting autoimmune responses are transient, just causing immunopathology during the infectious or trauma insult, and sometimes, but fortunately rarely, they result in self-perpetuating autoimmune disease [30], depending in part on the genetic variation of the affected individuals.

Final Remarks

The key physiological function of the immune system is to defend multicellular organisms against harmful non-self by transporting information from organs to secondary lymphoid tissues and mounting immune responses when this information indicates infection or tissue damage. Both overshooting and too cautious immune reactions lead to disease and are caused by the combination of the individual's genetic predisposition and the environmental conditions.

Perspectives

The immune system is evolution's answer to protect multicellular organisms against disease. Harnessing it by vaccination against infectious diseases [31], increasing its activity during immunotherapy of cancer [32], or dampening its function during autoimmunity [33] are already clinical practice. These applications might be broadened to multidrug-resistant bacteria and the induction of specific tolerance during autoimmunity and allergy.

Acknowledgments The Münz laboratory is financially supported by Cancer Research Switzerland (KFS-4962-02-2020), HMZ ImmunoTargET of the University of Zurich, the Cancer Research Center Zurich, the Sobek Foundation, the Swiss Vaccine Research Institute, the Swiss MS Society (2021-09), Roche, Novartis, the Vontobel Foundation, and the Swiss National Science Foundation (310030_204470/1, 310030L_197952/1 and CRSII5_180323).

Questions and Answers

Question 1 How does the thymus present the self-proteome of all organs for negative selection of autoreactive T-cells?

Answer 1 Thymic epithelial cells express the transcription factor autoimmune regulator (AIRE), which allows for leaky gene expression of organ-specific proteins that are degraded by cytosolic proteasomes for MHC class I presentation and delivered from the cytosol to lysosomes by autophagy for MHC class II presentation, allowing for negative selection of CD8$^+$ and CD4$^+$ T-cells, respectively [4].

Question 2 How are both intra- and extracellular proteomes represented by MHC class I and II ligands in order to mount CD8$^+$ and CD4$^+$ T-cell responses against pathogens in all cellular niches?

Answer 2 MHC class I ligands are products of mostly intracellular protein degradation by proteasomes. However, extracellular proteins can be endocytosed and then exit via ruptured endosomes to the cytosol for proteasomal degradation and can be imported into endosomes for vacuolar MHC class I loading. MHC class II ligands are mostly lysosomal products of endocytosed proteins, but intracellular proteins can also reach the MHC class II compartment (MIIC) via autophagy for lysosomal degradation and MHC class II loading [14].

Question 3 How do genetic and environmental risk factors interact to cause autoimmunity?

Answer 3 Most autoimmune diseases show concordance only in the minority of monozygotic twins, for example, 30% for multiple sclerosis (MS). Therefore, genetic predisposition seems to synergize with environmental factors to increase risk for autoimmune disease, for example, a particular MHC class II allele (HLA-DRB1*1501) with symptomatic primary infection with the Epstein Barr virus for MS development. This synergy between environmental and genetic risk factors for the development of autoimmune disease might allow for molecular mimicry between environmental and self-antigen, stimulate hyper-inflammation, and/or reduce immune control of an infectious trigger that then activates antigen presenting cells to stimulate autoimmune responses [34, 35].

References

1. Waterston RH, Lindblad-Toh K, Birney E, Rogers J, Abril JF, Agarwal P, Agarwala R, Ainscough R, Alexandersson M, An P et al (2002) Initial sequencing and comparative analysis of the mouse genome. Nature 420:520–562
2. Seok J, Warren HS, Cuenca AG, Mindrinos MN, Baker HV, Xu W, Richards DR, McDonald-Smith GP, Gao H, Hennessy L et al (2013) Genomic responses in mouse models poorly mimic human inflammatory diseases. Proc Natl Acad Sci USA 110:3507–3512
3. Medetgul-Ernar K, Davis MM (2022) Standing on the shoulders of mice. Immunity 55:1343–1353
4. Irla M (2022) Instructive cues of thymic T cell selection. Annu Rev Immunol 40:95–119
5. Jankovic M, Casellas R, Yannoutsos N, Wardemann H, Nussenzweig MC (2004) RAGs and regulation of autoantibodies. Annu Rev Immunol 22:485–501
6. Steinman RM (2012) Decisions about dendritic cells: past, present, and future. Annu Rev Immunol 30:1–22
7. Victora GD, Nussenzweig MC (2022) Germinal centers. Annu Rev Immunol 40:413–442
8. van de Pavert SA, Mebius RE (2010) New insights into the development of lymphoid tissues. Nat Rev Immunol 10:664–674
9. Schulz O, Hammerschmidt SI, Moschovakis GL, Forster R (2016) Chemokines and chemokine receptors in lymphoid tissue dynamics. Annu Rev Immunol 34:203–242
10. Zindel J, Kubes P (2020) DAMPs, PAMPs, and LAMPs in immunity and sterile inflammation. Annu Rev Pathol 15:493–518
11. Gong T, Liu L, Jiang W, Zhou R (2020) DAMP-sensing receptors in sterile inflammation and inflammatory diseases. Nat Rev Immunol 20:95–112
12. Broz P, Dixit VM (2016) Inflammasomes: mechanism of assembly, regulation and signalling. Nat Rev Immunol 16:407–420
13. Cresswell P (2019) A personal retrospective on the mechanisms of antigen processing. Immunogenetics 71:141–160

14. Pishesha N, Harmand TJ, Ploegh HL (2022) A guide to antigen processing and presentation. Nat Rev Immunol 22:751–764
15. Sallusto F (2016) Heterogeneity of human CD4$^+$ T cells against microbes. Annu Rev Immunol 34:317–334
16. Kapsenberg ML (2003) Dendritic-cell control of pathogen-driven T-cell polarization. Nat Rev Immunol 3:984–993
17. Reina-Campos M, Scharping NE, Goldrath AW (2021) CD8$^+$ T cell metabolism in infection and cancer. Nat Rev Immunol 21:718–738
18. Münz C (2009) Enhancing immunity through autophagy. Annu Rev Immunol 27:423–429
19. Münz C (2020) Autophagy in immunity. Prog Mol Biol Transl Sci 172:67–85
20. Steinert EM, Vasan K, Chandel NS (2021) Mitochondrial metabolism regulation of T cell-mediated immunity. Annu Rev Immunol 39:395–416
21. Garcin ED (2019) GAPDH as a model non-canonical AU-rich RNA binding protein. Semin Cell Dev Biol 86:162–173
22. Finlay D, Cantrell DA (2011) Metabolism, migration and memory in cytotoxic T cells. Nat Rev Immunol 11:109–117
23. Stertz S, Hale BG (2021) Interferon system deficiencies exacerbating severe pandemic virus infections. Trends Microbiol 29:973–982
24. Stockinger B, Omenetti S (2017) The dichotomous nature of T helper 17 cells. Nat Rev Immunol 17:535–544
25. Mills KHG (2022) IL-17 and IL-17-producing cells in protection versus pathology. Nat Rev Immunol 5:1–17
26. Golstein P, Griffiths GM (2018) An early history of T cell-mediated cytotoxicity. Nat Rev Immunol 18:527–535
27. Casanova JL, Abel L (2021) Lethal infectious diseases as inborn errors of immunity: toward a synthesis of the germ and genetic theories. Annu Rev Pathol 16:23–50
28. Wong LR, Perlman S (2022) Immune dysregulation and immunopathology induced by SARS-CoV-2 and related coronaviruses – are we our own worst enemy? Nat Rev Immunol 22:47–56
29. Schmitt EG, Cooper MA (2021) Genetics of pediatric immune-mediated diseases and human immunity. Annu Rev Immunol 39:227–249
30. Läderach F, Münz C (2022) Altered immune response to the Epstein-Barr virus as a prerequisite for multiple sclerosis. Cells 11:2757
31. Moore S, Hill EM, Dyson L, Tildesley MJ, Keeling MJ (2022) Retrospectively modeling the effects of increased global vaccine sharing on the COVID-19 pandemic. Nat Med 28:2416–2423
32. Oladejo M, Paulishak W, Wood L (2022) Synergistic potential of immune checkpoint inhibitors and therapeutic cancer vaccines. Semin Cancer Biol 88:81–95
33. van Loo G, Bertrand MJM (2022) Death by TNF: a road to inflammation. Nat Rev Immunol 15:1–15
34. Olsson T, Barcellos LF, Alfredsson L (2017) Interactions between genetic, lifestyle and environmental risk factors for multiple sclerosis. Nat Rev Neurol 13:25–36
35. Münz C, Lünemann JD, Getts MT, Miller SD (2009) Antiviral immune responses: triggers of or triggered by autoimmunity? Nat Rev Immunol 9:246–258

Sepsis

Jean-Charles Preiser and Jean-Louis Vincent

Introduction

The word "sepsis" (σήψη) comes from an ancient Greek word meaning putrefaction [1], which described well the clinical features of the condition, with little understanding at the time of the underlying cause or origin. Later, as the mechanisms of infection and sepsis began to be unraveled, it became clear that sepsis is a clinical syndrome, characterized by an inflammatory response that typically includes fever, abnormal white blood cell count, and increased concentrations of inflammatory markers such as C-reactive protein [2, 3].

Further knowledge about the pathophysiology of sepsis, including the concomitant presence of an anti-inflammatory response [4], led to the current definition of sepsis as a "dysregulated host response" to infection [5]. Administration of anti-inflammatory agents as adjunctive therapies in addition to the eradication of the underlying infectious focus with antibiotics and surgical intervention has not been associated with major improvements in outcomes of patients with sepsis. This lack of benefit with these general (non-individualized) approaches highlights the need for more personalized therapeutic modalities designed to impact the different clinical aspects of sepsis according to individual patient characteristics and disease status. Future therapeutic strategies are likely to involve interventions to control, limit, or augment individual components of the immune response, targeted according to patient phenotypes [4].

The overwhelming inflammation and in particular the oxidative stress induced by the activation of neutrophils, stress hormones, and proinflammatory cytokines (for example, tumor necrosis factor [TNF], interleukin [IL]-1 and -6) trigger the metabolic alterations of sepsis [6]. From a clinical standpoint, long-term sequelae, such as proteolysis leading to prolonged muscle weakness, persisting insulin resistance leading to type 2 diabetes, and osteopenia, are related to the prolonged sepsis-related catabolism [7]. However, many different clinical expressions have been reported, which are not always related to the intensity and/or the duration of the septic response. Novel approaches to evaluation of the sepsis response, including metabolomics, will hopefully improve understanding of the various aspects of the metabolic response, and thereby facilitate development of approaches targeting specific metabolic phenotypes.

J.-C. Preiser
Department of Internal Medicine, Erasme Hospital,
Université libre de Bruxelles, Brussels, Belgium

Department of Intensive Care, Erasme Hospital,
Université libre de Bruxelles, Brussels, Belgium
e-mail: jean-charles.preiser@hubruxelles.be

J.-L. Vincent (✉)
Department of Intensive Care, Erasme Hospital,
Université libre de Bruxelles, Brussels, Belgium
e-mail: jlvincent@intensive.org

© The Author(s), under exclusive license to Springer Nature Switzerland AG 2026
E. Lammert, M. Zeeb (eds.), *Metabolism of Human Diseases*,
https://doi.org/10.1007/978-3-031-96019-2_37

The metabolic changes occurring during sepsis that have been studied in small or large animal models or in human volunteers challenged with bacterial products have been revisited over the last few years. Several specific aspects of the metabolic response to sepsis have also been recently reevaluated, from whole-body alterations (energy expenditure [EE], use of energy substrates, change in body composition, protein wasting) to cellular and subcellular dysfunctions [6]. The influence of timing is also better appreciated, allowing distinction of successive metabolic phases, starting with an acute initial phase characterized by marked inflammation and catabolism and resistance to anabolic signals, followed by a hypometabolic phase comparable to hibernation, and then a hypermetabolic recovery phase [8, 9].

In this chapter, we will review the metabolic changes related to sepsis, starting from the whole-body level including EE and the use of energy substrates, before moving to the subcellular aspects including mitochondrial dysfunction and the potential dysfunction revealed by recently reported changes in metabolomes.

Whole-Body Metabolic Alterations

Energy Expenditure

The typical clinical signs of sepsis (fever, increased heart and respiratory rates) are usually associated with an increase in EE in uncomplicated sepsis [10, 11]. However, during the early phase of severe sepsis, EE measured by indirect calorimetry (IC), is usually lower than before the onset of sepsis; during the recovery phase, EE increases to values higher than in the pre-sepsis period [12–14] (Fig. 1). One should remember that EE is influenced by various physiological derangements (e.g., fever, hypothermia, changes in heart rate, shivering, agitation), and by frequently used therapeutic interventions (e.g., administration of sedative agents, non-selective β-blockers, active cooling). These various confounders make any equations unreliable to assess the time course of EE [8, 15]. Importantly, attempts to match EE are

no longer a therapeutic goal during the early phase of sepsis. Indeed, mobilization of energy substrates from the body stores continues, and can represent 50–75% of total EE [9, 16, 17]. The provision of additional energy substrates from food or medical nutrition is thus no longer recommended after several prospective trials reported the detrimental effects of high amounts of macronutrients during the early phase of sepsis [18]. Instead, early macronutrient restriction is now favored, an approach supported by the anorexia typically associated with sepsis [16].

Use of Energy Substrates

Adaptation of the body to a septic challenge requires mobilization of endogenous stores of fat from adipose tissue and of proteins from muscles to provide substrates for the endogenous production of glucose, mainly by the liver. However, in contrast to starvation, the rate of transformation of free fatty acids and lactate into easily useable substrates (glucose and ketones) can be downregulated during sepsis [19].

The flow of substrates is mediated by decreased insulin sensitivity of muscle and adipose tissues. Teleologically, this evolutionarily preserved adaptive mechanism allows increased lipolysis and proteolysis to mobilize glycerol and neoglucogenic amino acids to fuel the endogenous production of glucose. In parallel, the delivery of glucose derived from glycogenolysis and gluconeogenesis is predominantly redirected toward immune cells, brain, and erythrocytes, whose glucose uptake is not mediated by insulin [6, 8]. The adaptive nature of insulin resistance observed during the acute phase of sepsis can be viewed as a protective mechanism to overcome the depletion of adenosine triphosphate (ATP) [11]. During this acute phase, hyperglycemia is frequent and reflects the magnitude of insulin resistance [20]. The optimal blood glucose target in critically ill patients, including those with sepsis, is not well defined [44]. It is likely to vary during the course of sepsis, that is, higher during the early phase, and lower in the later stage. It may also depend on the patient's usual average

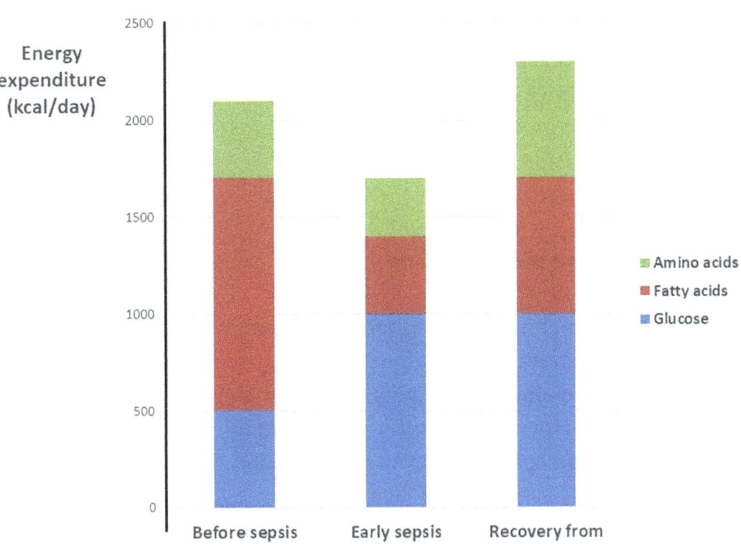

Fig. 1 Schematic representation of energy expenditure and use of energy substrates

blood glucose level, as estimated from the glycated hemoglobin [21]. Persistence of increased insulin resistance after the resolution of sepsis is not uncommon and is associated with a higher risk of subsequent type 2 diabetes (see chapter "Diabetes Mellitus") [22].

Clinical Consequences: Changes in Body Composition

The clinical phenotype of a critically ill patient with sepsis is characterized by an increase in extracellular water associated with capillary leakage and edema formation, and decreases in fat and muscle mass related to the mobilization of gluconeogenic substrates, an event described as "septic autocannibalism" [6, 23]. The functional consequences of the loss of fat have not been fully elucidated, but the protein wasting is related to intensive care-acquired weakness [24], resulting from the combination of a decrease in muscle protein synthesis, increased proteolysis by the ubiquitin–proteasome pathway, a lack of activation of autophagy, and alterations in the excitability of muscle membranes [25]. Changes in energy metabolism related to mitochondrial dysfunction and decreased oxidation of fatty acids also contribute to the muscle dysfunction [11].

Subcellular Changes

Mitochondrial Dysfunction

Mitochondrial dysfunction participates in the pathophysiology of sepsis and is associated with patient outcomes [26]. The key role of mitochondria in cellular homeostasis hints at a mechanistic explanation for the link between mitochondrial dysfunction and sepsis-related organ dysfunction. While the mitochondrial capacity for generating cellular ATP is decreased in sepsis, mitochondrial dysfunction is not associated with significant organ necrosis in human septic shock [27]. This may in part be explained by the move in EE from predominantly anabolic functions to functions necessary for short-term survival, such as ion homeostasis to preserve for example Na^+/K^+- and Ca^{2+}-ATPase pumps that are essential for cellular integrity [11, 28]. Despite mitochondrial dysfunction, cells may thus retain the ability to recover.

Mitochondrial dysfunction has been reported in human sepsis. Brealey et al. [26] showed low ATP concentrations and decreased complex I activity in biopsies of skeletal muscle. Later, associations between haplotypes of mitochondrial DNA and structural changes, reductions in the activities of the mitochondrial complexes, and sepsis survival were reported [29, 30]. Alterations in the biogenesis of mitochondrial components, changes in mitochondrial dynamics, and defective mitophagy have also been demonstrated, as recently reviewed [11].

Changes in the Metabolome

The recently developed metabolomic approach, enabling measurement of large numbers of metabolites in biological samples, is likely to change our understanding of sepsis, and facilitate more targeted therapies, similar to the evolution of treatments for cancer [4]. There is a need to stratify patients according to their clinical phenotype as evidenced by the concentrations of metabolites found in biological samples [32–34]. Results from metabolomic studies can be used to differentiate sepsis from other conditions. Patients with sepsis have an accumulation of lactate, pyruvate, and metabolites of the tricarboxylic acid (TCA) cycle, and altered oxidation of fatty acids, all consistent with the mitochondrial dysfunction observed using other experimental approaches. These patients also seem to have increased proteolysis, as evidenced by elevated levels of most amino acids [34, 35], and functional alterations of the urea cycle reflected by low ornithine, argininosuccinate, arginine, and citrulline levels [36]. They may also have decreased levels of high-, low-, and very low-density lipoproteins, and cholesterol [37, 38], and this could be involved in decreased endotoxin binding capacity, and altered mitochondrial respiration, cell function, and apoptosis [31, 37, 39].

Metabolomic studies may also help to identify the site of infection and the causative organism. For example, there are specific alterations including increased lysophosphatidylcholine levels in patients with urinary tract infection and intra-abdominal infections [40]. The presence of specific pathogens may also be reflected by the presence of specific metabolites [31].

Prognostication based on clinical phenotypes ("biomarkers") can represent an additional application of metabolomic techniques. For example, survival was lower in septic patients with high lactate and pyruvate, high carbohydrates, and low short- and medium-chain carnitines (reflecting mitochondrial and fatty acid oxidation dysfunction) than in other patients [31]. Specific metabolomic profiles were also found to be predictive of different organ dysfunction (lung and kidney in particular) [41].

Ultimately, the aim would be that metabolomics could be used to track response to treatment [42] and to characterize changes in the microbiome, another potential therapeutic target [43].

Perspectives

The metabolic response to sepsis has been profoundly revisited in the last decade. The ability to characterize temporal patterns in the use of energy substrates, to assess mitochondrial dysfunction, and to predict the response to therapy opens the way for a better understanding of sepsis and to more effective therapeutic approaches, tailored to individual patient needs [45]. New perspectives are clearly opened by these novel insights and understanding of the metabolic response to severe sepsis. Therapies targeting the clinical phenotype while preserving adaptive events can now be envisioned [46]. Likewise, a more rapid identification of causative microorganisms and foci of infection can be reached by the bedside application of metabolomic techniques.

Conflict of Interest The authors declare no conflicts of interest.

Questions and Answers

Question 1 How does energy expenditure (EE) relate to the severity of sepsis?

Answer 1 During the early phase, EE decreases with increasing severity of sepsis.

Question 2 Why should EE not be matched by an exogenous provision of energy substrates?

Answer 2 Because a non-inhibitable production of endogenous sources of energy occurs.

Question 3 Why is mitochondrial dysfunction viewed as adaptive?

Answer 3 Because functions necessary to short-term survival are privileged.

Question 4 How could metabolomic approaches change clinical management?

Answer 4 Stratification of clinical phenotypes to develop targeted therapies, earlier identification of causative microorganisms and foci of infection, and prognostication based on biomarkers.

References

1. Geroulanos S, Douka ET (2006) Historical perspective of the word "sepsis". Intensive Care Med 32:2077
2. Bone RC (1991) Sepsis, the sepsis syndrome, multi-organ failure: a plea for comparable definitions. Ann Intern Med 114:332–333
3. Vincent JL (2022) Evolution of the concept of sepsis. Antibiotics (Basel) 11:1581
4. Vincent JL, van der Poll T, Marshall JC (2022) The end of "one size fits all" sepsis therapies: toward an individualized approach. Biomedicines 10:2260
5. Singer M, Deutschman CS, Seymour CW et al (2016) The third international consensus definitions for sepsis and septic shock (Sepsis-3). JAMA 315:801–810
6. Wasyluk W, Zwolak A (2021) Metabolic alterations in sepsis. J Clin Med 10:2412
7. Rousseau AF, Prescott HC, Brett SJ et al (2021) Long-term outcomes after critical illness: recent insights. Crit Care 25:108
8. Preiser JC, Ichai C, Orban JC et al (2014) Metabolic response to the stress of critical illness. Br J Anaesth 113:945–954
9. Wernerman J, Christopher KB, Annane D et al (2019) Metabolic support in the critically ill: a consensus of 19. Crit Care 23:318
10. Fong YM, Marano MA, Moldawer LL et al (1990) The acute splanchnic and peripheral tissue metabolic response to endotoxin in humans. J Clin Invest 85:1896–1904
11. Preau S, Vodovar D, Jung B et al (2021) Energetic dysfunction in sepsis: a narrative review. Ann Intensive Care 11:104
12. Kreymann G, Grosser S, Buggisch P et al (1993) Oxygen consumption and resting metabolic rate in sepsis, sepsis syndrome, and septic shock. Crit Care Med 21:1012–1019
13. Uehara M, Plank LD, Hill GL (1999) Components of energy expenditure in patients with severe sepsis and major trauma: a basis for clinical care. Crit Care Med 27:1295–1302
14. Wasyluk W, Zwolak A, Jonckheer J et al (2022) Methodological aspects of indirect calorimetry in patients with sepsis – possibilities and limitations. Nutrients 14:930
15. De Waele E, Malbrain MLNG, Spapen H (2020) Nutrition in sepsis: a bench-to-bedside review. Nutrients 12:395
16. Preiser JC, Arabi YM, Berger MM et al (2021) A guide to enteral nutrition in intensive care units: 10 expert tips for the daily practice. Crit Care 25:424
17. Wolfe RR (2018) The 2017 Sir David P Cuthbertson lecture. Amino acids and muscle protein metabolism in critical care. Clin Nutr 37:1093–1100
18. Singer P, Blaser AR, Berger MM et al (2019) ESPEN guideline on clinical nutrition in the intensive care unit. Clin Nutr 38:48–79
19. Vandewalle J, Libert C (2022) Sepsis: a failing starvation response. Trends Endocrinol Metab 33:292–304
20. Lheureux O, Prevedello D, Preiser JC (2019) Update on glucose in critical care. Nutrition 59:14–20
21. Honarmand K, Sirimaturos M, Hirshberg EL et al (2024) Society of Critical Care Medicine Guidelines on glycemic control for critically ill children and adults 2024. Crit Care Med 52:e161–e181
22. Krinsley JS, Rule PR, Roberts GW et al (2022) Relative hypoglycemia and lower hemoglobin A1c-adjusted time in band are strongly associated with increased mortality in critically ill patients. Crit Care Med 50:e664–e673
23. Ali Abdelhamid Y, Kar P, Finnis ME et al (2016) Stress hyperglycaemia in critically ill patients and the subsequent risk of diabetes: a systematic review and meta-analysis. Crit Care 20:301
24. Knuth CM, Auger C, Jeschke MG (2021) Burn-induced hypermetabolism and skeletal muscle dysfunction. Am J Physiol Cell Physiol 321:C58–C71
25. van Gassel RJJ, Baggerman MR, van de Poll MCG (2020) Metabolic aspects of muscle wasting during critical illness. Curr Opin Clin Nutr Metab Care 23:96–101
26. Schefold JC, Wollersheim T, Grunow JJ et al (2020) Muscular weakness and muscle wasting in the critically ill. J Cachexia Sarcopenia Muscle 11:1399–1412
27. Brealey D, Brand M, Hargreaves I et al (2002) Association between mitochondrial dysfunction and severity and outcome of septic shock. Lancet 360:219–223
28. Hotchkiss RS, Swanson PE, Freeman BD et al (1999) Apoptotic cell death in patients with sepsis, shock, and multiple organ dysfunction. Crit Care Med 27:1230–1251

29. Buttgereit F, Brand MD (1995) A hierarchy of ATP-consuming processes in mammalian cells. Biochem J 312(Pt 1):163–167
30. Lorente L, Iceta R, Martin MM et al (2012) Survival and mitochondrial function in septic patients according to mitochondrial DNA haplogroup. Crit Care 16:R10
31. Lorente L, Martin MM, Lopez-Gallardo E et al (2016) Septic patients with mitochondrial DNA haplogroup JT have higher respiratory complex IV activity and survival rate. J Crit Care 33:95–99
32. Hussain H, Vutipongsatorn K, Jimenez B et al (2022) Patient stratification in sepsis: using metabolomics to detect clinical phenotypes, sub-phenotypes and therapeutic response. Meta 12:376
33. Rogers AJ, McGeachie M, Baron RM et al (2014) Metabolomic derangements are associated with mortality in critically ill adult patients. PLoS One 9:e87538
34. Johansson PI, Nakahira K, Rogers AJ et al (2018) Plasma mitochondrial DNA and metabolomic alterations in severe critical illness. Crit Care 22:360
35. Mierzchala-Pasierb M, Lipinska-Gediga M, Fleszar MG et al (2020) Altered profiles of serum amino acids in patients with sepsis and septic shock – preliminary findings. Arch Biochem Biophys 691:108508
36. Mierzchala-Pasierb M, Lipinska-Gediga M, Fleszar MG et al (2021) An analysis of urine and serum amino acids in critically ill patients upon admission by means of targeted LC-MS/MS: a preliminary study. Sci Rep 11:19977
37. Wijnands KA, Castermans TM, Hommen MP et al (2015) Arginine and citrulline and the immune response in sepsis. Nutrients 7:1426–1463
38. Barker G, Leeuwenburgh C, Brusko T et al (2021) Lipid and lipoprotein dysregulation in sepsis: clinical and mechanistic insights into chronic critical illness. J Clin Med 10:1693
39. Golucci APBS, Marson FAL, Ribeiro AF et al (2018) Lipid profile associated with the systemic inflammatory response syndrome and sepsis in critically ill patients. Nutrition 55-56:7–14
40. Amunugama K, Pike DP, Ford DA (2021) The lipid biology of sepsis. J Lipid Res 62:100090
41. Neugebauer S, Giamarellos-Bourboulis EJ, Pelekanou A et al (2016) Metabolite profiles in sepsis: developing prognostic tools based on the type of infection. Crit Care Med 44:1649–1662
42. Lin SH, Fan J, Zhu J et al (2020) Exploring plasma metabolomic changes in sepsis: a clinical matching study based on gas chromatography-mass spectrometry. Ann Transl Med 8:1568
43. Puskarich MA (2015) The challenge and the promise of studying mitochondrial dysfunction in humans with sepsis. Ann Am Thorac Soc 12:1595–1596
44. Bauermeister A, Mannochio-Russo H, Costa-Lotufo LV et al (2022) Mass spectrometry-based metabolomics in microbiome investigations. Nat Rev Microbiol 20:143–160
45. Scherger SJ, Kalil AC (2024) Sepsis phenotypes, sub-phenotypes and endotypes: are they ready for bedside care? Curr Opin Crit Care 30:406–413
46. Demailly Z, Tamion F, Besnier E et al (2025) Understanding metabolic remodeling in shock through metabolomics lenses. Mol Cell Endocrinol 600:112491

Kidney

Kidney: Overview

Markus Rinschen, Linus Völker,
and Paul Brinkkötter

Anatomy and Physiology of the Kidneys

The two human kidneys are localized in the retroperitoneal space of the abdomen. The adult size is approximately 11 × 6 × 4 cm. Anatomically, the kidney can be subdivided into three segments, the cortex, medulla, and pelvis (Fig. 1a). The latter gives rise to the ureter, which exits the kidney at the hilum.

Each kidney harbors approximately 1 million nephrons, its functional units, which are composed of glomeruli and tubules (Fig. 1b, c). The glomeruli are located in the renal cortex and are perfused by 4,000–5,000 L/day of blood. The glomerulus contains a capillary bed in which the primary urine is filtered into the tubules. The three-layered filter consists of fenestrated vascular endothelial cells, the glomerular basement membrane, and visceral epithelial cells, i.e., the podocytes, located at the outer aspect of the capillary loops (Fig. 1c, d). The latter form specific foot processes with a 40 nm gap in between neighboring foot processes bridged by a specialized cell–cell contact, the slit-diaphragm [1]. Plasma and plasma proteins are filtered based on their charge and molecular size (approximately <70 kDa) resulting in almost protein-free primary urine. The normal glomerular filtration rate (GFR) ranges from 130 to 180 L/day. The primary urine is then collected in Bowman's capsule and further transported via the proximal tubule followed by the loop of Henle into the medulla where it makes a hairpin turn and ascends back into the cortex before it enters the distal tubule and the connecting tubule into the collecting duct.

A prominent brush border at the apical, lumen-oriented plasma membrane of epithelial cells characterizes the proximal tubule (Fig. 1b, c, e). The loop of Henle consists of a thin descending, followed by a thin and thick ascending part (Fig. 1b, f). The distal convoluted tubule is similar to the proximal tubule; however, it lacks its prominent brush border. At the macula densa of the distal tubule (Fig. 1b, g), the nephron passes its own glomerulus at very close proximity (Fig. 1b, c). Several connecting tubules finally enter one of the collecting ducts (Fig. 1b, h), which transport the urine through the entire medulla and release it into the pelvis. Along the nephrons, the primary urine is concentrated to 1–3 L/day of terminal urine.

The blood flow of the kidney is maintained via the renal artery and renal vein. Continuous blood supply is of utmost importance to maintain high-capacity, energy-consuming transport processes (Fig. 1e–h) and other biological functions of the

M. Rinschen (✉) · L. Völker · P. Brinkkötter
Nephrolab Cologne, University Hospital of Cologne,
Köln, Germany
e-mail: markus.rinschen@uk-koeln.de;
linus.voelker@uk-koeln.de;
paul.brinkkoetter@uk-koeln.de

E. Lammert, M. Zeeb (eds.), *Metabolism of Human Diseases*,
https://doi.org/10.1007/978-3-031-96019-2_38

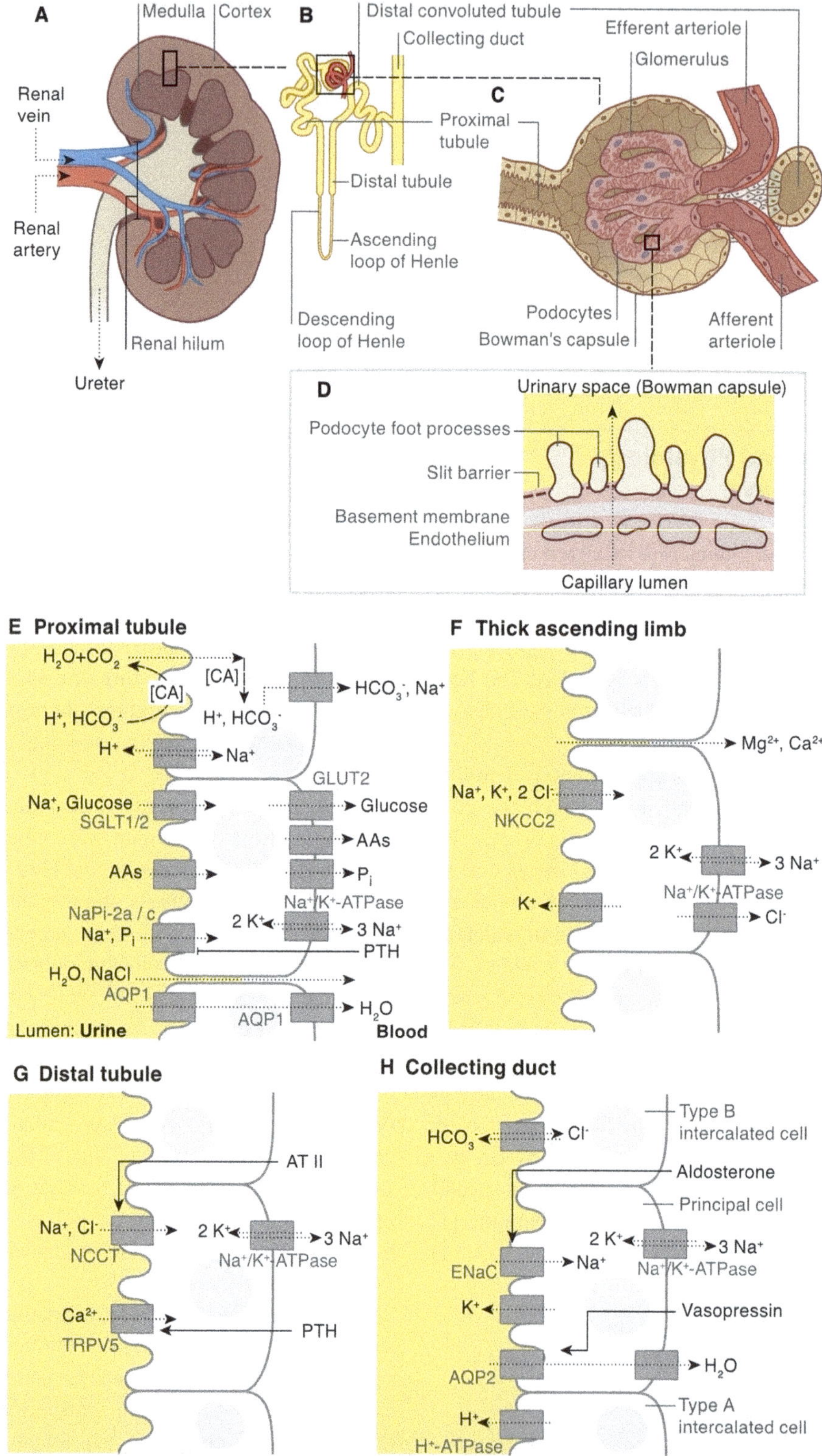

Fig. 1 Anatomy and basic physiology of renal reabsorption and secretion. (**a**) Cross-section through a kidney revealing renal artery and vein, and the ureter emerging from the renal hilum. Each kidney contains around 1 million nephrons each oriented with their glomeruli toward the renal cortex and descending toward the medulla.

entire kidney epithelium. The renal artery enters the kidney at the hilum and divides into main segment arterial branches and subsequently into interlobular arteries and afferent arterioles to the glomerulus as well as the vasa recta for the blood support of the inner medulla (Fig. 1a, c).

The kidney participates in the body homeostasis to maintain a relatively constant intracellular environment. Among its specific functions are (i) regulation of water (H_2O) and electrolyte metabolism, (ii) regulation of acid–base homeostasis, (iii) detoxification and excretion of waste products, (iv) reabsorption of glucose and other essential metabolites, (v) endocrine function, and (vi) synthesis of glucose under fasting conditions to a low extent.

Kidney-Specific Metabolic and Molecular Pathways and Processes

The GFR equals the sum of the filtration rates of all nephrons and is defined as the freely filtered plasma volume per time. It can be measured as inulin clearance or estimated as creatinine clearance. The clearance refers to the plasma volume, which is cleared from these substances within a given time. Its calculation requires measurement of urine concentration of creatinine, serum concentration of creatinine, and urine volume within a given time (usually 24 h).

The filtration fraction is strictly regulated by filtration pressure. This pressure depends on the hydrostatic pressure, which is dependent on the resistance of the afferent and efferent arteriole, and the oncotic pressure built up by the plasma protein concentration (see chapter "Liver: Overview"), as well as the hydrostatic pressure in the Bowman's capsule.

The main task of the kidney is to control water and electrolyte metabolism, to save and reabsorb several essential metabolites from the primary urine, and to excrete toxic metabolites. These processes occur in a segment-specific manner. The kidney tubular function largely depends on the unique distribution of transporters, channels, and pumps in various leaky or tight subsegments. The entire kidney tubule has a polarized epithelium with unique distribution of transport

---

Fig. 1 (continued) (**b**) The nephron is the functional unit of the kidney, consisting of glomerulus and tubular system (proximal tubule, descending and ascending loop of Henle, distal tubule, and collecting duct). Blood vessels (not shown) supply the glomerulus and also the parts closer to the renal medulla (vasa recta). (**c**) Ultrastructure of the glomerulus. The kidney filtration barrier consists of endothelial cells, their basement membrane, and slit diaphragms spanning between podocyte foot processes. The glomerulus is surrounded by Bowman's capsule collecting the filtrate and directing it toward the proximal tubulus. The distal convoluted tubule, a part of the distal tubulus, sends feedback to the glomerulus. (**d**) The filtration barrier consists of fenestrated endothelium, endothelial basement membrane, and foot processes of surrounding podocytes containing a slit diaphragm. (**e–h**) Detailed views of key transport processes in the nephron from the luminal side (yellow background) to the blood (white background). (**e**) Transport processes in the proximal tubule include trans- and paracellular water uptake, bicarbonate reabsorption, and sugar and amino acid (AA) reabsorption. The latter are mostly driven by secondary gradients of Na^+ built up by basolateral Na^+/K^+-ATPase. For reabsorption of glucose, sodium-glucose linked transporters 1 and 2 (SGLT 1/2) and glucose transporter 2 (GLUT2) are necessary (see chapter "Metabolic Syndrome"). If bicarbonate resorption is necessary, it occurs via CO_2, formed and disposed of by carbonic anhydrase (CA). Parathyroid hormone (PTH) is able to downregulate phosphate (P_i) reuptake. (**f**) In the thick ascending limb of Henle, urine is concentrated by massive resorption of ions. Mg^{2+} and Ca^{2+} are absorbed mainly paracellularly. Na^+ and K^+ are reabsorbed via the Na–K–Cl cotransporter (NKCC2) driven by the sodium gradient. (**g**) In the distal tubule, Na^+ and K^+ are further reabsorbed, and Ca^{2+} reabsorption can be upregulated by PTH. Angiotensin II (AT II) can directly increase sodium reabsorption and subsequent water retention by activating the sodium-chloride symporter, also known as Na^+-Cl^- cotransporter (NCCT). (**h**) The collecting duct consists of two major cell types, that is, principal cells responsible for water and urea reabsorption, and intercalated cells responsible for acid/base homeostasis. Sodium and water reabsorption can be upregulated by aldosterone and vasopressin (see chapters "Brain: Overview" and "Community-Acquired Pneumonia"), respectively. Note: Names of transporters and channels not specifically mentioned in the text or not regulated by signals mentioned in the book were omitted due to space constraints. *AQP* aquaporin, *ENaC* epithelial Na^+ channel, *TRPV5* transient receptor potential cation channel subfamily V member 5

proteins between the cell surfaces at urinary space (apical) and blood (basolateral). The Na^+/K^+-ATPase as a main pump sits on the basolateral side of the nephron and commonly serves as the driving force for direct or indirect electrolyte transport processes [2].

The proximal tubule is the main segment for Na^+, bicarbonate (HCO_3^-), and water reabsorption (Fig. 1e). Na^+ is mainly reabsorbed via several cotransporters. Importantly, under physiological conditions, the human proximal tubule reabsorbs virtually all of the filtered glucose (via the apically localized sodium-glucose linked transporters SGLT1 and 2, around 180 g/day) as well as all amino acids (50 g/day) [3]. The capacity for glucose reabsorption by the proximal tubule is approximately 200 mg/dL.

Water is transported both transcellularly via the water channel aquaporin-1 (AQP1) and paracellularly via leaky cell–cell contacts.

The brush border of the proximal tubule contains the enzyme carboanhydrase, which catalyzes the conversion of carbonic acid (H_2CO_3) to water (H_2O) and carbon dioxide (CO_2), which can diffuse freely through the epithelial cell membrane. CO_2 is converted back to HCO_3^- by carboanhydrase expressed in the proximal tubule cells. These processes facilitate the reabsorption of bicarbonate in the proximal tubule cells in order to compensate acidotic body conditions.

In addition, several metabolites are secreted or reabsorbed via various cation and anion transporters. The proximal tubule secretes exogenous metabolites such as penicillin, and organic ions such as choline and histamine [4]. Some of the excreted substances end up in the terminal urine, which serves the detoxifying function of the kidney. Ammonia (NH_3) is secreted to buffer luminal protons. Urate, a metabolite accumulating in gout (see chapter "Gout"), is reabsorbed in the proximal tubule.

Under fasting conditions, the epithelial cells of the cortical tubular system are able to generate glucose. This occurs to a lesser extent than in the liver (see chapter "Liver: Overview"). The kidney uses lactate and glutamine as substrates for glucose formation via gluconeogenesis. Proximal tubules do not perform glycolysis due to low expression of the hexokinase enzymes. There are many other enzymes actively involved in amino acid metabolism expressed in proximal tubule (transaminases, gamma-glutaryl-transferases) [5].

In the thin descending limb of Henle, water is mainly reabsorbed without the reabsorption of sodium resulting in a hyperosmolar urine, which is further concentrated in the thick ascending loop of Henle (Fig. 1f). Here, the Na–K–Cl cotransporter (NKCC2) on the apical cell surface is crucially involved in actively taking up all three mentioned ions [6]. Cells in the loop of Henle are therefore largely dependent on respiratory mitochondrial ATP generation and are the first cells to detrude during malperfusion of the kidney (e.g., in acute kidney failure). Since the epithelium of the ascending loop of Henle is nonpermeable to water, it releases hypotonic urine. The large export of sodium from this segment of the loop of Henle is responsible for the accumulation of salt and urea in the kidney medulla and generates a cortico-medullary osmolarity gradient. The diffusion of K^+ back to the lumen generates a positive transluminal potential. This induces paracellular Mg^{2+} and Ca^{2+} reabsorption.

The regulation of salt and water excretion takes place in the principal cells of the collecting duct (Fig. 1h). Upon stimulation with aldosterone, Na^+ is reabsorbed in exchange for K^+ [7]. Aldosterone increases both expression and activity of the apical sodium channel ENaC (epithelial Na^+ channel). Upon stimulation with antidiuretic hormone (ADH, or vasopressin), water is reabsorbed passively due to the cortico-medullary osmolarity gradient. In addition, urea is passively reabsorbed via specific urea transporters (UTs). Urea is taken up by the vasa recta and circulates between inner and outer medulla (recycling of urea) [8]. Urea diffusion is mediated via urea transporters UT-A1, UT-A2, and UT-A3, which are abundantly expressed in the medullary parts of collecting ducts, loop of Henle, and vasa recta. Both water and urea transport are crucial for the final concentration of urine and fine-tunes diuresis.

Interspersed between the principal cells in the collecting duct, intercalated cells regulate the acid–base metabolism (Fig. 1h) [9]. Type A intercalated cells actively secrete H^+ ions (via an apical H^+/K^+-ATPase), whereas type B intercalated cells actively secrete HCO_3^- (via a cotransporter). These processes are crucially regulated during disorders in metabolic and respiratory acidosis. For example, in acidic conditions, Type A intercalated cells increase their capacity to secrete protons into the urine.

Inside-In: Metabolites of the Kidney Affecting the Kidney Itself

Autoregulatory mechanisms maintain renal plasma flow and GFR almost constant over a wide range of renal arterial pressure. For example, at a low renal arterial pressure, GFR is kept constant by either increasing the diameter of the afferent arteriole, using prostaglandins and components of the kinin–kallikrein system released from vascular endothelial cells, or decreasing the diameter of the efferent arteriole, using angiotensin-II (AT II) acting via AT II receptors. Arteriolar resistance is also under intrinsic myogenic control meaning that arteriolar smooth muscle cells can contract autonomously if they are stretched by increased blood flow. In addition, it is regulated by the tubuloglomerular feedback, as well as norepinephrine and other hormones (see below). The tubuloglomerular feedback mechanism relies on specialized cells in the macula densa at the end of the thick ascending tubule where the loop of Henle passes its own glomerulus. These cells release adenosine and nitric oxide (NO) in response to a decrease in Cl^- reabsorption indicative of a decreasing GFR. Both signals lead to afferent arteriole dilatation, hence a rise in glomerular perfusion pressure [10].

In addition, the macula densa cells stimulate the release of the protease renin by the juxtaglomerular cells into the afferent arteriole in response to detection of hypo-osmolar urine. The release of renin and activation of the subsequent AT cascade (the renin–angiotensin–aldosterone system, see below) also result in an efferent vasoconstriction further increasing GFR. Both, the tubuloglomerular feedback and the release of renin maintain GFR and distal flow at a constant rate.

Inside-Out: Metabolites of the Kidney Affecting Other Tissues

The kidneys are active endocrine organs and produce several hormones to regulate miscellaneous functions, for example, systemic hemodynamics (modulation of blood pressure and body fluids via renin, prostaglandins, kinins), erythropoiesis (via erythropoietin; see chapter "Blood: Overview"), and mineral metabolism (via calcitriol; see chapter "Rheumatoid Arthritis" and Fig. 2).

As stated above, renin is secreted by the juxtaglomerular cells. In addition, a decrease in systemic or renal perfusion pressure activates cardiopulmonary baroreceptors leading to an increased activity of the sympathetic nervous system inducing the release of renin via binding of norepinephrine to $\beta1$-adrenergic receptors. In the plasma, renin cleaves angiotensinogen (synthesized by the liver) to AT I, which is subsequently cleaved by the angiotensin-converting enzyme (ACE) localized in pulmonary arterioles. The end product of this cleavage reaction, the octapeptide AT II increases blood pressure via direct vascular effects and via release of aldosterone from the adrenal gland (using AT_1 receptors).

Erythropoietin is a 34 kDa sized polypeptide hormone and is secreted by endothelial cells of peritubular capillaries in response to hypoxia. It acts on stem cells in the bone marrow to increase production of hemoglobin and erythrocytes. Chronic kidney failure is associated with normochromic, normocytic anemia (i.e., reduced number of normal red blood cells) due to reduced erythropoietin production.

With regards to vitamin D metabolism (see chapter "Teeth and Bones: Overview"), calcidiol is hydroxylated to calcitriol in proximal tubular cells via 12-α hydroxylase that is stimulated by parathyroid hormone (PTH). Calcitriol mediates calcium and phosphate uptake by the gut and enhances bone resorption of calcium and phosphate. In addition, calcitriol binds to specific receptors in the parathyroid gland to inhibit PTH release as part of a negative feedback loop (see also chapter "Teeth and Bones: Overview").

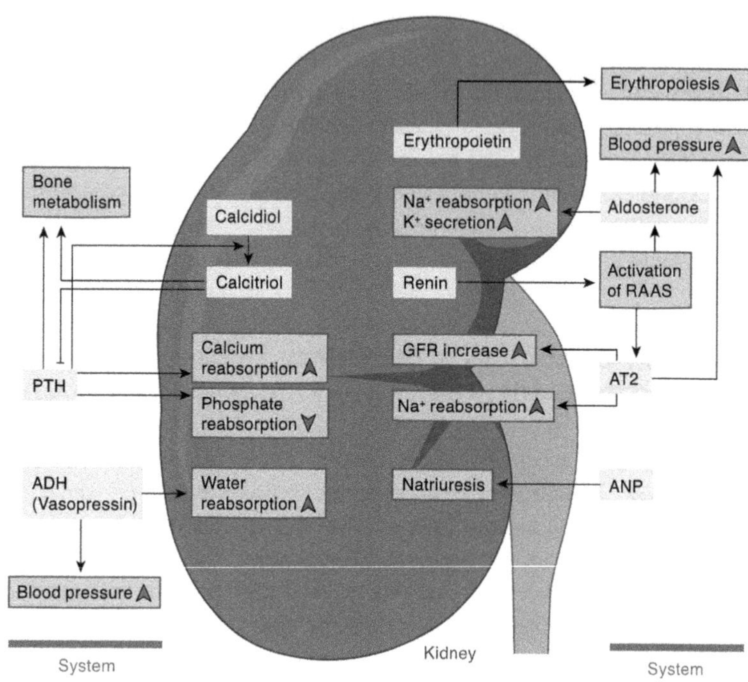

Fig. 2 Schematic overview of kidney metabolites affecting body homeostasis as well as body metabolites affecting the kidney. *ADH* antidiuretic hormone, *ANP* atrial natriuretic peptide, *AT2* angiotensin II receptor type 2, *GFR* glomerular filtration rate, *PTH* parathyroid hormone, *RAAS* renin–angiotensin–aldosterone system

Outside-In: Metabolites of Other Tissues Affecting the Kidney

Kidney function is under tight regulation by a variety of stimuli and fine-tunes water and salt metabolism as well as blood pressure (Fig. 2).

The atrial natriuretic factor (ANF), also known as atrial natriuretic peptide (ANP), is a peptide hormone secreted by the cardiac atria in response to an increase in atrial pressure (see chapter "Hyperlipidemia"). It directly relaxes the afferent arteriole of the glomeruli, thereby increasing GFR, and it induces an increase in sodium excretion, mainly via direct effects on the proximal tubule.

PTH is a key hormone of calcium metabolism. It is secreted by the parathyroid glands in response to decreased plasma calcium (Ca^{2+}) levels. It leads to massive calcium and phosphate release from the skeletal system. In the kidney, PTH increases calcium reabsorption in the distal tubule and decreases phosphate reabsorption in the proximal tubule via its G-protein-coupled receptor (called PTH receptor, Fig. 1g) [11]. In addition, it stimulates calcitriol production in the proximal tubule, which is responsible for increased Ca^{2+} and PO_4^{3-} absorption from the gut. Renal failure is associated with increased PTH levels, which become evident already in the early stages of chronic kidney disease, primarily as a consequence of phosphate retention and the associated rise in FGF-23, both of which suppress calcitriol synthesis and thereby contribute secondarily to decreased calcium levels.

AT II, the final effector of the renin–angiotensin–aldosterone system (RAAS), causes efferent arteriole constriction and thus an increase in GFR. In general, it increases blood pressure via its vasoconstrictive effect mediated by the AT_1 receptor. AT II directly acts on the distal part of the nephron to increase Na^+ and HCO_3^- reabsorption. In addition, AT II triggers release of the steroid hormone aldosterone from the adrenal glands, which increases Na^+ reabsorption via increased expression and apical localization of epithelial Na-channels in the distal tubules, and increases K^+ secretion via the renal outer medullary potassium channels [2]. It also leads to an increase in H^+ secretion. Adrenal insufficiency, as in Addison's disease, is associated with decreased aldosterone levels, decreased Na$^+$, increased K$^+$ levels and metabolic acidosis.

Vasopressin is a polypeptide hormone that is secreted by the posterior pituitary gland (see chapter "Brain: Overview") in response to increased plasma osmolarity and decreased blood volume. It binds to basolateral receptors on principal cells of the collecting duct activating vasopressin-2 (V2) receptor, which is a G-protein-coupled receptor, signaling through increased levels of cyclic adenosine monophosphate (cAMP) and subsequent activation of protein kinase A (PKA). This leads to increased translocation of the water channel aquaporin-2 (AQP2) to the apical cell surface (Fig. 1h). Water is then reabsorbed transcellularly via apical AQP2 and basolateral AQP3 and AQP4 channels [12]. Thus, vasopressin increases passive water reabsorption. Failure of this system leads to diabetes insipidus, a disease associated with urinary output of up to 20 L/day. Increased vasopressin secretion under pathophysiological conditions leads to dilutional hyponatremia. Besides its antidiuretic function, vasopressin directly increases blood pressure via arteriole constriction mediated by activation of the V_1a receptor on arteriolar smooth muscle cells.

Final Remarks

The kidney serves important functions for body homeostasis, including (i) regulation of water and electrolyte metabolism, (ii) regulation of acid–base homeostasis, (iii) detoxification and excretion of waste products, (iv) reabsorption of glucose and other essential metabolites, (v) endocrine function, and (vi) synthesis of glucose under fasting conditions. Perturbation of these functions is seen in a variety of pathophysiological conditions.

Questions and Answers

Question 1 What is the glomerular filtration rate (GFR), and how is it regulated?

Answer 1 The GFR is the volume of plasma filtered by the glomeruli per day (normally around 180 liters/day). It is primarily regulated by the hydrostatic and oncotic pressures in the glomerular capillaries, and the hydrostatic pressure in Bowman's capsule. The hydrostatic pressure depends on the resistance of the afferent and efferent arterioles.

Question 2 Which segment of the nephron is mainly responsible for glucose reabsorption, and how does it achieve this?

Answer 2 The proximal tubule is primarily responsible for glucose reabsorption. It contains sodium–glucose cotransporters (SGLT1 and SGLT2) on the apical (lumen-facing) membrane, which facilitate nearly complete reabsorption of the filtered glucose under normal conditions. The transport is secondary-active, meaning that glucose transport depends on the activity of the basolateral Na–K ATPase.

Question 3 What is the macula densa, and what role does it play in kidney function?

Answer 3 The macula densa is a group of specialized cells located in the distal tubule where the nephron passes its own glomerulus. These cells monitor the sodium and chloride content of the tubular fluid. In response to low levels of Cl$^-$ (indicating a decrease in GFR), they release factors like adenosine and nitric oxide that dilate the afferent arteriole, thus helping to increase glomerular perfusion and maintain a stable GFR.

Question 4 How does antidiuretic hormone (ADH, or vasopressin) influence water reabsorption in the kidneys?

Answer 4 ADH is released from the posterior pituitary gland in response to increased plasma osmolarity or decreased blood volume. It binds to receptors on the principal cells of the collecting duct, triggering the insertion of aquaporin-2 water channels into the apical membrane. This increases water reabsorption and helps concentrate the urine.

References

1. Pavenstädt H, Kriz W, Kretzler M (2003) Cell biology of the glomerular podocyte. Physiol Rev 83(1):253–307
2. Greger R (2000) Physiology of renal sodium transport. Am J Med Sci 319(1):51–62
3. Wright EM (2001) Renal Na(+)-glucose cotransporters. Am J Physiol Renal Physiol 280(1):F10–F18
4. Pritchard JB, Miller DS (1993) Mechanisms mediating renal secretion of organic anions and cations. Physiol Rev 73(4):765–796
5. Chrysopoulou M, Rinschen MM (2023) Metabolic rewiring and communication: an integrative view of kidney proximal tubule function. Annu Rev Physiol 86:405
6. Greger R, Schlatter E, Hebert SC (2001) Milestones in nephrology: presence of luminal K+, a prerequisite for active NaCl transport in the cortical thick ascending limb of Henle's loop of rabbit kidney. J Am Soc Nephrol 12(8):1788–1793
7. Lang F, Rehwald W (1992) Potassium channels in renal epithelial transport regulation. Physiol Rev 72(1):1–32
8. Knepper MA, Roch-Ramel F (1987) Pathways of urea transport in the mammalian kidney. Kidney Int 31(2):629–633
9. Al-Awqati Q (2013) Cell biology of the intercalated cell in the kidney. FEBS Lett 587(13):1911–1914
10. Schnermann J, Levine DZ (2003) Paracrine factors in tubuloglomerular feedback: adenosine, ATP, and nitric oxide. Annu Rev Physiol 65:501–529
11. Hernando N, Forster IC, Biber J, Murer H (2000) Molecular characteristics of phosphate transporters and their regulation. Exp Nephrol 8(6):366–375
12. Nielsen S, Frøkiaer J, Marples D, Kwon TH, Agre P, Knepper MA (2002) Aquaporins in the kidney: from molecules to medicine. Physiol Rev 82(1):205–244

Gout

Sonia Nasi and Alexander So

Introduction to the Disease

Acute gout is the inflammatory reaction provoked by monosodium urate (MSU) crystals when they form within a joint. Hyperuricemia is necessary for MSU crystal formation. Gout affects mainly males, due to their physiologically higher serum uric acid (UA) levels, and there is epidemiological evidence that the prevalence of gout and hyperuricemia is on the increase in western and Asian populations in both sexes and with aging. Based on data from an insurance database, Wallace estimated that between the years 1990 and 1999, the prevalence of gout increased by 60% in those aged over 65 and doubled in the population over 75 years of age [1]. In a study based on UK general practice data, the prevalence of gout in the adult population was estimated to be 1.4%, with a peak of over 7% in men aged over 75 years old [2]. Indeed, a strong association between hyperuricemia and the metabolic syndrome and age-related macular degeneration (see chapter "Metabolic Syndrome" and chapter "Age-Related Macular Degeneration") was observed. Potential explanations include lifestyle

and dietary changes brought about by increasing prosperity, increased life expectancy of the population, and the co-existence of multiple medical comorbidities and their treatments (i.e., hypertensive agents) that favor hyperuricemia in the elderly.

Gout starts with a first acute attack (flare) followed by an intercritical phase without attacks. Overtime, gout will become chronic. Diagnosis is based on clinical symptoms and the presence of MSU crystals in the joints.

Pathological Changes in Metabolism and Possible Outcomes

Uric acid in body fluid, at pH 7.4, exists in the urate form. It is generated by catabolism of purine nucleotides, which occurs mainly in the liver. The last steps consist of the conversion of xanthine into uric acid and is catalyzed by the enzyme xanthine oxidase (XO). Humans, as opposed to other mammals, lack the ability to further degrade urate to allantoin because the enzyme uricase is nonfunctional [3].

Hyperuricemia is defined as the level of serum urate that exceeds its plasma solubility. This favors crystal formation and deposition in soft tissues and around joints, leading to tophus (aggregates of MSU crystals) formation and

S. Nasi (✉) · A. So
Service de Rhumatologie, Centre Hospitalier Universitaire Vaudois and University of Lausanne, Lausanne, Switzerland
e-mail: sonia.nasi@chuv.ch;
AlexanderKai-Lik.So@chuv.ch

E. Lammert, M. Zeeb (eds.), *Metabolism of Human Diseases*,
https://doi.org/10.1007/978-3-031-96019-2_39

triggering an acute inflammatory reaction recognized as gout. In addition, increased urinary concentrations of urate can also lead to renal stone formation (see chapter "Kidney Stones"). Although the renal manifestations (see chapter "Kidney: Overview") are well documented in the literature, they are much less frequently encountered than gout, the most common clinical presentation. Hyperuricemia is either the result of excess formation of urate due to increased purine metabolism or the consequence of insufficient renal elimination of urate to maintain normal physiological values, or a mixture of both.

In the majority of cases, hyperuricemia is mainly explained by diet as well as idiopathic underexcretion of urate by the kidney. Studies confirmed that increased consumption of certain foods (meat, seafood, beer—because of their high purine content), liquor and sugar-sweetened soft drinks (because of their fructose content) increased the risk of developing gout [4–6]. Meat, seafood, and beer are risky because of their high purine content; sugar-sweetened soft drinks increase risk due to their high fructose content. Excess fructose (in the liver) and also alcohol deplete ATP levels and increase adenine degradation to uric acid.

An additional contributory factor to hyperuricemia, particularly in the older population, is drug-induced underexcretion due to use of thiazides and low-dose aspirin; indeed, they reduce the renal excretion of uric acid.

Data from genome-wide association studies (GWAS) and family studies showed that gout and hyperuricemia are polygenic traits [7]. In the kidney, urate undergoes glomerular filtration, tubular reabsorption, and then re-excretion. Most of the genes associated with hyperuricemia are implicated in either the excretion or reabsorption of urate in the renal tubule. The strongest association found is with SLC2A9, a urate transporter [8].

It is now established that one of the major mechanisms of gouty inflammation is the release of interleukin-1β (IL-1β) when MSU crystals are in contact with monocytes and neutrophils. In vitro, MSU crystals are capable of activating the NLRP3 (nucleotide-binding oligomerization domain-like receptor family, pyrin domain containing 3)-inflammasome in monocytes, macrophages, and dendritic cells to secrete large quantities of IL-1β [9]. The NLRP3 inflammasome is a cytoplasmic protein complex composed of NLRP3 (a protein of the NLRP family), an adapter protein ASC (apoptosis-associated speck-like protein containing a CARD), as well as caspase-1. Caspase-1 catalyzes the cleavage of pro-IL-1β and pro-IL-18 into their active forms (IL-1β and IL-18) leading to their secretion. Recently, protein kinase R (an RNA-dependent protein kinase) has been implicated in MSU-stimulated IL-1β release, as genetic deficiency in mice for this molecule blocked IL-1β secretion. Indeed, it interacts physically with the inflammasome to initiate caspase-1 activity [10].

In addition, MSU crystals can elicit inflammation in an inflammasome-independent manner (see chapter "Anatomy and Physiology of the Immune System") triggering at least two different pathways: one through crystals interacting with the cell surface (dendritic cells and macrophages) to initiate an intracellular signaling cascade that involves spleen tyrosine kinase (Syk), another via the release of pro-IL-1β into the extracellular space during cell activation or cell death, and its subsequent cleavage by serine proteases such as cathepsin-G, elastase, and proteinase-3 released by neutrophils at site of inflammation [11]. Finally, novel studies have also shown a role of IL-1α in the inflammatory process [12].

Since the body could not counteract hyperuricemia through a feedback mechanism or degradation of urate, people with hyperuricemia will remain with high urate levels for all their life, unless they are treated with urate-lowering agents.

Treatment of Gout

Gout therapy is based on two principal strategies: the control of hyperuricemia that predisposes to formation of crystals (urate-lowering therapies) and the control of gouty inflammation to calm the acute attack. All patients should be given dietary advice and general counseling of the importance of long-term treatment adherence.

The aim of the first group of therapies is to reduce the serum urate level to below the solubility threshold for crystal formation (<6 mg/dL or 360 umol/L). Inhibition of xanthine oxidase (XO) is the most widely used approach to control hyperuricemia and two inhibitors are currently available, allopurinol and febuxostat. Uricosurics, drugs that promote urine excretion of urate by inhibiting its reabsorption in the renal tubule, are less widely prescribed but are effective also. Examples of this category of drugs are probenecid and benzbromarone.

The treatment of acute gout aims to relieve pain and inflammation rapidly. Traditional approaches include nonsteroidal anti-inflammatory drugs (NSAIDs) like diclofenac or indomethacin, colchicine (a drug that inhibits tubulin polymerization), and corticosteroids. In most cases, these drugs are rapidly effective, but caution has to be exercised in some patients as NSAIDs can cause side effects such as renal dysfunction and raise blood pressure; the serum concentration of colchicine is influenced by cytochrome P450 3A4 (CYP3A4) and P-glycoprotein efflux pump (PgP) activity and both enzymes can have several potential drug interactions. The discovery of the IL-1 axis of gouty inflammation has led to studies that have evaluated the effectiveness of IL-1 inhibitors. They have been found to be effective, either in the prevention or in the treatment of an acute flare. The frequency of flares was halved using a monoclonal antibody against IL-1β (Canakinumab), and pain was significantly reduced [13, 14]. Similarly, an inhibitor of both IL-1α and β (Rilonacept) reduced gout flares by around 50% [15]. In uncontrolled studies, an interleukin-1 receptor antagonist called anakinra was effective in the treatment of acute gout in patients who had either intolerance or contraindications to standard therapy [16].

Finally, it is possible to lower serum urate by administration of exogenous uricase, the enzyme responsible for the oxidation of urate to allantoin, whose gene is no longer functional in man.

Influence of Treatment on Metabolism and Consequences for Patients

All drugs that reduce serum urate can cause acute flare when treatment is started. This is due to an alteration of the stability of MSU crystals when uric acid levels decrease suddenly. These evidences underline the importance of providing adequate prophylaxis for any patient who starts a urate-lowering therapy.

Allopurinol (an XO Inhibitor)

Allopurinol is a purine analog that is metabolized to active oxypurinol, a potent inhibitor of XO.

The dose of allopurinol required to reduce serum urate levels below the limit of solubility can vary. As the drug is eliminated by the kidney, attention has to be paid to renal function in determining the effective as well as safe maximal dose.

The most severe side effect is the allopurinol hypersensitivity syndrome, a Stevens-Johnson's type reaction that is characterized by fever, skin rashes and desquamation, liver function abnormalities, and a fatal outcome in a significant proportion of patients.

Febuxostat (Nonpurine XO Inhibitor)

Febuxostat is a nonpurine selective inhibitor of XO. In vitro studies showed that febuxostat is a potent ligand for, and inhibitor of, both the oxidized and reduced forms of XO, and clinical studies have confirmed its efficacy in reducing serum urate levels [17]. It leads to raised serum xanthine level but, clinically, this does not present a real problem.

Febuxostat shows a dose-dependent effect. Flares of gout were observed (in 64–70% of subjects), particularly after the cessation of gout prophylaxis therapy with naproxen (a NSAID) or colchicine.

The side effect profile of febuxostat is comparable to that of allopurinol [18]. The serious adverse events reported included liver function abnormalities and cardiovascular events. Rashes were observed in <2% of febuxostat-treated patients, and no patients developed a reaction that resemble the allopurinol hypersensitivity syndrome.

Uricosuric Drugs

Uricosuric drugs interfere with tubular mechanisms of urate reabsorption to enhance uricosuria inhibiting the renal/liver transporters SLC2A9 and SLC22A2. These drugs should not be used in patients with high urate excretion, as they can precipitate renal stones or urate nephropathy. Benzbromarone is the most powerful drug in terms of urate reduction but shows serious hepatotoxic side effects severely restricting its distribution. Currently, probenecid is the most frequently prescribed uricosuric.

As probenecid also increases urine calcium excretion, it is contraindicated in patients with a history of renal calculi. Probenecid can interfere with renal excretion of other drugs (penicillins, some antivirals), so a careful drug history is vital before initiating treatment.

Uricase

This therapy uses exogenous uricase that is chemically linked to polyethylene glycol (PEG) in order to prolong its half-life. Uricase rapidly reduces serum urate levels, can decrease tophus size, and significantly reduce the urate levels [19]. The resulting allantoin does not cause any problems but long-term uricase therapy can cause drug sensitization and allergic reactions, and this may be the major problem in chronic therapy.

Perspectives

Gout is a common medical condition that is well understood in terms of physiology, but its treatment remains suboptimal in many countries. Besides treatment of the acute attack, urate-lowering therapies are important to maintain a low serum urate level in order to prevent severe complications and frequent attacks. Currently, the treatment of asymptomatic hyperuricemia is not recommended, but this recommendation may be modified in the future, as there is strong evidence linking hyperuricemia to cardiovascular and renal morbidity.

Current research topics will bring new understanding of the genetics of gout and hyperuricemia as well as the mechanisms that regulate gouty inflammation, the renal and liver urate transporters, and the biochemical interaction between metabolic syndrome and hyperuricemia.

In addition, new XO inhibitors and IL-1 inhibitors, which are currently studied, could become future drugs for treatment (Figs. 1 and 2).

Questions and Answers

Question1 What are the risk factors for gout development?

Answer 1 The main risk factors for the development of gout encompass elevated serum uric acid (hyperuricemia), male gender, age over 65, renal insufficiency, hypertriglyceridemia, hypercholesterolemia, diabetes, obesity, hypertension, dietary factors, alcohol intake, and certain medications.

Question 2 What are the molecular mechanisms triggered by MSU crystals?

Answer 2 On the one hand, MSU crystals lead to the production of IL-1β by monocytes, macrophages, and dendritic cells in an NLRP3-inflammasome dependent manner. On the other hand, MSU crystals act in inflammasome-independent manners. These include activation of an intracellular signaling cascade involving Syk kinase, or release of pro-IL-1β from dead cells and its subsequent cleavage in mature IL-1β by neutrophil serine proteases.

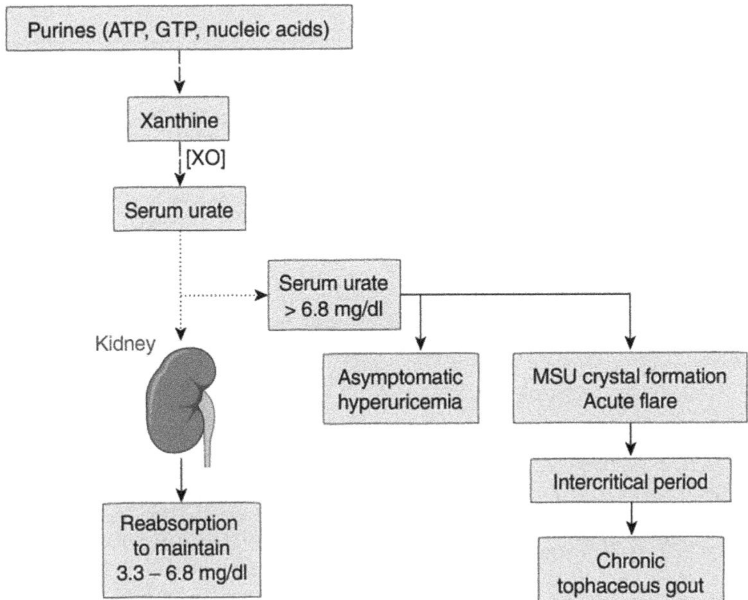

Fig. 1 Mechanisms and progression of gout. Uric acid is the end point of purine degradation, resulting from the conversion of xanthine by xanthine oxidase (XO), and is present in the blood as urate. Urate is filtered by the kidney and is reabsorbed to maintain a serum concentration below 6.8 mg/dL under healthy conditions. Serum urate concentrations can be increased by dietary factors and genetic disorders (not shown). When serum urate levels exceed its solubility (6.8 mg/dL), monosodium urate (MSU) crystals may form. Gout progresses through clinically distinct stages. Initially, the condition is asymptomatic (and can remain so). Crystals can be released into the joint space, triggering an interleukin (IL)-1-dependent inflammatory response, resulting in an acute flare, characterized by severe pain and fever. An initial flare usually resolves in 3–14 days. The periods between acute flares are termed intercritical periods, in which symptoms are absent, but urate crystals are still present in previously involved joints, stimulating low-grade inflammation. In untreated patients, continuing urate accumulation leads to chronic tophus formation

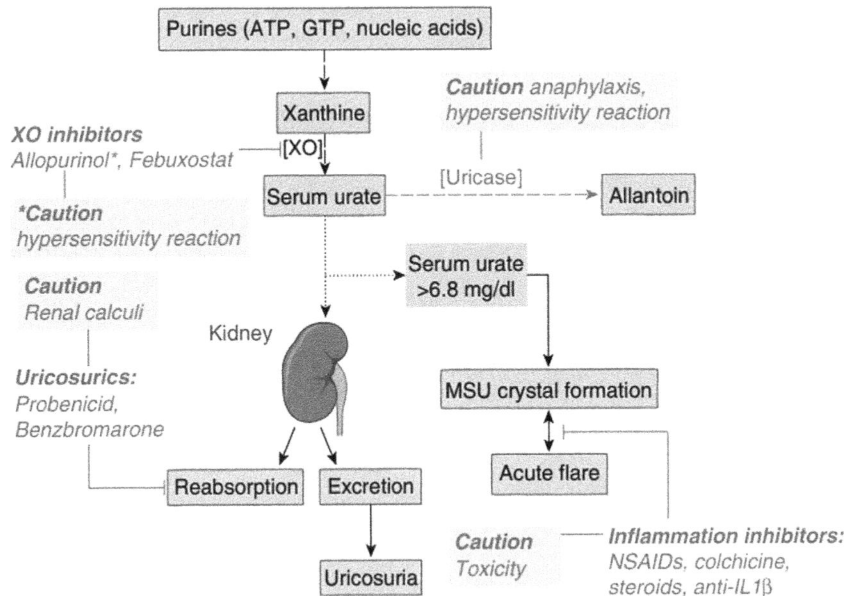

Fig. 2 Location of the different targets of currently available treatments of acute gout and hyperuricemia. Urate-lowering therapies have three major mechanisms of action: inhibition of XO, addition of uricase, or increased excretion of urate by the kidney (uricosurics). Inflammation inhibitors relieve the acute signs and symptoms of gout, usually arthritis. *IL* interleukin, *MSU* monosodium urate, *NSAIDs* nonsteroidal anti-inflammatory drugs

Question 3 What are the main therapeutic strategies to treat gout?

Answer 3 Gout therapy is based on two principal strategies: the control of hyperuricemia and the control of inflammation. Urate-lowering therapies include XO inhibitors and uricosuric. Anti-inflammatory drugs encompass NSAIDs, colchicine, corticosteroids, and IL-1 inhibitors.

References

1. Wallace KL et al (2004) Increasing prevalence of gout and hyperuricemia over 10 years among older adults in a managed care population. J Rheumatol 31(8):1582–1587
2. Mikuls TR et al (2005) Gout epidemiology: results from the UKUK General Practice Research Database, 1990–1999. Ann Rheum Dis 64(2):267–272
3. Choi HK et al (2005) Pathogenesis of gout. Ann Intern Med 143(7):499–516
4. Choi HK et al (2004) Purine-rich foods, dairy and protein intake, and the risk of gout in men. N Engl J Med 350(11):1093–1103
5. Choi HK, Curhan G (2004) Beer, liquor, and wine consumption and serum uric acid level: the Third National Health and Nutrition Examination Survey. Arthritis Rheum 51(6):1023–1029
6. Choi JW et al (2008) Sugar-sweetened soft drinks, diet soft drinks, and serum uric acid level: the Third National Health and Nutrition Examination Survey. Arthritis Rheum 59(1):109–116
7. Smyth CJ, Cotterman CW, Freyberg RH (1948) The genetics of gout and hyperuricaemia; an analysis of 19 families. J Clin Invest 27(6):749–759
8. So A, Thorens B (2010) Uric acid transport and disease. J Clin Invest 120(6):1791–1799
9. Martinon F (2010) Mechanisms of uric acid crystal-mediated autoinflammation. Immunol Rev 233(1):218–232
10. Lu B et al (2012) Novel role of PKR in inflammasome activation and HMGB1 release. Nature 488(7413):670–674
11. Joosten LA et al (2009) Inflammatory arthritis in caspase 1 gene-deficient mice: contribution of proteinase 3 to caspase 1-independent production of bioactive interleukin-1beta. Arthritis Rheum 60(12):3651–3662
12. Narayan S et al (2011) Octacalcium phosphate crystals induce inflammation in vivo through interleukin-1 but independent of the NLRP3 inflammasome in mice. Arthritis Rheum 63(2):422–433
13. Schlesinger N et al (2012) Canakinumab for acute gouty arthritis in patients with limited treatment options: results from two randomised, multicentre, active-controlled, double-blind trials and their initial extensions. Ann Rheum Dis 71(11):1839–1848
14. Schlesinger N et al (2011) Canakinumab reduces the risk of acute gouty arthritis flares during initiation of allopurinol treatment: results of a double-blind, randomised study. Ann Rheum Dis 70(7):1264–1271
15. Schumacher HR Jr et al (2012) Rilonacept (interleukin-1 trap) for prevention of gout flares during initiation of uric acid-lowering therapy: results from a phase III randomized, double-blind, placebo-controlled, confirmatory efficacy study. Arthritis Care Res (Hoboken) 64(10):1462–1470
16. So A et al (2007) A pilot study of IL-1 inhibition by anakinra in acute gout. Arthritis Res Ther 9(2):R28
17. Schumacher HR Jr et al (2008) Effects of febuxostat versus allopurinol and placebo in reducing serum urate in subjects with hyperuricemia and gout: a 28-week, phase III, randomized, double-blind, parallel-group trial. Arthritis Rheum 59(11):1540–1548
18. Becker MA et al (2010) The urate-lowering efficacy and safety of febuxostat in the treatment of the hyperuricemia of gout: the CONFIRMS trial. Arthritis Res Ther 12(2):R63
19. Ganson NJ et al (2006) Control of hyperuricemia in subjects with refractory gout, and induction of antibody against poly(ethylene glycol) (PEG), in a phase I trial of subcutaneous PEGylated urate oxidase. Arthritis Res Ther 8(1):R12

Kidney Stones

Nadine Wunder, Andreas Neisius,
and Glenn Michael Preminger

Introduction

Urinary stone disease is characterized by crystalline depositions, which are classified due to their location in the renal calyces, pelvis, or ureter (urolithiasis) and composition. With its rising prevalence, especially in countries with a high standard of living, urinary stone disease causes a significant health-care burden in the working-age population. High stone recurrence rates demand a proper metabolic workup and efforts to prevent new urinary stone formation.

In this chapter, a compact review of the pathophysiology of stone disease, its metabolic evaluation, and selective medical treatment, which is highly effective in preventing stone recurrence, is provided.

N. Wunder · A. Neisius
Department of Urology, Hospital of the Brothers of Mercy Trier, Medical Campus of the Johannes-Gutenberg University Mainz, Trier, Germany
e-mail: n.wunder@bbtgruppe.de;
a.neisius@bbtgruppe.de

G. M. Preminger (✉)
Division of Urologic Surgery, Department of Urology, Comprehensive Kidney Stone Center, Duke University Medical Center, Durham, NC, USA
e-mail: glenn.preminger@duke.edu

Pathophysiology of Kidney Stones

Epidemiology

Under the influence of geographic, climatic, ethnic, nutritional, and genetical aspects, the prevalence of urinary stone disease varies from 1 to 20% worldwide [1]. Prevalence rates were rising until the turn of the millennium in countries with a high standard of life probably due to dietary and environmental factors, along with enhanced diagnostics. As living conditions have improved, especially in the urban areas of the more developing countries, current data show similar prevalence figures to those previously described in countries with a high standard of living [2].

The prevalence of urinary stone disease had been greater in males (ranging from 8% to 19%) compared to females (ranging from 3% to 5%) [2]. Yet, over the past 30 years, there has been an increasing rate of stone development in women as the male-to-female ratio has decreased from 3.1 to 1.3 to 1.1 [3, 4].

Stone composition and treatment of a metabolic disease determine the risk of new stone formation [5]. Overall, after forming a urinary stone for the first time the risk of recurrence is 26% in 5 years [6]. Half of the patients with urinary stones will develop at least one recurrence during lifetime [7].

E. Lammert, M. Zeeb (eds.), *Metabolism of Human Diseases*,
https://doi.org/10.1007/978-3-031-96019-2_40

Etiology

Stone Composition

The urinary environment in stone patients is conducive to the crystallization of stone-forming salts, due to increased supersaturation of promoters or reduced inhibitor activity (Fig. 1). Promoting ions that participate in high concentration in the stone formation are calcium ions (Ca^{2+}), oxalate, sodium sulfate, and phosphate. Magnesium ions (Mg^{2+}) and citrate are building complexes with calcium ions (Ca^{2+}) and thus inhibiting the crystallization.

The stone composition is the foundation for initiating additional diagnostics and deciding on the right methods for secondary prevention.

About 80% of urinary stones are composed of calcium oxalate with a variable amount of calcium phosphate. Less than 20% of stones are non-calcium calculi. Calcium oxalate stones, calcium phosphate stones, and uric acid stones have a non-infectious origin whereas struvite stones are caused

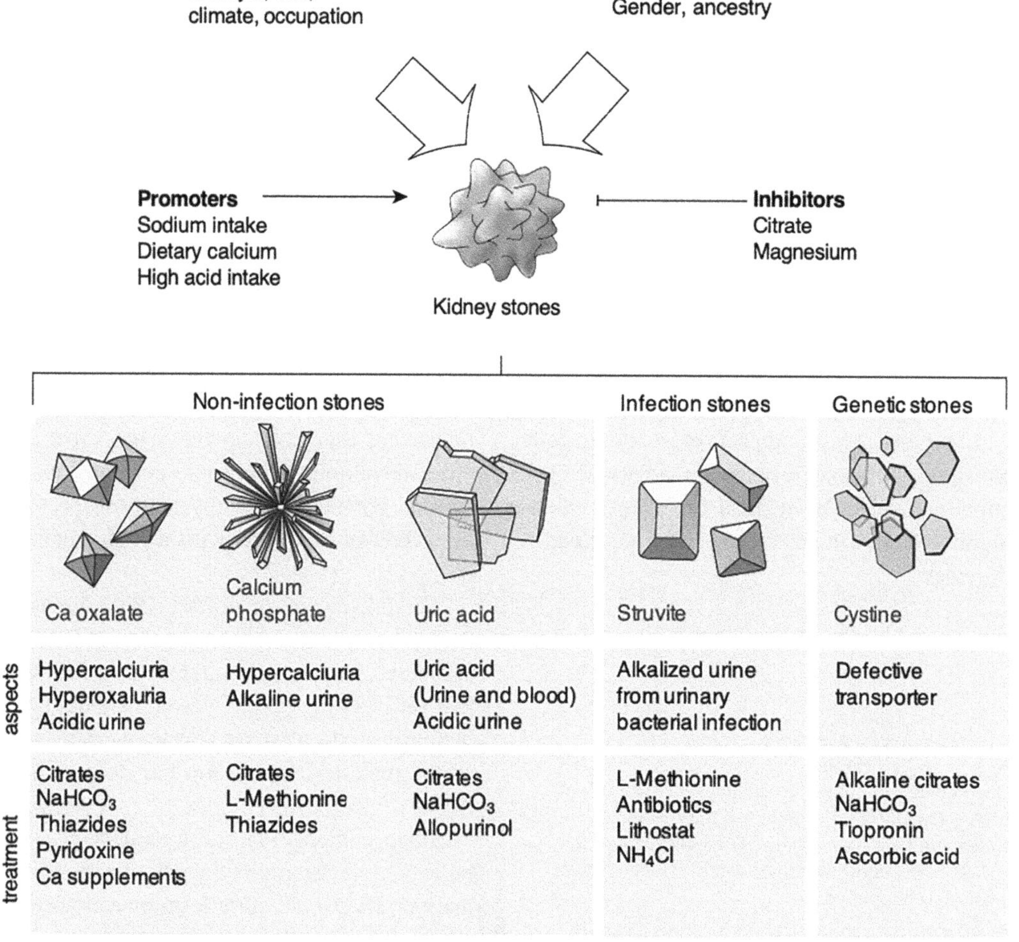

Fig. 1 Factors influencing formation of different stone types, their pathogenesis, and therapy. Risk factors for stone disease generally include lifestyle and genetic/epigenetic contributions. Formation of urinary stones is modulated by promoting or inhibiting metabolic factors (mainly concerning ion composition). Calculi can show different structures and, more importantly, chemical composition and are grouped into noninfectious, infectious, and genetic stones. Causes and/or diagnostic markers for the important subgroups are shown, as well as the most common treatment

by urinary tract infections. Genetic disorders can lead to the formation of cystine stones (Fig. 1). When active compounds of drugs crystallize in the urine or unfavorable changes in the urine composition occur under pharmacological treatment, drug stones can be formed in rare cases [8].

Risk Factors

Risk factors for forming urinary stones can be stratified by metabolic, environmental, genetic, and anatomic abnormalities. In 97% of patients with urinary stone disease, a metabolic or environmental etiology can be found [9], but in rare cases a genetic or anatomic disorder is causal.

Many metabolic risk factors are depicted within the metabolic syndrome (see chapter "Metabolic Syndrome"), which has a high correlation in developing urinary stone disease [10].

Single metabolic risk factors include high calcium (idiopathic hypercalciuria), high oxalate (hyperoxaluria), and high uric acid (hyperuricosuria) in the urine. Low urinary citrate (hypocitraturia) and an abnormal high or low pH-level of the urine (as seen in gouty diathesis) can contribute to the development of urinary stones.

Environmental triggers for stone formation include low urinary volume and dietary conditions that lead to low magnesium ion (Mg^{2+}) as well as high oxalate and sodium concentrations in the urine.

Cystinuria, primary hyperoxaluria, renal tubular acidosis (RTA type I), adenine phosphoribosyltransferase deficiency, and xanthine oxidase deficiency are rare genetic disorders leading to the development of urinary stones.

Some anatomical anomalies causing urinary stasis in the upper urinary tract allow urinary components to crystallize and accumulate to a new stone. Horseshoe kidney, ureterocele, ureteropelvic junction obstruction, ureteral strictures, vesico-uretero-renal reflux, calyceal diverticulum or cysts, and the medullary sponge kidney are common examples for anatomic conditions contributing to stone formation [11].

Critical care must be taken in determining the underlying mechanism of stone formation, when evaluating high-risk patients including children, patients with a family history of forming stones, and patients with chronic diarrheal or gastrointestinal malabsorptive states, mineral and bone disease, urinary tract infections, gout, or chronic kidney disease [11]. High-risk patients should undergo an extensive evaluation even if they are first-time stone formers.

Diagnostics

Imaging

For primary diagnostics, ultrasound is a widespread tool to detect stones in the upper urinary tract without the exposure to radiation in a very reasonable and quick way. Signs of obstruction, which presents as dilatation of the upper urinary tract, can lead to the diagnosis of a ureteral stone (ureterolithiasis).

Low-dose noncontrast-enhanced computed tomography (CT) is considered to be the gold standard for assessing the current stone burden in the entire upper urinary tract [12]. Besides the location and the size of the calculus, CT can provide information about the density of the stone, thereby suggesting the actual stone composition [13].

In cases where computed tomography is not available, an IVU (intravenous urography) or plain abdominal X-rays (KUB = plain kidneys, ureters, bladder—or abdominal tomograms) can be performed alternatively to visualize radiopaque calculi, which are the most common stones.

Metabolism

Stone analysis should be performed to determine the stone composition, which may help to determine the metabolic conditions for stone formation (Fig. 1). Often calcium calculi consist of mixed components, making stone analysis alone a poor option to determine underlying metabolic defects for recurrent stone formation.

Any patient with cystine, struvite, or uric acid stone should undergo a complete metabolic workup. This evaluation consists of analyzing a 24-h urine sample for various stone-forming risk factors. Urine should also be cultured for infection and analyzed for urine sediment [14]. Also the blood should be examined for creatinine, ionized calcium, potassium, sodium, uric acid, C-reactive protein, and the blood cell count [11].

A urine pH-level below 5.5 is a favorable environment in forming uric acid stones. Uric acid stones suggest gouty diathesis or any other cause of purine overproduction (see chapter "Gout").

A urine pH-level above 7.5 promotes the development of calcium phosphate stones. In addition, high urine pH may suggest bacterial growth and thus the development of infection calculi such as struvite stones. Also, the evidence of urea-splitting organisms in the urine culture suggests the presence of urinary tract infection with bacteria that lead to formation of struvite stones.

Cystine is poorly soluble in urine especially in the physiological pH-range and crystallizes spontaneously [11]. Cystinuria can be uncovered with a nitroprusside test, in which cystine reacts with cyanide in nitroprusside to form a colored product indicating the presence of cystine [15].

Calcium phosphate crystals and an elevated blood serum level of ionized calcium suggest primary hyperparathyroidism, as parathyroid hormone (PTH) releases calcium ions (Ca^{2+}) from the bones to increase ionized calcium levels in the blood (see chapter "Teeth and Bones: Overview" under part "Teeth and Bone") promoting the development of calcium phosphate stones in the urinary tract and osteoporosis (see chapter "Osteoporosis").

Calcium oxalate calculi, as the most common type of stone, can occur in a various number of conditions. As around 80–90% of oxalate is synthesized in the liver, the remainder comes from dietary oxalate or vitamin C ingestion as ascorbic acid. A metabolic or environmental increase of oxalate quickly leads to a supersaturation of oxalate in the blood and consequently causes calcium oxalate to crystallize in the urinary system. Hyperoxaluria can be either caused by an increased oxalate synthesis and excessive vitamin C (>2,000 mg/day) or oxalate-rich food ingestions [16]. In clinical practice the main cause of hyperoxaluria is enteric hyperoxaluria caused by chronic bowel diseases and/or bowel surgery and the development of malabsorption.

Treatment of Kidney Stones

Treatment of the Stone

Depending on the location of the stone, its size and symptoms (i.e., pain, obstruction, infection) therapeutic decisions are made individually, whether the stone can pass the urinary tract on its own conservatively or surgical intervention is needed.

The least invasive methods to treat calculi are chemolysis and shock wave lithotripsy (SWL). Oral alkali therapy with alkaline citrate, usually in the form of potassium citrate, or sodium bicarbonate can be used to dissolve uric acid stones. Currently, only uric acid stones are amenable to dissolution [11]. Shock wave lithotripsy (SWL) is another minimally invasive treatment modality for small or moderate-sized stones. Externally generated shock waves can disintegrate a stone into small fragments, which can pass through the urinary tract spontaneously.

Nowadays there are many equally effective minimally invasive surgical options to remove symptomatic renal or ureteral calculi with significantly decreased patient morbidity. The choice of an optimal stone removal procedure is best determined by stone size, composition, and location, combined with urinary tract anatomy. In general, endourological approaches such as ureterorenoscopy (URS) are preferred for stones <20 mm whereas percutaneous approaches such as percutaneous nephrolithotomy (PNL) are given preference to stones >20 mm [11].

Treatment of the Metabolism: Secondary Prevention

As stone recurrence rates may be high, continued medical stone management has been designed to suppress new stone formation or growth of existing calculi.

After the formation of the first stone, the patient's medical history and stone composition determine their risk group for recurrence of urinary stone disease. Baseline metabolic screening and general preventive measures are recommended for all patients, while high-risk patients require extensive metabolic evaluation, targeted therapy, and close monitoring, particularly if they are noncompliant with treatment [17].

A high fluid intake (at least 3 L per day), a balanced diet, and lifestyle optimization to control general risk factors are three general actions that can be taken to prevent stone development [11].

Therefore patients with low urine volumes must increase their urine output to 2–2.5 L/day, which conservatively can be managed by an increased fluid intake of 3 L/day [18], preferably water.

Also, a high-fiber diet that includes many fruits and vegetables and is limited to a maximum animal protein consumption of 1 g/kg bodyweight/day can help to suppress stone formation [19]. In addition, all patients are counseled to moderate their intake of sodium chloride (NaCl) to a maximum of 2 g/day [7] as high urinary sodium causes higher calcium ion (Ca^{2+}) excretion in the kidneys (see chapter "Kidney: Overview" under the part "Kidney") and a lower urinary citrate concentrations due to loss of bicarbonate [11]. For the same reason the calcium intake should not exceed 1.2 g/day [7]. Rhubarb, spinach, red beet, nuts, or other foods that have a high load of oxalate should be consumed in a moderate way to minimize hyperoxaluria.

It has been demonstrated that single-stone formers treated conservatively and solely with the avoidance of dietary excess and increased fluid intake can reduce stone recurrence by about 50% [18].

Lifestyle optimization that includes measures to avoid obesity [20], metabolic syndrome [21], and diabetes mellitus [22] by retaining a normal body mass index (BMI) level, being adequately physically active, and remaining on a balanced nutrition helps to lower the prevalence of urinary calculi.

Medical therapy is initialized once lifestyle and dietary optimization and fluid increase are not sufficient. The specific medical treatment is directed at the underlying metabolic risk factors that contribute to recurrent stone formation.

Prevention of Calcium Stones Due to Hypercalciuria

Thiazide diuretics are the recommended medication to treat patients with hypercalciuria as they limit the urinary calcium ion (Ca^{2+}) excretion and thus reduce the formation of calcium-containing calculi. By inhibiting the sodium-chloride symporter (Na^+-Cl^- cotransporter), sodium ions (Na^+) in the renal tubular cells are decreased and the basolateral sodium-calcium exchanger (Na^+/Ca^{2+} exchanger) is activated. Subsequently calcium ions (Ca^{2+}) are reduced in the cells, which increases the calcium reabsorption from the renal tubule.

Besides potential side effects of hypotension and hyperuricemia, secondary hypokalemia and hypocitraturia can occur via similar mechanisms. Thus potassium citrate is typically administered as a potassium supplement to patients taking thiazide diuretics [23].

Prevention of Calcium Stones Due to Hypocitraturia

Although hypocitraturia often coexists with hypercalciuria or hyperuricosuria, alkaline citrate is an efficient and safe treatment for patients with hypocitraturia [24]. By forming more soluble calcium-citrate complex, alkaline citrate decreases the effective concentration of free calcium ions (Ca^{2+}) and therefore the saturation and crystallization of urinary calcium oxalate.

Potassium citrate is preferred over sodium citrate, since increased sodium intake may also increase urinary calcium excretion [25]. Yet in patients with renal insufficiency, or with a high risk for hyperkalemia, sodium citrate or sodium bicarbonate may be administered [23].

For patients with distal renal tubular acidosis, the metabolic acidosis and hypokalemia can be corrected with potassium citrate, as citrate is partly metabolized to bicarbonate, which corrects the acidosis and thus reduces the urinary calcium ion (Ca^{2+}) excretion [26].

Prevention of Oxalate Stones Due to Hyperoxaluria

As described above, a general strategy to prevent formation of oxalate stones is dietary reduction of oxalate. The intestinal malabsorption of fat, as seen in patients with Crohn's disease, often presents with increased levels of oxalate in blood and urine after intestinal resection of jejunoileal bypass what causes enteric hyperoxaluria. The loss of fatty acids is combined with the loss of calcium and therefore disturbs the normal complex formation between oxalate and calcium in the bowel. Consequently, the increased oxalate absorption is what triggers the formation of oxalate stones.

For that reason, oral administration of calcium supplements has been recommended to enable calcium oxalate complexation in the intestine [11]. As many patients with Crohn's disease need to avoid calcium products due to intolerance to any calcium-containing nutriments, hyperoxaluria can also be reduced by decreasing the oxalate synthesis in the liver via pyridoxine (vitamin B6) that also can be administered orally [27].

Prevention of Uric Acid Stones

As an acidic urine, pH <5.5 is the primary cause of uric acid stones. The major goal in uric acid stone management is to decrease urinary saturation of uric acid by raising the urine pH-level to 6.5–7.0, providing alkali therapy. As the solubility for uric acid increases tenfold with a pH change from 5.0 to 7.0, urinary alkalization is by far the most important factor. Hyperuricemia or hyperuricosuria is rarely the cause of uric acid stone formation [11].

Prevention of Struvite Stones Due to Infection

Struvite stones occur in the setting of urinary tract infections. Some bacteria contain urease that splits urea into carbon dioxide and ammonia and increases the urinary pH-level [28, 29]. Chronic antibiotic treatment and, most importantly, complete surgical stone removal, along with urinary acidification, can significantly reduce the risk of new infectious stone formation. L-methionine [30], ammonium chloride [31], and urease inhibition via acetohydroxamic acid [32] are some drugs that can properly acidify the urine.

Prevention of Cystine Stones Due to Cystinuria

Main therapeutic options avoiding cystine crystallization are a reduced cystine excretion, an increased urinary pH-level >7.0 by alkaline citrates or sodium bicarbonate, and an increased urinary volume by elevating the hydration up to >3.5 L/day [11]. Tiopronin is an agent that splits disulfide bindings in cystine and forms a tiopronin–cysteine complex. As this tiopronin–cysteine complex has a higher solubility than cystine, it reduces the availability of cystine for cystine stone formation [33]. However, patients taking tiopronin may demonstrate extensive side effects, such as nephritic syndrome, which is why it is recommended in patients with a cystine excretion >3 mmol/day or when other options are insufficient.

Perspectives

Recurrence rates for stone formers can be high. Even though minimally invasive surgical stone removal is a successful solution and standard to treat already existing stones, selective medical therapy is an important adjunct to manage primary and recurrent stone disease. The need for repetitive surgical procedures can be significantly reduced by an effective prophylactic program that includes increased fluid intake, dietary modification, and selective medical therapy. Current research trends aim on understanding metabolic risk factors and dietary behavior to avoid urinary stone disease.

Questions and Answers

Question 1 What are 80% of kidney stones composed of?

Answer 1 Calcium oxalate (with a variable amount of calcium phosphate).

Question 2 What are the major tools to diagnose kidney stones?

Answer 2 Ultrasound and computed tomography (CT) of the upper urinary tract/kidney.

Question 3 What are the least invasive treatment methods?

Answer 3 Chemolysis and shock wave lithotripsy.

References

1. Trinchieri A (2003) Epidemiology. In: Segura JW, Pak CY, Preminger GM (eds) Stone disease. Health Publications, Paris
2. Trinchieri A (2008) Epidemiology of urolithiasis: an update. Clin Cases Miner Bone Metab 5(2):101–106
3. Lieske JC, Pena de la Vega LS, Slezak JM, Bergstralh EJ, Leibson CL, Ho KL et al (2006) Renal stone epidemiology in Rochester, Minnesota: an update. Kidney Int 69(4):760–764
4. Tundo G, Khaleel S, Pais VM Jr (2018) Gender equivalence in the prevalence of nephrolithiasis among adults younger than 50 years in the United States. J Urol 200(6):1273–1277
5. Keoghane S, Walmsley B, Hodgson D (2010) The natural history of untreated renal tract calculi. BJU Int 105(12):1627–1629
6. Ferraro PM, Curhan GC, D'Addessi A, Gambaro G (2017) Risk of recurrence of idiopathic calcium kidney stones: analysis of data from the literature. J Nephrol 30(2):227–233
7. Hesse A, Brandle E, Wilbert D, Kohrmann KU, Alken P (2003) Study on the prevalence and incidence of urolithiasis in Germany comparing the years 1979 vs. 2000. Eur Urol 44(6):709–713
8. Matlaga BR, Shah OD, Assimos DG (2003) Drug-induced urinary calculi. Rev Urol 5(4):227–231
9. Delvecchio FC, Preminger GM (2003) Medical management of stone disease. Curr Opin Urol 13:229–233
10. Taylor EN, Stampfer MJ, Curhan GC (2005) Obesity, weight gain, and the risk of kidney stones. JAMA 293(4):455–462
11. Skolarikos A, Neisius A, Petřík A, Somani B, Tailly T, Gambaro G et al (2023) In: European Association of Urology (ed) EAU guidelines on urolithiasis. EAU Guidelines Office, Arnhem
12. Niemann T, Kollmann T, Bongartz G (2008) Diagnostic performance of low-dose CT for the detection of urolithiasis: a meta-analysis. AJR Am J Roentgenol 191(2):396–401
13. Zilberman DE, Ferrandino MN, Preminger GM, Paulson EK, Lipkin ME, Boll DT (2010) In vivo determination of urinary stone composition using dual energy computerized tomography with advanced post-acquisition processing. J Urol 184(6):2354–2359
14. Williams JC Jr, Gambaro G, Rodgers A, Asplin J, Bonny O, Costa-Bauza A et al (2021) Urine and stone analysis for the investigation of the renal stone former: a consensus conference. Urolithiasis 49(1):1–16
15. Finocchiaro R, D'Eufemia P, Celli M, Zaccagnini M, Viozzi L, Troiani P et al (1998) Usefulness of cyanide-nitroprusside test in detecting incomplete recessive heterozygotes for cystinuria: a standardized dilution procedure. Urol Res 26(6):401–405
16. Smith LH, Fromm H, Hofmann AF (1972) Acquired hyperoxaluria, nephrolithiasis, and intestinal disease. Description of a syndrome. N Engl J Med 286(26):1371–1375
17. Skolarikos A, Somani B, Neisius A, Jung H, Petrik A, Tailly T et al (2024) Metabolic evaluation and recurrence prevention for urinary stone patients: an EAU guidelines update. Eur Urol 86(4):343–363
18. Hosking DH, Erickson SB, Van den Berg CJ, Wilson DM, Smith LH (1983) The stone clinic effect in patients with idiopathic calcium urolithiasis. J Urol 130(6):1115–1118
19. Dussol B, Iovanna C, Rotily M, Morange S, Leonetti F, Dupuy P et al (2008) A randomized trial of low-animal-protein or high-fiber diets for secondary prevention of calcium nephrolithiasis. Nephron Clin Pract 110(3):c185–c194
20. Siener R, Glatz S, Nicolay C, Hesse A (2004) The role of overweight and obesity in calcium oxalate stone formation. Obes Res 12(1):106–113
21. Chang CW, Ke HL, Lee JI, Lee YC, Jhan JH, Wang HS et al (2021) Metabolic syndrome increases the risk of kidney stone disease: a cross-sectional and longitudinal cohort study. J Pers Med 11(11):1154
22. Geraghty R, Abdi A, Somani B, Cook P, Roderick P (2020) Does chronic hyperglycaemia increase the risk of kidney stone disease? Results from a systematic review and meta-analysis. BMJ Open 10(1):e032094
23. Pearle MS, Goldfarb DS, Assimos DG, Curhan G, Denu-Ciocca CJ, Matlaga BR et al (2014) Medical management of kidney stones: AUA guideline. J Urol 192(2):316–324
24. Pak CY, Fuller C (1986) Idiopathic hypocitraturic calcium-oxalate nephrolithiasis successfully treated with potassium citrate. Ann Intern Med 104(1):33–37

25. Preminger GM, Sakhaee K, Pak CY (1988) Alkali action on the urinary crystallization of calcium salts: contrasting responses to sodium citrate and potassium citrate. J Urol 139(2):240–242
26. Preminger GM, Sakhaee K, Skurla C, Pak CY (1985) Prevention of recurrent calcium stone formation with potassium citrate therapy in patients with distal renal tubular acidosis. J Urol 134(1):20–23
27. Duffey BG, Alanee S, Pedro RN, Hinck B, Kriedberg C, Ikramuddin S et al (2010) Hyperoxaluria is a long-term consequence of Roux-en-Y Gastric bypass: a 2-year prospective longitudinal study. J Am Coll Surg 211(1):8–15
28. Griffith DP, Musher DM (1973) Prevention of infected urinary stones by urease inhibition. Investig Urol 11(3):228–233
29. Kramer G, Klingler HC, Steiner GE (2000) Role of bacteria in the development of kidney stones. Curr Opin Urol 10(1):35–38
30. Jarrar K, Boedeker RH, Weidner W (1996) Struvite stones: long term follow up under metaphylaxis. Ann Urol (Paris) 30(3):112–117
31. Wall I, Tiselius HG (1990) Long-term acidification of urine in patients treated for infected renal stones. Urol Int 45(6):336–341
32. Williams JJ, Rodman JS, Peterson CM (1984) A randomized double-blind study of acetohydroxamic acid in struvite nephrolithiasis. N Engl J Med 311(12):760–764
33. Lindell A, Denneberg T, Jeppsson JO (1995) Urinary excretion of free cystine and the tiopronin-cysteine-mixed disulfide during long term tiopronin treatment of cystinuria. Nephron 71(3):328–342

Reproductive System

Reproductive System: Overview

Matthew C. H. Rohn, Lyndsey T. Ellis,
and Danny J. Schust

Anatomy and Physiology of the Reproductive System

The gonads are the end organs of reproduction, represented by the ovary in females and the testis in males (Fig. 1). These organs contain the germ cells responsible for the production and release of gametes (oocytes in the ovary and spermatozoa in the testis), the fundamental cells central to human reproduction. The number of oocytes contained within the ovary peak in utero and steadily decline thereafter. At puberty the number declines further, with the production of mature oocytes that are released with each menstrual cycle, continuing until menopause. The testes produce spermatozoa, starting from the age of puberty and typically continuing until death. In addition to the formation of gametes, testes and ovaries produce sex hormones that affect the physiology of many, if not all, nonreproductive organs. The function of the gonads is mainly regulated by the hypothalamic–pituitary axis (Fig. 1). The human

M. C. H. Rohn (✉) · D. J. Schust
Division of Reproductive Endocrinology and
Infertility, Department of Obstetrics and Gynecology,
Duke University, Durham, NC, USA
e-mail: matthew.rohn@duke.edu;
danny.schust@duke.edu

L. T. Ellis
OB-GYN Services 17 Case Street Norwich,
Norwich, CT, USA

breast is a secondary reproductive organ that serves to feed the offspring in the first months to years of life.

Reproductive Organ-Specific Metabolic Pathways

Steroids are produced by several tissues, including the adrenal cortex, the gonads, the placenta, the brain, and the peripheral tissues, such as adipose tissue. All steroid hormones are derivatives of cholesterol, and production of sex steroids requires the expression of enzymes within the steroidogenic pathway (Fig. 1) [1].

The testes mainly produce androgens, such as androstenediol, androstenedione, testosterone, dihydrotestosterone (DHT), and small amounts of dehydroepiandrosterone (DHEA); these androgens are released from the Leydig cells, while the Sertoli cells in the testis convert testosterone to small amounts of estrogen, such as estradiol (E2), required for spermatogenesis. Estrogens in males are produced at a much lower rate than in females.

The ovaries produce estrogens, such as estrone (E1) and E2; progesterones, that is, progesterone (P4) and 17α-hydroxy-progesterone; and androgens (similar to the testis, except for dihydrotestosterone). Androgen production in females is much lower than in males. For example, levels of circulating testosterone in females are about

© The Author(s), under exclusive license to Springer Nature Switzerland AG 2026
E. Lammert, M. Zeeb (eds.), *Metabolism of Human Diseases*,
https://doi.org/10.1007/978-3-031-96019-2_41

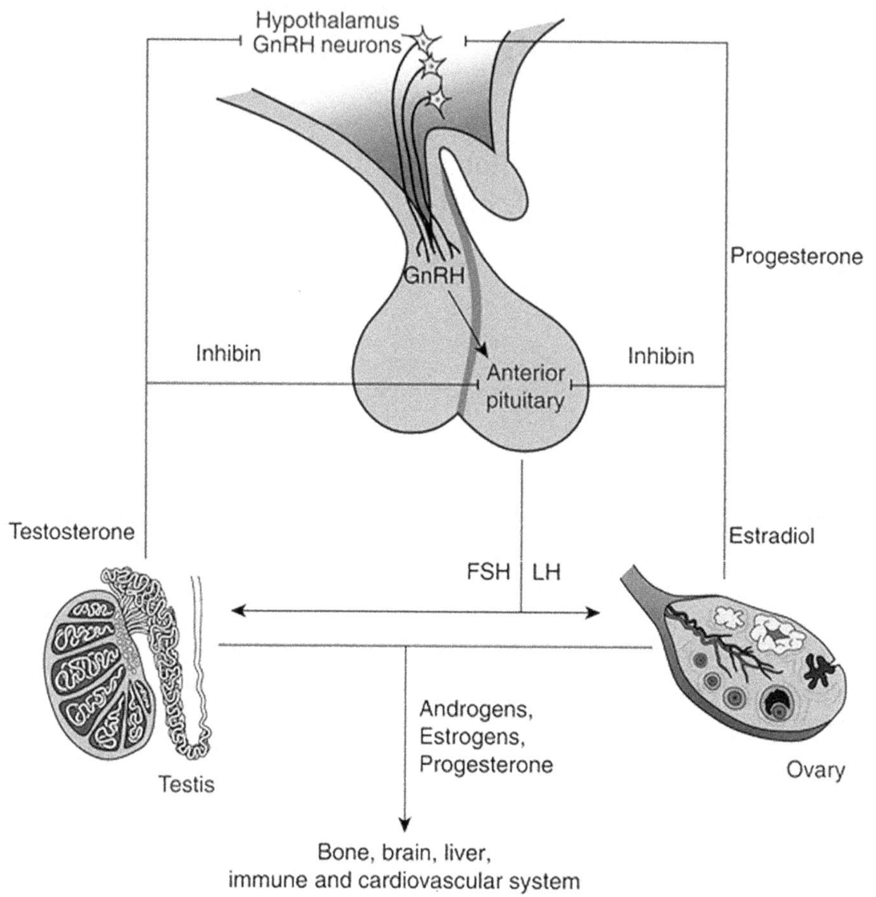

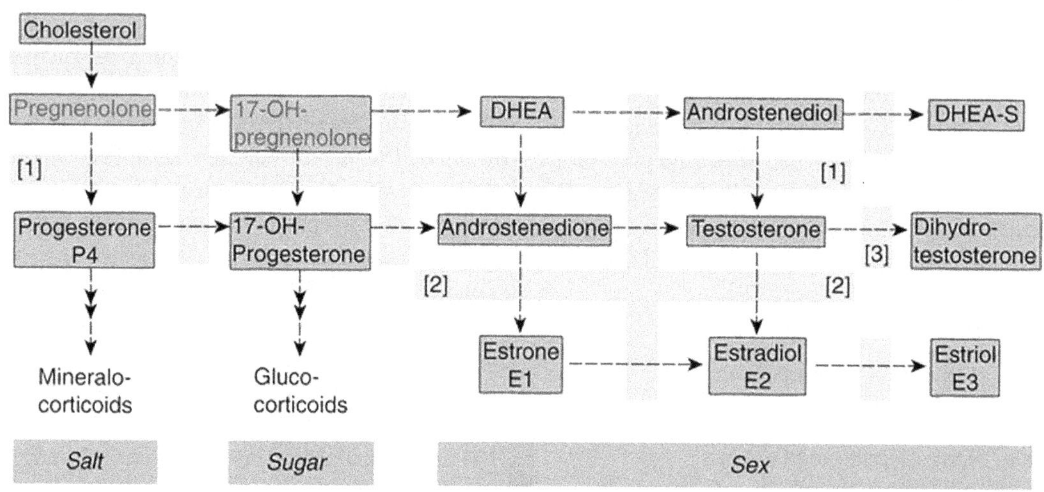

Fig. 1 Production pathways of sex steroids and the hypothalamic–pituitary–gonadal axis. *Upper Part*: Regulation of sex steroid production by the hypothalamus–pituitary axis and feedback mechanisms. *FSH* follicle-stimulating hormone, *GnRH* gonadotropin-releasing hormone, *LH* luteinizing hormone. *Lower Part*: Steroid hormone biosynthesis showing important intermediates and enzymes. *Gray boxes* indicate enzymes [1]. 3β- hydroxysteroid dehydrogenase / $\Delta^5\Delta^4$ isomerase; [2] aromatase; [3] 5α-reductase. Further, enzymes and molecules in *light gray* are not named in the figure but included for completeness. *DHEA* dehydroepiandrosterone, *DHEA-S* dehydroepiandrosterone sulfate

1/8th that of males. DHEA is found in fairly high concentrations in the circulation of females, mostly in its sulfated form, DHEA-S. Sulfation of DHEA occurs in the adrenal glands, the liver, and the intestines. The major function of DHEA is that of a precursor to other androgens and estrogens (Fig. 1). The weak estrogen, estriol (E3), is produced only during pregnancy, by the placenta.

The adrenal glands generally produce sex steroids as byproducts of glucocorticoids and mineralocorticoids in fairly small amounts when compared to the gonads. One important physiologic exception is that approximately half of the circulating androgens in females are of adrenal origin.

Inside-In: Signals in the Reproductive Organs

Development and release of each type of germ cell require delicately balanced interactions between several gonad-specific cell types (testicular Sertoli and Leydig cells in males and ovarian granulosa and theca cells in females) and the sex steroids produced by these cells.

Theca cells surrounding the ovarian follicle produce androgens, which are converted to estrogens by the granulosa cells of the ovarian follicle. Ovarian estradiol (E2) and its feedback interactions with the hypothalamus and pituitary gland direct the development of the ovarian follicle. Peptide hormones produced by the ovary (e.g., inhibin, activin, and follistatin) also participate in positive and negative regulation of the hypothalamic–pituitary–gonadal axis and function in controlling menstrual cyclicity.

Testicular androgens are mostly produced by the Leydig cells under the influence of pituitary luteinizing hormone (LH). The Sertoli cells of the testis support sperm proliferation and maturation and can convert androgens to estrogens via aromatase or to dihydrotestosterone via 5α-reductase (Fig. 1). They also secrete inhibin to regulate LH release (Fig. 1 and below). Most dihydrotestosterone, however, is synthesized locally in androgen-responsive peripheral tissues (such as hair follicle cells).

Outside-In: Hypothalamus/ Pituitary/Reproductive Organ Axis

The reproductive system is regulated by a series of activating and inhibiting hormone-mediated signals, starting centrally at the hypothalamus and pituitary, and extending peripherally to the testes and ovaries (Fig. 1).

The hypothalamus is divided into several types of nuclei that generate neural signals and have neuroendocrine capabilities. Primarily involved in gonad regulation are those nuclei that integrate olfactory, visual, emotional, and other signals The neuroendocrine signals consist of a set of peptide hormones that are either stored in the posterior lobe of the pituitary (and later released into the blood) or target the anterior lobe of the pituitary (Fig. 1). The primary hormones involved in regulating the gonads are released from the anterior pituitary in response to signals from the hypothalamus, which is itself regulated by those secreted by the testes and ovaries. This signaling circuit is the basis by which human reproduction is regulated, the details of which are discussed below.

Signals from the Hypothalamus

Gonadotropin-releasing hormone (GnRH) is released in a pulsatile fashion from the hypothalamus and stimulates gonadotropic cells in the anterior pituitary to produce follicle-stimulating hormone (FSH) and luteinizing hormone (LH). These gonadotropins are secreted by the same cell type; beyond the stimulation of FSH and LH release, the hypothalamus also regulates, via the pituitary, other reproductive systems such as the development of breast tissue and milk production and thyroid activity. Dopamine from the hypothalamus, also called prolactin inhibitory factor, suppresses prolactin production by lactotroph cells in the pituitary. Pituitary lactotrophs are also regulated by several other factors. Thyrotropic cells in the pituitary are triggered by thyrotropin-releasing hormone (TRH) from the hypothalamus to produce thyroid-stimulating hormone (TSH), which in turn promotes thyroid activity. Oxytocin and vasopressin (also known as antidiuretic hormone, ADH) are synthesized by the hypothalamus and stored in the posterior pituitary (see chapter "Kidney: Overview"). Oxytocin acts on uterine smooth muscle to cause contractions during labor, affects myoepithelial cells in the breast allowing for milk letdown, and has an important role in the reward processes involved in maternal-infant bonding.

Signals from the Pituitary Gland

Pulsatile release of follicle-stimulating hormone (FSH) and luteinizing hormone (LH) from the pituitary act on the ovaries to control the menstrual cycle. FSH induces ovarian granulosa cell proliferation, increases aromatase activity, and promotes recruitment of the follicular unit. Increases in aromatase activity are associated with increases in estradiol (E2) production. As E2 increases, FSH stimulates increased LH responsiveness. Relative abundance of E2 as compared to inhibin suppresses FSH release at the level of the pituitary (Fig. 1). Elevated and sustained levels of E2 cause a surge in LH [4]. This, in turn, results in release of a selected oocyte from the ovary (ovulation) and transformation of granulosa cells into the secretory luteinized cells of the corpus luteum. Luteal cells make large amounts of progesterone (P4), E2, and inhibin. Increased P4 and inhibin levels prevent E2 from stimulating another LH surge, and high E2 and P4 levels reduce the frequency of the GnRH pulse favoring LH over FSH secretion.

If pregnancy does not occur, these LH levels are not, however, sufficient to support the corpus luteum (for very long). As P4, E2, and inhibin levels drop, FSH levels rise. Menstruation begins, and the cycle repeats. In case of a pregnancy, luteal P4 supports the pregnancy until hormonal support is taken over by the placenta (typically at 7–9 weeks of gestation). The cyclic production of ovarian E2 and P4 directs the growth and differentiation of the uterine endometrium to prepare it for embryo implantation [5].

FSH and LH are similarly important in the regulation of the testis. FSH and LH increase the proliferation of Leydig cells, and LH upregulates 3β-hydroxysteroid dehydrogenase, which is responsible for the last step in testosterone formation (Fig. 1) [6].

Prolactin controls the initiation of lactation. In contrast to other pituitary hormones, it is not negatively regulated by classic feedback loops but rather by local autocrine and paracrine factors, neurotransmitters, and by peripherally produced steroid hormones. Ovarian E2 and pituitary thyrotropin-releasing hormone (TRH) are strong stimuli for prolactin production.

Growth hormone (GH), insulin-like growth factors (IGFs), IGF-binding proteins (IGFBPs), and IGF receptors of the somatotropic axis enhance ovarian steroidogenesis.

Inside-Out: Signals from Reproductive Organs Affecting Other Organs

The sex steroids result in several gender-specific health risks. The local and peripheral effects of steroid hormones depend on the presence of specific steroid hormone receptors in target tissues. These receptors are present in the cytoplasm or nucleus of the target cell and are expressed throughout the body. Upon ligand binding, steroid–receptor complexes commonly act as transcription factors by targeting steroid response elements that, in turn, regulate the expression of a remarkably wide variety of genes.

The presence of 5α-reductase in peripheral tissues allows for the local conversion of testosterone to the more potent dihydrotestosterone (DHT). While locally elevated testosterone levels direct the development of the male internal reproductive structures during fetal development, DHT directs appropriate development of the penis and scrotum in the fetus and masculinization at puberty in boys (increased penile length, testicular size, and growth of axillary, pubic, chest, abdominal, and facial hair). Estrogens and progesterone (P4) have little effect on the development of female internal or external genitalia during fetal development, but they are essential to pubertal development (breast and bone growth) and the onset of menstruation. Excesses in circulating androgens act on hair follicles and may lead to hirsutism and acne, an effect that is often most notable in females.

As a multitude of targets exist, female steroid hormone excess and deficiency, as they occur during pregnancy and menopause, respectively, or in pathological states, affect a vast array of tissue types (Fig. 2).

Pregnancy

Marked increases in circulating ovarian estrone (E1), estradiol (E2), and progesterone (P4) and placental estriol (E3) and P4 during pregnancy dominate the systemic steroid effectors in maternal blood. Estrogens and P4 increase maternal blood volume and cardiac output to supply the developing fetus with blood while maintaining or lowering maternal blood pressure. E1, E2, and E3 can all exert these effects via estrogen receptor binding; however, E3 comprises approximately 80% of circulating estrogens during pregnancy and therefore may exert the most significant effects. Elevated placental estrogens increase the production of angiotensinogen (by the liver), and estrogens and P4 increase release of renin (from the kidney), thus increasing the level of angiotensin and subsequently aldosterone; this promotes sodium and water retention (see chapter "Kidney: Overview" under the Part "Kidney"). Although the maintenance of maternal blood pressure in the face of increased blood volume is largely the result of the lower maternal responsivity to angiotensin II during pregnancy, the elevated levels of circulating P4 accentuate this adaptation by mediating smooth muscle relaxation throughout the gravid female's body. This latter effect of P4 also results in delayed gastric and gallbladder emptying, reduced bowel motility, and relaxation of the lower esophageal sphincter in the pregnant females, often experienced by the gravid female as reflux, nausea, vomiting, and/or constipation.

P4-related bronchial and tracheal smooth muscle relaxation may improve asthma symptoms in pregnancy [7, 8]. Other respiratory effects of elevated maternal P4 include an increase in tidal volume (lung volume), minute ventilation (the volume of gas inhaled or exhaled per minute), and respiratory rate (breathing frequency); this results in an overall decrease in arterial CO_2 gas tension via central nervous system changes that increase sensitivity to CO_2 [5].

Estrogens upregulate hepatic synthesis of hormone-binding proteins and several clotting factors promoting hypercoagulability and an increase in total circulating thyroid hormones. P4 and estrogens exert effects on maternal immune function causing susceptibility to certain viral infections (e.g., varicella) but allowing tolerance against the semi-allogenic fetus. Pregnant females with antibody-mediated autoimmune diseases often note an exacerbation of symptoms, while those with T-cell-based inflammatory conditions may experience improvement [6].

Menopause

At menopause, there is a cessation of ovarian follicular development with resultant marked decreases in circulating estrogen, inhibin, and progesterone (P4) levels. Ovarian androgen production remains relatively intact. This relative hormone deficiency has several physiologic effects. Decreased estrogen and inhibin release the negative feedback on the hypothalamus and pituitary gland (Fig. 1), and consequently follicle-stimulating hormone (FSH) levels rise. Alterations in these hormones have been linked to the hot flashes, insomnia, and depressed mood that are commonly found in peri- and early postmenopausal females. Estrogen deficiency leads to tissue atrophy and loss of elasticity in the breast, vagina, and skin. As estrogen antagonizes the effects of parathyroid hormone (PTH) on calcium mobilization in bone (see chapter "Teeth and Bones: Overview" under the part "Teeth and Bones"), and

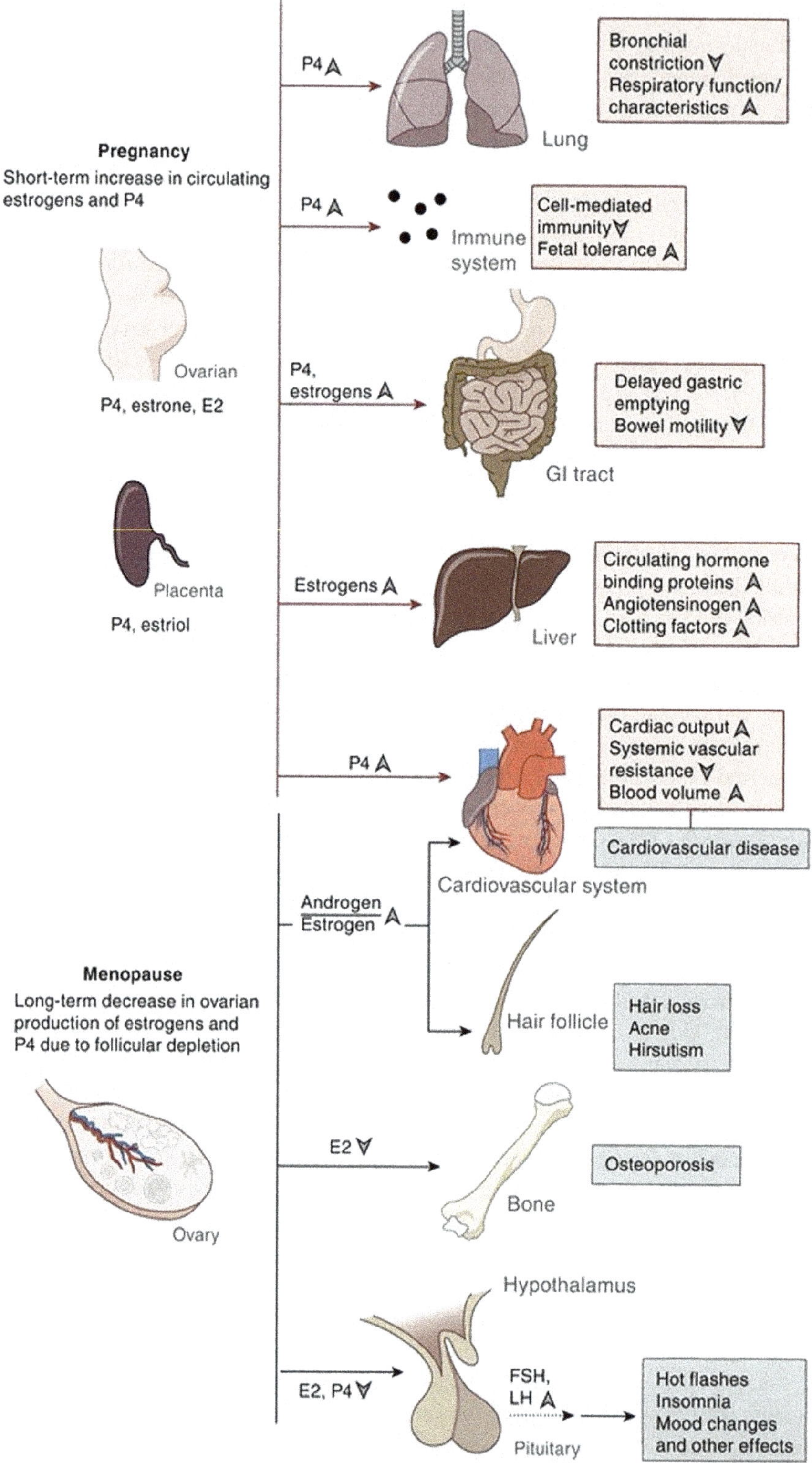

Fig. 2 Effects of female steroid hormones (exemplified by changes in these hormones during pregnancy and menopause). *P4* progesterone, *E2* estradiol

since estrogen deficiency increases osteoclast activity, postmenopausal females are at increased risk for osteoporosis as compared to premenopausal females (see chapter "Osteoporosis").

Estrogen increases circulating high-density lipoproteins (HDL) and decreases low-density lipoproteins (LDL) (see chapter "Hyperlipidemia"). Moreover, it decreases endothelial production of endothelin-1, a potent vasoconstrictor (see chapter "Blood Vessels: Overview" under the part "Blood Vessels"). These two protective cardiovascular effects of estrogens may help to explain the lower rates of cardiovascular disease seen in females when compared to age-matched males. Reduced estrogens, in combination with a relatively intact androgen production, cause a gradual loss of this protection after menopause. Interestingly, androgens are converted to dihydrotestosterone in hair follicles (see above), and continued production of this potent androgen often results in hirsutism and some degree of hair loss on the scalp of postmenopausal females. Adipose tissue is an important source of the enzyme aromatase that converts androgens to estrogen. Obesity can therefore increase levels of circulating estrogens in postmenopausal females, somewhat protecting these females from the effects of decreased ovarian estrogens.

Male reproductive hormones have equally dramatic effects on the vast array of tissues that express androgen receptors. For example, androgen receptors are expressed in skeletal muscle, and their activation stimulates the myogenesis responsible for the increased muscle mass noted in postpubertal males when compared to females [9]. The elevated levels of circulating androgens in males combine with genetic predisposition to make baldness much more common in males than in females (see above). Androgens increase bone mass through receptor-mediated direct and indirect effects on osteoclasts. Their effects on cardiovascular disease are complex, but androgens certainly promote less favorable lipid profiles by increasing LDL and triglyceride levels and decreasing HDL levels (see chapter "Hyperlipidemia") [10]. The prostate gland is unique among the male internal reproductive tract constituents in that dihydrotestosterone, rather than testosterone, is required for its development, growth, and maintenance. The nuclear androgen receptors in the prostate have a higher affinity for dihydrotestosterone, so adequate 5α-reductase activity is essential for prostate growth. Males lacking this enzyme have a poorly developed prostate gland and essentially no risk for subsequent prostate hypertrophy or prostate cancer (see chapter "Prostate Cancer"). Like females, males experience reproductive aging, termed andropause, although its onset and progression is less predictable and more gradual. Circulating testosterone levels decrease with age in males, thus increasing the risk of osteopenia and osteoporosis (see chapter "Osteoporosis"), decreased muscle mass, erectile dysfunction, and impaired sperm parameters [11]. Testicular Leydig cells also become less responsive to luteinizing hormone (LH) with aging. In response, LH levels increase to maintain androgen production (see above) but also increase the ratio of testicular estrogen to androgen secretion. In overweight males, circulating estradiol (E2) levels are increased. Estrogen-stimulated increases in hepatic sex hormone-binding globulin production effectively decrease levels of circulating free testosterone in these aging males.

Anatomy and Physiology of the Human Breast

The human breast is comprised of glandular and ductal tissues embedded in adipose tissue. Alveolar glandular tissues in the 15–20 lobes of the adult breast dump secretions into a converging series of excretory and lactiferous ducts that terminate in approximately 15 distinct orifices on the nipple surface [12]. These ducts are lined by stratified squamous epithelium.

Tissue-Specific Metabolic Pathways of the Breast

The breast is the only tissue that can secrete a fully life-sustaining product. Human milk is a fat emulsion in liquid phase that contains over 100 distinct substances. The main constituents are carbohydrates, including lactose (7%); fat (3–5%); proteins, including casein, α-lactalbumin, lactoferrin, immunoglobulin A, lysozyme, and albumin, (1%); and minerals (0.2%) [13]. Mature breast milk provides all of the essential components to support the growth needed in infancy.

Outside-In: Signaling of the Breast

At birth, the breast consists mainly of primitive ductal tissues, which remain mostly quiescent until the onset of puberty. In females, pubertal elevations in circulating ovarian estrogens and progesterone (mainly estradiol and progesterone) stimulate the resumption of growth and differentiation of the rudimentary ducts. The elevations in serum progesterone (P4) that accompany maturation of the hypothalamic–pituitary–ovarian axis to allow ovarian cyclicity also maintain ductal and initiate glandular growth and differentiation in the breast. Mammary development continues for several years after menarche but will not be fully complete until late in the first trimester of a woman's initial pregnancy. All aspects of mammary growth are supported by growth hormone (GH) and adrenal steroids.

The initiation of milk production is largely controlled by pituitary prolactin [14]. Serum levels of prolactin are elevated during the last trimester of pregnancy, but milk production is inhibited until after delivery by simultaneous elevations in serum estrogen. Oxytocin drives milk ejection. Suckling and the stimulation of additional sensorineural pathways control pituitary oxytocin secretion. Although basal serum prolactin levels return to pre-pregnancy levels and suckling-related spikes in prolactin secretion abate by approximately 2 months post delivery, the breastfeeding female may still be producing large amounts of breast milk, since milk delivery can be maintained by nipple stimulation and oxytocin alone. This can continue indefinitely.

Inside-Out: Signaling of the Breast

The suckling reflex describes the series of sensory impulses that travel from the nipple to the brain with breastfeeding. These impulses cause the release of prolactin to aid in continued milk production. Prolactin inhibits the secretion of FSH from the pituitary directly, but also indirectly through inhibition of the GnRH pulse generator and appropriate pulsatile secretion of GnRH. This typically prevents ovulation in the first months after delivery and can be contraceptive, though unreliably, in frequent and exclusive breastfeeders [15].

Perspectives

The gonads are responsible for the production of germ cells, and the female reproductive organs support embryonic and early postnatal development. As the cells within the reproductive tissues are highly proliferative, cancers of these tissues are common diseases, and the underlying hormonal signaling pathways are of critical importance for the development of both breast cancer (see chapter "Breast Cancer") and prostate cancer (see chapter "Prostate Cancer"). The gonads are the major producers of androgens and estrogens, steroid hormones that act on multiple targets throughout the body, mainly at the level of gene transcription. In addition to reproduction, they influence muscle strength, bone stability, hair growth, lipoprotein profiles, and many other physiologic functions. Furthermore, as the cells within the reproductive tissues are highly proliferative, cancers of these tissues are common diseases, and the underlying hormonal signaling pathways are of critical importance for the development of both breast cancer and prostate cancer.

Questions and Answers

Question 1 What are the functions of the Sertoli and Leydig cells?

Answer 1 The Leydig cells release androgens, including androstenediol, androstenedione, testosterone, dihydrotestosterone, and small amounts of dehydroepiandrosterone (DHEA). Sertoli cells convert testosterone to estrogen, such as estradiol (E2), required for spermatogenesis.

Question 2 How do the theca and granulosa cells work together to produce estrogens in the female body?

Answer 2 Theca cells in the ovary produce androgens. Enzymes within the granulosa cells of the ovary then convert the androgens into estrogen that can then act locally or enter circulation.

Question 3 What is the function of oxytocin in human reproduction?

Answer 3 Oxytocin plays a role primarily in the gravid female and after delivery of the infant. Oxytocin is released to stimulate contraction of the uterine smooth muscle during labor and delivery. It also acts on myoepithelial cells to cause milk letdown in the lactating female. Oxytocin and nipple stimulation alone can allow for indefinite breastfeeding of the offspring.

Question 4 What hormone causes ovulation, and what hormone prevents subsequent ovulation within the same menstrual cycle?

Answer 4 The LH surge caused by increased and sustained estradiol levels results in ovulation. Progesterone and inhibin are then released from luteal cells preventing release of an additional oocyte.

References

1. Jakimiuk AJ, Weitsman SR, Navab A, Magoffin DA (2001) Luteinizing hormone receptor, steroidogenesis acute regulatory protein, and steroidogenic enzyme messenger ribonucleic acids are overexpressed in thecal and granulosa cells from polycystic ovaries. J Clin Endocrinol Metab 86:1318–1323
2. Braak H, Braak E (1992) Anatomy of the human hypothalamus (chiasmatic and tuberal region). Prog Brain Res 93:3–14; discussion 14–16
3. Clifton DK, Steiner RA (2009) Neuroendocrinology of reproduction. In: Straus JF, Barbieri RL (eds) Yen and Jaffe's reproductive endocrinology, 6th edn. Elsevier, Saunders, pp 3–75
4. Shaw ND, Histed SN, Srouji SS, Yang J, Lee H, Hall JE (2010) Estrogen negative feedback on gonadotropin secretion: evidence for a direct pituitary effect in women. J Clin Endocrinol Metab 95:1955–1961
5. Hoffman BL, Schorge JO, Halvorson LM, Hamid CA, Corton MM, Schaffer JI (Eds) (2020) Reproductive endocrinology. In: Williams gynecology, 4th ed. McGraw Hill Education, New York
6. Lejeune H, Sanchez P, Chuzel F, Langlois D, Saez JM (1998) Time-course effects of human recombinant leuteinizing hormone on porcine Leydig cell specific differentiated functions. Mol Cell Endocrinol 144(1–2):59–69
7. Weinberger SE, Weiss ST, Cohen WR, Weiss JW, Johnson TS (1980) Pregnancy and the lung. Am Rev Respir Dis 121(3):559–581
8. Jackson DL, Schust DJ (2011) The role of the placenta in autoimmune disease and early pregnancy loss. In: Kay H, Nelson DM, Wang Y (eds) The placenta: from development to disease. Wiley-Blackwell Publishing, Inc, Chichester, pp 215–221
9. Sinha-Hikim I, Taylor WE, Gonzalez-Cadavid NF, Zheng W, Bhasin S (2004) Androgen receptor in human skeletal muscle and cultured muscle satellite cells: up-regulation by androgen treatment. J Clin Endocrinol Metab 89(10):5245–5255
10. Pederson L, Kremer M, Judd J, Pascoe D, Spelsberg TC, Riggs BL, Oursler MJ (1999) Androgens regulate bone resorption activity of isolated osteoclasts in vitro. Proc Natl Acad Sci USA 96:505–510
11. Crosnoe LE, Kim ED (2013) Impact of aging on male fertility. Curr Opin Obstet Gynecol 25(3):181–185
12. Rusby JE, Brachtel EF, Michaelson JS, Koerner FC, Smith BL (2007) Breast duct anatomy in the human nipple: three-dimensional patterns and clinical implications. Breast Cancer Res Treat 106:171–179
13. Jenness R (1979) The composition of human milk. Semin Perinatol 3:225–239
14. Motil KJ, Thotathuchery M, Montandon CM, Hachey DL, Boutton TW, Klein PD, Garza C (1994) Insulin, cortisol and thyroid hormones modulate maternal protein status and milk production and composition in humans. J Nutr 124:1248–1257
15. Konner M (1978) Nursing frequency and birth spacing in Kung hunter-gatherers. IPPF Med Bull 15:1–3

Breast Cancer

Tanja Fehm and Eugen Ruckhäberle

Introduction

Breast cancer (BC) is the most frequent female cancer worldwide with about 2.3 million newly diagnosed cases and approximately 685,000 deaths per year [1, 2]. Whereas incidence is rising, mortality has dropped in many countries, likely resulting from improved screening programs and better systemic adjuvant therapies (see below). Risk factors for BC are manifold and include genetics, high age, age at childbirth, short time of breast-feeding, obesity, early menarche and late menopause, and hormonal treatment.

BC diagnosis includes mammography and sonography, and, in case of suspicion, magnetic resonance imaging. Whenever any abnormality is seen, a core cut or stereotactic biopsy is necessary to diagnose the disease. For staging purposes, in cases of invasive breast cancer a bone scan as well as a computer tomography of the thorax and abdomen is recommended by the guidelines.

Breast cancer (BC) is a type of cancer originating from the epithelium of the mammary gland. As most cancers, it can be invasive or noninvasive. Noninvasive breast cancer comprises two distinct entities, lobular carcinoma in situ (LCIS) and ductal carcinoma in situ (DCIS). Invasive breast cancer is a collection of diseases defined by distinct pathological (e.g., ductal, lobular, mucinous) and molecular characteristics (e.g., estrogen receptor [ER] and progesterone receptor [PR] expression, Ki67, Human epidermal growth factor receptor (HER2) amplification, and more recently transcriptome-based classifications such as luminal, Her2-enriched, and basal-like cancers). Following the actual World Health Organization (WHO) Classification, the majority of invasive breast cancers are histologically classified as invasive ductal carcinoma of no special type (IDC; 75–80%) or invasive lobular carcinoma of classical type (ILC; 10–15%). Rare histological subtypes (mucinous, tubular, medullary, cribriform, invasive papillary cancer, secretory, adenosquamous carcinoma, as well as adenoid cystic cancers) make up the rest [3].

Furthermore, invasive BC is histologically subclassified according to the expression of the female hormone receptors for estrogen and progesterone (ER and PR), with 70% being positive (ER$^+$) and 30% being negative (ER$^-$ and PR$^-$), or the expression of the HER2 oncogene (the human epidermal growth factor receptor 2 [EGFR2]), with 20–25% Her$^+$ BCs and 75–80% Her$^-$ BCs. Just recently, there are newer reclassifications of the Her2 negative subgroup into a Her2 negative (Her2 0), a Her2 low (Her2 1 and 2), and even a

T. Fehm · E. Ruckhäberle (✉)
Department of Obstetrics and Gynecology, Heinrich Heine University Düsseldorf, Düsseldorf, Germany
e-mail: Tanja.fehm@med.uni-duesseldorf.de;
Eugen.ruckhaeberle@med.uni-duesseldorf.de;
direktion.frauenklinik@med.uni-duesseldorf.de

© The Author(s), under exclusive license to Springer Nature Switzerland AG 2026
E. Lammert, M. Zeeb (eds.), *Metabolism of Human Diseases*,
https://doi.org/10.1007/978-3-031-96019-2_42

Her2 ultralow cluster [5]. Of note, this new classification was triggered by a new group of systemic drugs called antibody drug conjugate (ADC). So far, discussion about the real existence of those subtypes is ongoing and not yet finished. An additional immunohistochemical marker that is used in clinical routine is the proliferation marker Ki67 [6]. Newest molecular markers in the classification of breast cancer include BRCA, PI3K (phosphatidyl-inositol-3-kinase), ESR1 (estrogen receptor-α), and the immune markers (tumor infiltrating lymphocytes or TIL and Programmed Death-Ligand 1 or PD-L1).

In general primary and recurrent breast cancer is considered as curable disease while metastatic disease is not curable. Nevertheless, in approximately 10–15% of metastatic cases long-term survival of at least 10–15 years has been observed with appropriate treatment.

Anatomy and Development of the Mammary Gland

The mammary gland is a highly dynamic organ that undergoes profound changes within its epithelium during embryogenesis, puberty, and the reproductive cycle. These changes are driven by dedicated stem and progenitor cells. Both short- and long-lived lineage-restricted progenitors have been identified in adult tissue as well as a small pool of multipotent mammary stem cells (MaSCs). While unipotent progenitor cells predominantly execute day-to-day homeostasis and postnatal morphogenesis during puberty and pregnancy, multipotent MaSCs have been implicated in coordinating alveologenesis and long-term ductal maintenance [7]. Female steroid hormones (estrogen and progesterone) and key regulators (e.g., Notch pathway, GATA binding protein-3 or GATA-3) are essential for normal development, and their deregulated expression is implicated in breast cancer. The microenvironment is also an important regulator of mammary gland development and tumorigenesis [8].

Today, there have been established two models of the postnatal development of the mammary epithelial differentiation. In model A, the stem cell compartment in the adult gland is heterogeneous and comprises quiescent mammary stem cells (MaSCs). These give rise to committed progenitors for the myoepithelial and luminal epithelial lineages. In model B, there are no adult MaSCs but rather embryonic multipotent stem cells that become restricted in their potential at around 16 gestational weeks. Thereafter, unipotent progenitors drive most aspects of postnatal development. In both models, there are independent ductal and luminal sublineages that produce ER^+ and ER^- cells, respectively.

The mammary ducts and alveoli of the adult human breast are lined by an inner layer of secretory luminal epithelial cells that produce milk during lactation and are surrounded by contractile myoepithelial cells for milk ejection, and basement membrane. During the pregnancy and lactation cycle, the mammary gland matures into a functional milk-secreting organ [8]. After pregnancy, these developments regress. This continuous remodeling is protective against breast cancer [9], particularly when it occurs before the age of 30 and with an interval of less than 14 years between menarche and the first pregnancy. Multiparity confers slightly more protection.

Pathophysiology of Breast Cancer

The recent knowledge of the differentiation of the mammary gland and the role of mammary epithelial hierarchy for normal physiological function and in the tumorigenesis of breast malignancy was recently extensively reviewed [7].

Intratumoral heterogeneity in breast cancer is caused by the fact that distinct epithelial breast cells serve as origin for the malignant transformation triggered by oncogenic driver mutations. Furthermore, the microenvironment plays a pivotal role in influencing tumor growth and evasion [10, 11].

Signaling downstream of the ER and HER are central to the development of breast cancer. Estrogen signaling is a key regulator of postnatal development of the mammary gland. ER signaling can affect many cellular processes such as cardiovascular protection (by beneficial estrogen effects on lipid metabolism, lipid-peroxidation, smooth-muscle-cell proliferation, hemostasis, and vasomotion; see chapter "Rheumatoid Arthritis"), bone preservation, neuroprotection (see chapter "Kidney Stones"), and proliferation of many cell types. However, deregulated ER signaling promotes carcinogenesis and cancer progression. ERα is associated with breast cancer initiation and progression, while ERβ function in breast cancer is still unclear [11].

Estrogen signaling includes two distinct pathways often referred to as genomic and nongenomic pathways, depending on whether the signaling initially alters gene transcription or protein activity. Whereas the genomic pathway is essential for breast carcinogenesis, the nongenomic pathways promote breast cancer invasion and metastasis. In contrast to its role in breast cancer initiation, estrogen signaling has a protective effect in later stages where the loss of ERα correlates with aggressive metastatic disease.

In the genomic pathway, binding of estrogen to intracellular ER induces its dimerization and translocation to the nucleus. Binding of the ER to an estrogen-responsive element in the promoter region of target genes subsequently enhances or represses transcriptional activity of these genes.

In the nongenomic pathway, ERs localized within caveolae (small, specialized invaginations in the plasma membrane) [12] interact and modulate a variety of adaptor proteins to exert rapid responses on protein level. Common targets include Src, HER2, mitogen-activated protein kinases (MAPK), and phosphatidyl-inositol-3-kinase (PI3K) and the respective signaling pathways (Figs. 1 and 2) [13, 14]. The G-protein-coupled ER in caveolae is the most likely candidate for rapid actions of estrogens [15].

Genomic and nongenomic pathways interact extensively. More specifically, transforming growth factor α and amphiregulin (AR), two genes induced by genomic ER signaling, can bind HER2 and consequently activate MAPK and Akt (also called protein kinase B) [16]. Moreover, nongenomic ER signaling activates the PI3K pathway, which then activates Akt and mammalian target of rapamycin (mTOR) and

results in decreased apoptosis. Additionally, nongenomic ER signaling activates the Ras–Raf–MAPK pathway, which increases release of matrix metalloproteinases (MMP) via Src (Fig. 1).

A second important pathway in BC is HER-signaling. The EGFR family includes four members, HER1-4, all of which are transmembrane receptor tyrosine kinases (RTK). Ligand binding induces dimerization and autophosphorylation. HER2 and HER3 rely on heterodimerization (or very high levels, in case of HER2) for activation, as HER2 does not have a ligand, and HER3 has no kinase activity. Yet, HER2 has the strongest kinase and signaling activity and is the dimerization partner of choice. Downstream signaling pathways are associated with cell proliferation, apoptosis, angiogenesis, and metastasis (Fig. 2). Thus, HER2 is associated with increased tumor invasiveness and metastasis. HER2 can also be activated by complexing with other membrane receptors such as insulin-like growth factor receptor 1 (IGF-1) [17]. Adding another level of interaction, HER2 and IGF-1 signaling can activate ER (so-called ligand-independent activation), for example, via MAPK-, PI3K/Akt-, or p38-signaling (see Fig. 1).

A field of increasing interest in breast cancer is the immune response of a human being to a malignant tumor. Knowledge of the pathophysiology has been developed in the last decades tremendously [18]. Key players in the model of immune response and immune checkpoint therapy are tumor cells, antigen-presenting cells (APC), and T-cells. In general, tumor cells release tumor antigens that are presented by APC. Throughout, these inactivated T-cells get activated and start to eliminate the tumor cells. As we have learned, in the interaction between T-cells and tumor cells there are activating receptors and inhibiting receptors (see chapter "Anatomy and Physiology of the Immune System"). Some tumors have the capacity to escape the immune response by activation of the inhibitory pathways. By inhibition of the inhibitory pathways, T-cells obtain back their capacity of detection of tumor cells and eliminate the tumor cells. The targeted treatment against

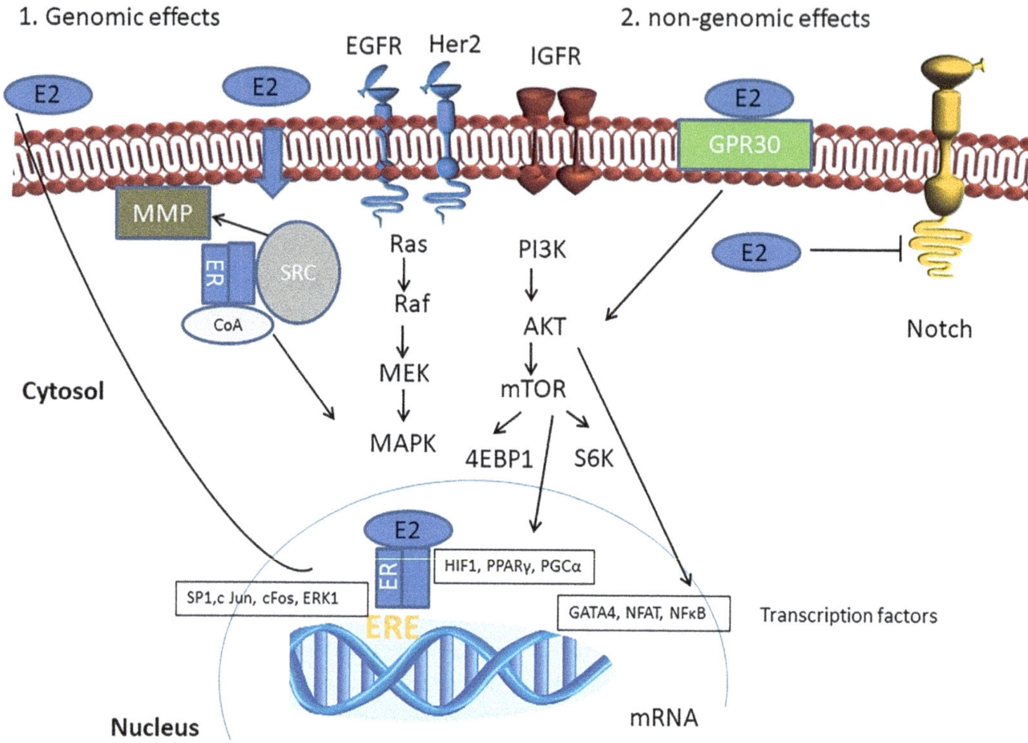

Fig. 1 Complexity of estrogen-mediated signaling in breast cancer. Estrogens exert their actions by binding to specific receptors, the estrogen receptors (ERs), which in turn activate transcriptional processes and/or signaling events that result in the control of gene expression. These actions can be mediated by direct binding of estrogen receptor complexes to specific sequences in gene promoters (1. genomic effects) or by mechanisms that do not involve direct binding to DNA (2. nongenomic effects)

PD-L1-PD-1 axis has gained highest interest in triple negative breast cancer and some drugs are already approved (Pembrolizumab and Atezolizumab).

Obesity is linked to cancer risk via various mechanisms [19–21]. Obesity-associated factors regulate metabolic pathways in both breast cancer cells and cells in the breast microenvironment, which provides a molecular link between obesity and breast cancer [18]. First, the insulin–cancer hypothesis attributes an important role to insulin resistance, which causes high insulin and glucose levels (see chapter "Alzheimer's Disease"). Increased activation of the insulin receptor on ductal cells stimulates cell division [22], a prereq-uisite for tumor formation, and high glucose concentrations may favor tumor cell proliferation and selection of malignant cells over nonmalignant ones [23]. Insulin also increases the level of sex-hormone-binding globulin and thus bioavailability of estrogens. Second, synthesis and bioavailability of sex steroids, most importantly estrogens, is increased, due to the expression of aromatase by adipose tissue (see chapter "Heart Failure" and below). Finally, obesity is characterized by a state of chronic low-grade inflammation due to pro-inflammatory adipokines, which link to cancer [24]. Furthermore JAK2/STAT3 is a regulator of lipid metabolism and promotes breast cancer cell "stemness" and chemoresistance [25].

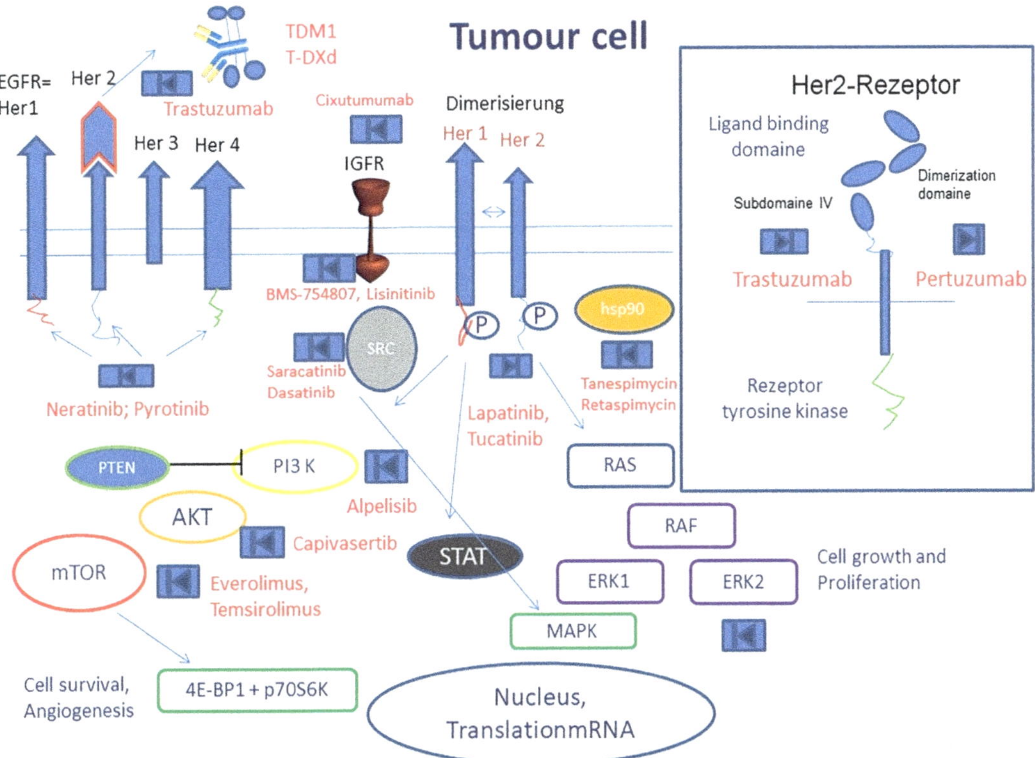

Fig. 2 Complexity of signal transduction in the EGF-Receptor family. EGF-Receptor signaling, and especially Her2-Receptor signaling, plays a major role in breast cancer pathophysiology. By understanding the complexity, several targeted therapy options started by Trastuzumab have become available in the last decades

Medical Treatment in Breast Cancer

Standard treatment modalities of noninvasive and invasive breast cancer include operative, systemic, and radiation treatment. In primary breast cancer, surgical treatment is still a treatment milestone; but systemic therapy has been increasingly used as induction treatment in luminal breast cancer or as neoadjuvant treatment in Her2-enriched and basal-like breast cancer.

Since circulating tumor cells are detectable in many patients at the time of breast surgery, systemic recurrence occurs commonly in the first 5 years after local treatment, thus arguing for systemic adjuvant (postoperative) endocrine or targeted therapy. Furthermore, there exists a medical need for biomarkers and multigene tests stratifying patients into groups with high or low risk of recurrence.

Three different systemic neoadjuvant therapies exist: (i) targeted therapy, such as endocrine therapy or treatment blocking the HER pathway and/or immune checkpoints and/or cell cycle as well as homologous recombination; (ii) systemic chemotherapy; and (iii) targeted therapy post neoadjuvant after primary surgery or post adjuvant after first adjuvant treatment line. Neoadjuvant therapy, a more recent treatment, uses the same substances and regimens like adjuvant treatment, but is given before surgery. Advantages of this concept include an in vivo sensitivity testing (response control in the breast and lymph node tumor under therapy), shrinkage of the tumor, and an increased number of cases where breast conservation becomes available. Furthermore, patients in need of post-neoadjuvant treatment are figured out by nonpathologic complete remission.

Targeted therapy depends on the BC subtype, but as most BCs are ER⁺, endocrine therapy, an antihormonal treatment, was the first and still is the most common adjuvant and inductive (presurgical short-term) treatment. Endocrine therapy aims to either reduce estrogen levels or to block ERα signaling. Traditionally, this was achieved via ovarian ablation. However, today drugs that target ER directly or indirectly are used. Three classes of antihormonal endocrine agents are used: selective estrogen receptor modulators (e.g., tamoxifen) that block the activity of ER; estrogen synthesis inhibitors (e.g., aromatase inhibitors such as anastrozole, letrozole, and exemestane); and selective estrogen receptor downregulators (e.g., Fulvestrant) that induce destabilization and degradation of ER.

Selective estrogen receptor modulators such as tamoxifen, toremifene, and raloxifene competitively inhibit binding of estrogen to ER and profoundly affect both the genomic and nongenomic activity of ERα. Tamoxifen causes a partial occlusion of the coactivator binding groove of ER, hampering interactions with activators and blocking transcription of ER target genes.

As estrogens (estrone, estradiol, and estriol) are produced from androstenedione and testosterone via aromatase (see chapter "Reproductive System: Overview"), aromatase inhibitors decrease estrogen levels. This effect is very relevant in post-menopausal women, in whom estrogen production occurs primarily by aromatization in peripheral tissues. Steroidal aromatase inhibitors (e.g., exemestane) bind irreversibly and nonsteroidal aromatase inhibitors (e.g., letrozole and anastrozole) bind reversibly to inhibit aromatase activity. Selective estrogen receptor degraders (SERDs) are antiestrogens that are designed to destabilize helix-12 or H12 of the estrogen receptor (ER) and function by binding to and inducing the degradation of ER, thereby inhibiting dimerization and abolishing the ER signaling pathway. One of the major advantages of SERDs seems to be the efficacy in cases with endocrine resistance to aromatase inhibitors. First representative was Fulvestrant. Meanwhile some other candidates of SERDs (Elacestrant, Giredestrant, Camizestrant) are tested in Phase III clinical trials.

Side effects of antihormonal therapy generally include hot flushes, sleeping disturbances, huffiness, and a decrease in bone density, as well as weight gain and an increase in cholesterol. Unfortunately, up to 40% of patients on endocrine therapy exhibit a primary (de novo) or acquired resistance [26]. Potential escape mechanisms include loss or modification of ER expression, crosstalk between pathways, upregulation of downstream pathways like PI3K, mTOR, AKT, and MAPK, an alteration in the expression of specific microRNA, and alterations in drug metabolism [27, 28]. To overcome endocrine resistance, new targeted drugs like cyclin D kinase 4/6 (CDK4/6) inhibitors, the mTOR inhibitor Everolimus, and the PI3 kinase inhibitor alpelisib were investigated and approved in the treatment of metastatic breast cancer. More targeted agents like MAP kinase (Capivasertib)—or mitogen activated protein kinase (MEK)—inhibitors are under investigation.

Major success in the treatment of hormone receptor-positive breast cancer was made by the group of inhibitors of the cyclin D kinases 4 and 6. CDK4/6 inhibitors target the cyclin D/CDK/retinoblastoma signaling pathway, inducing cell-cycle arrest, reduced cell viability, and tumor shrinking. As the cyclin D/CDK complex is activated downstream of estrogen signaling, the combination of CDK4/6 inhibitors with standard endocrine therapies represents a rational approach to elicit synergic antitumor activity in hormone receptor-positive BC. Three agents—palbociclib, ribociclib, and abemaciclib—are available.

Treatments against HER include monoclonal antibodies and tyrosine kinase inhibitors (Fig. 2). The first approved agent was Trastuzumab, a monoclonal antibody (thus the ending -mab) against HER2. It substantially improves outcomes in early-stage and metastatic BC and likely induces internalization and degradation of HER2, disrupting signaling via the PI3K–Akt pathway, thus resulting in apoptosis of the tumor cell (Fig. 2). Side effects are rare as Trastuzumab is specific for HER2. A second monoclonal antibody, Pertuzumab, binds to the extracellular dimerization domain II of HER2 and inhibits heterodimerization of HER2 with other HER family members, including EGF receptor (EGFR), HER3, and HER4. In combination with chemotherapy, those two antibodies demonstrated increase in the pathologic complete response (pCR) rate and improved prognosis in neoadjuvant treatment and an improvement in response rate and overall survival as first-line standard in Her2 positive metastatic breast cancer. However, de novo and acquired resistance pose problems as well. Explanatory escape mechanisms include signaling through alternative receptors (e.g., IGF-1R), shedding of the receptor, upregulation of downstream signaling pathways, and failure to elicit an appropriate immune response against the Trastuzumab-bound BC cells [29, 30]. Thus, new treatment concepts also target escape mechanisms [31]. For example, everolimus (Fig. 2), an mTOR inhibitor, demonstrated favorable results in ER$^+$ metastatic BC patients with acquired resistance to anti-HER2 treatment.

Tyrosine kinase inhibitors prevent phosphorylation of the cytoplasmic tyrosine kinase domain, usually of all HER kinases, and subsequent intracellular signaling [32]. Lapatinib, first in market tyrosine kinase inhibitor, targets HER1 and HER2. Side effects include diarrhea and rash due to unspecific binding. Future inhibitors will show improved targeting and administration. More recently Tucatinib, a more selective tyrosine kinase inhibitor of Her2 than Her1, was approved for metastatic breast cancer. Neratinib is an oral pan-HER inhibitor that irreversibly inhibits the tyrosine kinase activity of epidermal growth factor receptor (EGFR or HER1), HER2, and HER4. It is approved in Europe as post-adjuvant monotherapy of high risk Her2 and ER positive breast cancer.

Immunotherapy has gained interest in the last decade. Most of those drugs are targeted agents. The first drug that had demonstrated an effect on immune response was Trastuzumab. In the last decade, immune checkpoint inhibitors were investigated in nearly all subtypes of breast cancer. Meanwhile the PD-1 inhibitor Pembrolizumab is approved for neoadjuvant and post-neoadjuvant primary triple negative breast cancer, and the PD-L1 inhibitor Atezolizumab and the PD-1 inhibitor Pembrolizumab are approved for the first-line treatment of metastatic triple negative breast cancer. Those drugs show low response rates in breast cancer and are combined with chemotherapy. Side effects of the immune checkpoint inhibitors are less; immune-related side effects are hypo- or hyperthyroidism, colitis, pneumonitis, hepatitis, myositis, and very seldom hypophysitis.

A paradigm shift in the treatment of advanced breast cancer was made possible by a group of agents called antibody drug conjugates. Those agents comprise a monoclonal antibody conjugated to the cytotoxic payload via a chemical linker that is directed toward a target antigen expressed on the cancer cell surface, reducing systemic exposure and therefore toxicity. Moreover, the bystander effect of some ADCs, which kills cells surrounding the target tumor cells, is quite different from ADCs without this effect. Representative of an ADC without bystander effect is T-DM1, while T-DXd and sacituzumab govitecan demonstrated a stronger bystander effect. Those ADCs are approved for metastatic breast cancer and are under investigation in the early setting.

Besides endocrine treatment, chemotherapy belongs to the oldest treatment modalities. Chemotherapy has a significant impact on prognosis. Common chemotherapeutics (most of which are also used in other cancer types) include alkylating agents, antimetabolites, antibiotics, topoisomerase inhibitors, and plant alkaloids. Alkylating agents directly damage DNA to prevent the cancer cell from reproducing. Antimetabolites interfere with DNA and RNA synthesis by substituting for regular building blocks. Anthracyclines are antitumor antibiotics that interfere with enzymes involved in DNA replication. Topoisomerase inhibitors prevent separation of DNA strands in the M phase of the cell cycle. Plant alkaloids (e.g., taxanes, vinorelbine) can stop mitosis during M phase or inhibit enzymes from making proteins needed for cell reproduction.

Since consolidation of quality of life (QoL) and slowing of the disease are the major therapeutic aims in the metastatic setting, monochemotherapy should be the choice for these patients. Side effects of chemotherapy are mainly caused by the antiproliferative effect on normal, rapid-growing cell lines, such as hair and nail follicle cells, blood cells, and neuronal and enteric epithelial cells, and thus include alopecia (loss of hair), anemia, infections, and diarrhea.

Finally, the role of diet and physical activity in cancer prevention is complex. Regular activity is associated with reduced risk of BC by reducing sex hormones and adipokines, by preventing insulin resistance and chronic inflammation, and by improving immune function [33]. Long-term use of metformin (5 years) is associated with reduced risk of developing BC [34], probably via reduced insulin levels [35]. It has also been shown to reduce proliferation of most cultured BC cell lines [36].

As up to 70% of breast cancer patients with metastatic disease develop bone metastases [37], subsequent treatment to target the bone is necessary but cannot be discussed herein.

Perspectives and Future Directions

With improvements in technology (e.g., single-cell RNA sequencing [scRNA seq], transcriptomic and proteomic technologies) and increasing understanding of the development of the mammary gland and the carcinogenesis of BC, primary prevention strategies have come to the fore. Concepts of preventing or treating obesity with lifestyle changes, for example, diet, physical activity, and new agents (Vitamin A), are part of ongoing trials (e.g., Libre Trial). Current research on food and diet investigates anti-inflammatory effects, focusing on herbs and spices such as curcuma, ginger, and flavonoids [38, 39]. Further trials are needed to understand how obesity, diet, and physical activity may affect BC risk. A new field of investigation is the role of the microbiome in tumorigenesis and migration as well as response to various treatment modalities.

Early diagnosis allows breast-conserving surgery and preserves better prognosis. A new treatment era could start with the introduction of gene therapy for patients with gene mutations. Still, research to find novel or improve current BC treatments is bustling.

Other agents that are now tested target intracellular signaling pathways (e.g., IGF-1R, PI3K and AKT, MAPK [Fig. 2]) directly to complement current targeted therapy. The combination of targeted treatment and chemotherapy in one molecule (ADC) has been shown to be very successful. Several new antibodies and ADCs are in development. New protocols include ADCs in combination with other targeted agents like immune checkpoint inhibitors. Furthermore, new immune therapeutics like Vaccinations and Chimeric Antigen Receptor-T cells (CAR-T cells) are under investigation.

New targeted agents might even help to enhance the poor prognosis of basal-like BC patients, for whom currently chemotherapy is the only therapeutic option. Finally, the ultimate goal of all kinds of diagnosis and treatment of BC will be an individualized, tailored therapy (Table 1).

Questions and Answers

Question 1 What are the most frequent histological subtypes of breast cancer according to the WHO Classification?

Table 1 Intrinsic subtypes of breast cancer

Intrinsic subtypes in breast cancer and treatment				
	Luminal-A	Luminal-B	HER2-enriched	Basal-like
Estrogen/progesterone status	ER and/or PR positive	ER and/or PR positive	ER/PR negative	ER/PR negative
HER2-status	Negative	Negative or positive	Positive	Negative
Therapy	Endocrine	Chemo ± Trastuzumab + Endocrine	Chemo + Trastuzumab	Chemotherapy

Answer 1 According to the actual WHO Classification, invasive ductal carcinoma of no special type (IDC; 75–80%) and invasive lobular carcinoma of classical type (ILC; 10–15%) are the most common histological types.

Question 2 What cell type is responsible for milk production during lactation?

Answer 2 An inner layer of secretory luminal epithelial cells, so-called lactocytes, produce milk during lactation.

Question 3 What are important signaling pathways in breast cancer?

Answer 3 Estrogen- (genomic and nongenomic) and Her2-signaling pathways are the most important pathways in the pathophysiology of breast cancer. They include pathways involving PI3 kinase and MAPK. Immune checkpoints and their pathways have become of higher interest in breast cancer over the last decade.

References

1. Sung H, Ferlay J, Siegel RL, Laversanne M, Soerjomataram I, Jemal A et al (2021) Global cancer statistics 2020: GLOBOCAN estimates of incidence and mortality worldwide for 36 cancers in 185 countries. CA Cancer J Clin 71:209
2. Youlden DR, Cramb SM, Dunn NA, Muller JM, Pyke CM, Baade PD (2012) The descriptive epidemiology of female breast cancer: an international comparison of screening, incidence, survival and mortality. Cancer Epidemiol 36:237–248
3. Jenkins S, Kachur ME, Rechache K, Wells JM, Lipkowitz S (2021) Rare breast cancer subtypes. Curr Oncol Rep 23(5):54. https://doi.org/10.1007/s11912-021-01048-4
4. Zhang H, Karakas C, Tyburski H, Turner BM, Peng Y, Wang X, Katerji H, Schiffhauer L, Hicks DG (2022) HER2-low breast cancers: Current insights and future directions. Semin Diagn Pathol 39(5):305–312. https://doi.org/10.1053/j.semdp.2022.07.003. Epub 2022 Jul 9
5. Dowsett M, Nielsen TO, A'Hern R, Bartlett J, Coombes RC, Cuzick J, Ellis M, Henry NL, Hugh JC, Lively T, McShane L, Paik S, Penault-Llorca F, Prudkin L, Regan M, Salter J, Sotiriou C, Smith IE, Viale G, Zujewski JA, Hayes DF (2011) International Ki-67 in breast cancer working group. Assessment of Ki67 in breast cancer: recommendations from the international Ki67 in breast cancer working group. J Natl Cancer Inst 103(22):1656–1664. https://doi.org/10.1093/jnci/djr393. Epub 2011 Sept 29
6. Fu NY, Nolan E, Lindeman GJ, Visvader JE (2020) Stem cells and the differentiation hierarchy in mammary gland development. Physiol Rev 100(2):489–523. https://doi.org/10.1152/physrev.00040.2018. Epub 2019 Sep 20
7. Stingl J (2011) Estrogen and progesterone in normal mammary gland development and in cancer. Horm Cancer 2:85–90
8. Key TJ, Verkasalo PK, Banks E (2001) Epidemiology of breast cancer. Lancet Oncol 2:133–140
9. Ellis MJ, Perou CM (2013) The genomic landscape of breast cancer as a therapeutic roadmap. Cancer Discov 3:27–34. https://doi.org/10.1158/2159-8290.CD-12-0462
10. Olyak K, Kalluri R (2010) The role of the microenvironment in mammary gland development and cancer. Cold Spring Harb Perspect Biol 2:a003244. https://doi.org/10.1101/cshperspect.a003244
11. Osborne CK, Schiff R, Fuqua SA, Shou J (2001) Estrogen receptor: current understanding of its activation and modulation. Clin Cancer Res 7:4338s–4342s; discussion 4411s–4412s
12. Razandi M, Oh P, Pedram A, Schnitzer J, Levin ER (2002) ERs associate with and regulate the production of caveolin: implications for signaling and cellular actions. Mol Endocrinol 16:100–115
13. Migliaccio A, Piccolo D, Castoria G, Di Domenico M, Bilancio A, Lombardi M, Gong W, Beato M, Auricchio F (1998) Activation of the Src/p21ras/Erk pathway by progesterone receptor via cross-talk with estrogen receptor. EMBO J 17:2008–2018
14. Schiff R, Massarweh SA, Shou J, Bharwani L, Mohsin SK, Osborne CK (2004) Cross-talk between estrogen receptor and growth factor pathways as a molecular target for overcoming endocrine resistance. Clin Cancer Res 10:S331–S336
15. García-Becerra R, Santos N, Díaz L, Camacho J (2012) Mechanisms of resistance to endocrine therapy in breast cancer: focus on signaling pathways, miRNAs and genetically based resistance. Int J Mol Sci 14(1):108–145
16. Deroo BJ, Korach KS (2006) Estrogen receptors and human disease. J Clin Invest 116:561–570
17. Nahta R, Yuan LX, Zhang B et al (2005) Insulin-like growth factor-I receptor/human epidermal growth factor receptor 2 heterodimerization contributes to trastuzumab resistance of breast cancer cells. Cancer Res 65:11118–11128
18. Morad G, Helmink BA, Sharma P, Wargo JA (2021) Hallmarks of response, resistance, and toxicity to immune checkpoint blockade. Cell 184(21):5309–5337. https://doi.org/10.1016/j.cell.2021.09.020. Epub 2021 Oct 7
19. Kristy A (2021) Brown metabolic pathways in obesity-related breast cancer. Nat Rev Endocrinol

17(6):350–363. https://doi.org/10.1038/s41574-021-00487-0

20. Lee K, Kruper L, Dieli-Conwright CM, Mortimer JE (2019) The impact of obesity on breast cancer diagnosis and treatment. Curr Oncol Rep 21(5):41. https://doi.org/10.1007/s11912-019-078

21. Smith CJ, Ryckman KK (2015) Epigenetic and developmental influences on the risk of obesity, diabetes, and metabolic syndrome. Diabetes Metab Syndr Obes 8:295–302

22. Faulds MH, Dahlman-Wright K (2012) Metabolic diseases and cancer risk. Curr Opin Oncol 24(1): 58–61

23. Muti P, Quattrin T, Grant BJ et al (2002) Fasting glucose is a risk factor for breast cancer: a prospective study. Cancer Epidemiol Biomarkers Prev 11(11):1361–1368

24. Heikkila K, Ebrahim S, Lawlor DA (2007) A systematic review of the association between circulating concentrations of C reactive protein and cancer. J Epidemiol Community Health 61(9):824–833

25. Wang TY et al (2018) JAK/STAT3-regulated fatty acid beta-oxidation is critical for breast cancer stem cell self-renewal and chemoresistance. Cell Metab 27(1):136–150.e5

26. Davies C, Godwin J, Gray R, Clarke M, Cutter D, Darby S, McGale P, Pan HC, Taylor C, Wang YC, Dowsett M, Ingle J, Peto R (2011) Relevance of breast cancer hormone receptors and other factors to the efficacy of adjuvant tamoxifen: patient-level meta-analysis of randomised trials. Lancet 378(9793):771–784

27. Fedele P, Calvani N, Marino A, Orlando L, Schiavone P, Quaranta A, Cinieri S (2012 Nov) Targeted agents to reverse resistance to endocrine therapy in metastatic breast cancer: where are we now and where are we going? Crit Rev Oncol Hematol 84(2):243–251

28. Bianco S, Gévry N (2012) Endocrine resistance in breast cancer: from cellular signaling pathways to epigenetic mechanisms. Transcription 3(4):165–170

29. Hubalek M, Brunner C, Mattha K et al (2010) Resistance to HER2-targeted therapy: mechanisms of trastuzumab resistance and possible strategies to overcome unresponsiveness to treatment. Wien Klin Wochenschr 160:506–512

30. Pohlmann PR, Mayer IA, Mernaugh R (2009) Resistance to trastuzumab in breast cancer. Clin Cancer Res 15:7479–7491

31. Tsang RY, Finn RS (2012) Beyond trastuzumab: novel therapeutic strategies in HER2-positive metastatic breast cancer. Br J Cancer 106(1):6–13

32. Mendelsohn J, Baselga J (2003) Status of epidermal growth factor receptor antagonists in the biology and treatment of cancer. J Clin Oncol 21:2787–2799

33. Gallagher EJ, LeRoith D (2011) Diabetes, cancer, and metformin: connections of metabolism and cell proliferation. Ann N Y Acad Sci 1243:54–68

34. Bodmer M, Meier C, Krahenbuhl S, Jick SS, Meier CR (2010) Long-term metformin use is associated with decreased risk of breast cancer. Diabetes Care 33(6):1304–1308

35. Shaw RJ, Lamia KA, Vasquez D et al (2005) The kinase LKB1 mediates glucose homeostasis in liver and therapeutic effects of metformin. Science 310(5754):1642–1646

36. Alimova IN, Liu B, Fan Z et al (2009) Metformin inhibits breast cancer cell growth, colony formation and induces cell cycle arrest in vitro. Cell Cycle 8(6):909–915

37. Coleman RE (2001) Metastatic bone disease: clinical features, pathophysiology and treatment strategies. Cancer Treat Rev 27:165–176

38. Jungbauer A, Medjakovic S (2012) Anti-inflammatory properties of culinary herbs and spices that ameliorate the effects of metabolic syndrome. Maturitas 71(3):227–239

39. Baliga MS, Haniadka R, Pereira MM et al (2011) Update on the chemopreventive effects of ginger and its phytochemicals. Crit Rev Food Sci Nutr 51(6):499–523

Prostate Cancer

Ivan de Kouchkovsky, Eric Small,
and Rahul Aggarwal

Introduction to Prostate Cancer

Prostate cancer (PC) is the second most common cancer in men worldwide, with over 1,400,000 new cases per year [1]. The strongest risk factors for PC include age, genetic factors (which account for up to a third of PC cases) [2, 3], and metabolic parameters. Among men with one, two, or three first-degree relatives with PC, the risk of PC is increased 2-, 5-, and 11-fold, respectively. Obese patients (as defined by body mass index) have an increased risk of PC, including a more aggressive phenotype [4]. The underlying mechanisms are unclear; prior studies examining serum androgen, estrogen, and insulin levels have provided inconclusive evidence of an association of these hormones with elevated risk of PC.

For patients with disease that is confined to the prostate gland or pelvic lymph nodes at the time of diagnosis, PC can be curable with surgery and/or radiation therapy to the prostate gland. For patients with more advanced, metastatic cancer, or those with disease progression after prior local therapy evidenced by a climbing prostate specific antigen (PSA) level (termed biochemical relapse, see below), treatment is often applied systemi-

cally to delay subsequent progression, potentially prolong survival, and palliate symptoms.

Pathophysiology of Prostate Cancer

PC is unique among cancers in its exquisite dependence upon circulating androgens that drive tumor progression via activation of the androgen receptor (AR). Circulating androgens, synthesized from cholesterol precursors (see chapter "Reproductive System: Overview" under part "Reproductive System"), are derived primarily from the testes and adrenal gland. Activation of AR follows the common pathway of intracellular steroid receptors (see chapter "Reproductive System: Overview" under part "Reproductive System") and activates genes involved in cellular metabolism, cell cycle progression, and cellular proliferation [5].

Over the past two decades, a number of genetic events have been discovered that lead to the progression from benign prostate tissue to precancerous lesions and overt PC. These early inciting genetic events include loss of the tumor suppressor gene phosphatase and tensin homologue (*PTEN*) and fusions of the genes for *TMPRRS2* (transmembrane protease serine subtype 2) and a transcription factor of the ETS (E-twenty-six) family, which is observed in over half of PC tumors [6]. Activation of the AR increases transcription of this fusion product, driving PC proliferation.

I. de Kouchkovsky (✉) · E. Small · R. Aggarwal
Division of Hematology/Oncology, Department of Medicine, University of California San Francisco, San Francisco, CA, USA
e-mail: Ivan.deKouchkovsky@ucsf.edu;
Eric.Small@ucsf.edu; Rahul.Aggarwal@ucsf.edu

E. Lammert, M. Zeeb (eds.), *Metabolism of Human Diseases*,
https://doi.org/10.1007/978-3-031-96019-2_43

In more recent years, increased genetic sequencing of PC patients has identified a subset of PC cases driven by mutations in homologous DNA recombination repair pathway genes, such as *BRCA2*, *BRCA1*, *ATM*, and *CHEK2*. Germline mutations in these genes are detected in up to 12% of men with metastatic PC [7] and carry important cancer screening implications for patients and their living relatives. Acquired somatic alterations can also be seen in a subset of patients [8]. Whether inherited or acquired, the presence of homologous recombination repair gene alterations increases PC cells' dependence on alternative DNA damage repair pathways, and confers sensitivity to targeted therapy via inhibition of the poly(adenosine diphosphate-ribose) polymerase (PARP) enzyme [8] (see the "Nonhormonal Therapies" section below).

Androgen Deprivation Therapy and Mitigating Its Adverse Metabolic Impact

Downregulation of androgen signaling by "upstream" inhibition of androgen synthesis or blockade of AR downstream signaling triggers apoptosis of PC cells. Historically, lowering of circulating androgen levels was achieved by means of surgical orchiectomy (removal of the testes). This approach was shown to cause PCs to shrink, painful bone metastases to recede, and patients to live longer. Lowering circulating androgens to treat PC (termed androgen deprivation therapy, or ADT) remains the mainstay of primary systemic therapy for PC. Contemporarily, ADT is delivered medically rather than surgically with the use of gonadotropin-releasing hormone (GnRH) agonists or antagonists. In the case of GnRH agonists, initial stimulation of the anterior pituitary gland induces the release of follicle-stimulating and luteinizing hormones (FSH and LH), and a resulting flare in testosterone production prior to subsequent desensitization, and loss of testicular production of testosterone. Short-term androgen blockade with an AR antagonist just prior to the time of GnRH agonist initiation is often used to prevent worsening PC symptoms

associated with a testosterone flare. After several weeks of continued stimulation, however, GnRH agonists lead to suppression of the pituitary–gonadal axis and thus inhibit testosterone production from the testis (see chapter "Reproductive System: Overview" under part "Reproductive System"). The effectiveness of ADT relies upon the exquisite dependence on androgens in the vast majority of PCs.

However, by inducing a castrate state with a marked decline in circulating testosterone and therefore estrogen levels, ADT is associated with significant metabolic derangements, including decreased bone mineral density and increased risk of osteoporotic fractures (see chapter "Osteoporosis") [9], increased risk of insulin resistance and overt diabetes mellitus (see chapter "Diabetes Mellitus") [10, 11], dyslipidemia (see chapter "Hyperlipidemia"), increases in visceral fat (see chapter "Metabolic Syndrome"), sarcopenia, and potentially an increased risk of cardiovascular mortality (see Fig. 1) [12]. An ongoing phase 3 randomized clinical trial (NCT03031821) will address the potential benefit of metformin for the prevention of metabolic syndrome in patients initiating ADT. Interestingly, the development of insulin resistance and hyperinsulinemia is linked with increased risk of PC progression, through putative cross-activation of the insulin-like growth factor-1 (IGF-1) receptor by insulin, which is often upregulated in PC [13]. To mitigate the metabolic side effects of ADT, various treatment strategies have been developed (see below). Regular exercise has also been shown to reverse some of the metabolic side effects of long-term androgen deprivation therapy, and may result in increased survival among PC patients [14]. Several prospective studies are currently underway to evaluate the benefit of supervised exercise programs in patients with metastatic PC (NCT02613273, NCT04507698).

Intermittent ADT

Intermittent ADT usually consists of preplanned breaks in therapy after a duration of 6–12 months, to allow for testosterone recovery and partial mit-

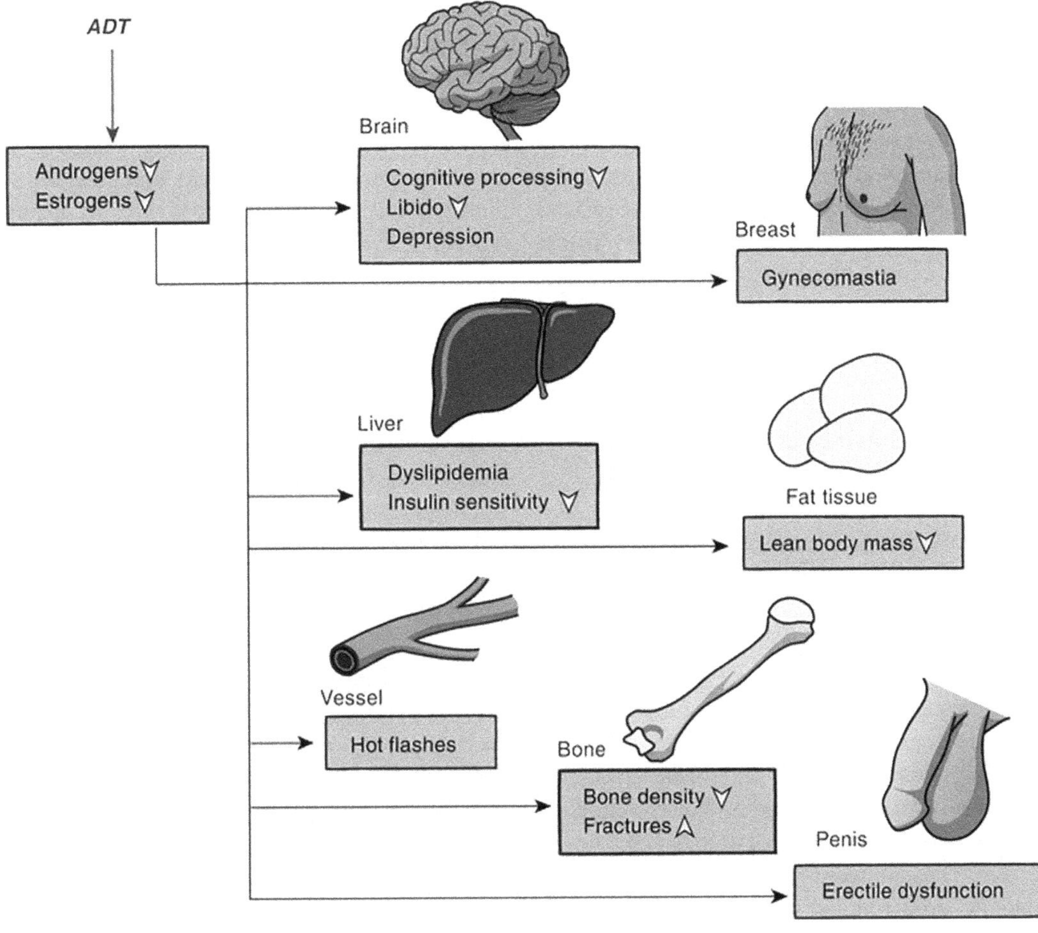

Fig. 1 Metabolic toxicities associated with androgen deprivation therapy. Gynecomastia is a significant breast enlargement in men. *ADT* androgen deprivation therapy

igation of side effects and metabolic impact. Prior studies have shown that during the "off" intervals, significant improvements in bone mineral density can be observed as the circulating testosterone and estrogen levels recover [15, 16]. An intermittent treatment approach may also delay the time to castration resistance (see below) when compared to continuous ADT [17]. Although most patients with metastatic PC benefit from continuous ADT intensified with an AR signaling inhibitor, intermittent ADT remains a standard treatment in men with non-metastatic PCs (i.e., with biochemical relapse following prior definitive therapy). Intermittent ADT may also be considered in carefully selected patients with metastatic PC, especially those with signifi-

cant metabolic comorbidities (mentioned above) [17, 18].

Peripheral Androgen Blockade

Although ADT is the backbone of systemic therapy in nearly all men with PC, peripheral androgen blockade—the use of AR-specific antagonists (e.g., bicalutamide, nilutamide) without concurrent castrating GnRH analogue therapy—may be considered as an alternative to ADT in carefully selected patients. AR antagonist monotherapy inhibits activation of the AR, thereby controlling cancer growth while potentially mitigating those toxicities of ADT related to depletions in circulating estrogen levels. For example, bicalutamide monotherapy leads to elevations in serum testosterone and estrogen levels via loss of negative feedback provided by AR activation and is associated with corresponding improvement in bone mineral density [19]. However, this approach is associated with significant rates of gynecomastia (which may warrant prophylactic breast irradiation), and long-term efficacy data are lacking relative to more standard ADT-based therapies.

Castration-Resistant Prostate Cancer and Approaches to Treat It

Despite the initial effectiveness of ADT in the vast majority of patients, the duration of response is highly variable, and disease progression on ADT is nearly universal. PC that has progressed on primary ADT (as described above) was formerly thought to be "hormone refractory" or "castrate resistant" (CRPC) and independent of signaling through the androgen receptor. However, while ADT markedly lowers circulating androgen levels, it does not reduce the levels to zero. The androgens that remain can still stimulate PC growth in the setting of "castrate resistant" disease. Furthermore, it was observed that discontinuing first-generation AR antagonist treatment at the time of disease progression can lead to declines in serum prostate-specific antigen (PSA), a clinical marker of PC progression, and regression of tumors (a phenomenon commonly referred to as anti-androgen withdrawal response) in a subset of patients with castrate-resistant PC [20]. This phenomenon is due to the small residual agonistic activity of first-generation AR antagonists, which may thus activate AR signaling even in the setting of ADT treatment. These observations have led to two key insights on the management of CRPC: (i) tumors often remain dependent on AR signaling in CRPC, and (ii) further manipulations to inhibit the androgen signaling axis have significant therapeutic potential in this setting.

Key cancer adaptations that allow disease progression despite ADT include: (1) increased expression of AR, often through *AR* gene amplification or activating mutations of the *AR* gene promoter and enhancers [6], (2) AR mutations, which increase promiscuity of ligand-mediated activation, for example, allowing estradiol or progesterone to bind [21], (3) upregulation of intra-tumoral androgen synthesis [22], (4) ligand-independent activation of the AR potentially via constitutively active AR splice variants (Fig. 2) [23], and (5) AR "bypass" mechanisms, including the development of AR-independent variants such as small cell/neuroendocrine prostate cancer. Secondary hormonal therapies targeting these adaptations, as well as non-hormonal approaches, have been developed, which have significant activity against PCs in this setting.

Androgen Synthesis Inhibition

Blockade of androgen synthesis from the adrenal gland and the PC cells itself has been developed as an effective treatment strategy. Ketoconazole is a nonspecific inhibitor of multiple enzymatic steps within the adrenal androgen hormone synthesis pathway (see chapter "Reproductive System: Overview" under part "Reproductive System"), with demonstrated clinical activity in PC [20]. Abiraterone acetate is a more recent androgen synthesis inhibitor developed to more selectively and potently target the cytochrome P450 17 enzyme, which shunts pregnenolone and progesterone precursors down the androgenic pathway [24]. One of the metabolic side effects of selective cytochrome P450 17 inhibition is elevation of mineralocorticoids (see chapter "Reproductive System: Overview" under part "Reproductive System"), including corticosterone and 11-deoxycortisol, which can cause hypertension, hypokalemia, and peripheral edema. These effects can be partially abrogated with the concomitant administration of low-dose corticosteroids, which block adrenal mineralocorticoid synthesis via feedback inhibition of the pituitary gland. Abiraterone acetate was first shown to improve overall survival in men with metastatic castrate-resistant PC [25]. Subsequent trials

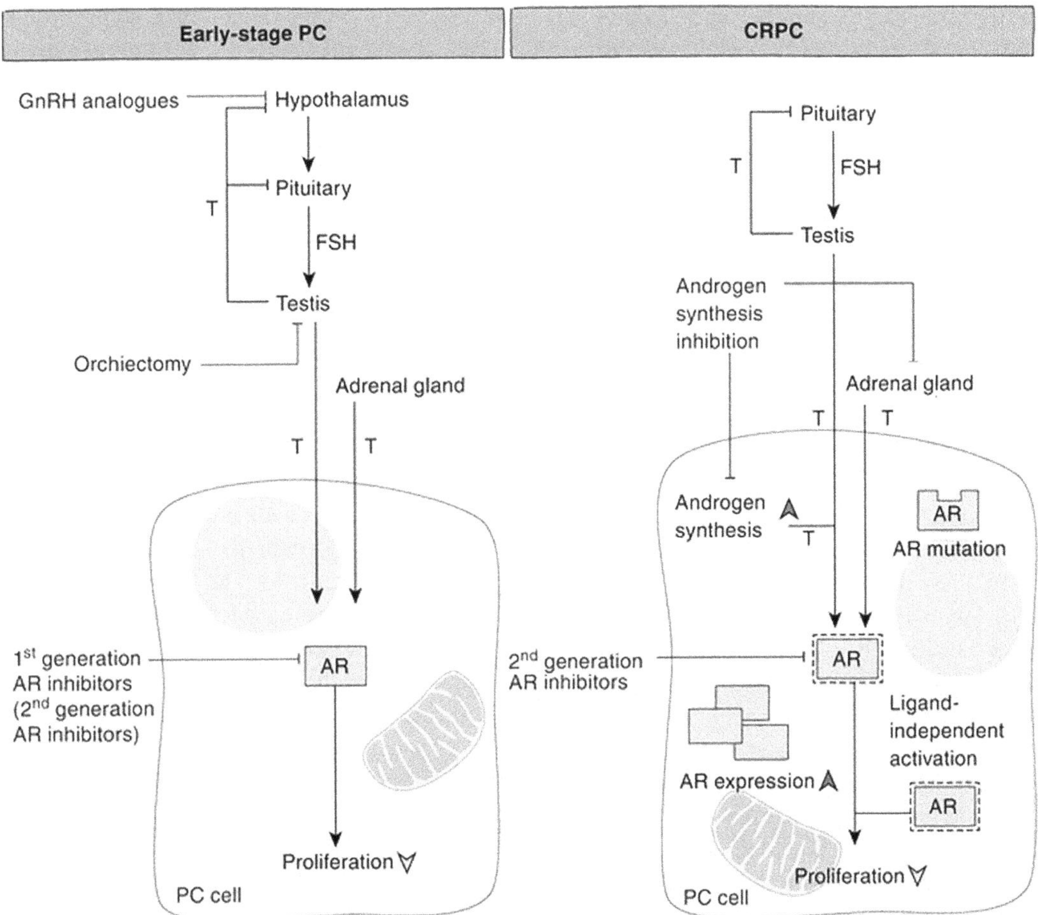

Fig. 2 Molecular mechanism and treatment approaches in early-stage and castrate-resistant prostate cancer (CRPC). Cellular adaptations that promote continued androgen signaling in castration-resistant prostate cancer (CRPC) are shown in the prostate cancer (PC) cell on the right. *AR* androgen receptor, *FSH* follicle-stimulating hormone, *GnRH* gonadotropin-releasing hormone, *T* testosterone

have also shown improved overall survival with the addition of abiraterone acetate to frontline ADT in men with metastatic hormone-sensitive PC, as well as patients with high-risk localized or regionally advanced PC receiving definitive radiation therapy [26, 27].

are ten times more potent than first-generation drugs and lack any agonistic activity. As with abiraterone acetate, use of second-generation AR antagonists in combination with ADT has been shown to prolong survival in men with both hormone-sensitive and castrate-resistant prostate cancer [29, 30].

Second-Generation AR Antagonists

Novel, second-generation AR antagonists (enzalutamide, apalutamide, darolutamide) were developed using prostate cancer models with AR overexpression, recapitulating the molecular characteristics commonly observed with CRPC tumors [28]. Second-generation AR antagonists

Non-hormonal Therapies

Various non-hormonal therapies have also been shown to prolong survival in patients with advanced prostate cancer, including taxane chemotherapies (docetaxel and cabazitaxel) [31, 32], the therapeutic vaccine sipuleucel-T, and radioligand therapies (radium-223 and lutetium-177-PSMA-617) [33–35]. Oral inhibitors of poly(adenosine diphosphate-ribose) polymerase (PARP) may additionally be used in the subset of patients with germline or somatic alterations in homologous DNA recombination pathway genes, such as BRCA1 and BRCA2 [8]. These therapies do not directly target AR signaling, are associated with fewer long-term metabolic side effects, but (with the exception of sipuleucel-T) higher incidences of fatigue and cytopenias.

Perspectives

Potent androgen receptor signaling inhibitors, initially designed for the management of advanced CRPC, have demonstrated a clear survival benefit when combined to upfront ADT in men with hormone-sensitive PC. Although a small subset of patients may be served by intermittent ADT alone, or non-castrating peripheral AR blockade, intensified ADT should be pursued in most patients for whom systemic therapy is indicated. However, the earlier introduction of combined testosterone and AR signaling suppression requires increased awareness and mitigation of the long-term metabolic consequences of hormone therapy.

Questions and Answers

Question 1 Consider the hormonal differences between androgen-deprivation therapy and peripheral androgen blockade, which may account for the lower incidence of osteoporosis and decreased libido, but increased gynecomastia seen with peripheral androgen blockade.

Answer 1 Peripheral androgen blockade results in increased levels of circulating testosterone via loss of negative feedback provided by AR activation in the pituitary gland. Reversible inhibition of AR by bicalutamide on non-cancer cells may be partially overcome by higher levels of testosterone [19]. Increased testosterone production also results in increased levels of circulating

estrogens via peripheral aromatization of testosterone. Increased estrogen levels promote osteoblastic activity and improve bone mineral density. However, the increase in estrogen signaling relative to testosterone on male breast tissue may also lead to significant gynecomastia.

Question 2 True or False: Androgen deprivation therapy with the GnRH agonist leuprolide should be discontinued upon the emergence of castration-resistance disease, and PSA should be monitored for a possible response to anti-androgen withdrawal.

Answer 2 False. Castration resistance often arise through upregulation of the AR signaling pathway. Although castration-resistant PC is able to progress despite low testosterone levels, it remains dependent on AR signaling and ADT should *not* be discontinued. A subset of patients treated with a first-generation AR antagonist (e.g., bicalutamide) in the hormone-sensitive setting may experience regression of their disease upon discontinuation of their AR antagonist (the so-called anti-androgen withdrawal effect). This effect is thought to be due to the small residual agonistic activity of first-generation AR antagonists, and is not seen with the discontinuation of GnRH agonists.

Question 3 Which of the following patients would be least appropriate for the addition of abiraterone acetate to standard ADT?

A. A patient with localized high-risk prostate cancer undergoing definitive radiotherapy
B. A patient with congestive heart failure and newly diagnosed metastatic prostate cancer
C. A patient with metastatic castration-resistant prostate cancer and insulin-dependent diabetes
D. A patient with newly diagnosed prostate cancer and hypertension well-controlled on three oral hypertensives

Answer 3 B. The addition of abiraterone acetate to standard ADT has been shown to prolong survival in patients with metastatic hormone-sensitive and castration-resistant PC, as well as in those with localized high risk or pelvic node positive non-metastatic PC undergoing definitive radiotherapy. Although the addition of abiraterone acetate (given with a low dose of daily prednisone) may lead to increased blood sugar levels and blood pressure levels, neither diabetes nor hypertension are an absolute contraindication to this. However, abiraterone acetate should be avoided in patients with congestive heart failure given the risk of mineralocorticoid excess and the resulting increased blood pressure, fluid retention, and hypokalemia.

References

1. Sung H, Ferlay J, Siegel RL et al (2021) Global cancer statistics 2020: GLOBOCAN estimates of incidence and mortality worldwide for 36 cancers in 185 countries. CA Cancer J Clin 71:209–249. https://doi.org/10.3322/caac.21660
2. Al Olama AA, Kote-Jarai Z, Berndt SI et al (2014) A meta-analysis of 87,040 individuals identifies 23 new susceptibility loci for prostate cancer. Nat Genet 46:1103–1109. https://doi.org/10.1038/ng.3094
3. Cancel-Tassin G, Cussenot O (2005) Genetic susceptibility to prostate cancer. BJU Int 96:1380–1385. https://doi.org/10.1111/j.1464-410X.2005.05836.x
4. MacInnis RJ, English DR (2006) Body size and composition and prostate cancer risk: systematic review and meta-regression analysis. Cancer Causes Control 17:989–1003. https://doi.org/10.1007/s10552-006-0049-z
5. Dai C, Heemers H, Sharifi N (2017) Androgen signaling in prostate cancer. Cold Spring Harb Perspect Med 7:a030452. https://doi.org/10.1101/cshperspect.a030452
6. Quigley DA, Dang HX, Zhao SG et al (2018) Genomic hallmarks and structural variation in metastatic prostate cancer. Cell 174:758–769.e9. https://doi.org/10.1016/j.cell.2018.06.039
7. Pritchard CC, Mateo J, Walsh MF et al (2016) Inherited DNA-repair gene mutations in men with metastatic prostate cancer. N Engl J Med 375:443–453. https://doi.org/10.1056/NEJMoa1603144
8. de Bono J, Mateo J, Fizazi K et al (2020) Olaparib for metastatic castration-resistant prostate cancer. N Engl J Med 382:2091–2102. https://doi.org/10.1056/NEJMoa1911440
9. Shahinian VB, Kuo Y-F, Freeman JL, Goodwin JS (2005) Risk of fracture after androgen deprivation for prostate cancer. N Engl J Med 352:154–164. https://doi.org/10.1056/NEJMoa041943
10. Keating NL, O'Malley AJ, Smith MR (2006) Diabetes and cardiovascular disease during androgen deprivation therapy for prostate cancer. J Clin Oncol Off J Am Soc Clin Oncol 24:4448–4456. https://doi.org/10.1200/JCO.2006.06.2497
11. Keating NL, O'Malley AJ, Freedland SJ, Smith MR (2010) Diabetes and cardiovascular disease during androgen deprivation therapy: observational study of veterans with prostate cancer. JNCI J Natl Cancer Inst 102:39–46. https://doi.org/10.1093/jnci/djp404
12. Nguyen PL, Je Y, Schutz FAB et al (2011) Association of androgen deprivation therapy with cardiovascular death in patients with prostate cancer: a meta-analysis of randomized trials. JAMA 306:2359–2366. https://doi.org/10.1001/jama.2011.1745
13. Aggarwal RR, Ryan CJ, Chan JM (2013) Insulin-like growth factor pathway: a link between androgen deprivation therapy (ADT), insulin resistance, and disease progression in patients with prostate cancer? Urol Oncol 31:522–530. https://doi.org/10.1016/j.urolonc.2011.05.001
14. Kenfield SA, Stampfer MJ, Giovannucci E, Chan JM (2011) Physical activity and survival after prostate cancer diagnosis in the health professionals follow-up study. J Clin Oncol Off J Am Soc Clin Oncol 29:726–732. https://doi.org/10.1200/JCO.2010.31.5226
15. Yu EY, Kuo KF, Gulati R et al (2012) Long-term dynamics of bone mineral density during intermittent androgen deprivation for men with nonmetastatic, hormone-sensitive prostate cancer. J Clin Oncol Off J Am Soc Clin Oncol 30:1864–1870. https://doi.org/10.1200/JCO.2011.38.3745
16. Tsai H-T, Pfeiffer RM, Philips GK et al (2017) Risks of serious toxicities from intermittent versus continuous androgen deprivation therapy for advanced prostate cancer: a population based study. J Urol 197:1251–1257. https://doi.org/10.1016/j.juro.2016.12.022
17. Crook JM, O'Callaghan CJ, Duncan G et al (2012) Intermittent androgen suppression for rising PSA level after radiotherapy. N Engl J Med 367:895–903. https://doi.org/10.1056/NEJMoa1201546
18. Hussain M, Tangen CM, Berry DL et al (2013) Intermittent versus continuous androgen deprivation in prostate cancer. N Engl J Med 368:1314–1325. https://doi.org/10.1056/NEJMoa1212299
19. Smith MR, Goode M, Zietman AL et al (2004) Bicalutamide monotherapy versus leuprolide monotherapy for prostate cancer: effects on bone mineral density and body composition. J Clin Oncol Off J Am Soc Clin Oncol 22:2546–2553. https://doi.org/10.1200/JCO.2004.01.174
20. Small EJ, Halabi S, Dawson NA et al (2004) Antiandrogen withdrawal alone or in combination with ketoconazole in androgen-independent prostate cancer patients: a phase III trial (CALGB 9583). J

Clin Oncol Off J Am Soc Clin Oncol 22:1025–1033. https://doi.org/10.1200/JCO.2004.06.037

21. Taplin M-E, Rajeshkumar B, Halabi S et al (2003) Androgen receptor mutations in androgen-independent prostate cancer: Cancer and Leukemia Group B Study 9663. J Clin Oncol Off J Am Soc Clin Oncol 21:2673–2678. https://doi.org/10.1200/JCO.2003.11.102

22. Montgomery RB, Mostaghel EA, Vessella R et al (2008) Maintenance of intratumoral androgens in metastatic prostate cancer: a mechanism for castration-resistant tumor growth. Cancer Res 68:4447–4454. https://doi.org/10.1158/0008-5472.CAN-08-0249

23. Guo Z, Yang X, Sun F et al (2009) A novel androgen receptor splice variant is up-regulated during prostate cancer progression and promotes androgen depletion-resistant growth. Cancer Res 69:2305–2313. https://doi.org/10.1158/0008-5472.CAN-08-3795

24. Ryan CJ, Smith MR, Fong L et al (2010) Phase I clinical trial of the CYP17 inhibitor abiraterone acetate demonstrating clinical activity in patients with castration-resistant prostate cancer who received prior ketoconazole therapy. J Clin Oncol Off J Am Soc Clin Oncol 28:1481–1488. https://doi.org/10.1200/JCO.2009.24.1281

25. Ryan CJ, Smith MR, de Bono JS et al (2013) Randomized phase 3 trial of abiraterone acetate in men with metastatic castration-resistant prostate cancer and no prior chemotherapy. N Engl J Med 368:138–148. https://doi.org/10.1056/NEJMoa1209096

26. Attard G, Murphy L, Clarke NW et al (2022) Abiraterone acetate and prednisolone with or without enzalutamide for high-risk non-metastatic prostate cancer: a meta-analysis of primary results from two randomised controlled phase 3 trials of the STAMPEDE platform protocol. Lancet 399:447–460. https://doi.org/10.1016/S0140-6736(21)02437-5

27. de Bono JS, Logothetis CJ, Molina A et al (2011) Abiraterone and increased survival in metastatic prostate cancer. N Engl J Med 364:1995–2005. https://doi.org/10.1056/NEJMoa1014618

28. Tran C, Ouk S, Clegg NJ et al (2009) Development of a second-generation antiandrogen for treatment of advanced prostate cancer. Science 324:787–790. https://doi.org/10.1126/science.1168175

29. Armstrong AJ, Szmulewitz RZ, Petrylak DP et al (2019) ARCHES: a randomized, phase III study of androgen deprivation therapy with enzalutamide or placebo in men with metastatic hormone-sensitive prostate cancer. J Clin Oncol Off J Am Soc Clin Oncol 37:2974–2986. https://doi.org/10.1200/JCO.19.00799

30. Chi KN, Agarwal N, Bjartell A et al (2019) Apalutamide for metastatic, castration-sensitive prostate cancer. N Engl J Med 381:13–24. https://doi.org/10.1056/NEJMoa1903307

31. Tannock IF, de Wit R, Berry WR et al (2004) Docetaxel plus prednisone or mitoxantrone plus prednisone for advanced prostate cancer. N Engl J Med 351:1502–1512. https://doi.org/10.1056/NEJMoa040720

32. de Bono JS, Oudard S, Ozguroglu M et al (2010) Prednisone plus cabazitaxel or mitoxantrone for metastatic castration-resistant prostate cancer progressing after docetaxel treatment: a randomised open-label trial. Lancet (Lond Engl) 376:1147–1154. https://doi.org/10.1016/S0140-6736(10)61389-X

33. Kantoff PW, Higano CS, Shore ND et al (2010) Sipuleucel-T immunotherapy for castration-resistant prostate cancer. N Engl J Med 363:411–422. https://doi.org/10.1056/NEJMoa1001294

34. Parker C, Nilsson S, Heinrich D et al (2013) Alpha emitter radium-223 and survival in metastatic prostate cancer. N Engl J Med 369:213–223. https://doi.org/10.1056/NEJMoa1213755

35. Sartor O, de Bono J, Chi KN et al (2021) Lutetium-177–PSMA-617 for metastatic castration-resistant prostate cancer. N Engl J Med 385:1091–1103. https://doi.org/10.1056/NEJMoa2107322

Cancer

Cancer Metabolism

Xue Zhang, Xin Guo, Catherine Lan Wang,
Saravana Gowtham Baskaran, Yahui Wang,
Zachary E. Stine, and Chi V. Dang

Introduction to the Disease

Cancer causes 10 million deaths per year globally, contrasting with a total of 7 million COVID-19 (coronavirus disease-2019) deaths throughout the pandemic, and cancer deaths are expected to increase by at least 50% over the next two decades. Through the accumulation of genetic mutations, cancer cells disobey the boundaries of normal tissue homeostasis and proliferate rapidly, acquiring metabolic needs that can differ from those of more slowly dividing non-cancerous cells [1, 2]. While many quiescent cells acquire homeostasis through maintenance metabolism to meet the energy demands required for processes such as maintenance of cell membrane potential and protein synthesis, cancer cells require proliferative metabolism that is needed to produce not only energy but also the cellular building blocks, particularly protein synthesis, required for rapid growth. Cancer cells reprogram their metabolism to provide the components for macromolecular biosynthesis, that is, nucleotides to produce new deoxyribonucleic acid (DNA), lipids to create new cell membranes, and ribosomes and amino acids required for increased protein production [1]. Cancer is challenging to treat due to its similarity to non-cancerous cells and the high amount of inter-tumor and intra-tumor heterogeneity, with particular similarities between cancer cells and the cells of the immune system, which are critical to the efficacy of many anti-cancer therapies, particularly cancer immunotherapy. While major advances have been made in the understanding of cancer metabolism, targeting cancer metabolism outside of nucleotide metabolism has largely failed to translate in the clinic due to metabolic plasticity of cancer cells and shared metabolic

X. Zhang
Department of Oncology, Bloomberg-Kimmel Institute for Cancer Immunotherapy, Department of Biochemistry & Molecular Biology, Johns Hopkins University, Baltimore, MD, USA

Ludwig Institute for Cancer Research, New York, NY, USA

School of Life Science and Biopharmaceutics, Guangdong Pharmaceutical University, Guangzhou, China

X. Guo · C. L. Wang · S. G. Baskaran · Y. Wang · C. V. Dang (✉)
Department of Oncology, Bloomberg-Kimmel Institute for Cancer Immunotherapy, Department of Biochemistry & Molecular Biology, Johns Hopkins University, Baltimore, MD, USA

Ludwig Institute for Cancer Research, New York, NY, USA
e-mail: xguo8@jhmi.edu; cwang210@jhmi.edu; sbaskar3@jhmi.edu; ywang915@jh.edu; cvdang@jhmi.edu

Z. E. Stine
Ludwig Institute for Cancer Research, New York, NY, USA

pathways between cancer and non-cancer cells in the tumor microenvironment.

Pathological Changes in Metabolism and Possible Outcomes

To transform into a cancer cell, somatic cells sustain a series of mutations, allowing for unlimited growth and replication. Traditionally, mutations in at least two major classes of determinants are required: oncogenes, whose constitutive activation deregulates growth of a cell toward cancer (e.g., through overexpression of growth factor receptor genes), and tumor suppressors, whose inactivation removes important anti-growth signals or protective pathways.

Constant cell division requires continuous production of cellular building blocks, and consequently, many cancerogenic mutations directly (or indirectly) affect cellular metabolism. In this respect, major changes in metabolism are discussed, followed by important examples of how mutations in oncogenes and tumor suppressors can reprogram metabolism. Finally, we discuss the challenges in targeting metabolism and potential interplay between cancer metabolism and the immune system.

Changes in Metabolism

Quiescent, healthy cells primarily convert glucose to pyruvate to enter the citric acid cycle, also called tricarboxylic acid (TCA) cycle, in the mitochondria to efficiently produce adenosine triphosphate (ATP) through oxidative metabolism. Only in the absence of oxygen do most noncancerous cells switch to converting pyruvate to lactate, enabling the generation of ATP anaerobically through glycolysis. Healthy cells can also use fatty acids for oxidation to produce ATP in a highly efficient manner. However, cancer cells often show increased glucose uptake and increased glycolysis even in the presence of oxygen [2], converting pyruvate to lactate to regenerate nicotinamide adenine dinucleotide (NAD^+)

required for sustaining glycolysis (Fig. 1). The phenomenon of glucose being converted to lactate in the presence of oxygen is termed the Warburg effect, also known as *aerobic glycolysis*, as it was first reported by Otto Warburg in the 1920s [2]. Many but not all cancers exhibit *aerobic glycolysis* [3]. Glycolytic conversion of glucose to pyruvate provides a rapid means of producing ATP. However, aerobic glycolysis produces much fewer ATP per molecule of glucose than oxidative glucose catabolism, but it allows glucose to contribute to biosynthetic pathways, such as de novo serine synthesis, production of NADPH (NAD phosphate) for fatty acid synthesis, and lipid glycerol backbone. The conversion of pyruvate to lactate regenerates NAD^+ that is necessary for redox homeostasis particularly for continued glycolysis. Thus, *aerobic glycolysis* shunts glucose carbons for biosynthesis and catabolizes glucose to lactate for rapid ATP production [4]. The avid uptake of glucose by cancer cells is used clinically to identify and follow tumors during cancer therapy. Using a labeled glucose derivative, ^{18}F-deoxyglucose, positron emission tomography (PET) allows localization and imaging of tumors in patients.

Glycolysis shares common intermediates with other anabolic pathways. For example, the pentose phosphate pathway (PPP; see also chapter "Diabetes Mellitus") shares glucose-6-phosphate as a starting molecule with glycolysis (Fig. 1) [5]. In the PPP, glucose-6-phosphate is converted to ribose by a multiple step pathway, the 5-carbon sugar required for nucleotide synthesis. Additionally, the PPP is a major source of NADPH, a reducing agent required for nucleic acid synthesis, fatty acid synthesis, and detoxification of reactive oxygen species (ROS). Therefore, the PPP provides much of the reducing power and ribose that cancer cells require for nucleotide biosynthesis.

In proliferating cells, extracellular serine or glucose, through de novo serine synthesis, serve as a donors for one-carbon metabolisms. To this end, glucose, or rather the glycolytic intermediate, 3-phosphoglycerate, is first converted to serine [6], which is then converted to glycine or used for protein synthesis (Fig. 1). The conver-

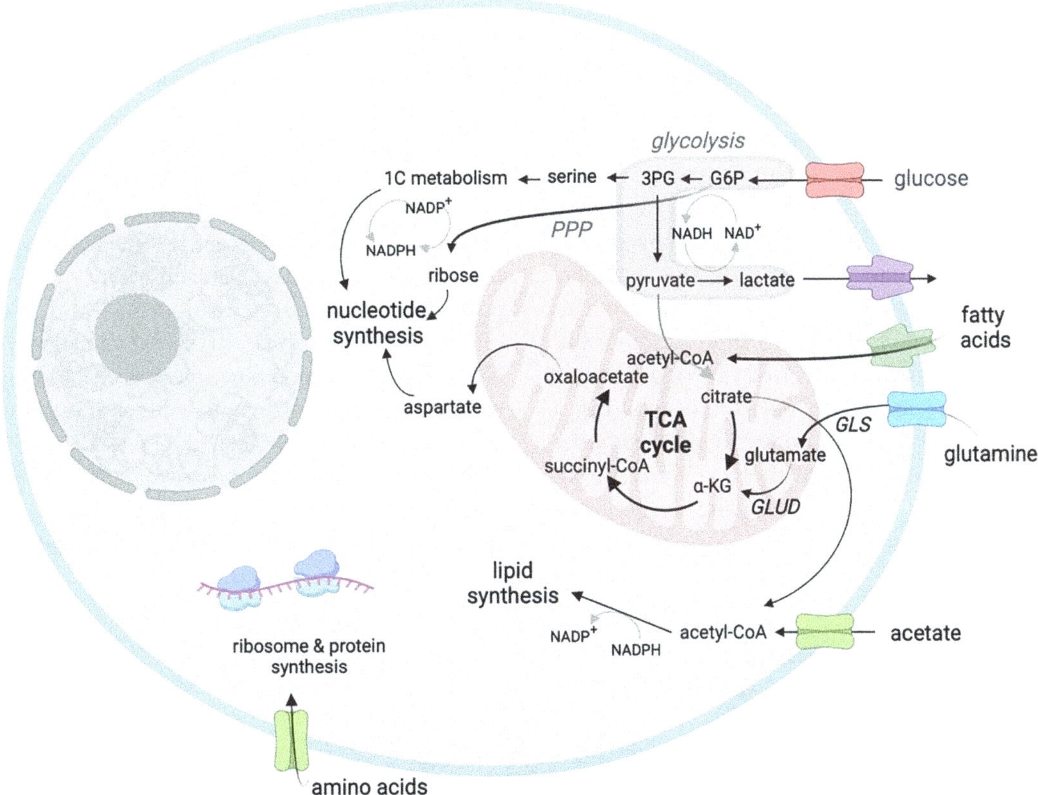

Fig. 1 Cancer cells reprogram metabolism to promote biosynthesis. Glucose and glutamine play a central role in cancer metabolism. Cancer cells show enhanced shunting of glucose to lactate and biosynthetic pathways. Amino acids are imported for protein synthesis. The pentose phosphate pathway (PPP) provides ribose for nucleotides and NADPH for biosynthesis. The serine/glycine synthesis pathway provides carbon donors for nucleotide synthesis. Cancer cells depend on glutamine to fuel the tricarboxylic acid (TCA) cycle for energy and biosynthetic precursors. Glutamine enters the TCA cycle following its conversion to α-ketoglutarate (α-KG). Citrate, a precursor for lipid synthesis, can be derived from either glucose or glutamine. *1C* one-carbon, *3PG* 3-phosphoglycerate, *G6P* glucose 6-phosphate, *GLS* glutaminase, *GLUD* glutamate dehydrogenase, *LDH* lactate dehydrogenase, *THF* tetrahydrofolate. (Created with Biorender.org)

sion of serine to glycine is an important step in one-carbon metabolism, which provides a carbon donor [7] and is used for nucleotide and methionine synthesis. Diversion of glucose into the serine/glycine synthesis pathways allows cancer cells to create building blocks for growth.

Warburg erroneously believed that glucose is robustly converted to lactate due to a lack of functioning mitochondria in all cancer cells. In fact, mitochondria are functional in many cancers [8]. In addition to providing ATP though oxidative phosphorylation, the TCA cycle provides precursors for cellular building blocks (Fig. 1). It provides citrate, which acts as a major cytoplas-mic acetyl-CoA source for fatty acid synthesis (Fig. 1). However, since the Warburg effect can divert a substantial portion of glucose to lactate, many cancer cells fuel the TCA cycle with other carbon sources including acetate, fatty acids, branched-chain amino acids, and glutamine (Gln) [9]. To enter the TCA cycle, Gln is first converted to glutamate by glutaminase and then to the TCA cycle intermediate α-ketoglutarate. Additionally, Gln plays a central role in the formation of the tri-peptide glutathione through its role in glutamate production and cystine import, which is critical for controlling reactive oxygen species (ROS) [10].

Proliferating cells need to generate sufficient lipids (in form of triglycerides and phospholipids) for synthesis of plasma membrane for growth and cell division, and for storage of energy [11]. While some lipids may be obtained exogenously, many cancers synthesize lipids de novo, and there are ongoing efforts to target the dependence of tumors on these pathways [12]. Fatty acid synthesis occurs in the cytoplasm through additions of 2-carbon groups from acetyl-CoA, which can be generated from citrate, as the latter can be exported from the mitochondrion whereas acetyl-CoA itself cannot (Fig. 1). In addition to anabolic lipid metabolism, some cancer cells catabolize lipids through β-oxidation in the mitochondria, producing ATP [11].

Common Mutations Affecting Metabolism

Many oncogenes have been shown to reprogram cellular metabolism in cancer [1] (Fig. 2). Activating mutations in oncogenes such as growth factor receptors lead to continuous induction of cell division irrespective of external signals. For example, constitutively active phosphatidylinositol-3-kinase (PI3K) or Akt (also called protein kinase B) signaling pathways are observed in numerous cancer types, leading to activation of the mammalian target of rapamycin complex 1 (mTORC1) and increased *aerobic glycolysis*. mTORC is a master regulator of cell metabolism, growth, mitochondrial biogenesis, lipogenesis, and protein synthesis [13]. Enhanced

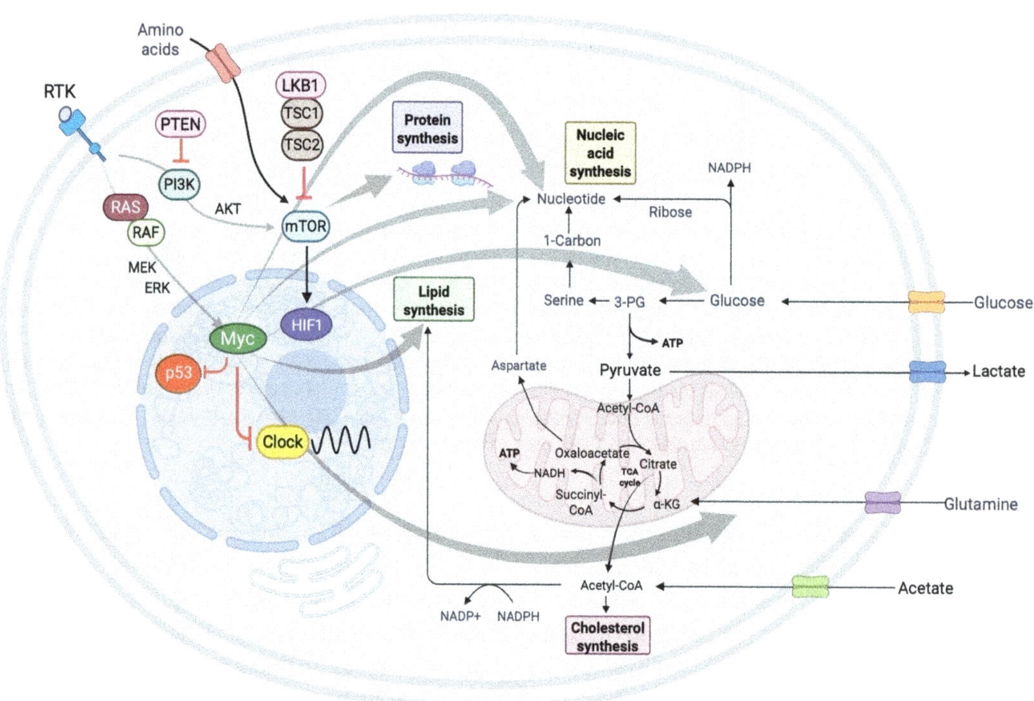

Fig. 2 Cancer therapies target cancer metabolism and the pathways that control them. Commonly found activating mutations of many oncogenes such as receptor tyrosine kinases (RTKs) and their downstream signaling factors phosphatidyl-inositol-3-kinase (PI3K), Kirsten rat sarcoma viral oncogene homolog (KRAS), protein kinase B (Akt), rapidly accelerated fibrosarcoma (RAF), or inactivating mutations of tumor suppressors (phosphatase and tensin homolog [PTEN], tuberous sclerosis complex [TSC] 1/2, liver kinase B1 [LKB1]) in cancer lead to mammalian target of rapamycin complex 1 (mTORC1) activation and downstream metabolic changes. Myelocytomatosis (Myc) protein is stabilized downstream of KRAS, blocks the molecular circadian Clock and stimulates proliferative metabolism. The hypoxia-inducible factor (HIF) is increased by mTOR or hypoxia and contributes to increased glucose fermentation. (Created with Biorender.org)

glycolysis is also partly due to non-hypoxic increase in hypoxia inducible factor 1α (HIF-1α) downstream of mTORC1.

Constitutively active PI3K signaling can also occur through loss-of-function mutations in phosphatase and tensin homolog (PTEN), a PI3K inhibitor and major tumor suppressor, or direct activating mutations in PI3K. Mutations in PI3K pathway signaling components are present in ~70% of breast cancers (see chapter "Breast Cancer") and 50% of colorectal cancers, which is likely to lead to mTORC1-driven cell growth [14, 15]. mTORC1 can also be constitutively activated through activating mutations of Kirsten rat sarcoma viral oncogene homolog (KRAS) or rapidly accelerated fibrosarcoma (B-RAF) mutations, or loss of the mTORC1 inhibitors Liver kinase B1 (LKB1) or tuberous sclerosis complex (TSC) 1/2 (Fig. 2).

Transcription factors that play a central role in cancer pathogenesis can also exert significant control over metabolism. c-Myc, one of the most commonly overexpressed genes in cancer, has been shown to enhance the Warburg effect and increase glutamine dependence of cells [16]. In addition to promoting mitochondrial and ribosomal biogenesis, c-Myc promotes the expression of glutaminase, the first gene in glutamine metabolism and glutamine transporters to promote the uptake of glutamine [16, 17]. Furthermore, c-Myc promotes expression of numerous genes involved in glycolysis, as well as many of the transporters and enzymes involved in glucose uptake [16]. In addition to oncogene activation, mutations in tumor suppressor genes have been shown to reprogram cellular metabolism. The tumor suppressor p53 is one of the most frequently mutated genes in cancer. p53 prevents tumorigenesis through its role in cell cycle arrest and apoptosis in response to DNA damage and other stressors. p53 mutations have also been shown to alter metabolism controlling glycolysis, the PPP, oxidative respiration, and glutamine and serine metabolism [18–21]. Due to rapid growth and poor tumor vascularization, many cancer cells have limited oxygen availability and thus activate the transcription factor HIF-1α [22]. HIF-1α controls transcription of key metabolic genes including glucose transporters and glycolytic enzymes to enhance the Warburg effect, optimizing tumor metabolism for the tumor microenvironment.

Mutations in some metabolic enzymes can cause the buildup of metabolites, termed oncometabolites, which are proposed to play a role in tumor progression. Mutated isocitrate dehydrogenase (IDH), a TCA cycle enzyme, creates the oncometabolite (2-hydroxyglutarate), which interferes with cellular differentiation by causing epigenetic changes through the inhibition of α-ketoglutarate-dependent chromatin and DNA methylation [23]. Mutant IDH inhibitors are now FDA (U.S. Food and Drug Administration) approved for the treatment of refractory acute myeloid leukemia (AML) harboring the mutation. Frequent mutations in the TCA cycle enzyme fumarate hydratase (also called fumarase) cause the accumulation of fumarate, proposed to disrupt metabolism and stabilize HIF-1α [24]. However, much more remains to be understood about how the accumulation of oncometabolites in a cell can contribute to cancer development.

Introduction to Treatment

Cancer treatment often focuses on the use of surgery, DNA-damaging radiation, or cytotoxic chemotherapy agents that damage rapidly dividing cells by inhibiting DNA synthesis, transcription, and cell division. More recently, cancer immunotherapy holds great promise for better clinical outcomes. As metabolism plays a key role in DNA synthesis, anti-metabolic therapies may sensitize cancer cells to cytotoxic chemotherapies. In turn, cytotoxic agents likely cause disruption of cancer metabolism.

The ideal cancer therapies target and kill cancer cells while sparing the non-cancerous tissues around them. Classical therapies aim to target the deregulated signaling pathways leading to malignant growth and proliferation. For example, the ATP analogue sunitinib blocks several important receptor tyrosine kinases (RTK), which favor proliferation. Similarly, vemurafenib, a Raf inhibitor, targets the mitogen-activated protein kinase (MAPK) pathway to induce cancer cell death. As cancer cells exhibit altered metabolism compared to quiescent cells, cancer metabolism has generated interest as a potential therapeutic target [25], for example, inhibiting *aerobic glycolysis* or glutamine metabolism may slow cancer growth [17]. 2-Deoxyglucose, a glucose analogue, is used to inhibit glycolysis and thus hamper cell growth. Cancer metabolism-based therapy seeks to disrupt one or more of the numerous metabolic changes that occur in cancer to slow cancer cell growth, induce cancer cell death, or enhance sensitivity to other treatments. However, metabolic cancer therapies are complicated by the metabolic flexibility of cancer cells, feedback loops, and the metabolic similarities between cancer cells and some proliferating cells. The first drugs to target cancer metabolism were anti-folate drugs in the 1940s [26], giving rise to the commonly used anti-folate drug methotrexate. Folate is an important donor of methyl groups during nucleotide synthesis. Other treatments, like the nucleoside analogue gemcitabine, also inhibit nucleotide synthesis. As glycolytic intermediate-derived glycine is a critical source of folate, anti-glycolytic therapy may sensitize cancer cells to anti-folate drugs.

A number of metabolic enzyme inhibitors have entered clinical trials over the decade with limited success [12]. Inhibitors of mitochondrial oxidative metabolism, glutaminase, and glycolytic metabolism have failed to advance to approval due to a combination of lack of efficacy and narrow therapeutic window. There are now multiple mTORC inhibitors approved for cancer. While the tumor autonomous inhibition of cancer metabolism has shown limited efficacy in treating cancer, there is a growing interest in immunometabolism corresponding to the expanded use of immune checkpoint inhibitors in the clinic. Manipulation of the availability of metabolites in the metabolic milieu may enhance anti-tumor immune response.

Influence of Treatment on Metabolism

Cytotoxic chemotherapeutic agents can have a wide array of side effects due to their effects on non-cancerous cells, including organ damage, immune problems, gastrointestinal defects, hair loss, fatigue, and cognitive alterations. Anticancer therapies have global metabolic effects in patients that are poorly understood. Many patients undergoing chemotherapy experience changes in weight and fat accumulation. While nausea and other side effects can cause weight loss in many cancers, women receiving adjuvant therapy (see chapter "Breast Cancer") often experience weight gain.

Perspectives

Cancer cells undergo profound metabolic changes, allowing for rapid proliferation and cell maintenance in a hostile tumor microenvironment. While *aerobic glycolysis* can fuel biosynthetic pathways, glutamine can fuel the TCA cycle to provide energy and other cellular building blocks. Hence, enhanced understanding of changes in cancer metabolism, metabolic plasticity, and the mutations that cause them should provide a potential therapeutic avenue for treating cancer. While inhibitors against oncogenic isocitrate dehydrogenase mutations [27, 28] have proven successful, targeting glycolysis, the mitochondria, and fatty acid metabolism remains to be established [12].

Questions and Answers

Question 1 Did more people die of cancer than COVID-19 in 2020 when no COVID-19 vaccine was broadly available?

Answer1 Even in 2020 more people died of cancer than COVID-19.

Question 2 Please describe the Warburg effect!

Answer 2 It is the conversion of glucose to lactate despite the presence of oxygen and this conversion often occurs in tumor cells (and is also called *aerobic glycolysis*).

Question 3 How is glucose uptake by many tumor cells used in the clinics?

Answer 3 Using ^{18}F-deoxyglucose (that cannot be metabolized), PET scans allow localization and imaging of tumors in patients.

References

1. Ward PS, Thompson CB (2012) Metabolic reprogramming: a cancer hallmark even Warburg did not anticipate. Cancer Cell 21(3):297–308
2. Dang CV (2023) Cancer metabolism historical perspectives: a chronicle of controversies and consensus. Cold Spring Harb Perspect Med. 13(12):a041530
3. Koppenol WH, Bounds PL, Dang CV (2011) Otto Warburg's contributions to current concepts of cancer metabolism. Nat Rev Cancer 11(5):325–337
4. Kelloff GJ, Hoffman JM, Johnson B, Scher HI, Siegel BA, Cheng EY et al (2005) Progress and promise of FDG-PET imaging for cancer patient management and oncologic drug development. Clin Cancer Res 11(8):2785–2808
5. Riganti C, Gazzano E, Polimeni M, Aldieri E, Ghigo D (2012) The pentose phosphate pathway: an antioxidant defense and a crossroad in tumor cell fate. Free Radic Biol Med 53(3):421–436
6. Possemato R, Marks KM, Shaul YD, Pacold ME, Kim D, Birsoy K et al (2011) Functional genomics reveal that the serine synthesis pathway is essential in breast cancer. Nature 476(7360):346–350
7. Jain M, Nilsson R, Sharma S, Madhusudhan N, Kitami T, Souza AL et al (2012) Metabolite profiling identifies a key role for glycine in rapid cancer cell proliferation. Science 336(6084):1040–1044
8. Wallace DC (2012) Mitochondria and cancer. Nat Rev Cancer 12(10):685–698
9. Altman BJ, Stine ZE, Dang CV (2016) From Krebs to clinic: glutamine metabolism to cancer therapy. Nat Rev Cancer 16(10):619–634
10. Son J, Lyssiotis CA, Ying H, Wang X, Hua S, Ligorio M et al (2013) Glutamine supports pancreatic cancer growth through a KRAS-regulated metabolic pathway. Nature 496(7443):101–105
11. Santos CR, Schulze A (2012) Lipid metabolism in cancer. FEBS J 279(15):2610–2623
12. Stine ZE, Schug ZT, Salvino JM, Dang CV (2022) Targeting cancer metabolism in the era of precision oncology. Nat Rev Drug Discov 21(2):141–162
13. Laplante M, Sabatini DM (2012) mTOR signaling in growth control and disease. Cell 149(2):274–293
14. The Cancer Genome Atlas Network (2012) Comprehensive molecular characterization of human colon and rectal cancer. Nature 487(7407):330–337
15. Cancer Genome Atlas Network, Comprehensive molecular portraits of human breast tumours (2012) Nature 490(7418):61–70
16. Dang CV (2012) MYC on the path to cancer. Cell 149(1):22–35
17. Wise DR, Thompson CB (2010) Glutamine addiction: a new therapeutic target in cancer. Trends Biochem Sci 35(8):427–433
18. Bensaad K, Tsuruta A, Selak MA, Vidal MN, Nakano K, Bartrons R et al (2006) TIGAR, a p53-inducible regulator of glycolysis and apoptosis. Cell 126(1):107–120
19. Hu W, Zhang C, Wu R, Sun Y, Levine A, Feng Z (2010) Glutaminase 2, a novel p53 target gene regulating energy metabolism and antioxidant function. Proc Natl Acad Sci USA 107(16):7455–7460
20. Jiang P, Du W, Wang X, Mancuso A, Gao X, Wu M et al (2011) p53 regulates biosynthesis through direct inactivation of glucose-6-phosphate dehydrogenase. Nat Cell Biol 13(3):310–316
21. Maddocks OD, Berkers CR, Mason SM, Zheng L, Blyth K, Gottlieb E et al (2013) Serine starvation induces stress and p53-dependent metabolic remodelling in cancer cells. Nature 493(7433):542–546
22. Semenza GL (2010) HIF-1: upstream and downstream of cancer metabolism. Curr Opin Genet Dev 20(1):51–56
23. Lu C, Ward PS, Kapoor GS, Rohle D, Turcan S, Abdel-Wahab O et al (2012) IDH mutation impairs histone demethylation and results in a block to cell differentiation. Nature 483(7390):474–478
24. Isaacs JS, Jung YJ, Mole DR, Lee S, Torres-Cabala C, Chung YL et al (2005) HIF overexpression correlates with biallelic loss of fumarate hydratase in renal cancer: novel role of fumarate in regulation of HIF stability. Cancer Cell 8(2):143–153
25. Tennant DA, Duran RV, Gottlieb E (2010) Targeting metabolic transformation for cancer therapy. Nat Rev Cancer 10(4):267–277
26. Farber S, Diamond LK (1948) Temporary remissions in acute leukemia in children produced by folic acid antagonist, 4-aminopteroyl-glutamic acid. N Engl J Med 238(23):787–793
27. Rohle D, Popovici-Muller J, Palaskas N, Turcan S, Grommes C, Campos C et al (2013) An inhibitor of mutant IDH1 delays growth and promotes differentiation of glioma cells. Science 340:626
28. Wang F, Travins J, Delabarre B, Penard-Lacronique V, Schalm S, Hansen E et al (2013) Targeted inhibition of mutant IDH2 in leukemia cells induces cellular differentiation. Science 340:622

Index

A

Abiraterone acetate, 390–393
Acetoacetic acid, 244
Acid–base homeostasis, 201, 203, 207, 345, 349
Acinar cells, 183–185, 187, 188
Acute gout, 351, 353, 355
Acute on chronic liver failure (ACLF), 215
Adipocytes, 9, 10, 20, 21, 23, 108, 115, 192, 194, 219–227, 232, 233, 269, 302–303
Adipokines, 192, 194, 219, 222–224, 227, 232–234, 283, 284, 287, 380, 384
Adipose tissue, 17, 99, 187, 192, 219, 232, 269, 283, 313, 336, 367, 380
Age-related macular degeneration (AMD), 80, 81, 83–88
Aldosterone, 274–276, 346–348, 371
Alveolar epithelium, 241, 242
Ammonia, 10, 202, 203, 206, 211–214, 346, 362
Amyloid plaques, 45–49
Analgesics, 133, 168, 308
Androgen deprivation therapy (ADT), 140, 388–392
Androgen receptors, 373, 387, 390, 391
Androgen receptor signaling, 392
Androgens, 140–142, 144, 225, 367, 369, 371, 373–375, 387, 388
Angiotensin, 276
Angiotensin-converting enzyme (ACE) inhibitor, 276, 277
Angiotensin-II (AT II), 347, 348
Angiotensin II receptor blocker (ARB), 255, 276
Angiotensin receptor-neprilysin inhibitors (ARNIs), 276, 277
Antacids, 157–159, 168
Anticholinergics, 157–159, 243
Antidiuretic hormone (ADH), 9, 210, 211, 346, 348, 349
Antioxidants, 18, 21, 22, 32, 49, 68, 96, 97, 270, 292
Antipsychotic drugs, 29, 30, 32
Antipsychotic medications, 32–34
Anti-VEGF-Therapy, 87
Apoptosis, 18, 59, 66, 108, 115, 117, 135, 194, 204, 269, 274, 275, 289, 290, 314, 328, 332, 338, 379, 382, 388, 401
Aquaporin (AQP), 344–345

Aqueous humor (AH), 75–77, 94, 95, 97, 98
Articular cartilage, 121–123, 126, 129, 131, 135
Articular tissues, 129, 131, 132
Atherosclerosis, 194, 224, 233–234, 314
Atrial natriuretic factor (ANF), 348
Atrial natriuretic peptide (ANP), 275, 348
Autoimmunity, 58, 151, 192, 212, 332, 333

B

Basal like, 377, 381, 384
B-cells, 53, 58–61, 141, 192, 328, 329
β-blockers, 99, 212, 213
β-cells, 35, 153, 183–188, 191–196, 206
β-galactosidase, 167, 168
Bile acids, 115, 150, 151, 154, 157, 174–176, 201, 203–207, 210, 312, 319
Blood biomarkers, 297
Blood pressure, 15, 20, 35, 80, 94, 95, 97, 99, 133, 209, 210, 212, 231, 243–245, 255, 263, 269, 283–286, 290–293, 316, 347–349, 353, 371, 393
Blood vessels, 14, 19, 20, 48, 49, 55, 75, 76, 80, 87, 88, 95, 192, 210, 246, 263, 283–285, 287, 289, 290, 297, 301, 306, 344–345, 373
Bone formation, 106–109, 113, 115–118, 122
Bone fragility, 113
Bone mineral density (BMD), 113, 115, 116, 118, 212, 388, 389, 392
Bone remodeling, 106–111, 113–118, 122, 129, 130, 135
Bone resorption, 106–110, 113–118, 122, 135, 347
Breast, 166, 214, 220, 225, 367, 371, 373–374, 377–385, 389, 390, 392, 401
Brown adipose tissue (BAT), 220, 221, 223, 226, 270

C

Calcium phosphate homeostasis, 105–110
Cancer prevention, 384
Capillaries, 76, 78, 79, 94, 183, 240, 241, 245, 263–266, 283, 284, 312–314, 337, 343, 347, 349
Cardiac cachexia, 269
Cardiac remodeling, 275

GPSR Compliance
The European Union's (EU) General Product Safety Regulation (GPSR) is a set
of rules that requires consumer products to be safe and our obligations to
ensure this.

If you have any concerns about our products, you can contact us on

ProductSafety@springernature.com

In case Publisher is established outside the EU, the EU authorized
representative is:

Springer Nature Customer Service Center GmbH
Europaplatz 3
69115 Heidelberg, Germany

www.ingramcontent.com/pod-product-compliance
Ingram Content Group UK Ltd.
Pitfield, Milton Keynes, MK11 3LW, UK
UKHW052351070726
473059UK00009B/2674